# ZOOLOGY

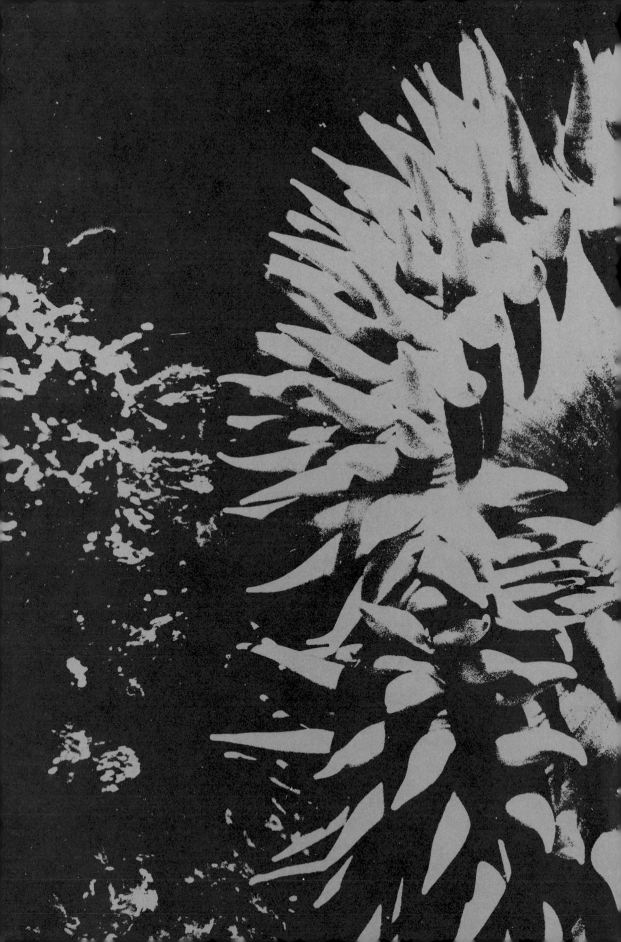

# ZOOLOGY
## An Introduction to the Study of Animals

### Richard A. Boolootian

Science Software Systems, Inc.
Director, Institute of Visual Medicine
Formerly, Associate Professor of Zoology
University of California, Los Angeles

Macmillan Publishing Co., Inc.
NEW YORK

Collier Macmillan Publishers
LONDON

**To my mother and father
Vivian and Vanig**

Copyright © 1979, Macmillan Publishing Co., Inc.

Printed in the United States of America

All rights reserved. No part of this book may be reproduced or transmitted in any form or by any means, electronic or mechanical, including photocopying, recording, or any information storage and retrieval system, without permission in writing from the Publisher.

A portion of this material has been reprinted from *College Zoology*, Ninth Edition, by Richard A. Boolootian and Karl A. Stiles, copyright © 1976 by Richard A. Boolootian and Estate of Karl A. Stiles. Earlier editions of *College Zoology* copyright 1912, 1926, 1931, 1936, 1942, 1951, © 1959, and copyright © 1969 by Macmillan Publishing Co., Inc. Renewal copyrights 1940 by Robert W. Hegner and 1954, 1959, and 1964 by Jane Z. Hegner.

Macmillan Publishing Co., Inc.
866 Third Avenue, New York, New York 10022

Collier Macmillan Canada, Ltd.

Library of Congress Cataloging in Publication Data

Boolootian, Richard A.
  Zoology.

  Includes index.
  1. Zoology. I. Title.
QL47.2.B66         591          78-1665
ISBN 0-02-312030-4

Printing:       7 8      Year:      4 5

ISBN 0-02-312030-4

# Preface

The purpose of this text is to introduce the student to the fascinating and diverse science of zoology. It is by no means an all-inclusive work, but represents what we consider to be the more basic aspects of zoology, together with discussions of current research findings. Given this perspective, it is hoped that students will become aware of the many areas of zoological study that remain open to new discoveries. This text can then serve as a matrix for students who wish to pursue the study of zoology in greater depths.

The approach is a balance between form and function, using representative animals as "types" to illustrate and underscore the systematic groups as they are studied. The unity of the animal kingdom is a basic thread running throughout the work. Similarities and differences between animal groups studied serve to highlight this concept of unity.

The text of 31 chapters is divided into five parts. Part One, "An Approach to Zoology," consisting of the first four chapters, introduces the student to the realm, scope, chemical and cellular basis, and phylectic organization of zoology. Part Two, "Animal Diversity: Invertebrates," containing eleven chapters, begins with the protists and continues through the more complex phyla of invertebrates. Particular focus is on basic body plans and the structure and function of representative invertebrates, with special attention to segmented worms and joint-footed animals. Levels of organization and comparison of structure culminate in this section with a look at those animals related to the vertebrates, the echinoderms.

The six chapters that comprise Part Three, "Animal Diversity: Vertebrates," begin with the primitive chordates and include representative animals from the lampreys, sharks and bony fish, frogs, reptiles, and birds. The last chapter of this section, on mammals, introduces this diverse and interesting class of vertebrates as a unique group exhibiting a wide range of adaptations.

The inherent unity of the animal kingdom, conveyed throughout the text, is reinforced from a form and function viewpoint. Part Four stresses, "The Design and Function of Organ Systems." The first chapter in this section focuses on the coordination of animal behavior, including both the history and current research findings in this field. The remainder of this section presents a review of the development of organ systems, with greater details of the functioning and interdependence of the integumentary, skeletal and muscular, digestive and excretory, circulatory and respiratory, and nervous systems.

Part Five, "The Continuity of Life," covers the basic requirements for continuity in the chapters on reproduction and development, and genetics and heredity. Following chapters on organic evolution and animal ecology give further insight into the history and continuing ecological factors influencing the existence of all life. The final chapter in this section and of the text, "The Human Species," gives a brief account of our species' development and places particular emphasis on our need to become more aware of the ecological impact of our actions on all forms of life.

Numerous photographs and illustrations appear in this text to facilitate understanding of the subject matter and bring it closer to the reader's experience. I wish to thank Science Software Systems, Inc., for providing numerous photographs which appear throughout the text.

In the preparation of the original manuscript, I have had the assistance of Santina Raspante, whose valuable contribution made the preparation of this text a rewarding experience.

A number of other colleagues were also instrumental in this project. Reviewers who offered detailed and very constructive criticism include: Professor Walter K. Taylor, Florida Technological University; Professor Charles Steinmetz, Southern Connecticut State College; Professor Ernest C. Anderson, Eastern Oregon State College; and Professor Willie J. Gore, Northeastern Junior College.

Through the years I have found that the best way of improving an already existing work is to solicit comments from those who use my texts. I look forward to receiving comments, criticisms, or questions relating to the information in this text and its presentation. It is with the help of those who want the best for their students that I can discover what is best for the ultimate reader of *Zoology: An Introduction to the Study of Animals*.

R.A.B.

# Contents

**Part One    An Approach to Zoology**

1  Introduction    3       2   Cell Chemistry and Morphology    8
3  Tissues and Organs    30       4   Taxonomy and Phylum Synopsis    40

**Part Two    Animal Diversity: Invertebrates**

5  Phylum Protozoa    53       6   Phylum Mesozoa    83       7   Phylum Porifera    85
8  Phylum Cnidaria (the Coelenterates) and Phylum Ctenophora    93
9  Flatworms    112       10   Rotifers and Roundworms    121
11  Bryozoans, Lamp Shells, and Arrow Worms    130
12  Segmented Worms    134       13   Joint-footed Animals    151
14  Mollusks    189       15   Echinoderms    203

**Part Three    Animal Diversity: Vertebrates**

16  Chordates    213       17   Lampreys, Sharks, and Bony Fishes    222
18  Frogs and Other Amphibians    242       19   Reptiles    267
20  Birds    286       21   Mammals    305

**Part Four    The Design and Function of Organ Systems**

22  Animal Behavior    329
23  Integumentary, Skeletal, and Muscular Systems    345
24  Digestion and Excretion    358
25  Circulatory and Respiratory Systems    370
26  Nervous System, Sense Organs, and Endocrine System    383

**Part Five    The Continuity of Life**

27  Reproduction and Development    405
28  Genetics and Heredity    420       29   Organic Evolution    440
30  Animal Ecology and Distribution    455       31   The Human Species    478

Index    491

# PART ONE

# An Approach to Zoology

# 1 Introduction

The early morning stillness is punctuated by the high-pitched plinks of water droplets, searching out the earth from their stalactite perches. A man, cloaked in animal skins, is roused from his sleep by the grunt of a vertebrate outside his rock shelter. Images traverse his awakening mind—images of mammals, birds, potential food, and foreboding enemies.

A lethargic eyelid lifts the veil of darkness, and the 6000 B.C. sunlight streams into the cavern, unevenly illuminating myriad animal drawings etched on the rock wall. The drawings were made with crude instruments, and the representations were imperfect—but how important these animals were to this man! He depended upon them for food, he feared them, he loved them, he was fascinated by them. This fascination has continued to this very day, as you, as a student of zoology, embark upon a journey through the animal kingdom.

The great variety of animals that comprises the kingdom Animalia existed long before our caveman ancestors came upon the scene. Our earliest dependence upon animals was for food; soon we began using them for shelter, clothing, and transportation as well. As our relationship with a greater variety of animals developed, we began to learn more about them. This knowledge accumulated and was not systematized until comparatively recently.

The first recorded classification of animals (**taxonomy**) was drawn up by Aristotle, as a pupil of Plato, in about 350 B.C. Aristotle was the first to trust data from observation, and his influence on successive generations was great. A well-known former pupil of Aristotle's, Alexander the Great, ordered his army to send his old master zoological specimens collected throughout their conquests.

## Our Relationship with Animals Throughout the Ages

Animals have served diverse purposes in the civilized world. They have been an artistic inspiration for centuries. History, poetry, music, and literature are enriched with references to animal life: Saint-Saëns composed "A Grand

Zoological Fantasy," Holmes philosophized about "The Chambered Nautilus," and Frost elucidated upon "The Need of Being Versed in Country Things."

Animals have been worshipped by cultures through the ages: the ancient Egyptians worshipped Hathor, the goddess of love, who was variously represented as having the horns, ears, or entire head of a cow. The Minoan culture of early Greece (about 200 B.C.) worshipped Minotaur, a god who was half-man, half-bull. In India the bull is still considered sacred; many people would kill a man found abusing a bull.

Animals have also served the sciences: the design of airplanes and jets were inspired by birds, and the shape of submarines by fish. Scientific research owes a great deal to the thousands of animals that have been used in studies which have benefited all living species. Practically every man-made medication has been tested on animals to determine its probable effect and toxicity.

Our study of zoology includes not only other animals, but the human species as well. We are a significant link in the evolutionary chain. Our ability to construct and use complex tools, vast communication skills, and to think abstractly separate us from the rest of the animal kingdom. But there is much that we can learn through observation of other animals, particularly how to live successfully within our environment.

## Scientific Method

The **scientific method** is an important part of any science, and one of the objectives of any course in zoology is to understand and experience this method of discovery.

The scientific method begins with **observation.** When a phenomenon occurs, its results are either directly or indirectly noted by the scientist, who seeks *knowledge*. If, upon observation, a **problem** is recognized, the scientist may decide to investigate it, for its implications might be of great relevance to life. It is implied in this decision that the problem's manifestations are *testable*.

Based on all his training, and frequently with a good measure of intuition (for creativity is the heart of science), the researcher puts forth a **hypothesis,** explaining what he expects is underlying the original observation or set of observations. Whether or not his hypothesis is correct is only a secondary consideration of the scientist, for its refutation is just as valuable as its confirmation.

Next, the hypothesis is **tested.** This may be done either through pure observations or by **experiment.** From the experiments, **evidence** is derived either for or against the hypothesis. A hypothesis may be refuted by one contrary case, but supportive evidence cannot prove a hypothesis; it can only lend credence to it.

When all the data are tabulated, the hypothesis is either accepted or rejected and a **conclusion** is reached. A major advantage of the scientific method of discovery is that it allows *repetition* to further validate findings of previous experiments.

Functioning within the strict confines of the scientific method requires the experimenter to possess a particular set of attitudes. These include intellectual honesty, open-mindedness, caution in reaching conclusions and vigilance in searching for possible flaws in hypotheses, theories, evidence, and conclusions. A major fault of the ancient philosopher-scientists was that broad, sweeping generalizations were the rule rather than the exception.

## General Characteristics of Life

The animal kingdom is almost limitless in its complex diversity. There are, however, some basic characteristics shared by all living organisms.

All animals contain **cytoplasm,** the substance that makes up **cells** (the basic units of life). The cell or cells that make up the living organism possess a constant **morphology** (form). Most zoologists agree that *form* and *function* are inseparable. As we will see throughout our study of animals, form is greatly determined by the physiological demands of the environment.

Animals and even their individual cells respond quickly to certain changes in the environment—this property is referred to as **irritability.**

Most animals are capable of **movement** of at least parts of their bodies. The majority of species are capable of movement involving the entire organism; this is **locomotion.**

The major factor separating living from nonliving matter is the ability of living matter to utilize materials from the environment by transforming them into active substances that can either build up cells, producing **growth,** or break them down, producing usable **energy.** This entire process is known as **metabolism.** The building-up process, or synthesis, is **anabolism;** the breaking-down process, or lysis, is **catabolism.**

Animals are usually capable of creating new individuals within their species by means of either asexual or sexual **reproduction.** Each new individual's primary concern is self-maintenance.

## Maintenance of the Individual

Animals must obtain food. They must be able to take this food in, either by ingestion or by diffusion. After the food is within the organism, basic physiological processes occur, resulting in digestion, transportation, and assimilation (anabolism) of the digested material. The food may now be thought of as potential energy. Other processes must exist to liberate the stored energy for a variety of activities. The elimination of waste materials is of equal importance, for many of these substances are poisonous (toxic).

Animals must also have a suitable habitat in which to live. **Marine** or ocean-dwelling animals are restricted in their habitats by such factors as water temperature and depth. The major habitats of marine organisms include beaches, open ocean, and the sea's bottom.

Other aquatic animals inhabit fresh waters (rivers, lakes, streams, swamps, pools, and ponds). Freshwater habitats differ from marine environments primarily in their lack of **salinity** (salt content).

The environment of terrestrial animals include the surface of the earth and areas below it.

**Parasitic** animals require **hosts** to support them nutritionally. Parasites' environments include the surface of the host, the digestive tract, the circulatory system, the muscles, and other body areas.

All animals must be able to protect themselves from organisms that prey upon them (predators), and from the environment. In our studies, we will see that several adaptations for the security of a particular animal's life are possible.

## Maintenance of the Species

The success of any species (a natural population potentially able to successfully and exclusively interbreed) is measured in terms of its capacity to produce new individuals which will manifest its traits (characteristics). Even if each member of a given species has equal potential for self-maintenance and reproduction, several limiting factors exist. There must be sufficient food to support each individual, and, almost as important, sufficient living space. A lack of either of these may cause the demise of the species or at least the migration of the affected population. Other considerations which compound this problem include competition among species for food and living space, and the effects on the species of predators and parasites.

**Figure 1-1** The main subdivisions of zoology with descriptions.

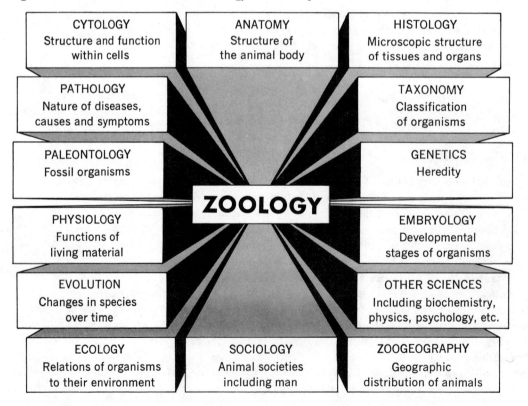

# The Realm of Zoology

Facts about animals have become so numerous that zoology has diversified into a myriad of allied sciences (Fig. 1–1). Each of these sciences has a direct relevance to every member of the animal kingdom, and it is the zoologist's job to coordinate his information with the data of allied disciplines and to help formulate a universal view of Animalia.

As you undertake the study of zoology, its value to your everyday life will become evident. In Chapters 5 through 21 the various groups of animals are discussed. **Representative animals** are used wherever possible to highlight the more important characteristics of a group. Chapters 22 through 31 are devoted to those areas of zoology that merit separate consideration. These include animal behavior, body systems, genetics, evolution, ecology, and a reflection on ourselves, the human species.

In your study of zoology, bear in mind the relevance of this science to human beings. Current world problems, such as population growth, pollution, famine, war, depletion of energy sources, and, most important, the survival of our species on this planet can be solved with help from the examples of those animals which, since the beginning of their existence, have lived peacefully within nature.

# 2
# Cell Chemistry and Morphology

Prehistoric man was aware that he was something more than a hollow receptacle for food. Since his time, striking advances have been made in understanding animal function. One of the most potent realizations began in 1663 when the English scientist Robert Hooke, in his contemplation of organic matter, found that one unit, the **cell,** seemed to encapsulate the essence of life. Almost 200 years later, in 1838, the German botanist Matthias Schleiden stated that the cell was the basic unit of plant matter. This concept was extended in 1839 by Theodor Schwann, who stated that both plants and animals were composed of cells. From this point on, progress in understanding cellular structure and function has been fantastic.

To understand the processes of life from the cell to the total organism, one must first acquire a working knowledge of the components of these living units. Such components—atoms and molecules—interact in characteristic ways. They may be formed into a muscle, participate in neural impulses to the muscle, or provide energy for the muscle to contract. Chemicals can maintain life or destroy it. These and an infinite number of other considerations all point to the inescapable relevance of chemical phenomena to the study of zoology.

## The Atom

Let us begin with a number of definitions to better understand basic cell chemistry.

A **proton** is a positively charged particle. **Neutrons** are essentially uncharged protons. A **nucleus** may be composed of one proton, a proton and a neutron, or several protons and neutrons. An **electron** is a negatively charged particle. An **atom** is composed of a central nucleus with one or more orbiting electrons. The **atomic weight** of an atom is calculated by adding the weights of its protons, neutrons, and electrons. The atomic weight of an electron is approximately 1/1800 that of a proton or neutron; its contribution to the total atomic weight is therefore minimal. The number of protons in a given atom remains constant. The number of neutrons and the number of electrons are each somewhat

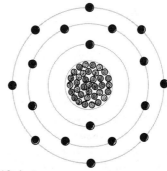

**Figure 2-1** The atom. Argon (Ar), one of the *inert gases*, has three complete shells of electrons, and is thus highly unreactive.

● 18 electrons
◉ 18 protons    ATOM (Argon)
◉ 18 neutrons

variable. The set of all atoms with the same number of protons is called an **element.** In all, there are about 103 different elements, 90 of which occur naturally. Each element is identified by a symbol and a common name, such as Fe and iron.

Electrons revolve around the nucleus, confined to specific orbits called **shells** (Fig. 2-1). Each concentric shell can hold a specific number of electrons. Atomic stability, the relatively unreactive or reactive state of an atom, is based largely on the completion or incompletion of its shells. Atoms with complete shells, such as the **inert** (or **noble**) **gases,** are highly unreactive and stable. Chemical reactions usually involve the outer electrons of given atoms. Atoms with incomplete shells, such as one with room for just one more electron, may be quite reactive. Likewise, if a shell has only one electron, the atom may tend to lose that electron, which also creates an unstable and reactive state.

Atoms, then, may have varying numbers of electrons. An atom has the same number of electrons as it has protons, so the positive and negative charges cancel and the atom is electrically neutral. When a reactive atom loses or adds electrons, the result is a charged **ion.** If electrons are removed, a positive charge is created and the ion is known as a **cation.** If electrons are added, a negative charge occurs, creating an **anion.**

An atom may also have a varying number of neutrons. The element carbon (C) is normally composed of six protons, six neutrons, and six electrons, symbolized as $^{12}C$ (the superscript indicates the total *nuclear* mass). However, other forms of carbon and certain other elements, called **isotopes,** have been observed. Isotopes (e.g., $^{13}C$ or $^{14}C$) have different numbers of neutrons (Fig. 2-2).

**Figure 2-2** Three isotopes of carbon. The extra neutrons increase the atomic mass, decreasing atomic stability.

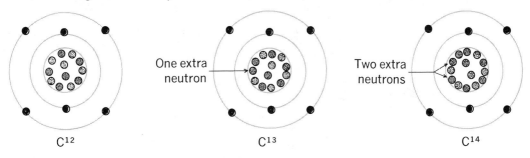

# The Chemical Bond

In living systems, it is rare to find an atom by itself. Atoms usually combine to form **molecules** or **compounds**. A molecule is the smallest unit that retains all the properties found in a given compound. These combined forms of atoms are held together by chemical **bonds**. There are two types of bonding, **ionic** and **covalent** (Fig. 2–3).

Ionic bonds are formed when an anion donates its extra electron to a cation. For example, if anionic chlorine, known as chloride ($Cl^-$) is in the presence of cationic sodium ($Na^+$), the two may be drawn together to form sodium chloride (NaCl), common table *salt*. Ionic bonds may occur among more than two atoms at a time. They tend to dissociate in aqueous (water) solution. When the ions are free in such a solution they are called **electrolytes.**

In covalent bonds the electrons are shared by participating atoms, becoming "community property." There are no direct transfers of electrons as in ionic bonding. Ninety-five percent of the chemicals in cells are bonded covalently. As in ionic bonding, multiple bonding is possible; sometimes even three or four bonds may be formed between two atoms.

Chemical bonds have **energy,** the ability to do work. During a chemical reaction, bonds may be either broken, formed, or both. For the bonds of a stable molecule to be broken, a certain amount of energy must be put into the system. Such **activation energy** is usually supplied in the form of heat. After the needed amount of activation energy is put into the system, the rest of the reaction is downhill. In fact, when the total energy of the system is decreased, the system becomes more stable. Generally, the higher the energy of a system,

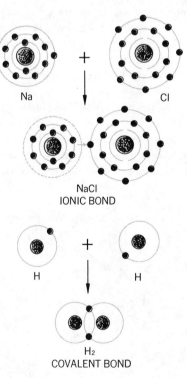

**Figure 2–3** Atomic bonds. Common table salt, NaCl, is *ionically* bonded through *transfer* of one Na electron to Cl. Molecular hydrogen, $H_2$, is *covalently* bonded; the electrons are *shared* by both atoms.

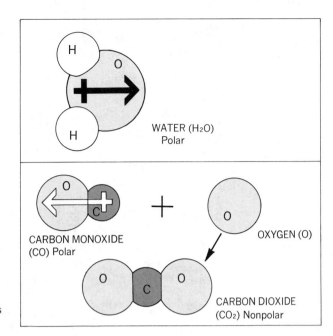

**Figure 2–4** Molecular polarity. *Top:* H$_2$O molecule; the positive portion of the dipole is on the hydrogen "end." *Bottom:* CO is polar, but CO$_2$'s dipoles cancel, to create a nonpolar molecule.

the less stable it is. A reaction will only occur if there is sufficient activation energy available. A reaction that requires energy is called an **endergonic reaction**. A reaction that releases energy is called an **exergonic reaction**.

A characteristic of some molecules which has direct relevance to zoology is the **dipole** (Fig. 2–4). In the water molecule, H$_2$O, the oxygen is more **electronegative** than the hydrogen and expresses its negativity by attracting some of the negative charge of the molecule to it. The hydrogen therefore becomes more positively charged, and this separation of charges is called the **dipolar moment**. The water molecule is said to be **polar**. Polar molecules tend to dissolve in polar solvents; because of this, water is a universal solvent. Nonpolar molecules have no dipole. We see that carbon dioxide, CO$_2$, is nonpolar because even though the oxygens are more electronegative than the carbon, the *net* effect of the electronegativity is zero. Polarity plays a major role in membrane permeability and so in part determines which substances may enter or leave a cell.

## Chemical Substances in the Cell

A chemical analysis of a typical animal cell shows that it is made up of the same elements found in nonliving materials and systems. The 20 elements in Table 2–1 are essential to the cell. These elements generally combine, to form compounds that are either **organic** or **inorganic**. Organic compounds contain carbon and are characteristic of living organisms. They consist primarily of proteins, fats, and carbohydrates. Inorganic compounds consist primarily of water and salts. The average percentages of each are as follows:

| Compound | Approximate Cellular Volume (%) |
|---|---|
| Water | 80.0 |
| Proteins | 12.0 |
| Fats | 3.0 |
| Nucleic acid | 2.0 |
| Carbohydrates | 1.0 |
| Inorganic salts | 1.0 |
| Steroids | 0.5 |
| Other substances | 0.5 |

**Table 2–1.** Essential Elements for Cell Life (percentages are slightly variable)

| Essential Element | Symbol | Percent in Cytoplasm |
|---|---|---|
| Oxygen | O | 63.00 |
| Carbon | C | 20.00 |
| Hydrogen | H | 10.00 |
| Nitrogen | N | 2.50 |
| Calcium | Ca | 2.50 |
| Phosphorus | P | 1.14 |
| Potassium | K | 0.11 |
| Sulfur | S | 0.14 |
| Chlorine | Cl | 0.10 |
| Fluorine | F | 0.10 |
| Sodium | Na | 0.10 |
| Magnesium | Mg | 0.07 |
| Iron | Fe | 0.01 |
| Copper | Cu | Trace |
| Cobalt | Co | Trace |
| Zinc | Zn | Trace |
| Silicon | Si | Trace |
| Manganese | Mn | Trace |
| Iodine | I | Trace |
| Nickel | Ni | Trace |

# Enzymes

A cell's existence and function rely completely on its ability to use accessible nutrients. Innumerable chemical changes occur in the cell. This ongoing process, referred to as **metabolism,** appears in two modes: **anabolic** and **catabolic. Anabolism** is "constructive" metabolism. Photosynthesis is an anabolic process; plants can construct necessary organic compounds from simpler inorganic substances, deriving the required energy from the sun. **Catabolism** is "destructive" metabolism, the breaking down of ingested organic compounds into simpler ones, resulting in a release of energy.

Every metabolic process is mediated by **enzymes,** which are the biological equivalents of catalysts. An enzyme is usually named by adding the suffix **-ase** to the name of the substance on which it acts (the **substrate**). Lipase, for example, acts on lipids. Effective in minute concentration, enzymes perform their catalytic functions over and over again without being depleted.

The immense enzyme molecules are very sensitive to temperature and pH; most decompose at temperatures over 30°C and some will not function beyond a very limited pH range. When homeostatic (steady-state) heat-regulating systems in cells and larger organisms are out of adjustment, enzymatic failure occurs, causing necrosis (death) to the cell or organism.

Although some enzymes are composed entirely of protein, most are complexes of an **apoenzyme,** made of protein, and a **coenzyme,** which may be either a metal or a vitamin. Coenzymes have two general functions. One is that of a **carrier** in a chemical reaction. Frequently, an atom, group of atoms, or a molecule must be transferred from one compound to another. The coenzyme is that portion of the enzyme that physically bonds to the transient group, carrying it to the recipient group. The coenzyme portion also provides a surface on which the molecules can react. The enzyme's physical and electrical configuration permit the molecules to come together more often than would occur by chance alone; its presence may increase the rate of reaction several thousandfold. After the reaction has occurred, the new molecule splits off, leaving the enzyme free to act again.

## Cellular Components

### SMALLER MOLECULES

**Water** ($H_2O$) is composed of two atoms of hydrogen and one of oxygen. It is the most abundant molecule of the cell, occupying 80 to 95 percent of its bulk. The importance of water lies in its ambifunctional character: it is used as a solvent, as a medium for material transport both inter- and intracellularly, and it takes an active role in several reactions, notably respiration.

**Ammonia** ($NH_3$) is a by-product of protein decomposition. It is a highly toxic substance; only five thousandths of 1 percent $NH_3$ in the blood of a rabbit will cause death. Consequently, ammonia must be disposed of by complexing with larger molecules. The result of this reaction is usually the formation of **urea,** a molecule of limited toxicity, which is eventually excreted as a major component of urine.

### LARGER MOLECULES

**LIPIDS.** Lipids, the first of a group of extremely important larger molecules, serve four major functions. Since they contain many high-energy bonds, they can be oxidized to produce large amounts of energy. They are also supportive structures, as presented in the discussion on membranes below. A particular group of lipids, the **waxes,** play a role in protecting the organism through their involvement in skin and fur formation and maintenance. Lipids in the form of four-ringed **steroids** act as chemical-coordinating agents.

There are three major forms of lipids: steroids (mentioned above), fats, and phospholipids. **Fats,** also important in energy storage, are made of three fatty-acid molecules and one glycerol molecule. The most frequently encountered fats in zoology are **palmitic** and **stearic** acids. (Since all fats contain the carboxyl function, $-C{\overset{O}{\underset{OH}{\diagup\kern-0.5em\diagdown}}}$ , they are all acids.)

**Phospholipids,** the third group, have a structure similar to that of fats, but one of the three fatty acids is here replaced by a nitrogen-containing base plus a phosphate group. This replacement creates the polar portion, which will be discussed below.

**CARBOHYDRATES. Carbohydrates** also play a major role in cellular chemistry. They are almost always composed of atoms of carbon, hydrogen, and oxygen in a 1:2:1 ratio, respectively. The simplest form of carbohydrates are the monosaccharide **sugars.** They have three to seven carbons. The most common monosaccharide sugar in animals is **glucose,** with the formula $C_6H_{12}O_6$, indicating six atoms of carbon, 12 of hydrogen, and six of oxygen.

Carbohydrates store energy in animals. **Glycogen,** composed of a long chain of glucose molecules, is used in animals for long-term energy storage.

**PROTEINS. Proteins** are the chief structural components of the cell, occurring in enzymes, hormones, chromosomes, membranes, and many other cellular

Table 2-2. The Important Amino Acids for Human Beings

| Name | Abbreviation | Notes (see key) |
|---|---|---|
| Alanine | Ala | |
| Arginine | Arg | |
| Asparagine | Asn | |
| Aspartic acid | Asp | |
| Cysteine | Cys | |
| Cystine | — | °° |
| Diiodotyrosine | — | °° |
| Glutamic acid | Glu | |
| Glutamine | Gln | |
| Glycine | Gly | |
| Histidine | His | |
| Hydroxyproline | — | °° |
| Isoleucine | Ile, Iln | ° |
| Leucine | Leu | ° |
| Lysine | Lys | ° |
| Methionine | Met | ° |
| Phenylalanine | Phe | ° |
| Proline | Pro | |
| Serine | Ser | |
| Threonine | Thr | |
| Thyroxine | — | °°° |
| Triiodothyronine | — | °°° |
| Tryptophan | Try | ° |
| Tyrosine | Tyr | |
| Valine | Val | ° |

Key:
° Essential amino acids that must be ingested.
°° Generally considered necessary for human life; may be formed from existing or ingested precursors in body.
°°° Thyroid gland hormones, which incorporate iodine and appear in the form of amino acids.

components. Particular organisms have characteristic proteins—a useful aid in determining similarity among organisms. Proteins contain carbon, hydrogen, oxygen, and nitrogen and sometimes sulfur, phosphorus, and/or iodine, assembled as strings of **amino acids.** Twenty amino acids are most important for protein formation. The single-starred amino acids in Table 2–2 are the **essential amino acids** for human beings. These cannot be synthesized by the organism; they must be ingested if survival is to continue. The ingestion may be in the form of a protein containing the particular amino acid required, and digestion may liberate the amino acid from its bound position in the protein. The other amino acids listed are synthesized by the organism from existing precursors. Generally, an amino acid has the structure

$$\begin{array}{c} H \\ \diagdown \\ N-C-C \\ \diagup \phantom{XX} | \phantom{XX} \diagdown \\ H \phantom{XX} H \phantom{XXX} OH \end{array} \begin{array}{c} R \\ | \\ \phantom{X} \\ \phantom{X} \end{array} \begin{array}{c} \phantom{X} \\ O \\ \diagup \\ \phantom{X} \end{array}$$

where R represents one of a number of **functional groups** composed of carbon, hydrogen, and possibly other elements. The $\begin{smallmatrix}H\\ \diagdown\\ N-\\ \diagup\\ H\end{smallmatrix}$ portion of the molecule is known as the *amino end*, and the $-C\begin{smallmatrix}\diagup O\\ \diagdown OH\end{smallmatrix}$ part is the *carboxylic end*.

When a protein is synthesized, the amino acids are bonded to each other by **peptide linkages.**

## Reaction Types

All chemical reactions share some characteristics. They occur at various rates, depending on temperature, concentration of the reacting substance, and other factors. Reactions occur in a particular direction, either one-way (indicated by a single arrow) or reversible (indicated by a double arrow). Most chemical reactions are reversible, as seen, for example, in the self-hydrolysis of water:

$$H_2O \rightleftharpoons H^+ + OH^-$$

At a given temperature and pressure, each reversible reaction has a preferred ratio of concentration of products (on the right side of a chemical equation) to concentration of reactants (on the left). That ratio, the **equilibrium constant,** is represented by the letter $K$:

$$K[H_2O] = \frac{[H^+] + [OH^-]}{H_2O}$$

Chemical equilibrium is established when the products and reactants have reached the characteristic equilibrium concentrations, so that for any reactants being transformed into products, an equivalent number of reactants are being re-formed *from* products.

There are four basic categories of chemical reactions constituting the majority of the metabolic activities of cells: transfer, synthesis, hydrolysis, and redox (reduction–oxidation).

**Transfer reactions** involve the removal of one portion of a molecule by a carrier and its transportation to another molecule. Commonly transferred substances are $H^+$ ions, amino groups, and phosphates. As mentioned earlier, coenzymes usually perform this function.

**Synthesis reactions** generally form a larger compound from smaller ones, and are usually accompanied by a simultaneous removal of water (**dehydration**).

**Digestion** is mostly the breaking down of carbohydrates, lipids, and proteins by a catabolic process involving **hydrolyses,** "water-dissolving" reactions, which usually leave the new fragments hydrated.

**Redox reactions** are actually coupled **reduction** and **oxidation** reactions. Such reactions *always* occur together and involve an exchange of electrons. When one atom or molecule gives up an electron, it is said to be **oxidized;** and when the other gains an electron, it is said to be **reduced.** Oxidative reactions are generally used for liberating energy, while reductive reactions are used to store energy.

## Acids, Bases, and Salts

For our purposes, an **acid** is any substance that releases hydrogen ions; an example is HCl, hydrochloric acid. A **base** is a substance that releases hydronium ions ($OH^-$); one such is NaOH, sodium hydroxide. A pH below 7 is acidic, 7 is neutral, and one above 8 is basic. Acids and bases are said to be *strong* if they are completely or nearly completely dissolved in aqueous solution. The fact that a compound is a strong acid does not mean that it is toxic or has corrosive effects.

A **salt** is an ionically bonded molecule composed of whatever ions are present in the particular solution. For example, if HCl and NaOH (a strong acid and a strong base) are put together into aqueous solution, $Cl^-$ and $Na^+$ ions will be liberated, and upon dehydration of the solution, NaCl (table salt) will be formed.

## Cell Morphology

Let us now turn our attention to the zoological implications of the chemical compounds and interactions that we have just discussed. *Cell morphology* is the study of the form and structure of the cell. Until recently, the term **protoplasm,** coined by Purkinje in 1840, was taken to be synonymous with "living substance." Protoplasm, a conglomeration of highly complicated subcellular components, embodies the prerequisites for life. Today we use the term only in a descriptive sense; the cell itself is considered to be the true living substance.

Most cells range in size from 0.5 μm (micrometer) to 20 μm (μ, the Greek lowercase letter mu, is the symbol for *micro*, equal to $10^{-6}$), but some bacteria are as small as 250 nm (nanometers; nano = $10^{-9}$). The lower limit for the diameter of an active, living cell varies around a mean of 225 nm. Upper limits are represented by some nerve cells in human beings, which may attain lengths of approximately 2 meters.

## SUBCELLULAR CONSIDERATIONS

With few exceptions, cells possess **subcellular organelles,** which taken together allow the cell's continued existence (Fig. 2–5). The most basic division of the cell is that between cytoplasm and karyoplasm (nucleoplasm). **Cytoplasm** is comprised of all extranuclear substances, while the remainder of the cell's contents within the nucleus makes up the **karyoplasm.**

Microscopically, protoplasm gives the appearance of a grayish jellylike substance in which one may see globules, droplets, granules, and crystals of various shapes and sizes. Cells are *stained* with various dyes so that structures that are normally colorless will be rendered more distinct. The materials in the cell form a **colloid,** which has the consistency of a syrup. Depending on the conditions, protoplasmic colloids may be in either a **sol** (liquid) or a **gel** state. Furthermore, colloids may vary between the sol and gel states.

In 1827, Robert Brown, an English botanist, noticed that when the cellular contents were in a sol state, particles seemed to be moving around. This **Brownian motion** is due to invisible particles striking against larger granules. Brownian movement also occurs in water, other liquids, and gases and is not necessarily a characteristic of life. The particles involved are estimated to range from $10^{-4}$ to $10^{-6}$ mm (millimeter; milli = $10^{-3}$) in diameter.

## MEMBRANES, DIFFUSION AND OSMOSIS, AND CELL NUTRITION

Cells and subcellular organelles are held together by **membranes.** Basically, membranes regulate substances entering and leaving the cell. The passage of substances through cell membranes may take place in various ways. The most common way is **diffusion** (Fig. 2–6A), whereby molecules move from an area of greater concentration to an area of lesser concentration. This is a natural tendency, based on random molecular movements.

Cellular and subcellular membranes are not completely permeable. If they were, *any* surrounding substances could diffuse into the cell and the cell could not remain a separate unit. Cells have **semipermeable membranes** which allow selected substances free access in and out.

**Osmosis** (Fig. 2–6B) is defined as the diffusion of water through a semipermeable membrane. If a single cell is placed in an **isotonic** solution (where the concentrations of solutes inside and outside the cell are equal) the cell will remain undisturbed (Fig. 2–7A). If, however, a cell is placed in a **hypotonic** solution (where the concentration of solutes outside is lower than that within the cell), the cell will swell considerably, taking in water to equalize the concentration (Fig. 2–7B). The osmotic movement into a cell in hypotonic solution is known as **deplasmolysis. Plasmolysis** occurs when a cell is placed in a **hypertonic** solution. In this case, the concentration of solutes outside the cell is higher than within, and water will leave the cell, seeking an isoosmotic state (Fig. 2–7C).

A membrane may be semipermeable by virtue of minute **pores,** submicro-

**18**
*An Approach to Zoology*

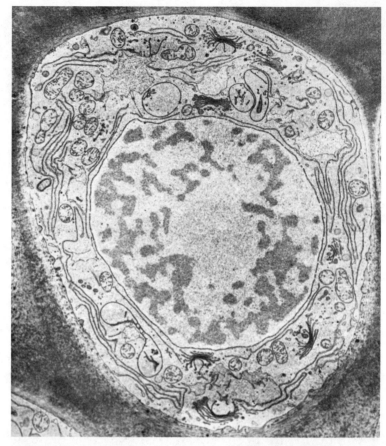

**Figure 2–5** The cell. *Top: Zea mays* (corn) cell. (Courtesy G. Hayley.) Approximately 2500×, shown at 50 percent reduction. *Bottom:* Diagram with subcellular organelles.

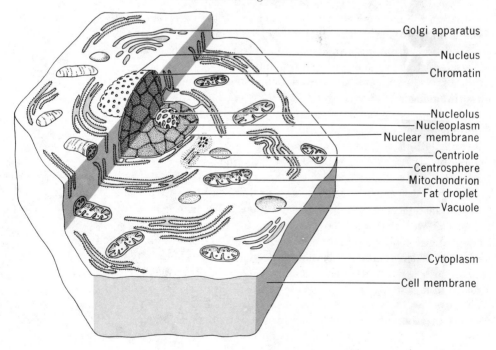

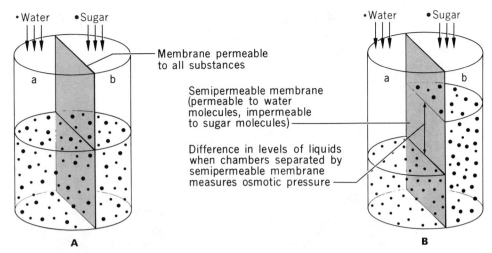

**Figure 2-6** (A) Diffusion. The permeable membrane offers no hindrance to the diffusion of water molecules or sugar molecules between chambers a and b, resulting in equal concentration on both sides. (B) Osmosis. The semipermeable membrane allows the passage of water molecules but hinders the passage of sugar. In accordance with the law of diffusion, water molecules move to an area of less concentration in chamber b.

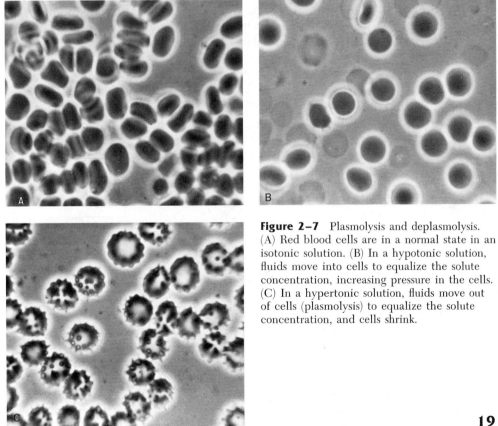

**Figure 2-7** Plasmolysis and deplasmolysis. (A) Red blood cells are in a normal state in an isotonic solution. (B) In a hypotonic solution, fluids move into cells to equalize the solute concentration, increasing pressure in the cells. (C) In a hypertonic solution, fluids move out of cells (plasmolysis) to equalize the solute concentration, and cells shrink.

scopic holes which physically limit the size of molecules that may pass through them. Generally, smaller molecules pass through cellular membranes more readily than larger ones. Lipids and lipid-soluble substances pass through more rapidly than do water-soluble substances. Membranes are **selectively** permeable. A solute's electrical polarity and other physical characteristics may greatly affect its ability to obtain nutrition and remove waste.

Cells may obtain nutrition by means other than diffusion and osmosis. One method is **phagocytosis** (also called **endocytosis**). This process is used when the ingested substance is too large to diffuse through the membrane. Impermeable *liquid* substances may enter or leave the cell by an analogous process called **pinocytosis.**

Another method of providing access in and out of cells is **active transport.** This method is used when there is a **concentration gradient** preventing solutes from entering a cell with an already high internal solute concentration. A substance may be chemically bonded to a **carrier** molecule (which is specific for that substance) and the carrier physically escorts the substance through the membrane. The energy required to overcome the concentration gradient and allow active transport to occur is assumed to come from adenosine triphosphate (ATP). ATP is the normal energy source for most cellular reactions and has been suggested as one type of carrier.

The last method of nonosmodiffusional access in and out of cells is **facilitated transport.** A molecule may be prevented from entering a cell because of its polarity. If there is a separation of charge in the molecule's structure that adversely interacts with the charges on a polar membrane, a carrier may again be employed. In this case the carrier's function is to neutralize the polarity of the entering molecule. Once the carrier deposits the molecule within the cell, it returns to repeat the process.

## MEMBRANE STRUCTURE

Membrane structure was first elucidated in 1935, by H. Davson and J. F. Danielli. Their conception is now known as the **unit membrane model,** named by D. Robertson, who substantiated it with the electron microscope. According to Davson and Danielli, a membrane is something like a sandwich. Under the electron microscope (Fig. 2–8) most stained membranes *do* appear as two parallel dark lines with a clear space between them. The dark stained lines are made up of proteins, which constitute 50 percent of the membrane's bulk. The other 50 percent is made of lipids, which have polar and nonpolar parts. The clear zone is made of the nonpolar fatty-acid side chains. The polar portions of the lipids reside in the protein layers in a linear arrangement. Although the unit membrane model has extensive application, recent studies have cast doubt on its universality, spurring investigators to look further for explanations more consistent with these new findings. Recent evidence indicates another strong possibility for membrane structure. In the **interspersed globule** model, a membrane is envisioned as again incorporating a bilayer of lipids. Interspersed in the bilayers are **micelles** (globules). Current evidence points to the likelihood that these micelles are made of a core of protein surrounded by lipid, or possibly protein alone. Various types of membranes may exist, each responding differently to experimental manipulation. Research has yet to provide substantial clarification of these theories.

Cell membranes are usually not bare. In animals, they are covered by a layer

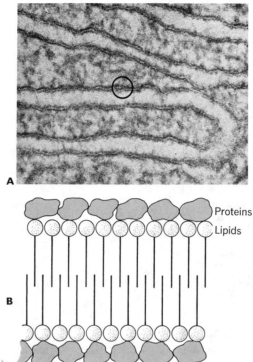

**Figure 2-8** Plasma membrane. (A) Electron micrograph of plasma membrane (circled). 400,000×. (Courtesy F. Sjostrand.) (B) Unit membrane model.

of complex carbohydrates and proteins or by a glycoprotein complex. Plant cells are frequently coated with the carbohydrate **pectin.**

## NUCLEUS AND NUCLEOLUS

The most obvious subcellular organelle is the **nucleus** (Fig. 2–5). Named by Robert Brown in 1833, the nucleus is the controller of the cell. Most nuclei are essentially spherical in shape and are bounded by two porous unit membranes. All cells except the mature red blood cells of most mammals and those in the kingdom Monera (the most primitive unicellular organisms) have nuclei. Within the nucleus are chromatin (also called **chromosomes**), which are threadlike strands of deoxyribonucleic acid (DNA), ribonucleic acid (RNA) and protein, and karyoplasm ("nuclear sap"). Glycoproteins and various hydrolytic enzymes, such as ribonuclease, alkaline phosphatase, and dipeptidase are also found in the nucleus.

A subcellular organelle of considerable interest is the **nucleolus,** which appears as a dark-stained body in microscopic preparations. A nucleus usually has two nucleoli, formed during the telophase of mitosis (see below). The nucleolus is apparently involved in the production of ribosomes.

## ENDOPLASMIC RETICULUM

A structure found traversing most of the cell is the **endoplasmic reticulum,** discovered by K. Porter, A. Claude, and E. Fullam in 1945 with the aid of the electron microscope. Essentially, the endoplasmic reticulum is a complex system of interlocking channels made up of unit membranes. Its exact function

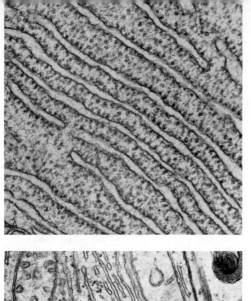

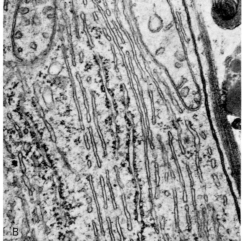

**Figure 2–9** Endoplasmic reticulum. (A) Rough ER from bat pancreas. (Courtesy K. Porter.) (B) Smooth ER, 24,000×.

is uncertain. Current theory is that it provides a means for orderly enzyme distribution and a route for the transmission of nervous impulses.

Two types of endoplasmic reticula have been identified. The first, known as rough endoplasmic reticulum (Fig. 2–9A), appears as a membrane with granules attached along its surface and length. These granules are **ribosomes,** the substrates for protein synthesis. This type of endoplasmic reticulum is especially prominent in cells with high rates of protein synthesis, such as the pancreas. The second type is called smooth endoplasmic reticulum (Fig. 2–9B). Ribosomes are not attached here. The function of smooth endoplasmic reticulum appears to be in the synthesis and transport of lipids and glycogen. Smooth endoplasmic reticula are found in liver cells, gland cells, mature leucocytes, spermatocytes, and various other structures.

Portions of the endoplasmic reticulum are expanded into **vacuoles.** Such areas may act as waste dumps and nutrient storage units for the cell. Most of the endoplasmic reticulum is flat in structure and the general term for these portions is **lamella.** The structure of endoplasmic reticulum varies according to cell type.

## GOLGI BODIES

**Golgi bodies** were discovered in 1898 by an Italian scientist, Camillo Golgi. Under the light microscope only the outer portions of the Golgi apparatus

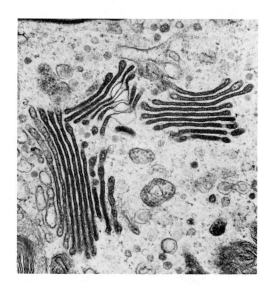

**Figure 2–10** Golgi bodies. Note forming and maturation faces. 11,500×. (Courtesy Alan Bell.)

were distinguishable with an osmic acid stain. The true morphology of Golgi bodies was discerned with the advent of the electron microscope. They appear as flattened stacked sacks with the edges forming a tubular network (Fig. 2–10). Some of these tubes are "dead ends"; others interconnect with other Golgi units. Golgi bodies are quite differentiated, consisting of a **forming face** and a **maturation face.** The maturation faces contain cavities that act as areas for accumulation, condensation, transformation, and synthesis of the Golgi product. It is thought that substances are transported to the Golgi bodies by means of endoplasmic reticular channels. Golgi vesicles have been observed to "bud off" the main body of the apparatus, move through the cytoplasm to the surface of the cell, and release their contents.

## LYSOSOME

A very important cell organelle, believed formed from portions of the Golgi bodies, is the **lysosome** (Fig. 2–11). It is essentially a bag of enzymes packaged

**Figure 2–11** Lysosome. 120,000×. (Courtesy F. Sjostrand.)

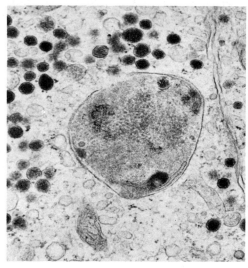

in the Golgi and sent out into the cytoplasm. There are about 30 enzymes found in the lysosome; the major ones, known as **lysosomal hydrolases,** are: acid phosphatase, ribonuclease, deoxyribonuclease, cathepsin, and $\beta$-glucuronidase. These enzymes catalyze the digestion of proteins, nucleic acids, some carbohydrates, and possibly fats. As mentioned earlier, a cell may take in solid food by phagocytosis. The resultant packet of food is known as a phagosome. Lysosomes fuse with phagosomes, forming **digestion vacuoles.** After digestion is completed, the digestion vacuole moves to the plasma membrane, fuses with it, and expels the nutrients into the cytoplasm.

In the process of internal fertilization, it is the lysosome that dissolves the structures surrounding the oocyte (egg) and allows spermal penetration. Certain cells that digest bone utilize the lysosome in the dissolution process. In fact, the word "lysosome" itself is derived from the verb "to lyse," which means to break up.

The lysosome is involved in several pathogenic conditions. In arthritis, joint inflammation may be aggravated by burst lysosomes. The drug **cortisone** stabilizes lysosomal membranes, thus making lysosomal enzyme release less likely and reducing inflammation.

If a cell is starving, the process of **autophagy** may occur. Here, the lysosome breaks down proteinaceous material of the cell itself, either from the cytoplasm in general, or sometimes destroying some of the cell's own organelles. This is the cell's last resort if no other energy-containing substances are available. If a lysosome's membrane is burst, the entire cell will "self-destruct."

## CENTRIOLES, MICROTUBULES, AND UNDULIPODIA

The **centrosome,** found near the nucleus of a cell, consists of two **centrioles,** which appear as tiny cylinders at 90° angles to each other. They usually have radiating fibers called **asters** and the centriole–aster unit is sometimes referred to as the **centrosphere.** Figure 2–12 shows a cross section of a centriole. Note the nine triplets of fiber running longitudinally, which appear as circles in cross section. These fibers are called **microtubules.** Centrioles may function during cell division by anchoring the **spindle fibers** (made of microtubules); this anchoring function remains somewhat in doubt, since plants lack centrioles but do have spindle fibers.

Similar in morphology to centrioles are the **undulipodia.** Often present in

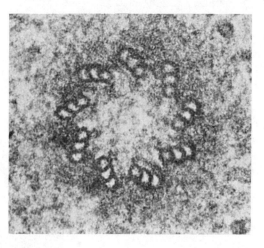

**Figure 2–12** Cross section of centriole with nine groups of three fibers. (Courtesy F. Sjostrand.)

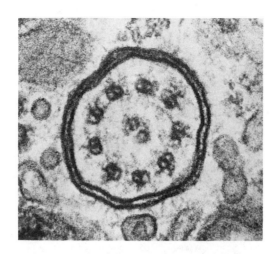

**Figure 2-13** Cross section of undulipodium. Note 9 + 2 arrangement of fibers. The inner two are not double. 82,000×.

specialized cells, the undulipodia consist of nine *double* pairs of microtubules *plus* an additional single set of two in the center (Fig. 2-13). Undulipodia are used for locomotion and sensation.

Owing to a discrepancy in length, undulipodia were originally thought to be two separate structures, **cilia** and **flagella**. Cilia range in length from 5 to 10 μm, while flagella may reach lengths of 150 μm. These terms are still used as quick indices of undulipodial length.

## MITOCHONDRIA

**Mitochondria** (Fig. 2-14) appear in various shapes, depending on the type of cell and intracellular conditions. They are bounded by double membranes, each layer approximately 4 nm thick. Mitochondria range in size from 0.5 to 3.0 μm in diameter and are up to 10 μm long. Within each mitochondrion is an intricate series of compartments with double membranes, **cristae mitochondriales** or simply **cristae**. Cristae are usually perpendicular to the long axis of the mitochondrion, although in the cells of some organisms (notably the human spermatogonium) they are parallel to the axis. The cristae are about 7 nm apart, and each is covered by a type of "fur" with balls of 7.5 to 10 nm at the

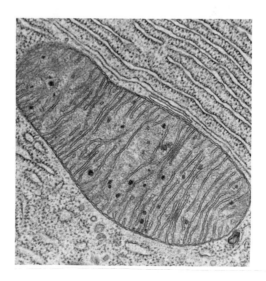

**Figure 2-14** Mitochondrion. Note transverse cristae and ribosomes. From bat pancreas. 53,000×. (Courtesy K. Porter.)

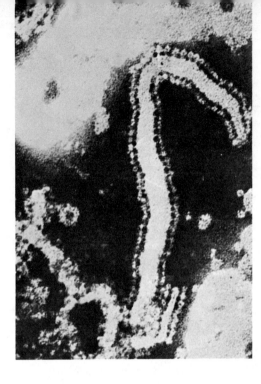

**Figure 2-15** Mitochondrial crista cross section. Electron micrograph exhibiting "ball" structure. 230,000×. (Courtesy D. Parsons, *Science* **142:** 1963.)

ends of the filaments (Fig. 2-15). In the interstices there is a matrix material which contains, among other things, ribosomes. Many authors believe that the mitochondria are attached to the nuclear membrane with a matrix of microtubules.

The mitochondrion is the powerhouse of the cell and is responsible for **cellular respiration.** This organelle accounts for almost all of the molecular oxygen ($O_2$) uptake of the cell, as well as 90 percent of the ATP formed. Mitochondria form the basis for all respiratory systems in animals.

Some students of the mitochondrion believe that it was once an independent organism, was taken into cells, thrived, and became a useful permanent adaptation. This theory is supported by the fact that each mitochondrion has its *own* circular strand of DNA (Fig. 2-16)—but no real proof exists for this theory.

Mitochondria are useful as cellular trauma indicators: if a cell is damaged, mitochondria may fragment, swell, or degenerate. Mitochondrial disintegration is a well-known sign of impending cell necrosis (death).

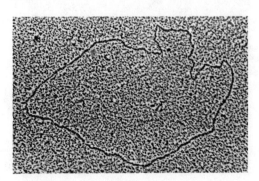

**Figure 2-16** Circular mitochondrial DNA. The isolation and electron microscopy was done by M. M. K. Nass. 38,400×. (*Proc. Nat. Acad. Sci.* **56:** 1215, 1966.)

# Cell Division: The Mitotic Process

With few exceptions (notably the nematodes), organisms grow by producing new cells. During the lives of most organisms, old cells die and must be replaced. With the exception of the nerve cell, all cells grow until they reach a certain size and then divide. This division is termed **mitosis.** The two new cells or daughter cells proceed to grow, and they in turn divide—the process continuing generation after generation. Many cells grow very little or not at all during the time between successive divisions. Why cells divide when they do is not known, but the relative quantities of karyoplasm and cytoplasm are maintained in each kind of cell. It has been suggested that when the ratio of cytoplasm to karyoplasm is upset, the cell divides.

Mitosis is somewhat arbitrarily broken up into several "stages." These are useful for understanding the process, but keep in mind that mitosis is a continuous process and nature does not stop at the end of each "stage," waiting for an indication to continue.

Figure 2–17 presents the basic features of the mitotic cycle. It will be useful to refer to it during this discussion.

**Figure 2–17** Animal mitosis.

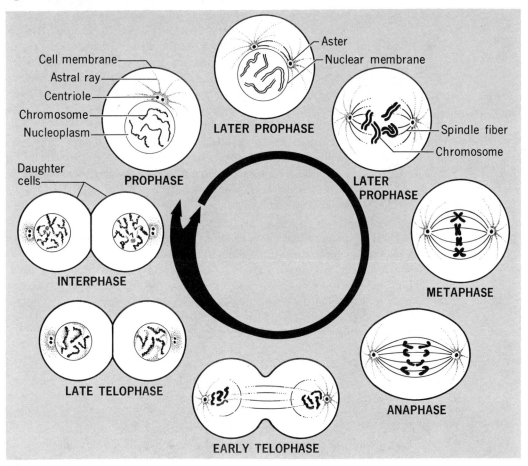

## THE CELL CYCLE AND INTERPHASE

This phase is commonly referred to as the "resting stage," but in this condition the cell by no means rests. Its normal metabolic activities are carried out during this time. During the $G_1$ period, the cell, in addition to performing its particular differentiated function, grows to normal size. Chromosome replication (Chapter 28) occurs during the S stage of the cell cycle. The $G_2$ period is essentially similar to the $G_1$, except that here the organism has replicated chromosomes. The "meat" of the mitotic cycle begins just after $G_2$ and ends with the beginning of another $G_1$. **Interphase** may be considered to consist of $G_1$, S, and $G_2$.

During interphase, the chromosomes may be visible as thin threads of DNA, randomly coiled, or they may not be visible at all. Nucleoli are visible within the intact nucleus, and next to the nucleus lies the centrosome. Thin filaments (the asters) may be visible radiating out from the centrioles.

## PROPHASE

In **prophase,** things begin to happen, sometimes rather rapidly. This phase may take varying lengths of time, depending on the organism and the type of cell itself. In the mesenchyme cells of a chick, at 39°C, prophase lasts between 5 and 50 minutes, usually taking at least 30 minutes.

An animal cell often rounds out somewhat during prophase and the cytoplasm becomes more viscous. The **chromosomes,** each now composed of a pair of **chromatids** of identical genetic constitution, are joined by a common **centromere.** The chromatids coil and shorten in this stage, making them thicker and thus more visible. Early in prophase the chromosomes migrate toward the nuclear membrane, which begins disintegrating. The **centrioles,** which originally made up the **centrosomes,** begin moving from their original position near the nucleus to opposite poles of the cell.

Eventually, the nuclear membrane completely disintegrates while the nucleoli become successively less distinct. About this time, in the middle of prophase, the centrioles begin **spindle** formation.

By the end of prophase, the centrioles have reached the opposite poles of the cell. Their spindle fibers appear to reach out toward the chromosomes, which at this time have almost completed migrating to the center of the cell.

## METAPHASE

When the chromosomes have reached the center line of the cell, known as the **metaphase plate, metaphase** officially begins. In the chick mesenchyme this stage takes from 1 to 15 minutes, usually from 2 to 10 minutes. As the chromosomes move to this orientation, the "arms" leading from the centromeres to the tips of the chromosomes look as if they are trailing along, being dragged by the centromere. The spindle fibers reach from their points of origin at the centrioles to the centromeres, where they attach.

## ANAPHASE

This stage is rather short, taking in chick mesenchyme cells from 1 to 5 minutes, usually 2 or 3 minutes. The main feature of **anaphase** is the splitting of the centromeres. The spindle fibers between the centrioles and the centromeres appear to shorten and pull the now-separated identical chromatids toward the opposite poles of the cell. Simultaneously, other spindle fibers appear between the centromeres, and some authors believe that these "push" the chromatids apart. As the chromatids move toward the poles, the cell begins a type of cleavage, known as **cytokinesis**, along the equatorial plane.

## TELOPHASE

In chick mesenchyme cells **telophase** usually takes from 70 to 180 minutes. At this point, the chromatids have reached the poles of the cell. The spindle filaments between centromeres are still visible in early telophase. As these fibers begin disintegration, cytokinesis continues and the nuclei of each daughter cell re-form. The chromatids replicate later, as do the centrioles. The chromatids are now considered chromosomes and begin to uncoil and become less distinct. Nucleoli reappear in the new nuclei. Cytokinesis continues to completion, and the result is two genetically identical **daughter cells**, each of about one-half the volume of the original mother cell. The $G_1$ segment of interphase follows and the cycle repeats.

Sex-cell (gamete) reproduction, called **meiosis,** will be discussed in Chapter 28. For now, it is only necessary to know that the mother cells, **spermatogonia** and **oogonia,** have the normal chromosomal complement of the rest of the cells of the organism (**diploid** chromosomal number). During meiosis, gametes are formed which have half the original number of chromosomes. This new chromosome complement is called **haploid.** When a sperm fertilizes an egg, a diploid **zygote** results.

# Summary

Protons, neutrons, and electrons, in various combinations, form atoms. Molecules are made up of two or more atoms linked by bonds, which may be either ionic or covalent. Chemical reactions involve bond breaking (requiring energy) or bond formation (releasing energy).

Enzymes are biological catalysts, and are composed of coenzymes and apoenzymes. They may be reused continuously.

Water, carbon dioxide, molecular oxygen, and ammonia are small molecules intimately involved in cellular metabolism. The larger molecules involved include lipids, carbohydrates, and proteins.

Most metabolic reactions are either transfer, synthetic, digestive, or redox reactions.

Acids involve proton loss, bases involve hydroxide loss, and salts are ionically bonded anions and cations.

Cells, the smallest units of life, contain a set of subcellular organelles, each performing a specific function to maintain the cell. Usually, the set consists of membranes, nucleus, nucleoli, endoplasmic reticulum, Golgi bodies, lysosomes, centrioles, microtubules, mitochondria, and undulipodia. The membranes enclosing other organelles are semipermeable.

Cells divide to produce new cells by the process of mitosis, which includes interphase, prophase, metaphase, anaphase, and telophase, in a continuous cycle.

# 3
# Tissues and Organs

According to the current theory of evolution, higher (more complex) animals evolved from the lower (less complex) ones. The first few species of animals that appeared in the warm waters of our planet were probably unicellular or precellular. When unicellular organisms grew to a certain size, they divided. More advanced forms of these lower animals became multicellular. Why did these first living cells act in these ways rather than growing into immense advanced bulbous cells?

## Tissues

The answer to the question we have posed lies in the relationship between cellular growth and division and the ratio of the surface area to the volume of a given cell. A cellular membrane must be able to take in nutrients and expel poisonous wastes for the entire volume of the cell. As the cell grows, its volume increases more rapidly (at the cube of the radius, $V = \frac{4}{3}\pi r^3$) than its surface area (at the square of the radius, $A = 4\pi r^2$) (Fig. 3-1). The membrane surface area of a cell is not capable of transporting substances in and out of the increased volume—and the cell divides. This division creates two new cells, each with a more favorable ratio of surface area to volume.

As organisms progressed in their evolutionary development, the number of cells per organism increased. These cells began to appear in layers, differentiating to perform specific functions in addition to universal metabolic functions. These layered, specialized groups of cells are known as **tissues**. A tissue may be formed of one or more types of cells. **Histology** is the study of tissues. **Organology** is the study of the organs that various tissues may comprise. There are several major types of tissue, including epithelial, muscular, connective, nervous, vascular, adipose (fat), and moist membrane (Fig. 3-1).

### EPITHELIAL TISSUE

Tissues that cover internal or external surfaces are known as **epithelia** (Fig. 3-2). These tissues function in protection, lubrication, absorption, secretion,

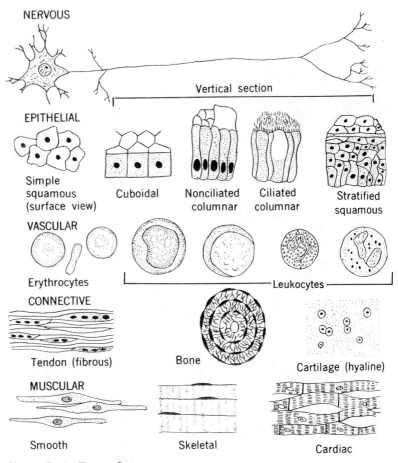

**Figure 3–1** Types of tissues.

and excretion. Most epithelia have one free surface and another which lies on vascular connective tissue. Often, epithelia are attached to a supportive **basement membrane,** or **basal membrane.**

TYPES OF EPITHELIA. There are several types of epithelial tissue, classified by the number of layers present and cellular morphology. **Simple epithelia** are found in many vertebrates. They are composed of one layer of flat (**squamous**) cells which give the appearance of somewhat irregular floor tiles.

**Cuboidal** epithelia, cubelike in shape, are usually found in glands and ducts. The kidney is comprised primarily of cuboidal epithelial tissue. **Columnar** epithelia resemble pillars in their construction and are found in the stomach and intestines.

Epithelia composed of more than one layer are **stratified** epithelia, found in the skin, sweat glands, and urethra. In the bladder, **transitional stratified epithelia** maintain an ability to change the number of layers through movement. The trachea has **pseudostratified** epithelia, with cells of different heights. Cells comprising epithelial tissue may also be ciliated or flagellated and may contain both receptors and effectors of the sensory system.

## MUSCULAR TISSUE

**Muscular tissue** (Fig. 3–3) is composed of cells specialized for contraction. A whole muscle is made up from groups of muscle fibers. These fiber cells are

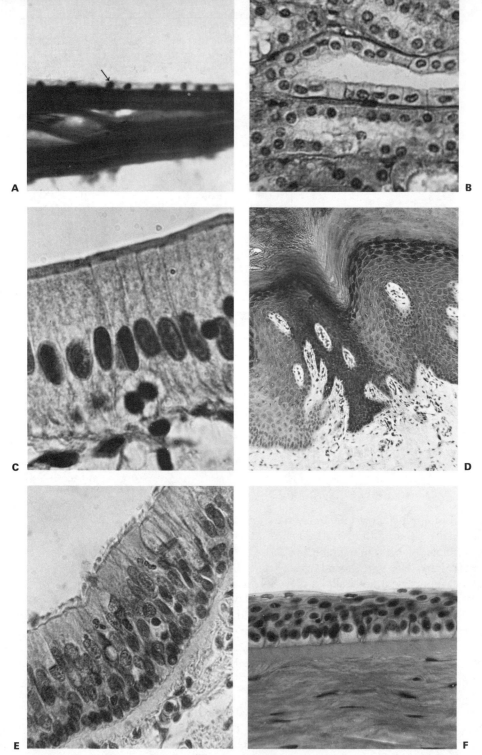

**Figure 3-2** Epithelia. (A) Simple squamous, from inside of cornea. Arrow indicates epithelial layer. 320×. (B) Simple cuboidal, from kidney. 800×. (C) Columnar, from intestine. 2000×. (D) Stratified squamous, from human skin. 250×. (E) Pseudostratified columnar, from trachea. 800×. (F) Stratified squamous and cuboidal, from outer surface of cornea.

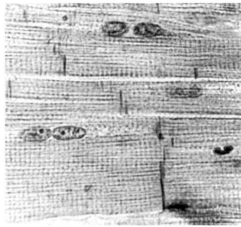

**Figure 3–3** Muscular tissues. *Top left:* striated voluntary. 384×. *Top right:* cardiac. 320×. *Bottom:* smooth involuntary. 240×.

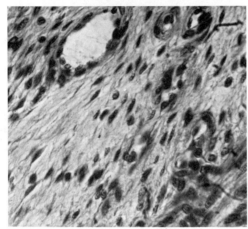

composed of an unspecialized cytoplasm known as **sarcoplasm,** which contains **myofibrils.** Myofibrils are contractive elements composed of successive **sarcomeres,** the units of muscular contraction.

Muscle fibers may be either **smooth,** with unstriped fibers, or **striated,** with fibers that are cross-striped. **Smooth involuntary** muscle tissue is found in the walls of the viscera and arterial blood vessels. There is no direct conscious (voluntary) control over its function. **Involuntary striated** or **cardiac** tissue is found in the heart. **Striated voluntary** muscles are under conscious, subconscious, and reflexive control exerted through the nervous system.

## CONNECTIVE AND SUPPORTIVE TISSUE

**Connective tissues** may be encountered in almost any part of the body. Their main functions are to form rigid structures capable of resisting shocks and pressures of various kinds and to bind together various parts of the body. Connective tissues which are closely packed are **parenchyma;** those of lower density, packed with a gelatinous material, are **collenchyma.** Connective tissues (Fig. 3–4) are generally composed of fibers, a matrix (a ground substance that supports and gives rise to fibers), and scattered cells. The fibers may be of three types: white (collagenous), found in parallel bundles that may be crossed or interlaced; yellow (elastic), which may be branched or bent and serve to

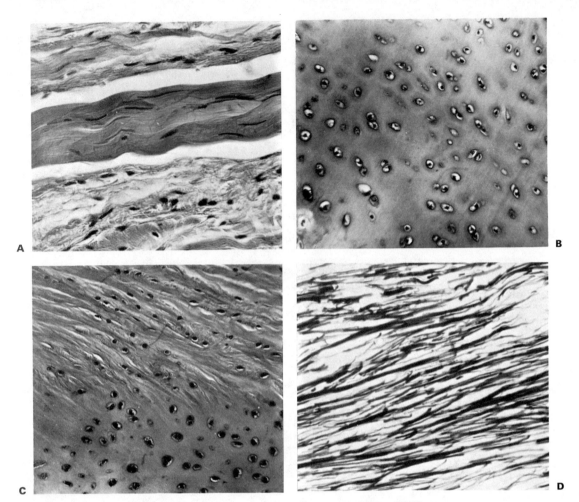

**Figure 3–4** Connective (supportive) tissues. (A) Tendon tissue. 500×. (B) Hyaline cartilage. 320×. (C) Fibrocartilage. 320×. (D) Ligament; note elastin. 500×.

bind organs to one another and skin to underlying muscles; and branched (reticular). **Tendons** contain collagenous fibers and attach muscles to bone. **Ligaments** contain collagenous (and sometimes elastic) fibers and attach bone to bone.

**Cartilage** is another type of connective tissue. It is made of firm specialized cells and an elastic matrix known as **chondrin**. The chondrin is enveloped by a thin fibrous coat of **perichondrium**. Three basic types of cartilage occur in vertebrate animals. **Hyaline cartilage** is a simple skeletal cartilage found in all vertebrate embryos. **Elastic cartilage** contains some yellow fibers and is found in the external ears and Eustachian tubes of mammals. **Fibrocartilage** is the strongest type of cartilage, containing mostly fibers and appearing in the body at areas of stress.

**Bone** or **osseous** tissue is found only in the skeletons of bony fish and land vertebrates.

### NERVE TISSUE

**Nerve tissue** is composed of **neurons**, units specialized for conductivity and ability to respond to stimuli (irritability). Three major types of neurons are

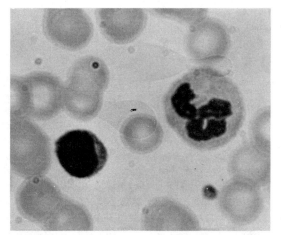

**Figure 3-5** Human blood vascular tissue. Note neutrophil (leukocyte with multilobed, dark-stained nucleus), basophil (leukocyte with dense nuclear material), thrombocytes or platelets (very small discoidal bodies), and erythrocytes (large discoidal bodies). 2000×.

recognized. **Sensory neurons** relay impulses that originate at **receptors** located at such sites as the taste buds, eardrums, skin, and retina to the central nervous system (CNS). **Motor neurons** transmit information to **effectors** (muscles or glands) and stimulate them to perform their functions. **Associate neurons** or **interneurons** may appear between neurons. Neurons are arranged in "chains," meeting at **synapses,** where impulses are chemically transmitted from neuron to neuron.

## VASCULAR TISSUE

Vascular tissue (Fig. 3-5) is a fluid tissue composed of white blood cells (**leucocytes**), red blood corpuscles (**erythrocytes**), blood platelets (**thrombocytes**), liquid **plasma,** and **lymph.** The tissue fluid is an accessory to the blood proper, arising from the blood by diffusion through the walls of the capillaries into the tissue spaces. It is the fluid medium in which the individual cells live.

## ADIPOSE (FAT) TISSUE

Adipose (fat) tissue is composed of either polygonal or rounded cells (Fig. 3-6), used for energy storage and insulation. The thin walls harbor fat droplets and an eccentric nucleus. There are two main types of adipose tissue. The

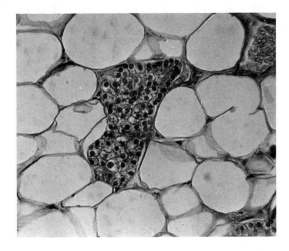

**Figure 3-6** White adipose tissue (light-colored bodies). 800×.

white variety makes up most of the body fat of an animal. **Brown** fat tissue (much less abundant than white) is found only in restricted spots within the organism.

### MOIST MEMBRANES

Formed of connective tissue and a modification of epithelium, **moist membranes** are constantly bathed in a thick mucous or thin watery secretion. They act as a defense against bacterial infection and provide a lubricating function. There are two types of moist membranes, mucous and serous. **Mucous membranes** are composed of either simple or stratified epithelium and are present in respiratory passages, genital tracts, sinuses, urinary tracts, alimentary canals, and other hollow organs. **Serous membranes,** usually composed of flat **mesothelium** (a thin membrane lining a body cavity), are supported by a thin layer of connective tissue. Moisture is provided by a thin substance, with free cells derived from blood and mesothelium. The portion of a serous membrane appearing on the exposed surface of the cavity is the **visceral** portion. Lining the cavity wall is the **parietal** portion. The pericardium, pleura, and peritoneum are serous membranes.

# Organs

An **organ** is a group of tissues cooperating in a common function. Most organs are composed of more than one type of tissue. The intestine (Fig. 3–7) is covered and lined by epithelia, its framework is of connective tissue, it is coordinated by nervous tissue, and vascular tissue provides for the transportation of nutrients. Organs usually have one type of tissue that is most important for that organ's particular function. Such a tissue is referred to as the **parenchyma.** Tissues of secondary importance, **stroma,** are frequently supportive. Several organs may work together to perform a major bodily function. There are ten major **organ systems.** An overview of these systems, with emphasis on their functions in human organisms, follows.

### ORGAN SYSTEMS

**BLOOD-VASCULAR LYMPHATIC SYSTEM.** The **blood-vascular lymphatic system** is a transportation system, carrying nutrients and wastes. The major trans-

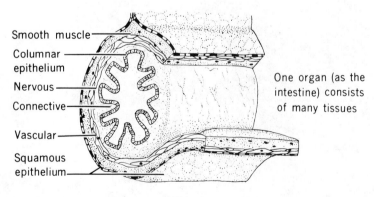

**Figure 3–7** One organ (such as the intestine) consists of many tissues that complement to form the functional whole.

ported substance is **blood.** Within the blood, oxygen from the lungs is carried to tissues, carbon dioxide from tissues to the lungs, food material to tissues, hormones and other internal secretions to various parts of the body, and metabolic wastes to the excretory organs. This system also helps to maintain a normal temperature in "warm-blooded" animals and aids in maintaining an internal fluid pressure. Blood is transported away from the heart by means of the arteries to the capillaries, through the veins, and back to the heart.

Some substances, known as **lymph,** diffuse out of the capillaries and are not immediately replaced. The **lymphatic** portion of this system picks up the lymph and redeposits it in the main circulatory system.

**RESPIRATORY SYSTEM.** The main organs of the **respiratory system** include lungs, gills, and air passageways. In human beings, air enters the respiratory system through the mouth and/or nostrils, then passes through the larynx, trachea, and bronchial tubes into the lungs, where **external respiration** occurs. Blood circulating through this system gains approximately 8 percent oxygen and loses approximately 7 percent carbon dioxide. **Internal respiration** involves the passage of oxygen and carbon dioxide at the capillary level, outside the lungs.

**DIGESTIVE SYSTEM.** The chief function of the **digestive system** is the ingestion, digestion, and absorption of food. Food enters the mouth and is masticated (chewed) by the teeth, with assistance from the tongue. Salivary glands provide saliva and some digestive enzymes. The food passes through the pharynx and esophagus into the stomach, where mechanical and chemical changes occur. Most of the nutrients are absorbed in the small intestine, while substantial water absorption takes place in the large intestine (colon). Undigested food material and other wastes are eliminated from the **rectum** at the end of the colon.

**ENDOCRINE SYSTEM.** The **endocrine system** is a major controlling factor in growth and development and exerts continuous control over most major metabolic activities. It is composed of secreting glands that lack ducts. The secretions (usually hormones or enzymes) emanate from the gland as a whole and are interdependent with the secretions of other endocrine glands. The thyroid, pituitary (**hypophysis**), adrenal, parathyroid, thymus, and other glands are constituents of the endocrine system.

**NERVOUS SYSTEM.** The brain, spinal cord, various ganglia, and nerves make up the **nervous system.** This system enables an organism to have awareness of itself and its environment through sensations: sight, smell, hearing, taste, and feel. It coordinates the various parts of the body, exerts control over the internal organs, and is responsible for human thought and conduct. The **central nervous system** (CNS) consists of the brain and spinal cord. The **peripheral nervous system** is comprised of the special sense organs and the autonomic and spinal nerves, which connect to the CNS. **Autonomic nerves** innervate the heart, some glands, and smooth muscular tissue, usually exerting their influence without the conscious knowledge of the organism. **Spinal nerves** extend from the spinal cord at regular intervals.

**SKELETAL SYSTEM.** The **skeletal system** serves both for support and for muscle attachment. It varies dramatically between organisms. Most higher organisms have an **endoskeleton,** contained within the animal. Lower organisms possess **exoskeletons,** which hold the organism together and provide protection.

**REPRODUCTIVE SYSTEM.** The **reproductive system** is concerned with the continuation of the species. The essential organs of reproductive systems are **ovaries**, in which eggs develop, and **testes**, in which sperms are formed. Collectively, these organs are termed **gonads**. Accessory organs include those that supply yolk and various secretions, ducts that carry the sperm or eggs, and copulatory organs necessary to ensure fertilization.

MUSCULAR SYSTEM. Motion and locomotion are the main functions of this system. Visceral, striated, and cardiac muscle are the gross components of the muscular system. In many organisms there are literally thousands of muscles. Each capillary, for example, has its own precapillary sphincter.

EXCRETORY SYSTEM. The **excretory system** concerns itself with the elimination of waste products of metabolism and digestion. **Urine**, consisting largely of water and urea, is extracted from the blood by the **kidneys**, passes through the ureters into the **bladder**, and is eliminated through the **urethra**. Other excretory systems include the **digestive tract**, which eliminates food and bile waste products; the **liver**, which excretes a toxic waste (bile) sometimes temporarily stored in the **gall bladder**; the **lungs**, which eliminate carbon dioxide from the blood; and the **skin**, which excretes substances through sweat glands.

INTEGUMENTARY SYSTEM. The **integumentary system** consists of the skin and its modifications—hair, scales, feathers, horns (a form of hair), and so on. Its functions, though primarily protective, include excretion and temperature regulation (through **perspiration**).

## General Considerations of Tissues and Organ Systems

The systems mentioned in this chapter are interdependent. Calling them separate systems is merely a convenient way of beginning to understand those interrelations. Tissues depend on their constituent cells, and organs are composed of one or many tissues, all of which must function cooperatively for the organ to be of use. In higher organisms, the nourishment of the circulatory system is essential to its organs. The circulatory system could not function without the muscular system, which includes the heart and arterial muscles. These muscles require stimulation from the nervous system in order to function. For the nervous system to function, energy must be available and the organism obtains that energy through the digestive system. All cells are aerobic (require oxygen), and so the respiratory system is indispensable to all other systems. The respiratory system, as well as all the others, would fail because of toxic buildup unless the excretory systems were there to eliminate wastes. The endocrine system is necessary to regulate the entire animal, to "keep it in tune." The skeletal system is necessary for, among other reasons, the production of red blood cells (from bone marrow). An organism must be protected from the environment by its integumentary system. Without it, everything would dehydrate and eventually the organism would die. Without a reproductive system, there would be no continuation of the species. The best means of understanding the complex relationships among these organ systems is to think of them collectively as one total system, the **organism**.

# Summary

Histology is the study of the structure and function of tissues. Tissues are composed of groups of cells cooperating in a particular set of functions.

Epithelial tissues cover surfaces and appear in various forms.

Muscular tissues are specialized for contraction and may be smooth or striated, voluntary or involuntary.

Connective tissue includes cartilage, ligament, and osseous tissue. It is made mostly of various types and configurations of fibers of variable strengths.

Nervous tissues, composed of neurons, serve to conduct electrical impulses throughout the organism.

Vascular tissue is fluid and occurs within the circulatory system. It is responsible for transport of wastes and nutriment and serves other functions.

Adipose (fat) tissue is used for storage of energy and insulation.

Moist membranes (either mucous or serous) function as defense against bacterial infection and in lubrication.

Organology is the study of organs. Organs are made of various types of tissues and perform many complex functions. The 10 major organ systems found in human beings and other organisms are: blood-vascular lymphatic, respiratory, digestive, endocrine, nervous, skeletal, reproductive, muscular, excretory, and integumentary. These systems are interdependent and will be discussed in detail in Chapters 23 to 27.

# 4
# Taxonomy and Phylum Synopsis

**Taxonomy**, the identification and classification of organisms, provides a systematic framework for studying the various life forms that inhabit our earth. Taxonomy not only establishes order from chaos, but also provides a system for naming animals.

## HISTORY OF TAXONOMY

The first person to collect and organize knowledge of animal classification was Aristotle (384–322 B.C.). His method, classifying animals by "their way of living, their actions, their habits, and their bodily parts" was somewhat haphazard and limited.

The English biologist John Ray (1627–1705) devised a system of classification utilizing structural similarity as a basis. Following Ray was the famed Carolus Linnaeus (1707–1778), a Swedish scientist who was the first to truly systematize taxonomy. He identified 236 different animal species, although his classification scheme was based primarily on morphology.

The need for taxonomy is universal. In the mountains of New Guinea, the Papuans classified 137 of 138 avian species without ever having heard of "taxonomy." In fact, all "primitive" people develop intricate naming systems for the life around them.

After the appearance of Charles Darwin (1809–1882), evolutionary considerations were worked into contemporary taxonomic studies, complementing the purely morphological differentiation. Ernst Haeckel (1834–1919) further refined the taxonomic system in 1886 by representing phylogeny by means of a branching diagram. With further sophistication, genetic, ecological, geographical, and morphological characteristics were taken into account, giving taxonomy a greater biological perspective.

Recent innovative suggestions include R. Sokal's numerical taxonomy (1966). Theodore Jahn (1966) expanded upon this idea with the suggestion that the numerical categories be computerized.

## PURPOSE OF TAXONOMY

It is estimated that 1,900,000 different animal species have been described in the literature of zoology, and that each year about 10,000 new ones are identified. Taxonomy's necessity is evident when one imagines trying to gather information on a particular unfamiliar organism. The objective of taxonomy is to provide a useful, convenient universal system with which all information about a given organism or group of organisms may be retrieved. The system is hierarchical, and in order to function, several prerequisites must be met. It must be able to discriminate among different forms of organisms, provide the criteria for such discriminations, and have the capacity for grouping smaller **taxa** (*plural*) into larger, more inclusive taxa. A **taxon** (*singular*) is any unit of the taxonomical system.

## The Taxonomic System

Taxonomy is dependent upon a **natural classification** system, using as many physical characteristics as possible to determine the true genetic relationships between animals. Systems that group animals on the basis of superficial resemblances (color, habitat) are more **artificial.** The natural classification system and the principle of evolution form the base for current taxonomical classification.

Basically, the smallest taxon is the **species,** a group of similar or identical organisms which interbreed. One or more species with considerable similarities form a **genus.** Species within a genus are assumed to have common phylogenetic origin. Each genus is distinct from all other genera. Genera with considerable similarities form a **family.** Similar families form an **order,** similar orders form a **class,** similar classes form a **phylum,** and the phyla collectively form a **kingdom.** As we get higher and higher in the taxonomical sequence, the categories get more and more inclusive and therefore more general.

### RULES OF NOMENCLATURE

In 1901, the International Congress of Zoology organized an International Commission on Zoological Nomenclature, which has served since that time. The adopted system of nomenclature is derived from Linnaeus's *Systema Naturae* (10th edition, 1758). The rules cover formation, derivation, and correct spelling of zoological names, the author's name, and any proposed but rejected names. The specimen first used to establish a specific name, termed the **type specimen,** always retains the original name even if the species is renamed. If a new genus is described, a representative type specimen is also decreed and follows the same rules as a species type specimen. Animal family names are formed by adding *-idae* to the stem of the name of the type genus (the names of a subfamily are formed by adding *-inae*). A particular organism is named by the **binomial nomenclature** system, which consists of the organism's generic and specific names. Generic names are capitalized and followed by the specific epithet; the entire name is either italicized or underlined. *Caiman sclerops,* for example, is the scientific name for a type of alligator. All scientific names are either in Latin or latinized.

If a species is divided into two or more subspecies, a **trinomial** nomenclature system may be employed. For example, two very similar types of robins are found in the southern and eastern parts of the United States. The southern form, duller in color than the eastern form, is termed *Turdus migratorius achrusterus,* the word "achrustera" meaning "duller color."

The complexity of the taxonomical system may be seen in this rather complete classification of a species of crayfish, *Cambarus bartoni:*

| | |
|---|---|
| Kingdom: | Animalia |
| Phylum: | Arthropoda |
| Subphylum: | Mandibulata |
| Class: | Crustacea |
| Order: | Decapoda |
| Family: | Astacidae |
| Genus: | *Cambarus* |
| Species: | *bartoni* |

## VARIATIONS IN TAXONOMICAL CLASSIFICATION

The taxonomic scheme presented in this text is a synthesis of current thought on the science. Unfortunately, a truly universal system does not yet exist. A few major and several minor differences occur on an international, national, regional, and even institutional level. Taxonomy is, in fact, a synthetic construct placed upon the **vivicum** (all living things) to better understand existing relationships. Variations in classification obviously have no effect on the actual numbers and characteristics of species.

There are several reasons for classification variation. The original name given to a species has priority over any subsequent names (termed **synonyms**) despite the possibly greater accuracy of the new name. A **law of priority**, enforced by the International Commission on Nomenclature, should be followed by zoologists when synonymous names exist. Zoologists sometimes find it impractical to use the original name given to a species, and will use a synonym for greater clarity. Hence, throughout literature, there may be several synonyms for the same organism.

Animals may differ on very fine points, and distinctions made between them become highly subjective when sufficient data are not available. The diversity in the level of inclusiveness insisted upon by various taxonomists also creates inconsistencies in taxonomical classification. Taxa may be grouped in too general categories or be subdivided to extremes; between lies what Aristotle called the "Golden Mean."

Before we examine the traits that link animals together in various grouping, we need some definitions.

# Definitions of Morphological Terms

When referring to parts of an organism, several morphological terms are used to designate positions, planes, or axes of the organism. The **anteroposterior axis** or **longitudinal axis** runs from the nose (**anterior** or **rostral end**) to the tail (**posterior** or **caudal end**), while a **transverse axis** is perpendicular to the longitudinal axis. A **sagittal plane** runs from the **dorsal** (top or back) to the **ventral** (bottom or front) extents of the structure. An infinite number of such sagittal planes are possible, the most common being the **midsagittal plane** (running directly along the midline of the structure, necessarily including the anteroposterior axis). A **transverse plane**, also called a **cross section**, intersects the longitudinal axis at only one point and is perpendicular to it. Perpendicular to both the sagittal and transverse planes is the **frontal plane**. Various **oblique** planes may be identified; they are not perpendicular to any of the axes mentioned. **Medial** means "toward the center or midline"; **lateral** means "away from the center or midline." In cases where the anterior–posterior designation is inappropriate, the terms **oral** ("mouth") and **aboral** ("not mouth") are used.

## Discriminational Bases of Taxonomy

Myriad animal characteristics are taken into account in the process of zoological classification. Animals are generally considered more related to each other if they share particular common traits. We shall consider next some of the more important ones.

### FORMS OF SYMMETRY

**Symmetry** refers to the arrangement of parts in relation to planes and centralized axes. In a **symmetrical** animal, one or several planes will divide the organism into parts with essentially equivalent geometrical designs. Animals are either symmetrical or **asymmetrical**. The common amoeba, *Amoeba proteus*, is an asymmetrical organism—no plane will consistently divide it into equal parts. The amoeba's body has no stationary form or arrangement of parts. In addition to the amoeba, many protists and most poriferans are asymmetrical. There are four basic types of symmetry (Fig. 4–1).

RADIAL SYMMETRY. A **radially symmetrical** animal may be divided into a number of similar parts, called **antimeres**, which are arranged about a central **longitudinal axis**. It is possible to draw a number of planes through this axis, dividing the body of these radially symmetrical animals into equal parts. The hydra (Fig. 4–1) is an example: its cylindrical shape is divided by similar tentacles that radiate out from the central mouth. Some simple sponges, the majority of coelenterates, and adult echinoderms are radially symmetrical. Radial symmetry seems best suited to **sessile** (attached or at least immobile) animals, since the similarity of the antimeres enables these animals to obtain food or repel enemies from all sides.

BIRADIAL SYMMETRY. A set of three mutually perpendicular axes may be imposed on a **biradially symmetrical** animal such as the ctenophore *Mnemiopsis* (a comb jelly). Two planes, one passing through the longitudinal and sagittal axes and the other through the longitudinal and transverse axes, will divide the animal into symmetrical halves, each plane producing its own *different* set of halves. Biradial symmetry is also found in some coelenterates of the class Anthozoa. This form of symmetry is best suited to floating organisms.

SPHERICAL SYMMETRY. Only organisms that are spherical in shape have **spherical symmetry**, as does the *Volvox*. This symmetrical arrangement is somewhat disadvantageous since only an indefinite type of locomotion (usually free-floating) can take place.

BILATERAL SYMMETRY. An animal exhibiting **bilateral symmetry** is so constructed that the main organs are generally arranged in pairs on either side of an axis which passes from the anterior end to the tail (posterior end). A dorsal surface and a ventral surface are recognizable, as are right and left sides. There

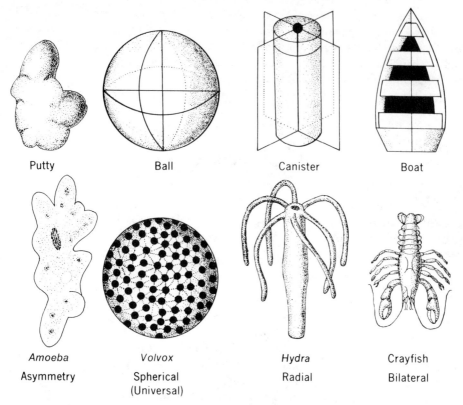

**Figure 4–1** Patterns of symmetry in animals.

is only one plane which will divide the body into two similar parts, the midlongitudinal, or midsagittal, plane. Bilateral symmetry is characteristic of most animals living today, including all vertebrates and many invertebrates. Such morphology stresses forward movement and the differentiation of a head end.

It is doubtful that perfect symmetry is to be found anywhere in the animal kingdom. Although the human form is considered a good example of bilateral symmetry, the right and left sides of the human body are not identical. The human heart, for instance, is eccentric. Nevertheless, the zoological concept of symmetry is of great importance in the study of animals; this will become evident as the different groups are studied.

## METAMERISM

**Metameric** animals have bodies composed of more or less similar parts, or they have organs arranged in a linear series along the main axis. Each part is called a **metamere, somite,** or **segment.** When segmentation occurs only in the external parts (and possibly the body wall) of an animal, it is **superficial metamerism.** Many animals, including all vertebrates, have **internal (metameric)** metamerism. In vertebrates, the vertebrae of the backbone, the ribs, and nerves are metamerically arranged. The earthworm has both superficial and metameric metamerism. The body consists of a great number of similar segments, and the ganglia of the nerve cord, the body cavity, blood vessels, muscles and excretory organs are segmentally arranged (Fig. 4–1). When the segments of an organism are similar, acting as semiindependent units (unable to function in isolation from other segments, but having independent nervous, excretory, and reproductive organs, as in the earthworm), the animal has **homonomous segmentation.** The crayfish, on the other hand, is a **heteronomous** animal, since division of labor has resulted in dissimilarity of the metameres of different regions of the body (Fig. 4–1).

## CEPHALIZATION AND POLARITY

Bilaterally symmetric animals have a definite anterior (oral) end and a posterior (aboral) end and are said to be **polar.** A concentration of nervous tissue is found within a **head** at the oral end; this morphological state is known as **cephalization.** A cephalic, polar organism has differentiation along its longitudinal axis. Cephalization and polarity are adaptive; they represent the most efficient means of reacting to most environments. As organismal complexity increased, animals tended to move unidirectionally. The advantages of having the major sense organs and most neural integrational material concentrated at the anterior end of an organism (the pole that meets the oncoming environment first) are obvious. In addition, for erect or semierect terrestrial animals, the visual sweep is vastly increased over that of animals lower to the ground.

## THE COELOM

The presence or absence of a **coelom** (pronounced *see'-lome*) is another animal characteristic useful in taxonomy. A coelom is a true body cavity, usually located in the area between the digestive tract and the outer wall of an animal. The coelom's evolutionary significance lies in the fact that it permits relatively large size and complexity in an organism by creating greater surface-area exposure, thereby allowing diffusional and osmotic processes more opportunity to occur.

**Acoelomata,** animals without a coelom include the jellyfish and the flatworms. Those with a coelom, the **eucoelomata,** may be thought of as "a tube within a tube." The coelom develops between somatic (outer) and visceral (inner) mesoderm. It is lined with mesodermal epithelium, called **peritoneum.**

A coelom may appear in various forms. The mollusks, arthropods, onychophoran, and others have a **hemocoel,** a great cavity within which the blood sloshes around (a rather primitive circulatory system). Some animals, such as the rotatorians and nematodes, have an unlined coelom. A peritoneumless coelom is a **pseudocoel.**

The coelom of the earthworm and of most annelids is divided into chambers by **septa.** Sometimes the vertebrate coelom is divided into **thoracic** and **abdominal** cavities by the diaphragm. These vertebrates also have a separate pericardial cavity.

## OTHER MEANS OF DISTINCTION

**Homologous** organs have common ancestral origins and identical embryological development and are thus fundamentally similar in structure. They may have similar functions, as do the hindlegs of a horse and the legs of a human being; or they may have very different functions, as do the wings of birds and the arms of human beings. Homologous structures show evolutionary relationships.

Organs that do not have a common embryological origin and are fundamentally different in structure but do have similar functions are **analogous.** The wings of a butterfly and those of a bird are analogous because they are both used for flight, but they do not share common structure or origin.

**CELL NUMBER.** Cell number is a quick indication of gross similarity or difference and has become more important in recent years. Unicellular organisms carry out life processes much differently from multicellular organisms.

**DIGESTIVE-TRACT PRESENCE AND VARIATION.** A digestive tract may occur in several forms, each being a more-or-less gross indication of taxonomic similarity. The protozoans and sponges lack a digestive system. The other group, enterozoans, possess digestive systems. Some organisms, such as coelenterates, have only a mouth; this type of system is incomplete. Other multicellular organisms have a complete digestive tract; both mouth and anus are present.

**EMBRYONIC DEVELOPMENT.** Another helpful means of classification is represented by the various embryological differences among organisms. The **biogenetic law** states that **ontogeny recapitulates phylogeny.** This means that evolutionary development that took hundreds of thousands of years (phylogeny) is "compressed" into the relatively short time span in which an organism develops from a fertilized egg into a new individual (ontogeny). Actually, the biogenetic law must be amended: Ontogenetic development is only roughly parallel to phylogenetic development. Many gaps, inclusions and omissions, occur in their comparison. While a multicellular organism is developing prenatally, it goes through changes that become successively more individualized over time. Initially, most animals are practically indistinguishable. The human fetus, for example, has a rudimentary tail for a short time during its development.

A major embryological point of dissimilarity among animals is the site where the mouth forms on the **blastula** (the "ball" of cells occurring early in ontogenesis) (Chapter 27). In the **protostomes** (generally lower animals, such as annelids, arthropods, and flatworms), the mouth is formed from the **blastopore**, the opening of the gastrula. In **deuterostomes** (higher forms, including the hemichordates and chordates), the anus is developed from the blastopore while the mouth develops elsewhere. The pattern of **cleavage** is also quite variable, with animals exhibiting like characteristics generally tending to cleave in similar fashion. **Holoblastic** cleavage, where the egg is *completely* divided, and **meroblastic** cleavage, where the division is incomplete, form an index of egg yolk concentration: The higher the concentration, the greater the probability for meroblastic cleavage.

The number of "germ layers" is also important. For multicellular organisms, either two or three germ layers occur during embryological development. The coelenterates are **diploblastic,** having two germ layers (**endoderm** and **ectoderm**), while all higher multicellular animals are **triploblastic,** with three germ layers (endoderm, ectoderm, and, between them, **mesoderm**).

**APPENDAGES.** A body part substantially protruding into the extraorganismal medium is termed an **appendage.** Different types of appendages serve different purposes. Locomotion, feeding, and sensory functions are usually performed by appendages. Tentacles, antennae, legs, fins, and arms are some of the different appendages serving various functions, such as locomotion, feeding, and sensation. The number of appendages an organism has is often a characteristic used in taxonomical classification, especially among the arthropods.

**SKELETAL SYSTEM.** An organism's position in a taxonomic sequence may be indicated by its skeleton, or lack of one, and the presence or absence of a **notochord** (primitive supporting structure). The skeleton may be on the outside of the body (exoskeleton) or on the inside of the body (endoskeleton). Some animals, such as turtles, have both an endoskeleton and an exoskeleton.

SEX. Animals vary considerably in their sexual characteristics. **Dioecious** or **gonochoristic** organisms are unisexual. **Monoecious** or **hermaphroditic** organisms have both male and female sexual organs in the same individual.

LARVAL FORMS. This is a rather important means of species identification. One or several **larval** forms may occur in some organisms after birth and before attainment of the **adult** form. Often, the larval form indicates evolutionary relationships between animals.

## Synopsis of Major Phyla

Throughout this text, we will be learning in some detail the characteristics of the various animal phyla. The purpose of this synopsis is to give you a sense of the diversity within the animal world. New forms of living species are constantly being named; thus all estimates of the number of animal species must be regarded as tentative. For further information, consult the appropriate chapters.

### KINGDOM PROTISTA

PHYLUM PROTOZOA. This phylum represents the most important "protozoic" animal-like organisms in the kingdom Protista. Collectively, somewhere between 15,000 and 50,000 species exist, but this figure is probably too low, for it is thought that most of the unicellular protists have yet to be discovered. There are three major subphyla in the phylum Protozoa.

*Subphylum Sarcomastigophora.* Many of these organisms are asymmetrical. Amoebae and flagellates are examples. Locomotion is accomplished by pseudopodia, undulipodia, (see p. 25) or both. Reproduction is asexual or sexual. Sarcomastigophora have only one type of nucleus.

*Subphylum Sporozoa.* These are asymmetrical organisms and are all parasitic. They lack locomotor organelles. Sporozoans have a single type of nucleus and spores are formed.

*Subphylum Ciliophora.* These asymmetrical organisms have undulipodia or sucking tentacles in at least one stage of the life cycle. Ciliophores have two kinds of nuclei and may reproduce asexually or sexually. Ciliates and suctorians are examples.

### KINGDOM ANIMALIA

PHYLUM PORIFERA (SPONGES). There are about 5,000 species of sponges, exhibiting either radial or no symmetry. Their body forms vary, most being globular, vaselike, or treelike. The poriferans lack tissues and are riddled with **ostia** (pores), chambers and canals that allow water passage throughout the body. Poriferans reproduce both asexually and sexually.

PHYLUM CNIDARIA OR COELENTERATA. The coelenterates, numbering about 10,000, may have radial or biradial symmetry and are either sessile or free-swimming. These are the first organisms with some form of digestive area, composed of a gastrovascular cavity. They possess primitive nervous and muscular systems. Reproduction may be either asexual or sexual. Most coelenterates are dioecious. All coelenterates have stinging organelles, **nematocysts.** They are all aquatic; the majority occupy a marine (saltwater) habitat. Hydroids, jellyfish, sea anemones, and corals are examples.

**PHYLUM CTENOPHORA.** There are about 100 species of ctenophorans, all with biradial symmetry. The organ systems of the ctenophorans are very similar to those of coelenterates. Ctenophorans usually possess eight rows of **comb plates** on their external surfaces which are used for locomotion. These marine organisms include sea walnuts, sea gooseberries, and comb jellies.

**PHYLUM PLATYHELMINTHES.** These are the most primitive of the bilaterally symmetrical animals. They are commonly called **flatworms** because their bodies are flattened in a dorsoventral relation. Although they exhibit a triploblastic body, the platyhelminths are, nonetheless, acoelomates. Most contain an incomplete digestive system. The basic structure of the excretory system is the flame cell. A primitive muscular system and a relatively advanced nervous system (including ganglia) exist. They have no circulatory, respiratory, or skeletal system. Most forms are monoecious and many are parasitic. Platyhelminths inhabit the sea, fresh water, land, and various **hosts.** They include planarians, flukes, and tapeworms.

**PHYLUM ASCHELMINTHES.** These organisms number about 12,000 species. They are unsegmented, triploblastic, and exhibit bilateral symmetry. An unlined body cavity (**pseudocoel**) is present, as is a nervous system. Aschelminthes may have **ciliated pit** sense organs, eyespots, bristles, or papillae. Usually having separate sexes, these organisms reproduce utilizing a system of ducts and gonads. The rotifers and nematodes are representatives of these free-living and parasitic animals.

**PHYLUM ANNELIDA.** The approximately 13,000 species of annelids are the lowest organisms to have metameric segments. These worms have a true coelom within their triploblastic bodies. They exhibit bilateral symmetry. Annelids have an advanced, closed circulatory system as well as a complete digestive system. The basic structure of the excretory system is the nephridium. Most annelids have **setae,** which are used primarily in locomotion. A ventral nervous system is present. Oligochaetes and leeches are monoecious whereas polychaetes are dioecious. Most are dioecious; some forms may bud. Annelids are primarily marine, but some may be terrestrial or parasitic. Examples include earthworms, sandworms, and leeches.

**PHYLUM ARTHROPODA.** The arthropods are about three times more numerous in described species (currently about 800,000) than all other phyla combined. They have bilateral symmetry with triploblastic, metameric bodies. All possess chitinous exoskeletons and metameric appendages. Arthropods have open circulatory systems (**hemocoels**). Their muscular system is highly developed, composed primarily of **striated muscles.** The respiratory system may utilize the body surface, gills, **tracheae,** or a type of breathing apparatus known as a **book lung.** A well-developed excretory system composed of **Malpighian tubules** and/or **coxal glands** is present. The nervous and sensory systems are also highly developed and sexes are usually separate. Arthropods may go through **molting** and **metamorphosis** (Chapter 13). This phylum includes lobsters, crayfish, insects, spiders, ticks, and mites.

**PHYLUM MOLLUSCA.** The 90,000 living molluscan species are usually bilaterally symmetric with a triploblastic body. Most have a dorsal or dorsoventral **mantle** which secretes a **shell,** a form of exoskeleton. Locomotion is accomplished in most mollusks with a ventral muscular foot. Mollusks have advanced circulatory and nervous/sensory systems. Digestive and respiratory systems are also present. Mollusks may be either monoecious or dioecious, having a system of ducts with one or two gonads. Fertilization occurs internally or externally. They are mostly marine; examples include clams, octopuses, squids, oysters, and snails.

**PHYLUM ECHINODERMATA.** The echinoderms number about 6,000 living species. Their radial symmetry is usually expressed in **pentamerous** divisions (divisions of five). These triploblastic organisms have an endoskeleton of calcareous ossicles. Locomotion is accomplished by **tube feet,** controlled hydrostatically by the unique **water-vascular system.** The coelom is definitive, with **amoebocyte**-containing intercoelomic fluid. (An amoebocyte is a body cell capable of individual movement by means of pseudopodia.) Most echinoderms are dioecious; all are marine. Examples include sea stars, urchins, sand dollars, and sea cucumbers.

**PHYLUM CHORDATA.** The chordate species number about 66,000. They have bilateral symmetry and are triploblastic. Chordates are the highest developed animals; all possess a **notochord** for at least a period of time (in higher chordates it is replaced by the development of a **vertebral column**). Most chordates show some form of metamerism and an endoskeleton is usually present. Organ systems reach their highest state of development in this phylum. Physiological similarity between species is greater here than in any other phylum. Characteristics of chordates distinguishing them from lower phyla include a **postanal tail** (in some vertebrates), paired **pharyngeal gills** (present in the embryos of all chordates), a **ventral heart,** and a **hepatic portal system.** A **neck** is found in most lung-breathing members of this phylum. Chordates inhabit all of the biosphere; they include bony fish, amphibians, reptiles, birds, and mammals.

# Summary

Taxonomy, the ongoing study of classification of organisms, is most useful for its systematic description of the evolutionary, adaptational, and anatomical relationships within the vivicum.

Aspects of symmetry, cephalization, polarity, metamerism, coelomic presence and form differences, homology, analogy, cell number, digestive tract presence and anatomy, skeletal system presence and form, sexual type, larval form presence, and type are all attributes taken into consideration in the taxonomic process.

Kingdoms, phyla, classes, orders, families, genera, and species are the major taxonomical categories (taxa) placed upon living organisms in respectively decreasing order of generality.

Various characteristics of the animal-like organisms of phylum Protozoa of the kingdom Protista and the major phyla of the kingdom Animalia were presented in an introductory manner. You are now prepared to approach these phyla in some detail. The discovery of **patterns** in the following chapters will serve to unify your view of zoology.

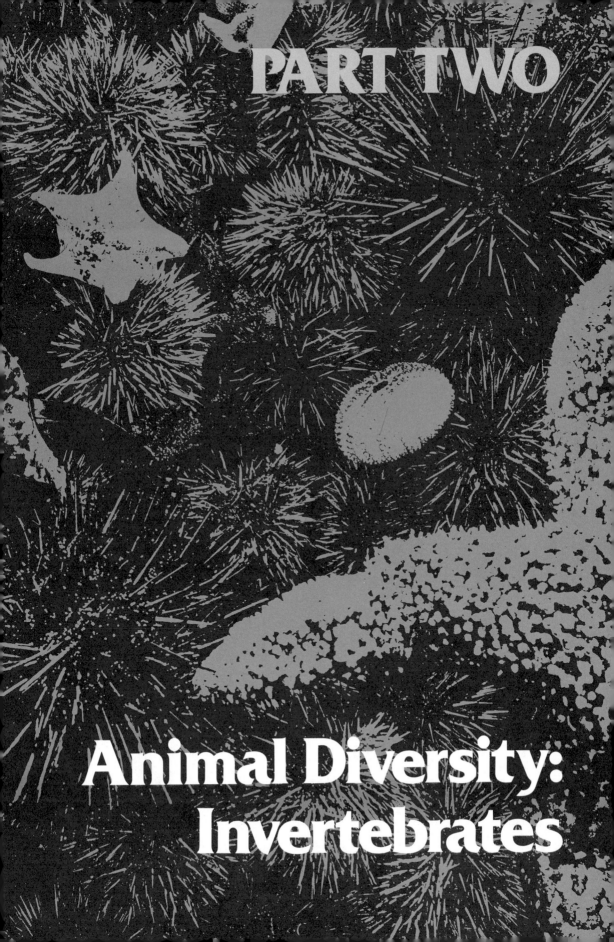

# PART TWO

# Animal Diversity: Invertebrates

# 5

# Phylum Protozoa

Protozoans are the least complex organisms that display animal-like traits such as **motility** (spontaneous motion) and ingestion of food. As the name protozoan (Gr. *protos*, first + *zoon*, animal) implies, these organisms of the kingdom Protista are considered to resemble more closely than any other organisms the original forms of animal life on earth. Protozoans are highly successful in an evolutionary sense; their numbers, as individuals, far exceed those of all other animals combined.

These one-celled animals are studied for a variety of reasons. They are important to us medically because many cause diseases, such as malaria and sleeping sickness. Their rapid rate of reproduction makes them ideal subjects for genetics experiments. Protozoans are simple prototypes of more complex organisms. Progressing from them to the increasingly complex organisms will give you a feeling for the general stream of evolution reflected in the classification scheme.

Structurally, a single-celled protozoan is in some respects comparable to an individual cell of a multicellular animal. Physiologically, the protozoan, which carries on all the basic life functions in its single cell, is comparable to the *entire* body of a multicellular animal. Some biologists prefer to call protozoans **acellular** (as opposed to unicellular) to emphasize their function as complete units. A few protozoans are composed of groups of cells, but these never differentiate into tissues.

Active protozoans are unable to inhabit dry areas, but are ubiquitous in water and moist places. They abound in freshwater ponds, lakes, and streams; billions live in the sea; moist soil often teems with them; and large numbers live on or within other animals.

The following major taxonomical divisions will be used in discussing protozoans:

Phylum: Protozoa
   Subphylum I. Sarcomastigophora
      Class 1. Mastigophora (flagellates)
      Class 2. Opalinata (opalinids)
      Class 3. Sarcodina (amoebas)
   Subphylum II. Sporozoa
   Subphylum III. Cnidospora
   Subphylum IV. Ciliophora (ciliates)

# Sarcomastigophora

Sarcomastigophores move by means of false feet (**pseudopodia**) or by undulipodia. They consist of three classes. The first, **mastigophores** (Fig. 5–1A), are small protists that usually possess a definite shape. They have long, whiplike undulipodia (flagella), which usually arise from the anterior end of the organism. Besides their locomotor function, flagella are used to capture food and as sense receptors. Mastigophores are abundant in puddles, ponds, and swamps. They fall into two distinct categories: plantlike organisms with chloroplasts, and animal-like organisms without chloroplasts. Green mastigophores, such as *Euglena viridis*, are an important part of our food chain. These are eaten by various fish and insect larvae. A large, colonial species of animal-like mastigophores, *Volvox* (Fig. 5–2), may live in either fresh or salt water.

Together with algae, mastigophores are a source of several vitamins, including A and D. Many mastigophores are parasitic in human beings, other animals, and plants.

The second class of Sarcomastigophora, **opalinids** (Fig. 5–1B), are so named because of their opalescent color. All members of this group are **symbiotic** (live in conjunction with other organisms), occurring mostly in the alimentary tracts of frogs and toads. Opalinids are oblong-shaped organisms covered with oblique rows of short undulipodia (cilia). A typical specimen may have from one to hundreds of nuclei. Opalinids are dependent upon a cyst stage in their life cycle for survival.

**Sarcodines** (Fig. 5–1C), the third class of Sarcomastigophora, are commonly characterized by the amoeba. They have one type of nuclei and move about by extensions of the cytoplasm called pseudopodia. Sarcodines live in a variety of habitats, including fresh water, the sea, soil, and as parasites within animals.

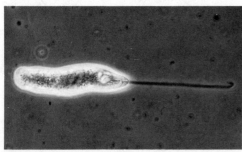
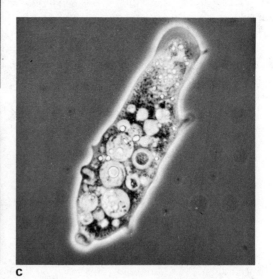

**Figure 5–1** Sarcomastigophores. (A) Mastigophore. (B) Opalinid. (C) Sarcodine.

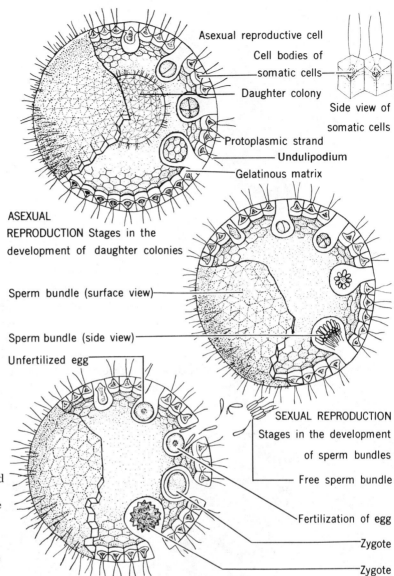

**Figure 5–2** Asexual and sexual reproduction of *Volvox*, the largest of the colonial mastigophores, consisting of 500 to 40,000 cells. The whole organism attains the size of a pinhead.

## Sporozoa and Cnidospora

All members of the subphyla **Sporozoa** and **Cnidospora** are parasites. During a portion of their life cycles most members of these subphyla develop into **spores,** which can survive outside a host. (This spore state is similar to the cyst stage of the opalinids.) This development into a spore state and the absence of locomotor organelles in the adult (**trophozoite**) stages distinguish sporozoans and cnidospores from all other protozoans. Sporozoans are quite varied in appearance, as we shall see. One representative from this group is shown in Fig. 5–3.

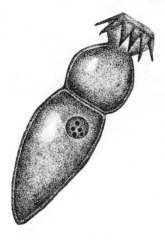

**Figure 5–3** Sporozoan trophozoite, *Actinophilus*, from gut of a centipede.

## Ciliophora

The final protozoan subphylum is the **Ciliophora** (ciliates) (Fig. 5–4). When compared to other protozoans, this group is the largest and most complex structurally. They are distinguished from other protozoans in possessing *short* undulipodia (cilia), two different types of nuclei, (a macronucleus and one or more micronuclei) and a unique form of reproduction. Most ciliates are free-living in fresh and salt water. Some, however, are important parasites of man and other animals. The ciliates, like other protists, play a considerable role in aquatic food chains. Many species ingest dissolved nutrients in the water and serve as food for small multicellular animals, which are, in turn, eaten by larger animals.

**Figure 5–4** Photomicrograph of euplotes 250×.

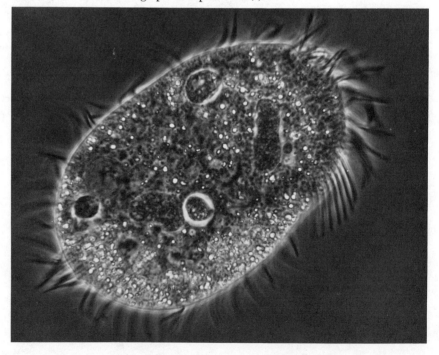

Let us now see what some members of these groups look like and how they have met the problems of survival.

## Euglena viridis—A Mastigophore

*Euglena viridis* is a common representative of the genus *Euglena*, represented by over 150 species. Euglenas are found in freshwater ponds, especially in clusters of weeds. In sufficient numbers, they impart a greenish tinge to the water.

### MORPHOLOGY

The *Euglena viridis* (Fig. 5–5) is 0.1 mm or less in length, blunt at the anterior end and pointed at the posterior end. Surrounding the peripheral layer of cytoplasm is a thin elastic membrane, the **pellicle.** Its striated appearance is due to the presence of parallel spiral thickenings. Near the anterior end is a funnel-shaped depression, the cell mouth (**cytostome**), leading into the cell gullet (**cytopharynx**). The cytopharynx is enlarged at the base to form a vesicle called the **reservoir**; adjacent to it is a **water-expulsion vesicle** (contractile

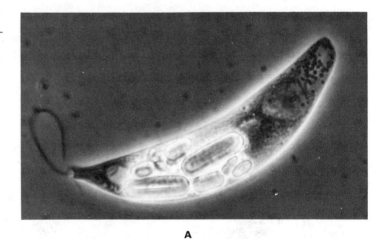

**Figure 5–5** *Euglena viridis.* (A) Photomicrograph. (B) Diagram showing internal structure.

vacuole), which collapses, discharging its contents into the reservoir and out through the cytopharynx.

Near the anterior end of the body is an orange-red eyespot (**stigma**), which is part of a light-sensitive organelle. A single undulipodium arises from the blepharoplast and extends to the outside through the cytostome. Near the center of the euglena is an oval or spherical **nucleus** containing a central body, the **endosome**. Suspended in the cytoplasm and radiating from a central point are a number of slender **chromatophores** or **chloroplasts,** green in color owing to the presence of **chlorophyll.** Each chloroplast contains a **pyrenoid,** where it is believed a starchlike substance, **paramylum,** is formed. This substance is food reserve material produced by a photosynthetic process. Paramylum bodies may also be freely suspended in the cytoplasm in the form of disks, rods, and links.

### NUTRITION

As in green plants, euglenas obtain their nutrition largely by photosynthesis, an internal process (**holophytic nutrition**). Euglenas do, however, require some materials which they cannot manufacture, such as vitamin $B_{12}$ and several minerals; these are absorbed from surrounding water. In the dark, euglenas can live on organic compounds dissolved in the water (**saprophytic nutrition**). Under these conditions the chloroplasts and pyrenoids degenerate and disappear. Although euglenas do not capture and eat other organisms (**holozoic nutrition**), *animal*-like mastigophores do eat protozoans, algae, and diatoms.

### LOCOMOTION

Swimming is accomplished with the aid of flagellum, causing the organism to gyrate through the water in a helical path. Euglenas are characterized by

**Figure 5–6** Reproduction and euglenoid movement in *Euglena viridis*.

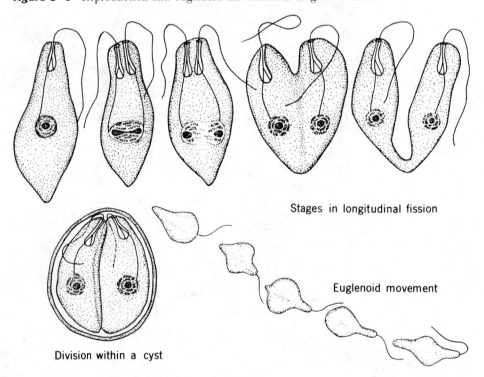

Stages in longitudinal fission

Euglenoid movement

Division within a cyst

their wormlike movements, involving waves of contraction known as **euglenoid movement** (Fig. 5-6).

## REACTIONS TO LIGHT

Euglenas (Gr. *eu*, good + *glene*, eye) are easily stimulated by changes in the direction of light. Most species swim toward an ordinary light source but away from direct rays of sun, which in long exposures can kill them.

By shading various portions of the body of a euglena, it has been found that the region of the eyespot is especially sensitive. When a euglena is swimming through the water, it is this anterior end that first encounters regions of different light intensity.

## REPRODUCTION

Reproduction in *Euglena* takes place by binary longitudinal fission (Fig. 5-6). The nucleus divides in two by mitosis; then anterior organelles are duplicated. The animal then divides longitudinally (in an anteroposterior position), splitting the cell into two equal parts. The old undulipodium may be retained by one half, while a new one is developed by the other.

Longitudinal division often takes place while the organisms are encysted; in this condition the organisms are almost spherical and surrounded by a gelatinous wall they have secreted. One cyst usually contains two euglenas, although further multiplication by longitudinal division may produce 4, 16, or 32 young euglenas in a single cyst.

# Other Mastigophores

*Chilomonas* is a species common in nature and laboratory cultures. It constitutes a large part of the food of *Amoeba proteus*. *Chilomonas* are about 35 micrometers long, with two undulipodia at the anterior end. It does not possess chromatophores, but absorbs nutriments through the surface of the body.

*Gymnodinium* is a dinoflagellate of which one species (*brevis*) may occur in such great numbers as to cause a poisonous "red tide" in coastal waters. "Red tide" is a popular misnomer for the brownish-amber discoloration of seawater caused by this microscopic mastigophore. Under certain conditions, *Gymnodinium* reproduces at a fantastic rate; 60 million have been counted in a single quart of seawater.

Another dinoflagellate, *Gonyaulax*, causes waters to appear a rusty red when found in great numbers. *Gonyaulax catenella* is known to be disastrously poisonous. Several kinds of shellfish along the Pacific Coast feed on them, making the shellfish poisonous for human consumption.

The genus *Mastigamoeba* includes species that live in fresh water or in the soil. They possess both undulipodium and pseudopodium, the latter aiding in the ingestion of food particles.

## MASTIGOPHORES IN WATER SUPPLIES

Drinking water may be contaminated by protozoans of fecal origin and may also be unpalatable because of the multiplication of free-living mastigophores under natural conditions. Among several protozoans known to make water undesirable for drinking, *Uroglenopsis* is probably the most important, since it imparts a fishy odor similar to cod-liver oil. Other mastigophores producing foul odors in drinking water are *Synura, Peridinium, Dinobryon,* and *Ceratium. Chlamydomonas* and *Mallomonas* are less objectionable, having aromatic odors. *Cryptomonas* gives off a "candied violet" aroma.

All these odors are due to oils produced by the organisms during growth, which are liberated upon death and decomposition. Treatment of water supplies with copper sulfate is standard procedure for control of bad-smelling mastigophores, but appears to illustrate the cliché "the cure is worse than the disease."

## PARASITIC MASTIGOPHORES

Parasites play an important role in limiting the population size of their hosts in much the same way predators do. Studying the biological responses of parasites to various environmental conditions will give you a greater perspective on the limitations and requirements of the biological world.

Blood- and tissue-inhabiting mastigophores are of two important types: the **trypanosomes** and the **leishmanias.** The genus *Trypanosoma* is widespread in nature and may be found in the blood of many common mammals, birds, reptiles, fishes, and amphibians. *Trypanosoma gambiense* (Fig. 5–7) and *Trypanosoma rhodesiense* cause two types of African sleeping sickness in man, transmitted by the bite of the tsetse fly. The trypanosomes pass by way of the fly's proboscis into the human host. Early stages of the disease are characterized by fever and swelling of the lymph nodes in the neck. Invasion of the nervous system results progressively in drowsiness, coma, emaciation (through inability to take food), and finally death. Medicinal treatment is fairly successful, especially if begun early in the course of the disease.

South American trypanosomiasis or Chagas's disease, caused by the *Trypanosoma cruzi,* occurs quite differently. It becomes serious only after the acute stages, when heart damage occurs; the nervous system is not invaded. The trypanosomes are transmitted by blood-sucking kissing bugs. Infection takes place by the rubbing of the bug's feces into the puncture wound created by its blood feeding. Besides the undulipodiated form, which may be found in the bloodstream, a spherical, nonundulipodiated form is frequently encountered in the tissues, particularly in the muscular tissue of the heart. Death from cardiac failure is common. This disease is found in human beings; of the other animals invaded, the armadillo is of particular importance.

Other infections caused by parasitic mastigophores include kala-azar or dumdum fever, caused by *Leishmania donovani* (Fig. 5–8); oriental sore, caused by *Leishmania tropica;* and espundia, caused by *Leishmania brasiliensis.* All types of *Leishmania* are carried by sandflies.

Four types of mastigophores live in the digestive tract: *Trichomonas tenax*

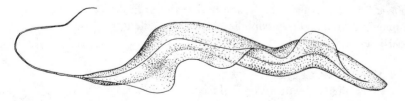

**Figure 5–7** *Trypanosoma gambiense,* the parasitic mastigophore causing African sleeping sickness in human beings.

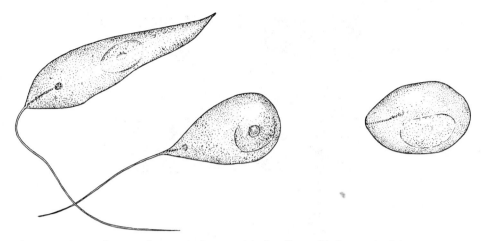

**Figure 5–8** *Leishmania donovani*, the parasitic flagellate of kala-azar in Asia.

inhabits the tartar of the human mouth; *Chilomastix mesnili, Trichomonas hominis,* and *Giardia lamblia* (Fig. 5–9) inhabit the intestine. *Trichomonas vaginalis* causes inflammation in the urinary tract and vagina of women.

There are many more parasitic mastigophores, found in various animals, including bats, rats, mice, sheep, rabbits, turkeys, and cattle.

**Figure 5–9** Intestinal mastigophores of human beings.

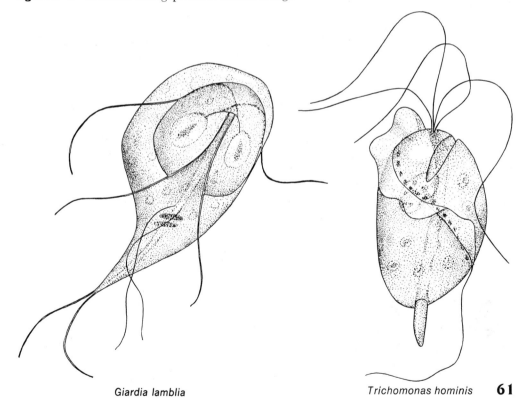

Giardia lamblia

Trichomonas hominis

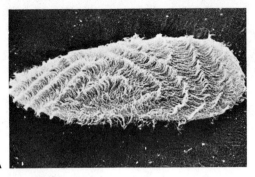

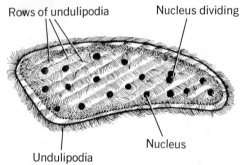

**Figure 5-10** *Opalina*. (A) Photomicrograph. (B) Diagram showing structure.

## Opalina—An Opalinid

*Opalina* (Fig. 5-10), a representative from the second class of Sarcomastigophora, is also a parasite, found in frogs and toads.

### REPRODUCTION

From summer until early the following spring, the multinucleated *Opalina* reproduces by fission (Fig. 5-11A). Later in the spring the rate of fission increases, yielding smaller opalinids, which encyst. When the host frog enters water to spawn, the cysts are discharged from the frog's intestinal tract into the water and swallowed by tadpoles. In the tadpole intestine each *Opalina* emerges from its cyst and divides into **anisogametes** (two unequal-sized gametes). The gametes unite in pairs (**syngamy**) and form a diploid **zygote** (Fig. 5-11B). The zygote may encyst and be discharged into the water to infect

**Figure 5-11** Reproduction in *Opalina*. (A) Fission (B) Syngamy (fusion of gametes).

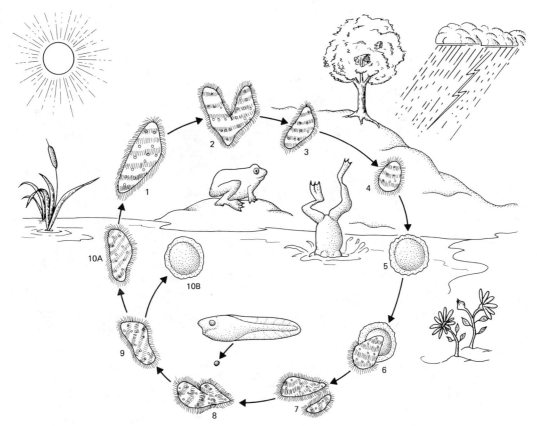

**Figure 5-12** Lifecycle of *Opalina ranarum*. 1-3, trophozoite stages; 4 and 5, formation of cysts; 6, cyst hatches; 7, formation of gametes; 8, copulation; 9, zygote; 10a, encystment of zygote and hatching of young asexual individual; 10b, cyst discharged into water and swallowed by tadpole.

another tadpole and begin its asexual development, or, if it does not encyst the zygote, may become a **trophozoite**, the active adult feeding form (Fig. 5-12).

### LOCOMOTION

*Opalina* has cilia, which beat in continuous helical waves.

The opalinids were once classified with the ciliates and the flagellates. As more information accumulated concerning their biology, particularly details of reproduction, they were placed in their own class.

## Amoeba proteus—A Sarcodine

*Amoeba proteus* (Gr. *amoibe*, change + Proteus, a sea god in classical mythology who had the power of changing shape) (Fig. 5-13) lives in freshwater ponds and streams, where it can often be found on the undersides of dead lily pads and other vegetation in shallow water. They exhibit activities necessary for self-maintenance which are comparable to those of more complex

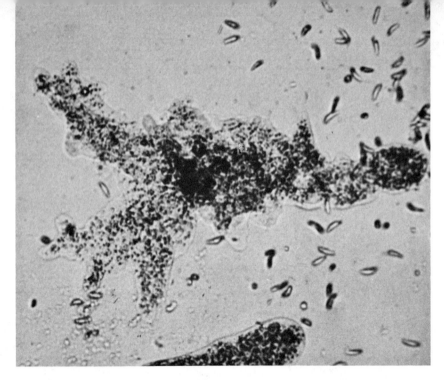

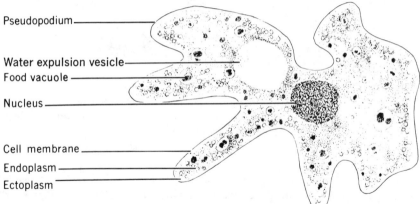

**Figure 5-13** *Amoeba proteus.* *Top:* Photograph. *Bottom:* Diagram showing main morphological features.

animals: movement; the capture, ingestion, and digestion of food; egestion of undigested matter; absorption and assimilation of the products of digestion; secretion and excretion of various substances; respiration; growth; reproduction; and response to changes in its environment.

## MORPHOLOGY

*Amoeba proteus* (Fig. 5-13) is about 0.25 mm in length. It appears under the microscope as an irregular, grayish mass of animated jelly, continually changing its shape by pushing out and withdrawing fingerlike extensions of the cytoplasm called **pseudopodia.** Two areas of the cytoplasm are recognizable: a thin area of clear protoplasm (**ectoplasm**) which surrounds a more centrally located granular protoplasm (**endoplasm**). Part of the endoplasm, the **plasmagel,** is relatively rigid and sheaths a more liquid part, the **plasmasol.** Within the endoplasm several cytoplasmic inclusions larger than the ordinary granules are visible. One of these, the **nucleus**, is disk-shaped, filled with chromatin granules. The nucleus plays an important part in such fundamental activities as growth, manufacture, and use of food and formation of new cells. In an amoeba cut into two pieces, only the part containing the nucleus will survive.

A clear bubblelike body, the **water expulsion vesicle**, can often be seen lying near the nucleus. At more or less regular intervals, it is carried to the surface, where it ruptures and collapses under cytoplasmic pressure, forcing its fluid contents out of the body. Food vacuoles are also visible in the endoplasm. These are temporary structures containing food bodies (**phagosomes**) in the process of digestion.

## LOCOMOTION

Amoebas move from place to place, capture other organisms, and ingest solid particles with their pseudopodia. These pseudopodia may arise at any point on the surface, appearing first as blunt projections consisting of ectoplasm. Granular ectoplasm flows into the projection and the entire amoeba moves forward in the direction of the pseudopodia. Several pseudopodia may form at the same time; one is usually large and effective while the others are smaller and soon disappear. Amoebas can move in this manner (**amoeboid movement**) at the rate of 1 inch per hour. The rate may vary with temperature changes, increasing up to about 30°C and ceasing at 33°C.

## INGESTION, DIGESTION, AND EGESTION

The amoeba feeds principally on minute organisms. A distinct selection of food particles is evident (Fig. 5-14); *Tetrahymena* and *Paramecium* are preferred foods. Food is usually engulfed in that part of the body closest to the direction in which the amoeba is moving (Fig. 5-15). Pseudopodia enclose the food particle from all sides; thin sheets of cytoplasm cover the top and bottom, completing the food cup formation. Active prey, such as the *Tetrahymena*, are captured by the formation of a large food cup which surrounds the organism without touching it; in this way a dozen or more flagellates or ciliates may be ingested in one food cup. The ingestion process takes 1 minute or longer, depending on the character of the food and the temperature. Ingestion increases up to 25°C and decreases to zero at about 33°C.

Digestion takes place in the **food vacuoles.** Digested material diffuses out of the vacuoles into the cytoplasm. Indigestible matter may be egested from any point on the amoeba's body; the matter is usually heavier than the cytoplasm

**Figure 5-14** *Amoeba proteus* exhibiting food selection. Arrows indicate the direction of movement of the endoplasm in the pseudopodia.

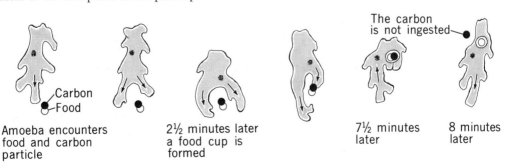

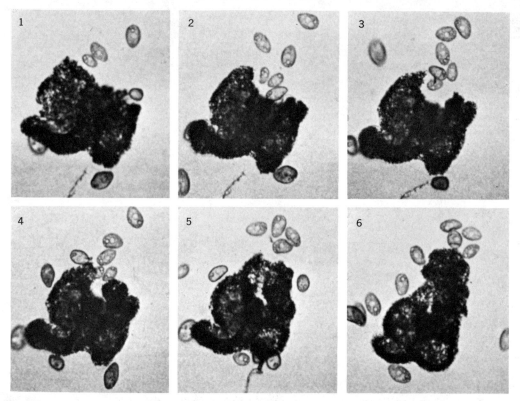

**Figure 5–15** *Amoeba* capturing and ingesting *Tetrahymena*, from 16-mm motion picture. About 200×.

and is left behind as the amoeba moves forward. Oxygen from surrounding water diffuses through the amoeba's cell membrane, allowing respiration to occur. Waste products of digestion, such as carbon dioxide and ammonia, are excreted out into the environment by a similar diffusion process.

## REPRODUCTION

Protoplasm is built up through **assimilation** until the amoeba reaches its full size. At this point it reproduces by **binary fission** (Fig. 5–16), a mitotic process, and two amoebas are formed. During mitosis the amoeba becomes spherical, covered with small pseudopodia.

Development in the amoeba is simply a matter of growth. The rate of growth is rapid just after division, gradually decreasing until the full size is reached again, averaging about 3 days.

## BEHAVIOR

All the activities discussed thus far are part of the amoeba's behavior. Changes in the environment (**stimuli**) are also part of the amoeba's behavior, producing a reaction (or **response**). The amoeba responds to a number of stimuli, including contact, light, temperature, chemicals, and electricity. Movement toward a stimulus is a **positive reaction;** movement away from a stimulus is a **negative reaction.** The data thus far obtained indicate that factors present in the behavior of the amoeba are comparable to the habits, reflexes, and automatic activities of higher organisms.

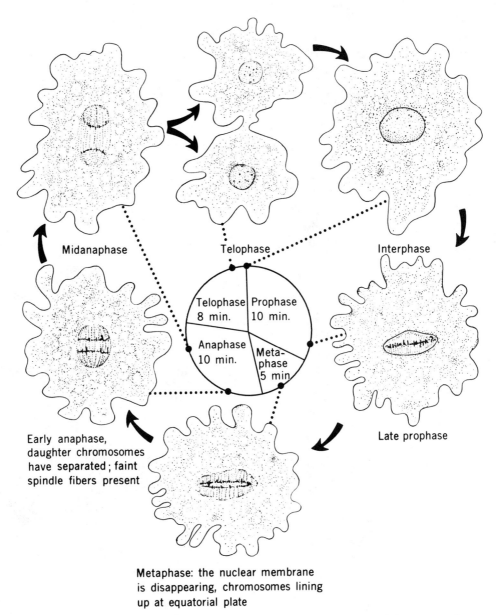

**Figure 5–16** Reproduction of *Amoeba* by binary fission. Begin study of this continual cycle at the interphase stage and end at telophase.

## Other Sarcodines

Many amoebas live in fresh water, salt water, and soil. Some of these have been placed in the genus *Amoeba;* the rest have been assigned to other genera.

*Pelomyxa palustris* is a large species, reaching a diameter up to 2 mm. It contains many nuclei and moves along without definite pseudopodia. Another large species, *Pelomyxa carolinensis,* may reach a length of from 2 to 5 mm and can be seen without the aid of a microscope. This amoeba contains from 300 to 400 nuclei and generally divides into three individuals during mitosis.

Several types of common freshwater sarcodines are protected by shells. *Arcella* secretes its organic shell; *Difflugia* builds its inorganic shell from minute grains of sand. Both thrust pseudopodia out of a small opening in the shell.

## PARASITIC SARCODINES

*Entamoeba histolytica*, the causative agent of amoebic dysentery, is the only serious disease-producing parasite in this group. Fortunately, most of those infected (estimated at 10 percent of our population) are only carriers; the entamoebas are present but do no damage. Occasionally, however, they do invade the intestinal wall and form abscesses which later rupture, becoming persistent ulcers, resulting in diarrhea and dysentery. Medicinal treatment against this parasite has proved effective.

*Entamoeba gingivalis* lives in the human mouth and is sometimes associated with pyorrhea. Kissing is the commonest method of contraction; over 50 percent of our population is infected.

Other intestinal and cavity-inhabiting forms exist; these are usually harmless or do little damage.

## SARCODINES AND GEOLOGY

Most of the 8000 or more sarcodines live in the sea. The Foraminifera and Radiolaria (Fig. 5-17) are large groups of sarcodines that have existed from

**Figure 5-17** Various representatives of sarcodines. (A) Radiolarians. (B) Foraminiferans.

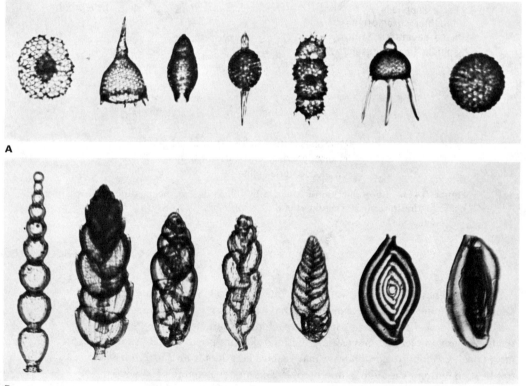

early times; their skeletal structures have been preserved as fossils in a great number of distinct rock strata.

The skeletons of these and other ocean-dwelling sarcodines continually sink to the bottom, creating evergrowing layers of ooze. The greater portion of the Atlantic Ocean (approximately 20 million square miles) is covered with an ooze formed from skeletons of the foraminiferan *Globigerina*. The White Cliffs of Dover are chalk deposits composed mostly from the shells of foraminiferans. The stones used in Egyptian pyramids are made up of skeletons from very large forms of the genus *Camerina*.

Many zoologists believe that mastigophores evolved from green algae and sarcodines from a mastigophore or mastigophorelike organism. Probably not all sarcodines arose from mastigophores; some may have evolved from more primitive sarcodines.

## Monocystis lumbrici—A Sporozoan Parasite of Earthworms

*Monocystis lumbrici*, a parasite of earthworms, illustrates many of the characteristics of sporozoans and cnidosporans, the two protozoan subphyla characterized by **spore** formation (Fig. 5–18). Its life cycle (Fig. 5–19) begins with the intake of spores into the earthworm's digestive tract, where the **sporozoites** are set free. Each sporozoite penetrates a bundle of developing sperm cells in the testis of the earthworm and is then termed a **trophozoite.** Here it lives at the expense of surrounding cells.

The sperms of the earthworm, deprived of nourishment, slowly shrivel up, becoming tiny filaments on the surface of the trophozoite (making it resemble a ciliated organism). The trophozoite grows and migrates to a seminal vesicle, where it joins with another trophozoite. The two are surrounded by a cyst wall where each divides, producing a number of small cells (**gametes**). The gametes unite in pairs to form **zygotes.** Each zygote becomes lemon-shaped and secretes a thin hard wall about itself. It is now known as a **spore,** completing the life cycle. Spores most likely remain within the worm throughout its lifetime and death; when the dead worm decays the freed spores may be ingested by another worm.

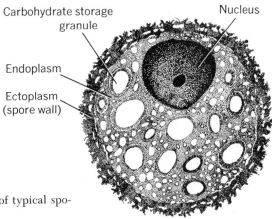

**Figure 5–18** Morphology of typical sporozoan.

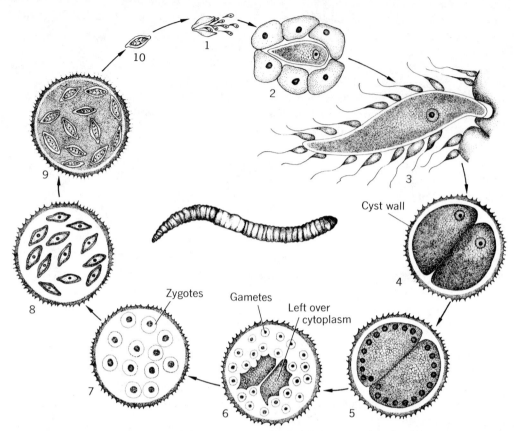

**Figure 5–19** Life cycle of *Monocystis*, a sporozoan that lives in the earthworm. (1) Sporocyst eaten by earthworm, sporozoites released. (2) Young trophozoite in sperm morula. (3) Trophozoite grows into worm; earthworm sperm develops from morula cells and attaches to trophozoite. (4) Trophozoites pair and form cyst wall (gametocytes). (5) Nuclei of gametocytes divide. (6) Gametes are produced. (7) Gametes fuse (fertilization) to form zygotes. (8) Zygotes secrete spore walls (sporocysts). (9) Nuclei divides, forming eight sporozoites in each spore. (10) Spores liberated when earthworm dies, or through intercourse, or disintegration of worm cocoon.

## Plasmodium—A Sporozoan Parasite of Human Beings

Of the blood and tissue parasites, the human malarial organism is by far the most important. Malaria is probably the most devastating disease, killing or afflicting large numbers of people each year. Three major types of human malaria exist: benign tertian malaria, caused by *Plasmodium vivax* (Fig. 5–20B), characterized by fever every 48 hours; quartan malaria, caused by *Plasmodium malariae*, usually producing fever every 72 hours; and pernicious malaria, caused by *Plasmodium falciparum*, which is the greatest killer, manifested by an irregular temperature curve with a sometimes continuous fever. The life cycles of all three species are essentially similar and may be summarized as follows.

Malaria is transmitted through the bite of the female mosquito. In human beings, only the genus *Anopheles* can transmit the disease. When the insect's mouthparts pierce the skin, saliva containing anticoagulants passes into the

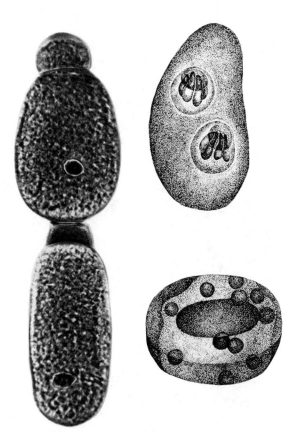

**Figure 5-20** Parasitic sporozoans. *Left:* Two attached trophozoites of *Gregarina blattarum*, from cockroach intestine. *Right:* Schizont of *Plasmodium vivax* in human red blood cell.

wound, allowing any malarial parasites present to enter the human bloodstream. This infective stage of the organism is a **sporozoite**. The sporozoites do not invade blood corpuscles directly but penetrate certain tissue cells (liver), where they grow, multiply, and go through at least two cycles. Relapses of the disease may occur from this reservoir in the tissue cells. *Plasmodium* parasites are eventually released into the bloodstream, where they proceed to enter red blood cells (**erythrocytes**), one parasite per cell. The plasmodium first assumes a ring shape, then an irregular form that soon fills the cell; these are the **trophozoite** stages. The time required for development depends upon the particular species. The mature trophozoite eventually divides (asexual reproduction by multiple fission, called **schizogony**) into **merozoite** cells, which are then released into the bloodstream by means of rupture of the erythrocyte. Each merozoite invades a new erythrocyte, where it becomes a trophozoite. Each time the cycle is repeated, a larger number of erythrocytes are involved; when these burst, chills and fever occur.

A few merozoites, instead of becoming trophozoites, develop into **gametocytes,** potential gametes that remain inactive while inhabiting a human host. Gametocytes that pass into a mosquito's stomach become active at once. Female gametocytes develop into a single spherical egg or **female gamete;** males undergo exflagellation, producing slender **male gametes** (**sperms**) which grow out radially from the cell surface and eventually float free. Fertilization occurs, producing a **zygote** that has the ability to move about and is therefore called an **ookinete**. The ookinete migrates to the stomach wall, where it

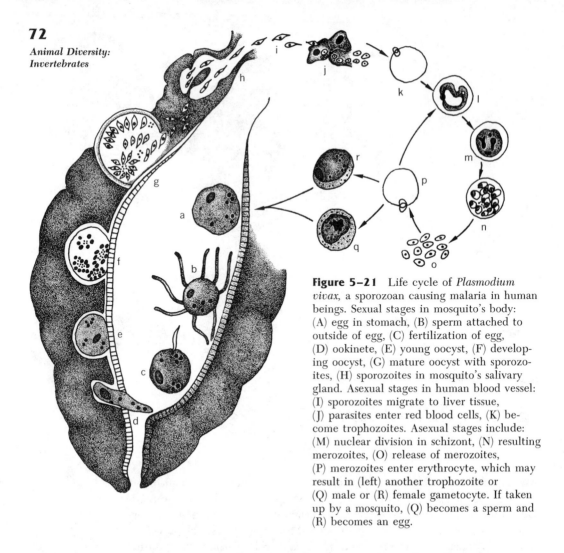

**Figure 5–21** Life cycle of *Plasmodium vivax*, a sporozoan causing malaria in human beings. Sexual stages in mosquito's body: (A) egg in stomach, (B) sperm attached to outside of egg, (C) fertilization of egg, (D) ookinete, (E) young oocyst, (F) developing oocyst, (G) mature oocyst with sporozoites, (H) sporozoites in mosquito's salivary gland. Asexual stages in human blood vessel: (I) sporozoites migrate to liver tissue, (J) parasites enter red blood cells, (K) become trophozoites. Asexual stages include: (M) nuclear division in schizont, (N) resulting merozoites, (O) release of merozoites, (P) merozoites enter erythrocyte, which may result in (left) another trophozoite or (Q) male or (R) female gametocyte. If taken up by a mosquito, (Q) becomes a sperm and (R) becomes an egg.

undergoes nuclear division, resulting in a great number of sporozoites contained within a swelling called an **oocyst.** The oocysts finally rupture, and the sporozoites are released into the insect's body cavity, where they make their way to the salivary glands. The next time the mosquito takes a blood meal, sporozoites are injected into the human body. The life cycle of *Plasmodium vivax* is summarized in Fig. 5–21.

Attempts to control malaria have involved three approaches: (1) Treatment of infected persons by antimalarial drugs; (2) protection of uninfected individuals by use of screens, nets, gloves, and application of repellent solutions to skin; and (3) various attempts to eliminate infective mosquitos through chemicals or genetics.

## Other Sporozoans

Sporozoans exhibit a variety of shapes, ranging from spherical to elongate forms. Three classes and seven orders are usually recognized. Two types easily

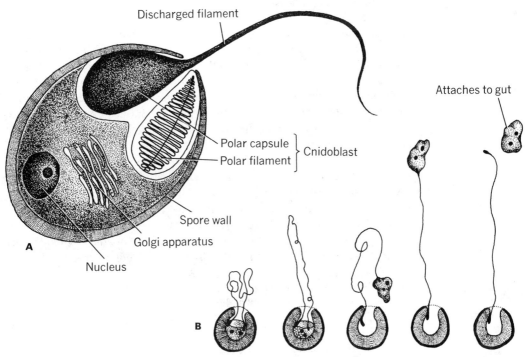

**Figure 5-22** *Myxidium*, a cnidosporan spore. (A) Within each polar capsule is a tightly wound filament which straightens out and temporarily attaches the spore to its host. (B) Discharge: emergence of filament and sporoplasm from *microsporidian* spore.

obtained for study are: **gregarines** (Fig. 5-20, left), found in the intestines of grasshoppers, cockroaches, and mealworms; and **coccidians**, most easily obtained from rabbits.

## Cnidosporan Parasites

Cnidosporan parasites invade only lower vertebrates and invertebrates. Some of their hosts are of considerable economic importance, notably honeybees, silkworms, and commercial fish. They are distinguished from sporozoans by the presence of polar capsules and polar filaments (Fig. 5-22), and the method of spore formation.

## Paramecium caudatum—A Freshwater Ciliate

Paramecia are commonly found in pond water containing large amounts of decaying vegetation. There are 10 well-known species, differing in shape, size, and structure.

## MORPHOLOGY

*Paramecium caudatum* (Figs. 5–23 and 5–24) ranges from 0.15 mm to 0.3 mm in length. The anterior end is blunt, the posterior end more pointed. Its greatest width is behind the center of the body. A depression extends from the anterior end obliquely downward and posteriorly into the endoplasm. An oral groove designates one side as oral and the opposite side aboral. Fine hairlike undulipodia (cilia) are regularly arranged over the surface, those at the posterior end being slightly longer than the rest. As in the amoeba, the cytoplasm can be divided into an outer, clear **ectoplasm** and an inner granular **endoplasm**. A distinct elastic membrane, the **pellicle**, is present on the outer surface of the ectoplasm. A large **water expulsion vesicle** is situated near each end of the body, close to the aboral surface; and a variable number of **food vacuoles** may be seen. A large **macronucleus** controls vegetative functions and a smaller **micronucleus** is important in reproduction. The two nuclei are suspended in the endoplasm near the mouth opening. A temporary opening called the **cell anus** can be observed only during the discharge of undigested particles; it is

**Figure 5–23** *Left:* spiral path of a free-swimming paramecium, *Right:* internal structure of *Paramecium caudatum*. Numbered food vacuoles show progress of digestion and absorption.

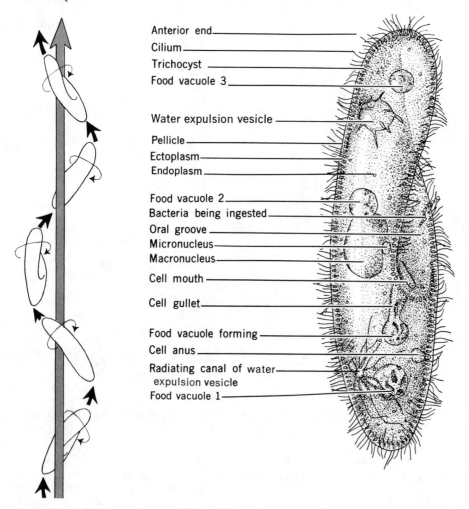

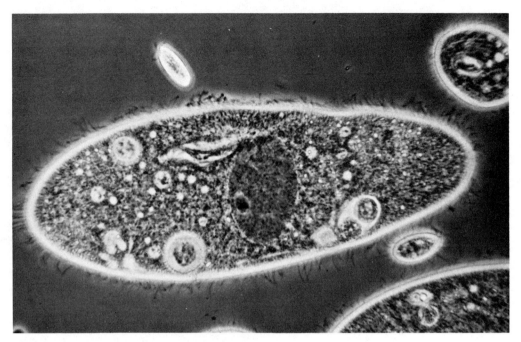

**Figure 5-24** Photomicrograph of a paramecium.

situated posterior to the oral groove and always re-forms at the same point.

The fluid endoplasm contains large food reserve granules and occupies the central part of the body. Under the higher powers of a microscope the pellicle may be observed to consist of a great number of hexagonal areas produced by subsurface ridges. From the center of each hexagonal area one undulipodium arises (Fig. 5-25).

**Figure 5-25** Pellicle structure of *Paramecium*. The carrot-shaped trichocysts are attached by delicate threads to the ridges of the hexagonal areas.

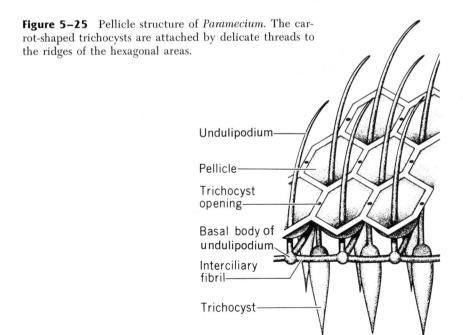

Physiological processes similar to those of the amoeba occur in the paramecium: capture and ingestion of food, digestion, build up of protoplasm, egestion, stimulus response, respiration, and reproduction.

## LOCOMOTION

Undulipodia of ciliates are short (5 to 10 μm) in comparison to those of mastigophores. Both, however, have a 9 + 2 fiber arrangement structure and function by beating in helical waves. The combined helical beats of the undulipodia move the organism forward. To reverse direction, undulipodia do not change the pattern of their beat but the direction of the axis of the helical waves. Another component of forward motion may be due to waves of the entire undulipodial blanket, called **metachronal waves.**

As a paramecium moves forward it spirals around a relatively straight course. The gyrations result from two factors: rotation about the long axis of the body due to obliquely oriented undulipodia and a swerving away from the oral side, caused by the faster beating of the oral groove's undulipodia. The beats of the undulipodia appear to be initiated by the action of certain ions. Electric potential or the action of cations can cause the axis of the helical waves to shift, causing the ciliate to reverse direction.

## OFFENSE AND DEFENSE

The paramecium possesses carrot-shaped structures (**trichocysts**), embedded in the ectoplasm just beneath and perpendicular to the body surface. A small amount of iodine or acetic acid added to a drop of water containing paramecium causes the ejection of trichocysts, followed by a halo of long threads. The function of trichocysts in *P. caudatum* is not known but they appear to be used (rather ineffectively) as a defense mechanism against larger protozoans and small metazoans that prey upon paramecia. Trichocysts may possibly secrete salts, thereby serving an osmoregulatory function.

## INGESTION, DIGESTION, AND EXCRETION

Selected bacteria, yeasts, small protozoans, and algae are captured with the aid of undulipodia in the oral groove. By the direction of their beating, undulipodia produce a current that drives a steady stream of water toward the cytostome (cell mouth). Food particles swept into the cytostome are carried down the cytopharynx and gathered together near the posterior end into a **food vacuole.** When this vacuole has reached a certain size it is released into the surrounding cytoplasm and another vacuole begins to form.

Digestion takes place as the food vacuole is carried along a course by the rotary streaming of the endoplasm (**cyclosis**). The course begins just behind the cytopharynx, passes posteriorly, then forward and aborally, and finally posteriorly toward the oral groove (Fig. 5–26). Food is digested by enzymes from the endoplasm and food vacuoles become smaller. Digested food is stored, used for vital activity, or built up into protoplasm. Some paramecia have a cell anus (cytopyge) through which undigested material is egested.

Most of the waste products appear to diffuse through the pellicle; nitrogenous wastes (urea and ammonia) may also be excreted by the water-expulsion vesicles.

## REGULATION OF WATER CONTENT

The two water-expulsion vesicles occur in the inner layer of ectoplasm and communicate with a large portion of the body by means of a system of

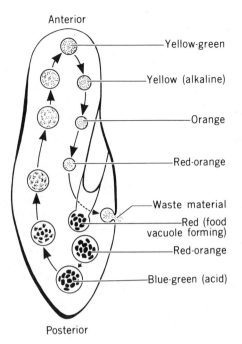

**Figure 5–26** Cyclosis and digestion in *Paramecium caudatum*. Arrows indicate the path of the food vacuole within the body. Red-stained yeast cells change to blue-green, indicating that the vacuole became acid soon after formation. As the vacuole with yeast cells circulates through the body, it changes back to a red-orange color, indicating less acidity. Experiments have shown that in digestion of food the vacuole is first acid, then alkaline.

ampullae or **radiating canals,** numbering between 6 and 11. These ampullae fill with liquid, then discharge their contents, forming the vesicles, which in turn expel the liquid to the exterior. After each emptying a new vesicle is formed.

The rate of expulsion varies with the activity of the organism, the temperature, and the concentration of salts in surrounding water. Water continually enters the paramecium because of its semipermeable membrane and the greater concentration of water molecules on the outside.

## RESPIRATION

Oxygen, dissolved in the water, diffuses through the surface of the body into the protoplasm. Carbon dioxide diffuses out of the body through the surface; some is probably discharged by the water expulsion vesicles.

## BEHAVIOR

As in the amoeba and euglena, changes in the environment serve as stimuli to which paramecia respond in several ways. One of the most common responses is the **avoiding reaction (negative response)** (Fig. 5–27). When a free-swimming paramecium encounters a harmful chemical (such as strong salt), it may swim backward for a short distance; then its rotation decreases in rapidity and it swerves toward the aboral side more strongly than under normal conditions. Its posterior end becomes a pivot upon which the organism swings in a circle. During this revolution, samples of the surrounding medium are brought into the oral groove. When a sample no longer contains the stimulus for the negative response, the organism moves forward again.

The paramecium reacts to other stimuli besides chemicals, including ultraviolet light, contact, changes in temperature, and electric current. If the anterior end (most sensitive part) is touched with a glass rod, the avoiding reaction is given. Frequently a slow-swimming paramecium will come to rest with its undulipodia in contact with an object; this **positive response** often

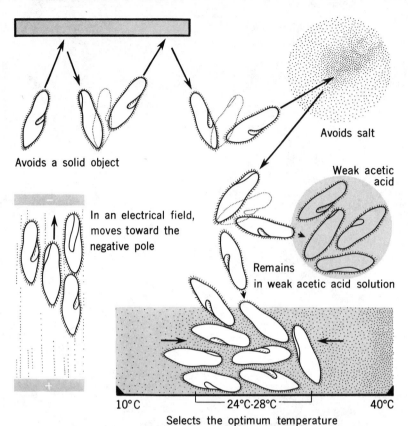

**Figure 5-27** Behavior of the paramecium in various environmental conditions.

brings it into an environment rich in food. Paramecia placed on a slide heated at one end have been observed to give the avoiding reaction, swimming in all directions until through trial and error they move toward the cooler end.

A paramecium may be stimulated in various ways at the same time. The **physiological condition** of a paramecium determines the character of its response. This physiological state is a dynamic condition, changing continually with the ongoing metabolic process.

## REPRODUCTION

The paramecium usually multiplies by simple binary fission. This asexual process is interrupted at intervals by a temporary sexual union (conjugation) of two individuals, with subsequent mutual nuclear fertilization.

In binary fission the organism divides transversely (Fig. 5-28) into two daughters. The micronucleus undergoes mitosis, the macronucleus elongates and divides by amitosis (without spindle formation), other duplicate organelles are formed, and the organisms finally split along a deep cleavage formed in the parent organism. The entire process takes about 2 hours.

The sexual process of conjugation involves the apposition of the oral surfaces of two paramecia joined by a protoplasmic bridge (Fig. 5-29). As soon as the union is effected, the micronuclei pass through a series of stages, resulting in a 50 percent reduction of chromosomes in each micronucleus. During conjugation there is an interchange of micronuclei (Fig. 5-30). The migratory micronucleus is smaller than the stationary one and may be considered comparable to the nucleus of a male germ cell. Its fusion with the stationary micronucleus

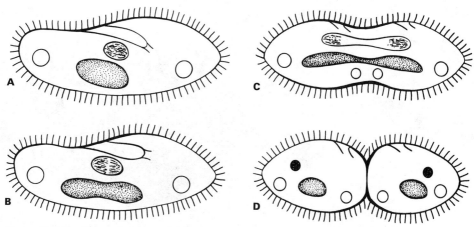

**Figure 5–28** Binary fission of *Paramecium caudatum*.

**Figure 5–29** Conjugation in *Paramecium caudatum*. This diagram is simplified and does not include all stages. For clarity, the macronuclei are omitted from the third stage; they do not actually disappear completely until after conjugating animals have separated.

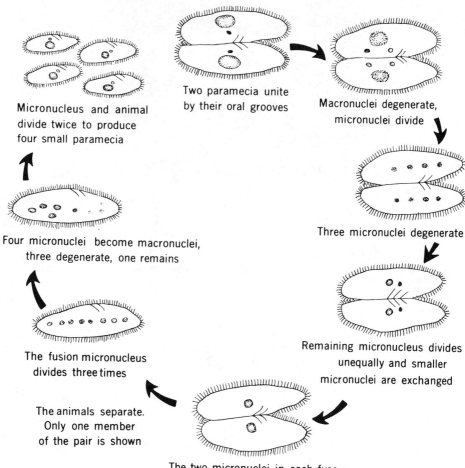

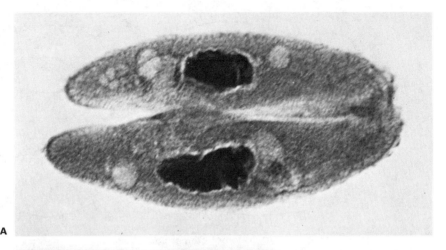

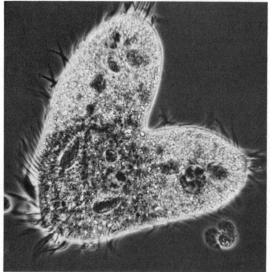

**Figure 5–30** (A) Photomicrograph of two paramecia in conjugation. An exchange of micronuclei is taking place. (B) Another ciliate species, *Euplotes*, in conjugation.

resembles the fusion of male and female nuclei in the eggs of higher animals during fertilization. Conjugation is similar to fertilization; however, after conjugation, organisms recommence asexual reproduction. In organisms with true fertilization, only sexual reproduction occurs.

A complex system of mating types exists within each species. *Paramecium aurelia* (similar to *P. caudatum*) has at least two mating types. These types will conjugate with each other but not with their own type. The discovery of mating types is a relatively recent finding.

Paramecia may also undergo a process of self-fertilization (**autogamy**) at regular intervals. Some ciliates do not undergo autogamy or conjugation for a period of years; in these cases, nuclear reorganization is essential.

## Other Ciliates

Ciliates are widespread in nature. Many species live in fresh water; others occur in the sea, soil, or as parasites (Fig. 5–31).

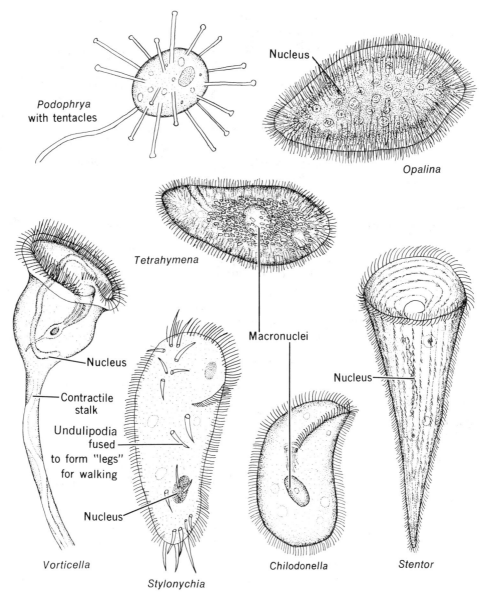

**Figure 5–31**  Representative ciliophorans.

*Tetrahymena* is a small ciliate that grows in a culture medium free from all other microscopic organisms and has about the same nutritional requirements as man. This suggests that even a single-celled organism may have physiological processes nearly as complex as those of man. This tiny organism is playing an increasingly important role in physiological and genetic research.

*Stentor* is trumpet-shaped, bluish in color, with a bearded macronucleus and undulipodia spirally arranged around the "mouth"; it may be free swimming or attached. A complicated disk of undulipodia appear at the anterior end.

*Stylonychia* has a flattened body and groups of undulipodia fused to form cirri, used as "legs" in creeping. *Vorticella* resembles an inverted bell attached to a contractile stalk.

## PARASITIC CILIATES

The ciliate *Balantidium coli* is the largest protist found in the human intestinal tract. This motile organism penetrates the membrane lining of the colon, where it frequently causes ulcers. It is a definite disease-producing parasite and may cause symptoms resembling acute amoebic dysentery. Infected persons pass large spherical cysts, easily identified by microscopic examination. This parasite also infects pigs and chimpanzees, both of which probably serve as reservoirs for human infection.

Many domesticated and wild animals, both terrestrial and aquatic, are parasitized by ciliates, most of which live in the digestive tract.

# Origin of Ciliata

Organisms of the subphyla Ciliophora and class Mastigophora are considered to be more closely related to each other than to other protozoans because of the presence in both groups of undulipodia (also present in *Opalina*), the homology of basal granules, and certain biochemical similarities. The greater complexity of ciliates suggests that this group arose from the simpler mastigophores.

# Summary

Protozoans are acellular animal-like organisms of the protist kingdom. Sarcomastigophores are a protozoan subphylum consisting of mastigophores (*Euglena viridis*), opalines, and sarcodines (*Amoeba proteus*). The subphyla Sporozoa and Cnidospora are all parasites, distinguished by their formation of spores and lack of locomotor organelles. Ciliophora, the final subphylum contains the largest and structurally most complex protozoans, distinguished by their short undulipodia, as in the ciliate *Paramecium caudatum*. Representative organisms were discussed in detail, highlighting various characteristics typical of their group.

# 6

# Phylum Mesozoa

In Chapter 5 we looked at ways in which acellular organisms have met the problems of survival. These protozoans are animal-like members of the kingdom Protista. We now enter the kingdom Animalia, the realm of **metazoans** or *multi*cellular animals.

The increasing levels of complexity seen in organisms of the protist kingdom continues in the animal kingdom. It will be useful to consider the organisms discussed throughout this text in terms of an organizational spectrum rather than as members of isolated groups. Keeping in mind the continuum of organizational levels these animals display, we now turn to those found in the phylum Mesozoa.

The name **Mesozoan** implies that these organisms are intermediate between protozoans and metazoans, which they may indeed be. They are the simplest organisms of the animal kingdom—small, slender, and parasitic. Their simplicity may be partly the result of modifications necessary for a parasitic existence. Most mesozoans live in the excretory organs of **cephalopods** (squid and octopuses). Some inhabit flatworms, sea stars, annelids, and other invertebrates.

## Morphology

A mesozoan body resembles a worm; it is long (0.5 to 10 mm) and cylindrical and is bilaterally symmetrical. The body consists of **axial** reproductive cells, enclosed by an outer layer of ciliated epithelial cells (the **somatic** cells). There are from 16 to 42 somatic cells present; anteriorly, eight of these cells form a polar cap which holds the animal to the renal organ (kidney) of its host.

## Physiology

*Dicyema* (Fig. 6-1), a parasite of cephalopods, most likely absorbs its nutrients direct from the host's urine. Several factors point to this conclusion:

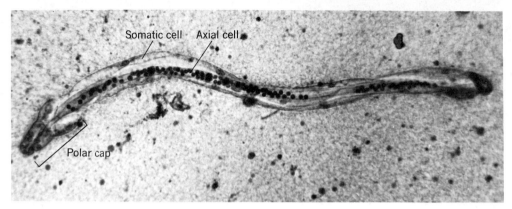

**Figure 6-1** *Dicyema*, a representative mesozoan and parasite of cephalopods.

*Dicyema* lives well in cephalopod urine but does not live for long in seawater; the urine contains inorganic and organic molecules that could be used by the parasite. Furthermore, if the parasite were to derive nutrients from the tissues of the host, there would be damage to the renal organs; no such damage occurs.

*Dicyema* must respire anaerobically, since there is little or no oxygen in cephalopod urine. Digestion and excretion appear to take place in each individual somatic cell.

## Life Cycle

Repeated asexual fission of axial cells in the adult *Dicyema* gives rise to new individuals that share the host. When the population of a host reaches a certain density, some adults give rise to hermaphroditic individuals, which remain in the parent and produce dispersal larvae. These larvae are undulipodiated and escape from the parent and the host into the sea. The fate of these larvae is unknown.

## Origin of Mesozoa

Mesozoans are either intermediate between unicellular and multicellular organisms or else degenerate forms of Platyhelminthes. Some of the best authorities believe their characteristics are chiefly primitive and not the result of parasitic degeneration.

## Summary

Mesozoans are the simplest organisms in the animal kingdom. They are small endoparasites of some marine invertebrates. Their evolutionary origin is not yet firmly established.

# Phylum Porifera

The phylum Porifera (Latin *porus*, pore + *ferre*, to bear) consists of the sponges. For centuries sponges were thought to be plants, probably due to the green color of some freshwater sponges that contain photosynthetic algae. Their animal nature was recognized by zoologists in 1765.

Sponges are thought to have diverged from the main line of animal evolution quite early. Except for mesozoans, they are the least complex of the Metazoa. They are of particular interest in their exhibition of a multicellular organization that is intermediate between protists and typical metazoans.

Most of the 5000 species are marine (inhabit saltwater); 150 species comprise the freshwater family Spongillidae. Sponges exhibit great diversity in shape, size, structure, and geographic distribution. Adult sponges are sessile (attached) and usually asymmetric; a few have radial symmetry. **Cellular specialization** exists as a result of division of labor among somatic cells, but there is no grouping or coordination of these cells to form definite tissues.

Three classes of sponges are recognized: **Calcarea,** comparatively simple shallow-water species with calcareous spicules; **Hexactinellida,** slightly more complex deep-sea sponges with siliceous spicules; and **Demospongiae,** the most dominant and complex type of sponge, with a skeleton composed of siliceous spicules and/or spongin fibers or neither.

## Leucosolenia—A Simple Sponge

*Leucosolenia* (Fig. 7–1) is a simple sponge, white or yellowish in color, which consists of long, narrow tubes that project upward from connecting horizontal tubes. The upright tubes have an opening (**osculum**) at the top; the inside of the tube is the **central cavity** or **spongocoel.** The body wall consists of an outer epidermis (**dermal epithelium**) of thin, flat cells and an inner lining of undulipodial collar cells (**choanocytes**). The choanocytes are in loose contact with each other and create a current of water through constant undulipodial (flagellar) beating.

Between the two layers of the body wall is a gelatinous mesoglea containing various amoeboid wandering cells (**amoebocytes**), which give rise to reproductive and several types of somatic cells, including those for pigment and food

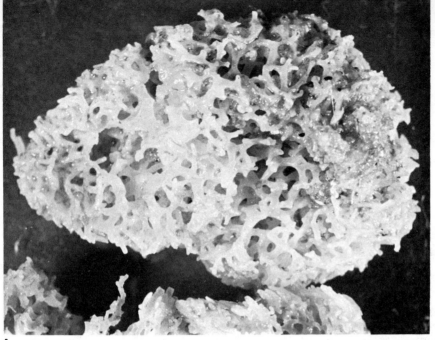
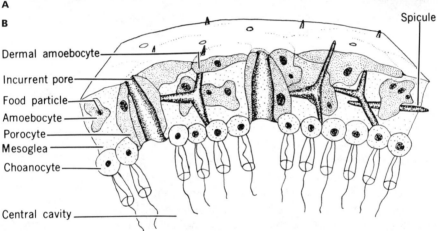

**Figure 7–1** *Leucosolenia,* a simple sponge. (A) Large colony. (B) Cross section, showing cellular structure of body wall.

storage. Many three-pronged (triradiate) **spicules** are interlaced in the soft tissues of the body wall, serving to strengthen the body and hold it upright.

A colored dye placed in the water will be drawn into the *Leucosolenia* through minute **incurrent pores** in the body wall and pass out through the **osculum** (an exhalant opening). Sponges are the only animals in which the large opening serves as an exit, rather than an entrance, for water.

## Canal Systems and Skeletal Structures

If it had not been for the development of elaborate canal systems, sponges would have remained in the simple asconoid condition of *Leucosolenia* and

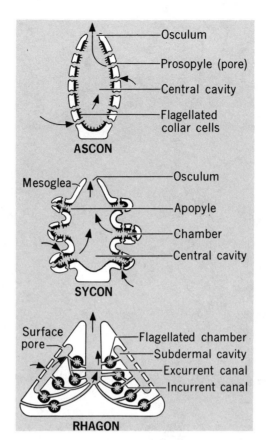

**Figure 7–2** Canal systems of sponges. The sycon type is thought to be a derivative of the simple ascon type. Note how each chamber of the sycon sponge resembles the single chamber of the ascon sponge. The rhagon type, like the bath sponge, is a more complex system, with canals and flagellated chambers. Arrows indicate the water flow through sponge types.

would never have been able to become massive in size. The canal system furnishes an avenue for food through the body and for transportation of excretory matter out of the body. Three types are usually recognized (Fig. 7–2): the simplest or ascon type, as in *Leucosolenia;* the sycon type, as in *Scypha;* and the leucon (rhagon) type, in which there is a number of small chambers lined with choanocytes.

The skeletons of sponges consist of calcium carbonate or silica (a mineral substance akin to glass) in the form of **spicules,** or of **spongin** in the form of protein fibers more or less closely united (Fig. 7–3). Spongin is secreted by flask-shaped cells (**spongoblasts**). Spicules are deposited in cells, and more than one cell may take part in the formation of a single spicule.

## Scypha—The Syconoid Sponge

### MORPHOLOGY

*Scypha* (Fig. 7–4) is a more complex marine sponge in comparison to *Leucosolenia*. It attaches permanently to rocks and other solid objects and varies in length between 12 mm to almost 25 mm. Its shape resembles a slender vase, bulging slightly in the middle. The osculum is surrounded by a ring of straight spicules. Smaller spicules protrude from other parts of its body. The body wall is riddled with numerous incurrent pores.

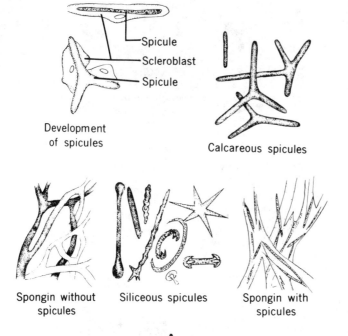

**Figure 7-3** (A) Spicules and spongin from various genera of sponges. (B) Photomicrograph of calcareous spicules.

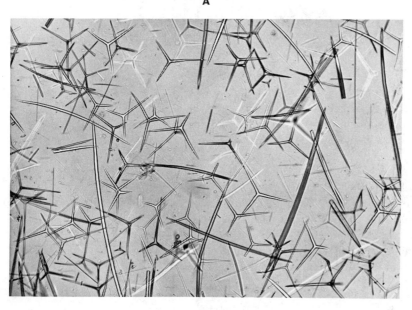

One large **central cavity** (**spongocoel**) leads from the base of the sponge up to the osculum. Around the central cavity, the thick body wall is built of elongated, sac-shaped **radial canals.** Each canal lies perpendicular to the central cavity and has a large exhalant opening (**apopyle**). Undulipodia lining the canals beat constantly, drawing water into the many inhalant canal chambers.

In the walls of the canals are inhalant openings (**prosopyles**), mesenchyme (a jellylike material), spicules, and a variety of amoeboid cells. There are three

**Figure 7-4** *Scypha*, the comparatively simple sycon sponge.

types of amoeboid cells: **porocytes** or pore cells, which surround pores and may close them in a musclelike manner; **scleroblasts,** which secrete spicules of calcium carbonate; and **archeocytes,** which receive, digest, and transport food, and produce other cells, particularly reproductive ones (ova, sperm, gemmules).

The soft body wall is supported and protected by a skeleton composed of many calcium carbonate spicules. Four varieties of spicules are present: (1) long, straight monaxon rods, which guard the osculum; (2) short straight monaxon rods, surrounding the incurrent pores; (3) triradiate spicules, embedded in the body wall; and (4) T-shaped spicules, lining the central cavity.

## PHYSIOLOGY

*Scypha* lives on fine particles and minute planktonic organisms, drawn into it by the current created by choanocyte undulipodia. Some digestion occurs within the choanocytes, but is for the most part carried out by amoebocytes. As in protozoans, digestion is intracellular; nutrients are diffused through the cells with the aid of amoeboid archeocytes, which also serve as food-storage sites.

Excretory matter is discharged through the general body surface, probably assisted by amoeboid wandering cells and possibly by choanocytes. Respiration, likewise, takes place through the cells of the body wall in the absence of any specialized organs.

Sponges are usually considered to be very quiet and sluggish but are actually among the most active and energetic of all animals, working night and day to create the currents of water that bring food and oxygen into the body and carry away wastes. The amount of water that flows through the body is tremendous; an average-size sponge draws about 45 gallons of water through its canal system in a single day.

True nerve tissue in sponges has not been demonstrated; single cells, how-

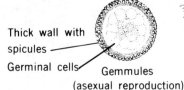
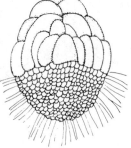

**Figure 7-5** *Left:* gemmules of asexual, freshwater sponge. *Right:* amphiblastula formed during sexual reproduction of *Scypha*.

ever, do respond to certain stimuli. A finger placed in the osculum may be forcibly squeezed due to responsive contractile cells (**myocytes**) surrounding the opening.

In many sponges an individual cut into pieces will grow into several normal sponges; this process is known as **regeneration.**

## REPRODUCTION

*Scypha* reproduces both sexually and asexually. Asexual reproduction involves the formation of **buds** near the point of attachment. These eventually break free and take up a separate existence.

Sexual reproduction involves the mesenchyme cells. Both eggs and sperm occur in a single individual. The fertilized egg segments by three vertical divisions into a pyramidal plate of eight cells. A horizontal division cuts off a small cell from each of the eight, resulting in a layer of eight large cells crowned by a layer of eight smaller cells. These arrange about a central cavity, producing a blastulalike sphere. The small cells multiply rapidly and develop undulipodia, while the large cells become granular, partially growing over the other cells to form an **amphiblastula** (Fig. 7-5). The amphiblastula escapes from the parent as a larva and swims about for several days until it attaches to a solid object and begins growth as a young sponge.

# Other Sponges

Sponges vary in structure from simple, thin-walled tubular structures to massive irregularly shaped forms (Fig. 7-6). Forms exhibited by certain species may vary according to their development in shallow or deep water. *Microciona* found in shallow water forms a thin encrustation on rocks, while in deep water it forms massive colonies that may be up to 6 inches high.

Sponges may be branched like trees or shaped like gloves, cups, or domes; the majority are irregular and amorphous. They vary in size from those no larger than a pinhead to those 8 feet in diameter. They are also highly variable in color, some being white or gray and others yellow, orange, red, green, blue, purple, or black.

Bath sponges, with which we wash cars and clean kitchens, and which the

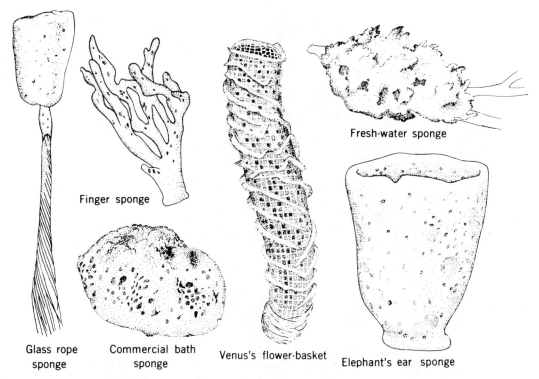

**Figure 7-6** Assorted freshwater and marine sponges.

ancient Greeks and Romans used for drinking and painting, are actually the supple skeletons of certain sponges.

Boring sponges live in shallow water near shores all over the world. They form irregular masses and are a bright sulfur yellow. Their name refers to their habit of attaching to oyster and clam shells which they bore full of holes, eventually killing the animal and breaking up the entire shell.

Some sponges are poisonous, and contact with them can produce skin irritations for human beings. Other live sponges give off an unpleasant odor, and many contain sharp spiny spicules; because of such devices they are seldom preyed upon. Other organisms use sponges for concealment and protection.

Venus's-flower-basket (*Euplectella*) is a sponge of ornamental interest, with a beautiful cylindrical skeleton resembling spun glass (Fig. 7–7). Other sponges are of economic importance owing to large flint deposits formed from siliceous spicules.

## FRESHWATER SPONGES

All freshwater sponges belong to the family Spongillidae. They are commonly yellow, brown, or green and are found in clear water, encrusted on stones, sticks, and plants.

These sponges reproduce by formation of **gemmules** (Fig. 7–5), consisting of a number of archeocytes from the middle layer of the body wall gathered into a ball and surrounded by a chitinous shell reinforced by silica spicules. Gemmules are used to identify sponges. They also allow the sponge to survive unfavorable environmental conditions.

There are about 150 species of Spongillidae, 20 occurring in the United States; the most abundant is *Spongilla lacustris*.

**Figure 7-7** Photograph of the intricate venus's flower basket. These marine sponges attach themselves to the mud sea floor by the long threads at one end.

## Origin of Sponges

Although there is little specialization of cells, the sponges represent a considerable advance over even complex protozoans. Despite the fact that sponges are multicellular and suggest the tissue level of organization, their lack of organs or digestive cavity indicates that they may have developed from a protozoan group, the Choanoflagellata (a mastigophoran). They resemble these colonial protozoans in many ways, notably the formation of skeletal spicules by single cells.

Although sponges are well enough adapted to their environment to have lived their primitive way of life for at least 600 million years, it is apparent that they have diverged far from the main line of metazoan evolution.

## Summary

Sponges are the least complex of the metazoans except for the Mesozoa. They are all sessile in the adult stage, usually asymmetrical and vary greatly in size and complexity of structure. Their primitive multicellular organization is an intermediate between the protists and more typical metazoans.

# 8

# Phylum Cnidaria (the Coelenterates) and Phylum Ctenophora

Phylum Cnidaria (Gr. *knide*, nettle) consists of over 10,000 species of various interesting animals, including hydras, jellyfish, and corals. All cnidarians are aquatic and most are marine. A smaller, exclusively marine group comprise the phylum Ctenophora. Ctenophorans number approximately 100 species and closely resemble jellyfishes; they will be discussed at the end of the chapter.

## General Characteristics of Coelenterates

Cnidarians (or **coelenterates**) were most probably derived from protozoans; most form larvae that closely resemble some ciliates. In Chapter 7 we saw that many sponges also form such larvae, and it is thought that sponges and coelenterates arose from a common ancestor.

In the evolutionary spectrum, coelenterates are higher than poriferans, exhibiting a definite **tissue level** of organization and cells that are more highly specialized and integrated. Coelenterates are also more receptive to stimuli and are capable of a great variety of responses. The organizational differences between poriferans and cnidarians may have been a result of divergent feeding habits: filter feeders (such as sponges) require less complicated biological machinery than do carnivores (such as coelenterates), since the latter must capture and digest active prey.

There are three classes in the phylum Cnidaria: **Hydrozoa**, represented by the *Hydra* and the colonial *Obelia*; **Scyphozoa**, large jellyfish, represented by *Aurelia*; and **Anthozoa**, characterized by the sea anemone *Metridium* and the coral *Astrangia*.

All three classes have certain basic coelenterate characteristics. They are either **radially** or **biradially symmetrical** with a principle axis extending orally–aborally (mouth to base). The body wall consists of three layers: an external **epidermis**, an internal **gastrodermis**, and between them a noncellular substance, the **mesoglea**. Within the body is a single **gastrovascular cavity** or

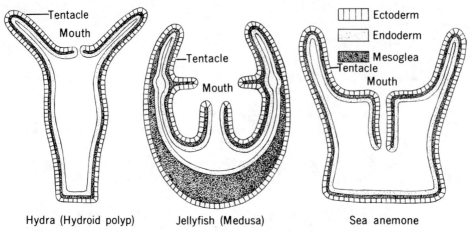

**Figure 8–1** Basic body plans of representatives from the three classes of coelenterates. The mouths of the hydra and sea anemone are directed upward, but the jellyfish swims with mouth down. For purposes of comparison, all three are shown with mouth up.

enteron. The coelenterates possess sting capsules (**nematocysts**), occurring in one or both layers of the body wall.

The two principal body forms of coelenterates are the **polyp,** an upright sessile form as in the hydra and sea anemone, and **medusa,** a free swimming form as in the jellyfish. These body structures share similar features, but differ notably in the large quantity of mesoglea present in medusa (Fig. 8–1).

Other basic coelenterate features include a diffuse network of neurons comprising the nervous system and a sometimes-present skeleton, composed of either calcium carbonate or horny material. Respiratory, circulatory, and excretory systems are lacking. Coelenterates exhibit a remarkable degree of **polymorphism:** various body forms may be present in a species.

We will now look at examples from the three classes of this diverse and often beautiful group of animals.

## Hydra—A Freshwater Hydrozoan

Hydras, named after the mythical nine-headed dragon slain by Hercules, are simple coelenterates, commonly found in freshwater ponds and streams (Fig. 8–2). They range in size from 2.5 to 3 cm and are either white or brown in color (with the exception of the green *Chlorohydra viridissima*). There are nine known species in the United States.

A complex organization is exhibited in hydras, but with little division of labor; work performed by organs in higher animals is done by tissues and individual cells. As with most "simple" animals, a study of these cells and tissues reveals the true complexity of their nature.

### MORPHOLOGY

The body of the hydra (Fig. 8–3) resembles an elastic tube and may be extended to a length of 2 cm. At the oral end is a circlet of tentacles. The

*Phylum Cnidaria (the Coelenterates) and Phylum Ctenophora*

**Figure 8–2** *Hydra*, showing a bud. (Courtesy Ward's.)

tentacles of some hydras are capable of remarkable extension, stretching from small blunt projections to very thin threads 7 cm or more in length. They move independently, capturing food and bringing it to the mouth. The number of tentacles varies considerably (usually six or seven, in some species up to 10), increasing with the size and age of the animal.

The tentacles surround a conical elevation, the **hypostome.** When the mouth is contracted (during rest or digestion), it is a minute circular pore, but when swallowing objects, the mouth and the surrounding hypostome can dilate to a relatively large diameter. Below the mouth is the **gastrovascular cavity** or **enteron,** and beneath this is a **stalk;** together these structures comprise the **body column.**

At the aboral end is a **foot** or **basal disk,** which serves in anchorage and locomotion. An **aboral pore,** located at the center of the disk, closes when the hydra attaches itself to an object and opens upon its release. A sticky substance is secreted by cells in the epidermis of the foot, aiding its anchoring and locomotion functions.

## HISTOLOGY

The body wall of hydra (Fig. 8–4) consists of two cellular layers: an outer thin **epidermis** and an inner **gastrodermis,** about twice as thick as the outer layer. Both are composed primarily of **epitheliomuscular cells.** A thin, noncellular **mesoglea** separates the layers throughout most of the body but is lacking in the center of the basal disk. The mesoglea supports the epithelial cells and serves as a place for attachment of their muscle processes.

The epitheliomuscular cells of the epidermis have polygonal outer surfaces that are fused together to form a continuous membrane over the hydra, interrupted only where stinging or sensory cells are in contact with the surface. The epitheliomuscular cells of the gastrodermis line the entire wall of the

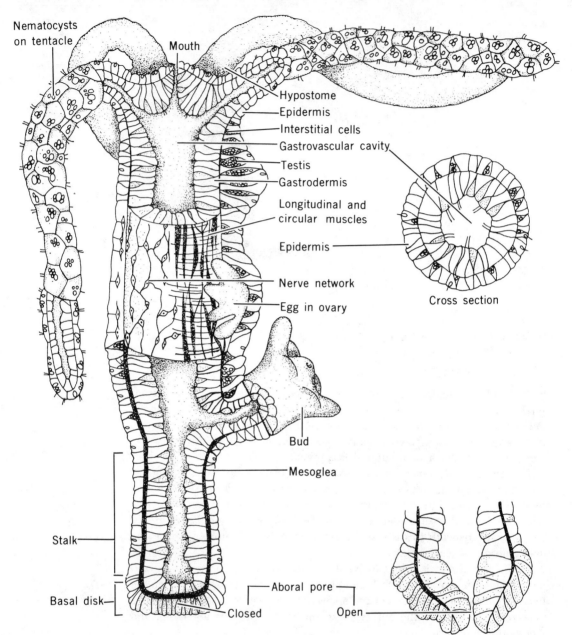

**Figure 8-3** *Hydra*, with parts cut away and sectioned to show structure.

gastrovascular cavity. They are concerned with digestion or absorption of food and are therefore called **nutritive cells.**

At the center of the hypostome are gastrodermal cells which are alternately filled with small secretion granules or have a fine, spongy texture. These are the **mucus-secreting gland cells;** their contractile fibers form a sphincter around the mouth which aids in swallowing food.

The **interstitial cells** are small and round, with clear cytoplasm and a relatively large nucleus, containing one or two nucleoli. These cells can differentiate into any of the specialized cells of the hydra, and are the chief agents in reconstructing tissues in growth and regeneration. They also form the

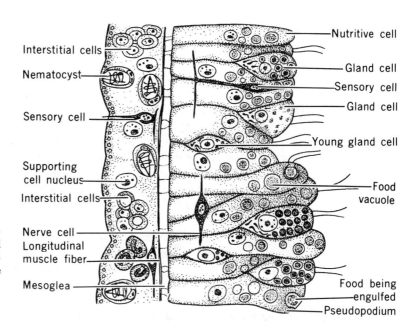

**Figure 8-4** Longitudinal section of body wall of *Hydra*, highly magnified. Epidermis is to the left of the mesoglea, gastrodermis is to the right.

primordial germ cells of the gonads. Without interstitial cells, the hydra cannot survive.

## MUSCULAR SYSTEM

The muscular system of the hydra consists primarily of two layers of contractile fibers. The outer muscle layer is longitudinal and is formed by the contractile fibers of the epidermal cells, while the inner layer is circular and derived from the contractile fibers of the gastrodermal cells. The circular muscle layer contracts slowly, performing peristaltic-like movements while the external longitudinal layer is capable of rapid contraction.

## NERVOUS SYSTEM

The coelenterate nervous system is relatively primitive. That of the hydra consists of two types of cells: **conducting** or **motor nerve cells** and **sensory cells.** These are distributed throughout the body, forming a network in the epidermis (**nerve net** or **nerve plexus**). Similar nervous elements are present in the gastrodermis, forming another, lesser developed nerve network. The two are connected by fibers passing through the mesoglea.

Most of the conducting and motor nerve cells of the hydra have several processes; they conduct impulses in any direction (nonpolar) and thus differ from the neurons of higher animals. The greatest concentration of nervous elements occurs around the hypostome, where the fibers pass in a circular direction to form a loosely organized **nerve ring.** Another, somewhat similar concentration of nerve fibers appears in the foot.

The sensory cells consist of slender, threadlike specialized nerve cells, lying in both epidermis and gastrodermis, between the epitheliomuscular cells. At their base they divide into two or more fibers, which connect with either the nerve plexus or with muscle fibers. The sensory cells are most abundant toward the foot region, on the hypostome, in the inner parts of the tentacles, and at basal disk. These parts of the hydra are the most sensitive to external stimuli.

**Nematocysts** (stinging capsules) are present on all parts of the hydra's epidermis except the basal disk, and are most numerous on the tentacles. Each nematocyst is formed inside a modified interstitial cell, called a **cnidoblast.** On the general body surface, cnidoblasts are usually wedged between the outer edges of supporting cells; on the tentacles and hypostomes, they lie within epitheliomuscular cells which are then known as **host cells.** Host cells on the tentacles are large and contain one or two large nematocysts (penetrants), surrounded by a number of smaller ones.

Four kinds of nematocysts occur in the hydra: **penetrants** (Fig. 8–5), which are the largest; **volvants,** small and pear-shaped; and **oval glutinants** and **small glutinants.** Penetrants and volvants are of special help in capturing prey; oval and small glutinants secrete a sticky substance possibly used in anchorage and locomotion as well as in obtaining food.

Projecting from the cnidoblast, near the outer end of the nematocyst, is a hairlike process, the **cnidocil.** The cnidocil is known as the "trigger" since it was once believed to be the cause for discharge of the nematocysts; these capsules are actually triggered by a combined influence of chemical and mechanical stimuli. The process is not fully understood, but involves a change in the permeability of the capsule wall. Pressure in the capsule may build up (owing to the presence of more water), and a contraction of the cnidoblast may then burst open the lid, allowing the nematocyst to spring out.

An animal "shot" by nematocysts is immediately paralyzed and sometimes killed by **hypnotoxin,** a poisonous substance injected into it through the tube.

**Figure 8–5** Discharged penetrant nematocyst of *Hydra.*

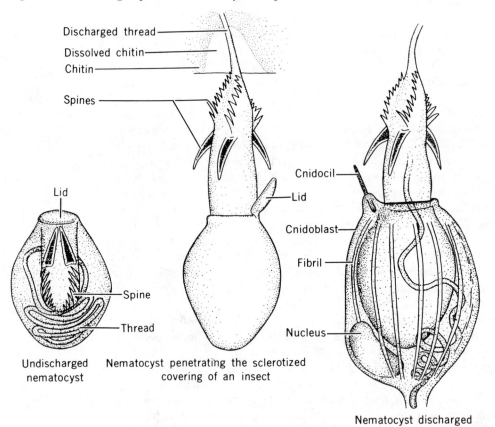

# PHYSIOLOGY

**INGESTION, DIGESTION, AND EGESTION.** Hydras feed primarily on small aquatic crustaceans. Larger hydras may ingest animals as big as tadpoles and young fish. Hydra usually rests with its basal disk attached to some object, its body and tentacles extended into the water, thus occupying a large hunting territory. Any small animal swimming within touch of a tentacle is at once shot full of penetrants, affixed by glutinants, or grappled by volvants.

The tentacle that has captured the prey is assisted by other tentacles (which may use nematocysts to quiet the victim) in bringing the prey to the mouth. The edges of the mouth, moist with mucus, slip smoothly over the helpless prey as it slides slowly into the gastrovascular cavity. The body wall contracts behind the food and forces it downward. Frequently, organisms many times the size of hydras are successfully ingested.

Immediately after the ingestion of food, gland cells in the gastrodermis become very active, forming enzymes which are discharged into the gastrovascular cavity, where they immediately begin the digestive process. Undulipodia extending out into the central cavity also aid in the breakdown of food by creating currents. This method of digestion differs from that of the amoeba, paramecium, and sponge in that it is carried on outside the cell (**extracellularly**). **Intracellular** digestion also takes place in the hydra; pseudopodia thrust out by the gastrodermal cells engulf particles of food, which are then digested within the cells. This digested food is absorbed and stored by the gastrodermis. All indigestible material is egested from the mouth, accomplished by a sudden squirting which discharges the debris some distance.

**RESPIRATION AND EXCRETION.** Oxygen diffuses into the cells from the water in which the hydra lives. Carbon dioxide diffuses out of the cells into the surrounding environment. Metabolic waste products are excreted through the general body surface.

# BEHAVIOR

**SPONTANEOUS MOVEMENTS.** All voluntary movements of hydra result from contraction of the contractile fibers, and are produced by either internal (spontaneous) or external stimuli. Spontaneous movements may be observed in a resting hydra: at intervals of several minutes, the body, tentacles, or both contract suddenly and rapidly, then slowly expand in a new direction.

**LOCOMOTION.** Hydra move by several methods (Fig. 8–6), one being a gliding movement, where the basal disk slowly slides over the object to which the animal was attached. Another is an inchworm-type movement, in which the hydra bends over and attaches its tentacles to a region, slides its basal disk up close toward them, and then releases the tentacles, assuming an upright position. They may also move about by continual somersaults, each time releasing the basal disk and attaching it to a new object. Another method of locomotion, seldom observed, is an upside-down position where the hydra uses its tentacles as legs. To rise to the surface, a gas bubble may form on its basal disk, helping to carry the hydra upward.

**CONTACT, LIGHT, AND OTHER STIMULI.** Hydra may react to mechanical stimulation in two ways: **nonlocalized** response, causing a rapid contraction of a part of or the entire animal (caused by agitation of its environment or **local** response, caused by direct stimulation of a body part. If one tentacle is stimulated with a fine glass rod, the other tentacles (and possibly the body) may

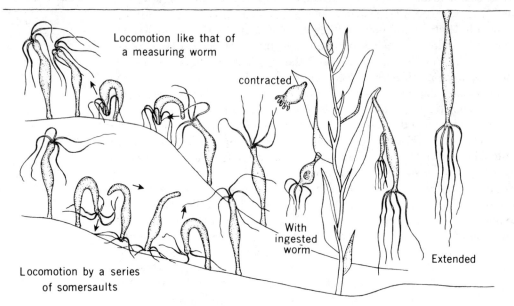

**Figure 8–6** *Hydra* feeding, and methods of locomotion.

contract, indicating that some transmission of impulses takes place through the nervous system.

The hydra has an optimum with regard to light. If a dish containing hydras is placed so that illumination is not equal on all sides, the animals will collect in the brightest region. If the light becomes too strong, they will move to a less illuminated area through trial and error.

Reactions to changes in temperature are indefinite, although in many cases they will move away from a heated region.

The overall physiological condition of an animal determines to a large extent the kind of reactions produced not only spontaneously, but also by external stimuli.

## REPRODUCTION

Asexual reproduction (by budding) and sexual reproduction (producing fertilized eggs) take place in the hydra and may occur at the same time. *Budding* is a common occurrence; several buds often appear on a single hydra. The bud first appears as a slight bulge in the body wall. This pushes out rapidly as a projection, which soon develops a circlet of blunt tentacles around its outer end. The cavities of both the stalk and tentacles of the bud are connected to the parent until it is full grown (requiring about 2 days). The bud then detaches and leads a separate existence.

**Sexual reproduction** involves the formation of a fertilized egg. In some hydra, sperm cells (spermatozoa) and egg cells (ova) are formed in one monoecious individual. Most hydra are dioecious, producing either egg or sperm cells. Both ova and spermatozoa develop from interstitial cells.

The male germ cells of the hydra are formed in small conical or rounded elevations (**testes**) which project from the surface of the body. Through a series of complicated divisions, spermatozoa are formed. In most hydras, definite nipples with small fissures are formed on the testes through which mature spermatozoa escape. As many as 20 or 30 testes may occur on a single hydra. Eggs develop within an ovary and are formed from several oocytes (enlarged interstitial cells) (Fig. 8–7). As the egg grows it becomes scallop-shaped, as a result of confinement between supporting cells. Attaining full growth, it becomes spherical but is still surrounded by epidermal cells which stretch enormously to cover the egg and remain rooted to the mesoglea. When the egg is mature, an opening appears on the epidermis, freeing the egg on all sides except for a point of attachment to the parent. At this point, fertilization usually occurs. Several sperm may penetrate the egg membrane, but only one enters the egg itself.

**Cleavage** now begins in the embryo, forming a **blastula** which resembles a hollow sphere with a single layer of epithelial cells composing its wall. These cells may be called the **primitive ectoderm.** By mitotic division, they form endoderm cells which drop into the cleavage cavity, completely filling it. The early **gastrula,** therefore, is a solid sphere of cells differentiated into a single outer layer, the ectoderm, and an irregular central mass, the endoderm. The ectoderm secretes two envelopes around the gastrula: the outer is a thick chitinous shell which may be covered with sharp projections; the inner is a thin gelatinous membrane. Different species of hydras can be identified by the structure of their embryonic shells.

In hatching, the embryo separates from the parent and falls to the bottom, where it remains unchanged for several weeks. The outer chitinous shell is eventually broken and the embryo escapes. Mesoglea is now secreted between the ectoderm and endoderm cells, and these layers then differentiate to the adult epithelial tissues: the epidermis and gastrodermis. A circlet of tentacles arises at one end, and a mouth appears in their midst. The young hydra formed grows into the adult condition without going through a larval stage.

**Figure 8–7** Developmental stages of *Hydra*.

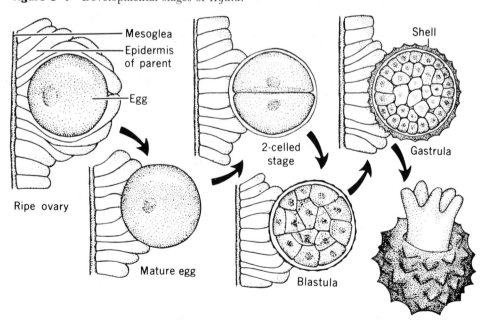

## REGENERATION AND GRAFTING

The ability of animals to regenerate lost parts was first discovered in 1740 by A. Trembley in his experiments with the hydra. Trembley found that if hydras were cut into two, three, or four pieces, each would grow into an entire animal. If a hydra were split longitudinally into two or four parts, each would become a perfect polyp; and if the oral end were split and kept slightly separated, a "two-headed" animal would result. Pieces of the hydra measuring only $1.2 \times 10^{-3}$ cm in diameter are capable of becoming entire animals.

Parts of one hydra may easily be grafted upon another, producing some very bizarre results. Parts of hydras from two different species have also been successfully united.

The ability to regenerate lost parts is obviously beneficial to animals. Regeneration takes place in some manner in all animals; in man, for example, new epidermal cells are continuously produced to replace old ones. Lost parts, however, are not restored, since our growing tissues are not coordinated for regeneration to occur; a decrease in regenerative power seems to be correlated with the increase in complexity of animal types.

# Other Hydrozoans

### *Obelia*—A COLONIAL HYDROID

*Obelia* (Fig. 8–8) is a marine hydrozoan that lives along the seacoasts of the Atlantic and Pacific oceans. This animal forms a **colony** of polyps that are specialized for feeding (**hydranths**) or reproduction (**gonangia**). The colony is attached to a substratum by a rootlike mass (**hydrorhiza**) from which arise upright branches (**hydrocauli**) that support the hydranths and gonangia.

The hydranths resemble the hydra somewhat in structure and function but are specialized for feeding, having about 30 solid tentacles. The gonangium has a central axis (**blastostyle**) which gives rise to buds that develop into **medusae**; these escape and produce either eggs or spermatozoa. The fertilized egg (**zygote**) develops into a ciliated free-swimming larva (**planula**), which soon becomes attached to a stone and grows into a polyp type of colony that reproduces asexually by budding.

METAGENESIS. The alternation of a generation that reproduces asexually by division or budding with a generation that reproduces sexually by means of eggs and sperms, as in *Obelia*, is known as **metagenesis**. Metagenesis occurs in many coelenterates. The polyp and medusa stages are not equally prominent in all hydrozoans; the medusa stage in the *Obelia* is degenerate and inconspicuous and in the hydrozoan jellyfish *Gonionemus* (Fig. 8–9) the polyp stage is only slightly developed.

### *Physalia*

*Physalia*, the Portuguese man-of-war, is a colony of hydrozoan polyps specialized for feeding, sensation, and reproduction; the polyps are supported by a gas-filled float. The surface of the float shimmers with beautiful iridescent colors.

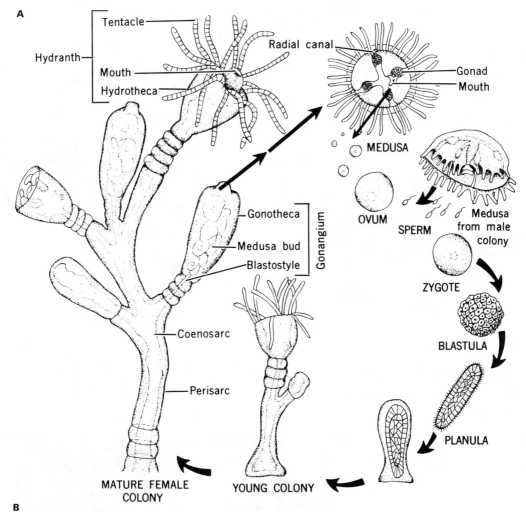
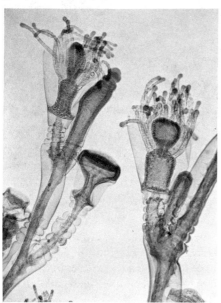

**Figure 8–8** (A) Life cycle and structure of *Obelia*, a colonial marine hydroid. Specialized polyps for feeding (hydranths) and asexual reproduction (gonangia) and specialized medusas for sexual reproduction illustrate the metagenesis typical of coelenterates. (B) Photomicrograph of *Obelia*.

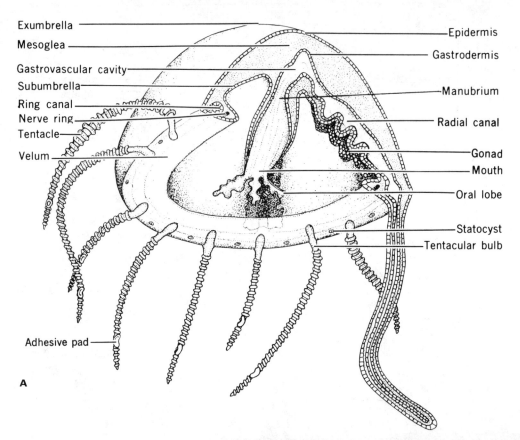

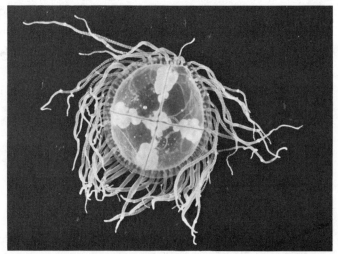

**Figure 8–9** (A) Diagram of hydrozoan medusa with part cut away to show internal structure. (B) Medusa.

## Jellyfishes—Class Scyphozoa

Most of the larger jellyfishes belong to the class Scyphozoa (Gr. *skyphos*, cup + *zoön*, animal). They usually range from 2.4 cm to 1 meter in diameter and are found floating near the surface of the sea. *Aurelia* (Fig. 8–10) is white or bluish, with pink gonads.

**Figure 8-10** Life cycle of the jellyfish, *Aurelia*. Longitudinal section for gastrula stage, vertical section of adult.

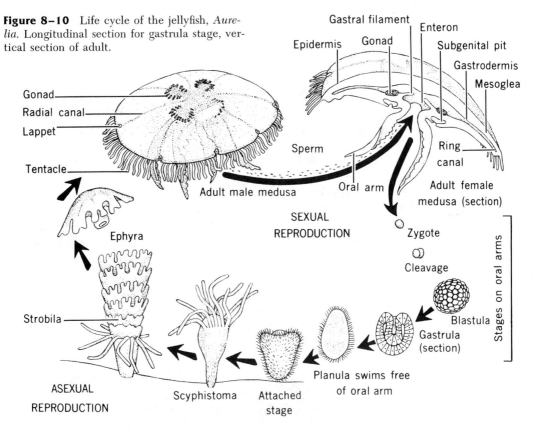

Oral arms hang down from a square mouth, which opens into a short **gullet**; this leads into a rectangular, central **enteron**. **Gastric pouches** extend laterally from four sides of the enteron. Within each gastric pouch is a gonad and a row of small **gastric filaments** bearing nematocysts.

Numerous radial canals, some of which branch several times, lead from the enteron to a **ring canal** at the margin. The eight sense organs of *Aurelia* lie between the **marginal lappets** and are known as **tentaculocysts**; these are considered to be organs of equilibrium. In addition, each tentaculocyst bears a pigment spot that is sensitive to light.

The food of *Aurelia* consists of small particles carried along radial canals by the beating of undulipodia of some gastrodermal cells. Ingestion occurs in the gastrodermal cells; the physiological processes of *Aurelia* are in general similar to those of the hydra.

The **gonads** are frill-like organs lying in the floor of the gastric pouches. The egg develops into a free-swimming **planula** which becomes attached to an object and develops into an elongated and deeply constricted polyp (**scyphistoma**). It then asexually divides into disks (a stage called **strobila**); each disk develops tentacles and swims away as a young medusa (**ephyra**). The ephyra gradually develops into an adult jellyfish.

## Sea Anemones and Corals—Class Anthozoa

Anthozoans (Gr. *anthos*, flower + *zoön*, animal), the final class of cnidarians, are sessile flowerlike animals. The hydrozoan *Obelia* has both polyp and medusa stages in its life cycle, while the scyphozoans emphasize the medusa stage (at times eliminating the polyp stage altogether); conversely, all anthozoans are polyps.

### Metridium—A SEA ANEMONE

A common representative of the class Anthozoa is the sea anemone. *Metridium dianthus* (Fig. 8–11) is an anemone that fastens itself to piles of wharves and solid objects in tidal pools along the North Atlantic and Pacific Coasts. It is a cylindrical animal with a crown of hollow tentacles arranged in a number of circlets about a slitlike mouth. They are usually brown- or yellow-toned, with tough skin.

At either side of the gullet (**stomodeum**) is an undulipodiated groove (**siphonoglyph**) through which water descends. The internal body cavity consists of six **radial** chambers separated by thin double partitions (**primary septa** or **mesenteries**). Water passes from one chamber to another through pores (**ostia**) in these septa. Smaller, **secondary septa** project out from the body wall into the chambers but do not reach the gullet; **tertiary septa** lie between the primaries and secondaries. The septa vary considerably in size, position, and number.

The free edges of the septa below the gullet in the **enteron** are expanded into

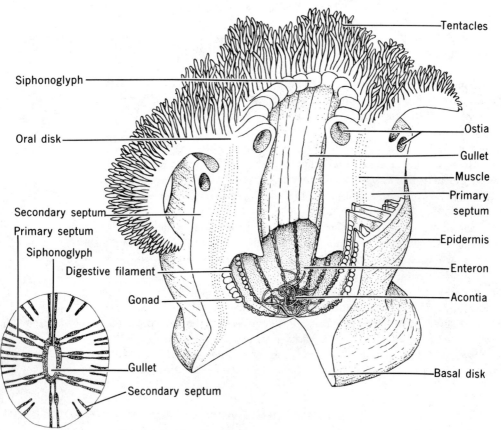

**Figure 8–11** *Metridium*, a sea anemone representative of class Anthozoa. *Left*, cross section through gullet shows arrangement of septa. *Right*, part of body cut away to show internal structure.

thickened **mesenteric filaments,** bearing glands that secrete digestive enzymes. Near the base of these filaments are borne long, delicate **acontia,** armed with gland cells and nematocysts.

Gonads are situated near the end of the septa; reproduction is both asexual and sexual. Asexual forms of reproduction include budding, fragmentation at the edge of the basal disk, and longitudinal fission. Sexual reproduction takes place in the water. *Metridium* and most other anemones are dioecious; sperm and egg cells exit through the mouth and unite to form a zygote that develops into a swimming, feeding larva. The larva eventually attaches to an object and matures.

Sea anemones are among the most beautiful inhabitants of the waters; when expanded they form a sea garden filled with flowerlike crowns of many colors. In their natural habitat, sea anemones are far from flowerlike; they serve as death traps for any small animal within reach of their tentacles or mouth (which is capable of grasping food). They may be beautiful in color but they wield their batteries of stinging nematocysts with deadly effect, paralyzing prey, which are carried through the mouth, down the gullet, and into the enteron. Food is digested by enzymes and absorbed by the gastrodermis. Undigested wastes are ejected through the mouth.

### *Astrangia*—A CORAL POLYP

*Astrangia danae* is a white coral polyp that inhabits the waters of North Atlantic coasts. Another species, *A. insignifica,* occurs along the Pacific Coast; the polyps of this species are orange and the coral is red. A number of individuals live together in colonies attached to rocks near the shore (Fig. 8–12). Each polyp resembles a small sea anemone and secretes a calcareous skeleton within which the animal rests; coral used for jewelry is actually the skeleton of a coral polyp.

**Figure 8–12** Colony of *Astrangia*. These corals secrete limestone cups into which the delicate polyps can retract.

**108**

*Animal Diversity: Invertebrates*

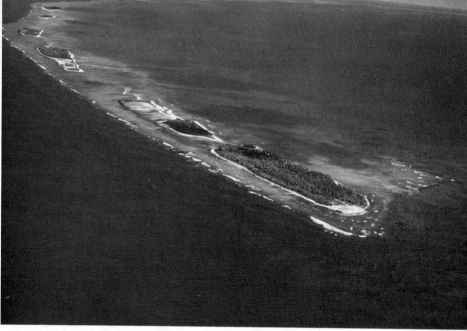

**Figure 8-13**  Coral atoll as seen from an aircraft.

Coral polyps build up various types of reefs, atolls, and islands (Fig. 8-13). The best known include the Fiji Islands of the Pacific Ocean and Bermuda, a Bahama Island where even some houses are built of coral blocks mined from certain areas. Coral reefs have been built up by the epithelial cells of countless numbers of small polyps, each one secreting its cup-shaped skeleton. The polyps die and new generations secrete their skeletal cups upon the old ones; only the surface of the coral mass is alive (Fig. 8-14).

## Origin and Relations of the Cnidaria

The origin of coelenterates is not firmly established. Some believe they evolved from colonial protists, others claim they arose from a primitive

**Figure 8-14**  Some representative corals.

flatworm (phylum Platyhelminthes). A comparison between embryological development of flatworms and coelenterates, however, shows them to be quite different.

Coelenterates are of considerable economic importance to man. Precious corals, usually bright red or pink, are used as jewelry; a belt of coral reefs, known as the horseshoe atoll, in west Texas is a prolific source of oil production—over 300,000,000 barrels produced to date.

Cnidarians have also aided the sciences: the interstitial cells of the hydra provide an invaluable tool for research in developmental biology, and the annual growth bands of certain corals which date over a period of 400 million years have helped keep astronomers informed of our planet's rotational speed and history.

## Comb Jellies—Phylum Ctenophora

Ctenophores (Fig. 8–15) are iridescent in sunlight and luminescent at night. They are commonly called sea gooseberries or sea walnuts, because of their shape, or **comb jellies** because of comblike locomotor organs (comb plates) arranged in eight rows. Their blue-green light is produced near nerve tracts beneath the comb plates. Ctenophores are **biradially symmetrical,** since the parts, although radially disposed, lie on either side of a median longitudinal plane.

The mouth (Fig. 8–16) is situated at the oral end, and a sense organ (**statocyst**) at the aboral end. The sense organ is a balancing device which may also be photosensitive. It consists of a small granular **statolith** which sits upon eight tufts of undulipodia; these are arranged to correspond with the comb plates so

**Figure 8–15** *Mnemiopsis*, a comb jelly.

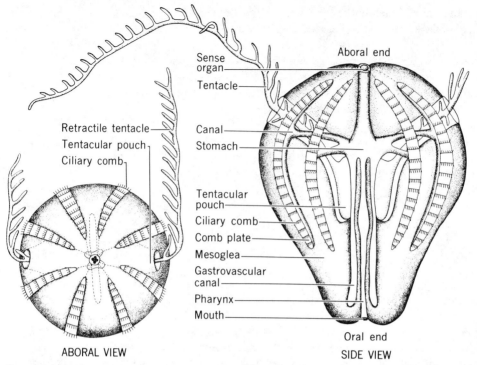

**Figure 8–16** Structure of typical ctenophores. *Left:* aboral view of *Pleurobrachia*. *Right:* side view of *Hormiphora*.

that undulipodial beating in a particular tuft can initiate beating in their corresponding comb plate.

Most ctenophores possess two solid, contractile **tentacles** which emerge from blind pouches opposite each other. The tentacles possess glue cells (**colloblasts**), which are modified epidermal cells. A colloblast is attached to the tentacle by a basal coiled spring. Secreted mucus is ejected against prey, firmly anchoring them to the tentacles. Ctenophores are carnivorous; their food consists of fish eggs, molluscan larvae, and small invertebrates. Feeding and digestion are very similar to cnidarians.

As in cnidarian jellyfishes, the cellular layers of ctenophores constitute a very small part of the body, which is primarily composed of a transparent jellylike **mesoglea**. A thin undulipodiated epidermis, derived from the ectoderm, covers the exterior and lines the pharynx (**stomodeum**). A gastrodermis, derived from the endoderm, lines the stomach and associated gastrovascular canals.

Scattered cells and muscle fibers lie in the mesoglea; this "layer" of cells in ctenophores resembles that of certain cnidarians but represents a higher grade of development. All the systems of ctenophores are of tissue grade in construction except for the indication of reproductive ducts in some forms.

Ctenophores are hermaphroditic; the ova and spermatozoa are formed on the walls of the digestive canals just beneath the undulipodial bands. The eggs and sperms pass by way of the mouth to the outside, where they are fertilized. Most develop directly into adult forms; in some, cydippid larvae form an intermediate stage in development. No nematocysts are present in organisms of this phylum. It is probable that ctenophores evolved from primitive coelenterate ancestors.

# Summary

Coelenterates exhibit a definite tissue level of organization. Nerve cell fibers and sensory cells are characteristic structures. The two principal body forms of coelenterates are the polyp, exhibited by the hydra, and medusa, exhibited by the jellyfish. Polyps and medusae are similar in basic structure but notably different in the enormous quantity of mesoglea present in the medusa. Digestion is intra- and extracellular; respiration and excretion are performed by the general body surface. Nervous tissue and sensory organs provide for various perception of stimuli and conduction of impulses. Reproduction is both asexual and sexual.

Ctenophores resemble cnidarians in some ways but are more complex, having undulipodiated bands, aboral sense organs, mesenchymal muscles, more definite organization of the digestive system and pronounced biradial symmetry.

# 9
# Flatworms

Animals of the phylum Platyhelminthes (Gr. *platys*, flat + *helmins*, worm) have dorsoventrally flattened bodies and hence are known as **flatworms**. Flatworms are noted for their lack of a true body cavity (coelom) and are referred to as acoelomates. In Chapters 10 and 11 we will see the transition from acoelomates to animals that possess body cavities.

Flatworms may be free-living, dwelling principally in fresh or salt waters, or parasitic. The parasitic flatworms are primarily flukes (or trematodes) and tapeworms (or cestodes). They are widely distributed among human beings and other vertebrates, often causing diseases and sometimes death in their hosts.

Many advances are exhibited by flatworms over coelenterates and ctenophores. They are the first phylum built on an **organ-system level** of complexity and have a distinct head with sense organs and central nervous system (cephalization). They are bilaterally symmetric, and possess a third embryonic tissue, **mesoderm**. Hence they are **triploblastic**: their structures are derived from ectoderm, endoderm, and mesoderm.

There are three classes of Platyhelminthes: **Turbellaria**, which are mostly free-living; and **Trematoda** and **Cestoda**, which are totally parasitic. Some of these parasites, such as tapeworms, are well known. Parasitic species are of great interest and very important economically. Their mode of life has brought about various morphological and physiological specializations, including increased powers of reproduction and extremely complicated life cycles. The influences of some species on human health constitute a large part of economic and medical zoology.

## Planaria—A Freshwater Turbellarian

The most common freshwater planarian in the United States is *Dugesia tigrina*. It lives on water plants in ponds; along the shores of ponds, lakes, and rivers; and in small streams under stones. Its upper surface is black, brown, or mottled and irregularly spotted with white; its undersurface is white or grayish. The body is broad and blunt at the anterior end and pointed at the posterior end, and may reach a length of 15 to 18 mm.

The anterior or head end (Fig. 9-1) has at each side a sharp projecting

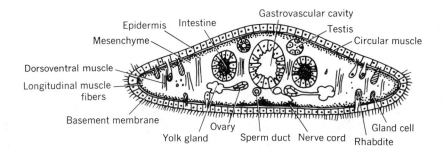
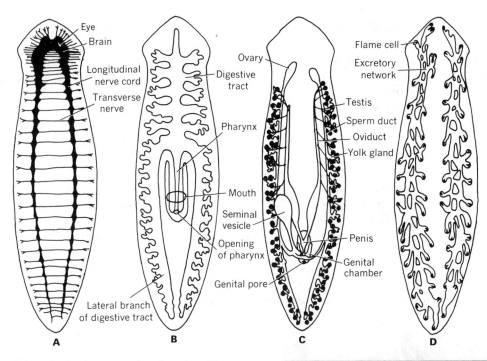

**Figure 9–1** Anatomy of a planarian. *Top:* cross section through pharyngeal region. *Bottom:* (A) nervous system, (B) digestive system, (C) reproductive system, (D) excretory system, (E) free-living planarian.

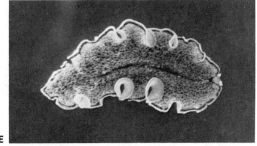

**auricle** containing olfactory sense cells. A pair of eyes is present on the dorsal surface between the auricles. The **mouth** is not on the head but near the middle of the animal on the ventral side. It opens into a cavity surrounded by a muscular tube, the **pharynx**, attached only at its proximal end. The pharynx consists of a complex of muscle layers and many gland cells. By means of the muscles, the pharynx can be thrust out of the mouth some distance when feeding. On the ventral side, posterior to the pharynx is a smaller opening, the **genital pore** (present only in sexually mature animals). The ventral surface of

113

the body is covered with undulipodia and mucus-secreting glands. The undulipodia beat in the mucus "slime road" and the animal moves forward.

The tissues of **mesodermal** origin (lying between the body wall and the intestine) consist of mesenchyme cells that are either joined (fixed) or free (with amoeboid movement); these are embedded in a fibrous mesh to form a connective tissue called **parenchyma.** The well-developed muscular, nervous, digestive, and reproductive systems are constructed in such a way as to function without the coordination of a circulatory system, respiratory system, coelom, or anus. The **digestive system** of *Dugesia* consists of a **mouth,** a **pharynx,** and an **intestine** with three main trunks, each having many small lateral branches.

Planaria feed on insect larvae, annelids, crustaceans, and other small animals, living or dead. The auricles are used to locate food; once located, the planarian inserts its pharynx into the prey's body cavity and secretes **endopeptidase** throughout, breaking up organs and tissues, which are then sucked out. Digestion occurs within cells lining the planarian intestine. There is only one opening to the digestive cavity; as in Cnidaria, undigested matter is egested through the mouth. Digested food is circulated within the branches of the digestive system and in the fluid-filled spaces in the parenchyma.

There is a complex network of small vessels on each side of the animal, from which many **flame cells** branch. A flame cell is large and hollow, with a cluster of flickering undulipodia extending into its central cavity; these create a current, forcing collected fluid through tubules, which open on the surface by several minute pores.

The **muscular system** consists principally of three sets of muscles, a **circular layer** just beneath the epidermis, a **longitudinal layer** immediately below the circular layer, and **dorsoventral muscles,** lying in the parenchyma.

The **nervous system** is well developed, consisting of an inverted V-shaped mass of tissue, a **brain,** and two ventral **longitudinal nerve cords** connected by

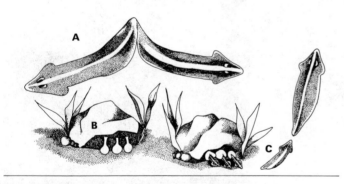
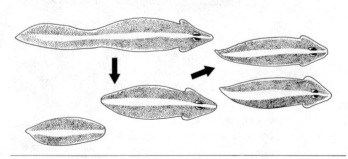

**Figure 9–2** Planarian reproduction. *Top:* sexual reproduction. (A) copulation, (B) cocoon under rock, (C) emergence and growth of young planarian. *Bottom:* asexual reproduction. The planarian divides by fission and the pieces develop and grow into adults.

**transverse nerves.** Nerves pass to various parts of the anterior end of the body. The pigmented eyes are sensitive to light but do not form an image.

Reproduction is asexual or sexual (Fig. 9–2). In asexual reproduction (fission), planarians divide transversely, each part reorganizing into a complete animal. Sexual reproduction is by cross-fertilization only, despite the presence of both male and female sex organs in each (hermaphroditic) planarian. There is no larval stage in development.

Planarians show remarkable powers of **regeneration.** If an individual is transversely cut in two, the anterior end will regenerate a new tail and the posterior end, a new head. Grafting pieces of one planarian to another is easily accomplished.

Planarians have been used in experiments to elucidate the mechanisms of learning (see Chapter 22).

# Other Platyhelminthes

Flatworms differ greatly among themselves; turbellarians exhibit the typical organization of the phylum, while trematodes and cestodes are modified considerably for a parasitic existence. Trematodes and cestodes do not have an epidermis but are covered with a thick surface cuticle. Cestoda also lack a mouth or intestine, absorbing nutrients through their general body surface. Trematoda, on the other hand, resemble Turbellaria in possessing a mouth and saclike intestine (and some have secondarily developed an anus). In the simplest forms of trematodes the intestine is unbranched; in others, branches occur and may penetrate to all parts of the body, rendering a circulatory system unnecessary. In certain families, channels filled with fluid occupy a considerable part of the body; the fluid surging back and forth as a result of muscular contractions may, in effect, serve as a transport system.

Both cestodes and trematodes possess excretory systems. These endoparasites have less need for osmoregulatory machinery than their free-living relatives, the turbellarians.

### LIVER FLUKE (*Fasciola hepatica*—A TREMATODE)

*Fasciola hepatica* or **sheep liver fluke** (Fig. 9–3) occurs in the bile ducts of the livers of sheep, cows, pigs, and many other herbivores. Human infections are rare, but may occur through consumption of aquatic plants (notably watercress) infested with encysted flukes.

The **mouth** of *Fasciola* lies in the middle of a muscular disk, the **oral sucker.** The **ventral sucker** (acetabulum) serves as an organ of attachment. Between the mouth and the ventral sucker is a **genital pore** through which eggs pass to the exterior. The **excretory pore** lies at the extreme posterior end of the body.

The body of the liver fluke is covered with a thick, heavy elastic cuticle. The **parenchyma** is a loose tissue lying between the body wall and the digestive tract; within it are embedded various internal organs. The **digestive system** consists of a mouth, muscular pharynx, short esophagus, and intestine with two main branches. There is a complex **excretory system** of flame cells draining into collecting ducts and then into two main looping ducts that fuse, forming a muscular epithelioid bladder connected to the excretory pore.

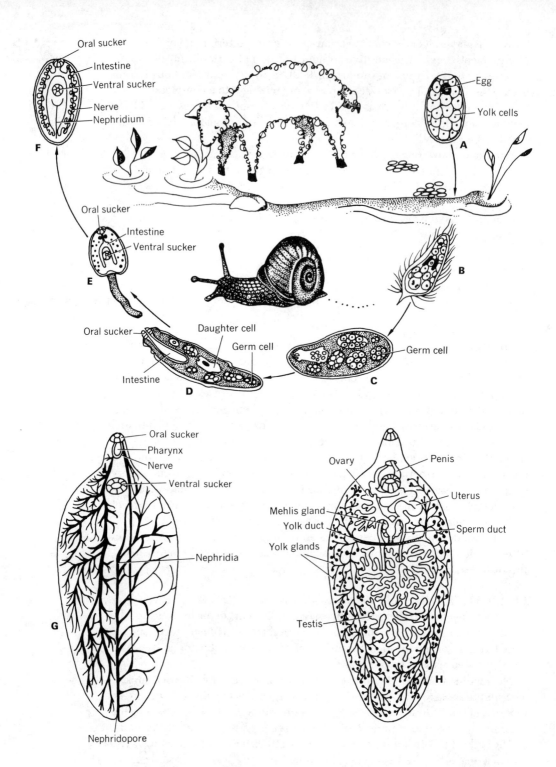

The **nervous system** consists of a small ganglion (at the anterior end) which gives off a series of longitudinal nerves with interconnecting nerve fibers. Sense organs are indefinitely present. Complex muscle layers lie just beneath the cuticle.

Most trematodes are hermaphroditic, with well developed and complex

116

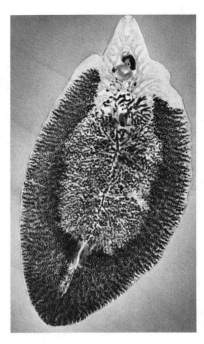

**Figure 9-3** Life cycle and structure of sheep liver fluke, *Fasciola hepatica*, (A) Eggs leave sheep's body through feces. (B) In water, eggs develop into miracidia that burrow into a snail. (C) Sporocyst from snail. (D) Redia larvae from sporocyst. (E) Cercaria larvae leave snail and encyst on vegetation. (F) Encysted metacercaria. If vegetation is eaten by a sheep, the metacercaria infect the sheep and develop into adults. (G) Adult, showing digestive system on the left and excretory system on the right. (H) Adult showing reproductive system. (I) Photograph of sheep liver fluke.

male and female **reproductive organs** present in the adult. The maturation process of eggs begins in the uterus of the fluke; these pass through the bile ducts of the sheep into its intestine, and are eventually carried out with the feces. Eggs that encounter water develop for 15 days or more, then hatch, releasing undulipodiated larvae (**miracidia**), which swim about until encountering a particular species of freshwater snail, into which they burrow. Within weeks, each larvae becomes a saclike body or **sporocyst**. A complex process resulting in the production of long-tailed larvae (**cercariae**) ensues. The cercariae leave the snail, swim about for a short time, and then encyst on a leaf or blade of grass. The encysted cercaria is called a **metacercaria**. If the leaf or grass is eaten by sheep, the metacercariae escape from their cyst wall in the small intestine, and make their way to the bile ducts in the liver of the sheep, where they develop into mature flukes in about 6 weeks.

One liver fluke may produce as many as 500,000 eggs; since the sheep's bile ducts may contain more than 200 adult flukes, there may be as many as 100 million eggs formed in one parasitized animal. The great number of eggs produced by a single fluke is necessary because many eggs do not reach water; the majority of the larvae do not find the particular species of snail necessary for further development; and the metacercariae to which the successful larvae give rise have little chance of being devoured by a sheep.

### PORK TAPEWORM (*Taenia solium*—A CESTODE)

The pork tapeworm (Fig. 9-4) lives (as an adult) in the digestive tract of humans. It is long (up to 4 meters), consisting of a knoblike "head," the **scolex**, and a great number of similar parts, the **proglottids**, arranged in a linear series. The animal clings to the inner wall of the intestine by means of **hooks** and **suckers** located on the scolex. Behind the scolex is a short **neck** followed by a string of proglottids, which increase in size and age from the anterior to the posterior end; a worm may contain as many as 800 or 900 proglottids.

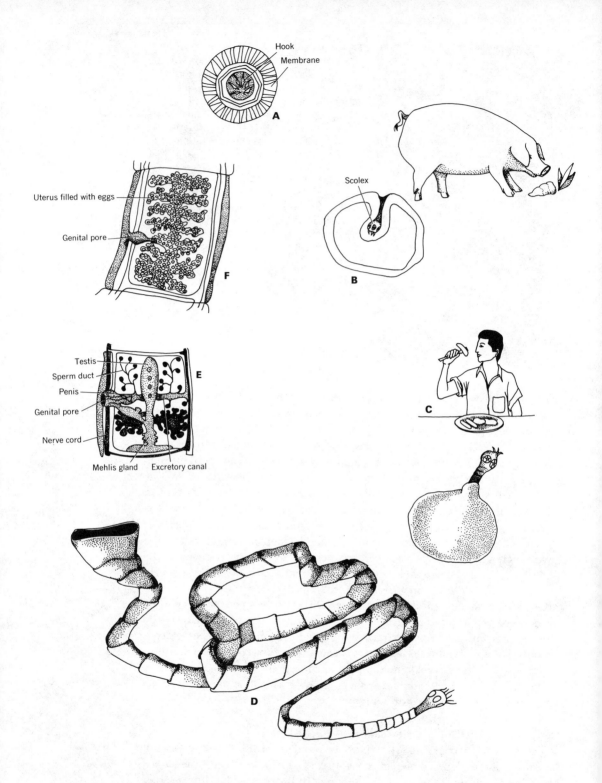

**Figure 9–4** Life cycle and structure of pork tapeworm, *Taenia*. (A) Embryo from ripe proglottid in human feces. Pig ingests embryo with food, (B) bladderworm, (C) raw or partially cooked pork eaten by man; Cysticercus everts, attaches to intestine, and grows into (D) adult. (E) Mature proglottid with reproductive organs. (F) Gravid proglottid.

No digestive tract is present; digested food in the intestine of the host is absorbed through the body wall. The **nervous system** is similar to that of planarians and liver flukes, but not as well developed. Longitudinal and cross-connecting **excretory canals** have many branches, ending in flame cells which carry metabolic waste out of the body.

A mature proglottid is almost completely filled with **reproductive organs**. A **six-hooked embryo** (**hexacanth**) develops from each egg while still within the gravid ("ripe") proglottid. These proglottids are egested from the host through the feces and may then be ingested by a pig. Here the hexacanth larvae escape from their envelopes, boring through the wall of the pig's intestine into the blood or lymph vessels. They are eventually carried to the voluntary muscles, brain, or eyes, where they form cysts. A scolex is developed from the cyst wall and the larva (known as a **bladder worm** or **cysticercus** at this stage) may then be ingested by human beings (through pork that is insufficiently cooked). The bladder is digested off and the released scolex fasten to the human intestine wall, where a series of proglottids are developed.

## Ribbon Worms—Phylum Nemertinea

**Ribbon worms,** (Fig. 9–5) are so called because of their long and dorsoventrally flattened shape. They range in length from less than 2 cm to over 30 meters and are mostly marine; a few inhabit fresh water and land. Animals in this phylum were once included in the phylum Platyhelminthes; they resemble the free-living flatworms in having bilateral symmetry, flame cells, unseg-

**Figure 9–5** Ribbon worm.

mented and contractile bodies, and in lacking a true coelom and respiratory system.

Anatomical features distinguishing nemertines from platyhelminths and other lower organisms include a long retractile **proboscis,** which lies in a sheath just above the digestive tract and may be used as a tactile and defense organ; a **blood-vascular system,** usually consisting of a median dorsal and two lateral trunks; and a **complete digestive tract** with both **mouth** and **anal openings.** They have a less complex reproductive system than the platyhelminths and are usually dioecious.

Nemertines are usually found coiled up in burrows in mud, sand or under stones; some frequent patches of seaweed. They feed on living or dead animals. Locomotion is effected by the undulipodia that cover the body surface, by contractions of body muscles or by the attachment of the proboscis and a subsequent drawing forward of the body.

The adults have great powers of **regeneration,** and some reproduce (in warm weather) by fragmentation of the body. *Lineus socialis*, which is only 100 mm in length, may be cut into as many as 100 pieces, each regenerating into a minute worm within 4 or 5 weeks. These minute worms may again be cut into pieces that regenerate, the process repeated until miniature worms less than 1/200,000 of the volume of the original worm result.

## Summary

Flatworms are characterized by an unsegmented, triploblastic body that is flattened dorsoventrally and bilaterally symmetric. These animals are the first group built on an organ-system level of complexity and exhibit cephalization—a major advancement over groups previously studied.

Turbellaria are free-living flatworms. Trematodes and cestodes are parasitic species of great economic and medical interest.

Nemertines are closely related to flatworms, but their proboscis, blood-vascular system, and digestive tract with mouth and anal openings places them in a separate phylum.

All the animals discussed in this chapter lack a true body cavity (coelom) and are thus known as acoelomates.

# 10
# Rotifers and Roundworms

In Chapter 9, we saw that the flatworms are acoelomates—they lack a body cavity (coelom). The animals we will study in this chapter are known as pseudocoelomates—they possess an unlined body cavity (pseudocoel) and anticipate the development of a peritoneum-lined or true coelom, to be studied in Chapter 11.

Rotifers and roundworms are classes of pseudocoelomate worms (Phylum Aschelminthes). Rotifers are usually free living, whereas roundworms can be either free living or typically parasitic. This chapter will present detailed characteristics of both groups, with particular emphasis on parasitic roundworms, which are of great agricultural and medical importance. Other pseudocoelomate organisms will be briefly discussed near the end of the chapter.

## Rotifers

**Rotifers** (or Rotatoria) (Fig. 10–1) are commonly known as "wheel animals," referring to the beating of cilia at their anterior ends that suggests the rotation of wheels. Rotifers, smallest of the metazoans, possess fantastic forms and brilliant colors, thus attracting the attention of many amateur microscopists. Most are inhabitants of fresh water, some are marine, and a few are parasitic. The body is usually cylindrical, covered by a shell-like cuticle. An often-forked, tail-like **foot** adheres to objects by means of a secretion from the cement glands.

Food is swept by the undulipodia through the **mouth** into the **pharynx**, the lower end of which forms a very characteristic grinding organ; and the **mastax**, where **chitin-like jaws** are constantly at work breaking up the food. A short **esophagus** leads into a **glandular stomach** or a **stomach** and **intestine**, depending on the species; here the food is digested. Undigested particles pass through the **intestine** into the **cloaca.** The excretory system consists of **flame cells** appearing at intervals alongside two long tubes which open into a **bladder** that contracts periodically, forcing the contents into the cloaca and out of the **cloacal opening** ("anus"). The bladder also maintains a proper water balance in the animal, just as the water expulsion vesicle does in the protists.

The sexes of rotifers are separate. Males are known in only a few species;

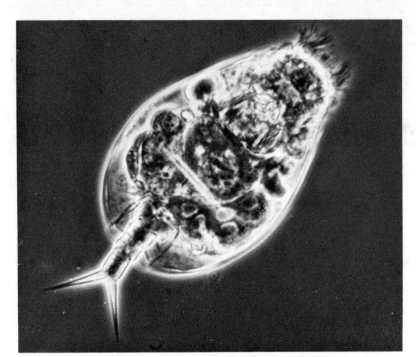
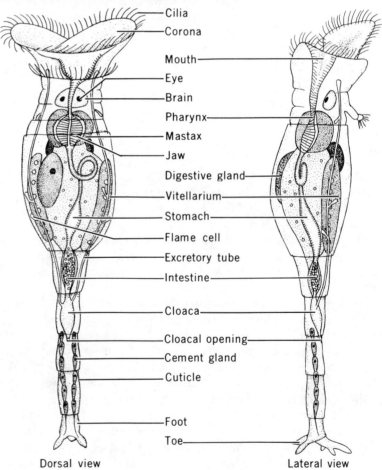

**Figure 10–1** Class Rotifera. *Top:* photomicrograph. *Bottom:* general structure of female rotifer.

where found, they are usually smaller and less developed than the females. Two kinds of eggs are laid: thin-shelled **summer eggs,** which develop **parthenogenetically** (unfertilized) and come in two sizes, the larger producing females and the smaller males; and thick-shelled **winter eggs,** which are fertilized and produce only females.

A characteristic of rotifers is their ability to resist dehydration. Certain species, if dried slowly, secrete gelatinous envelopes that prevent further drying; in this condition they may survive seasons of drought and severe extremes of temperature.

# Roundworms

**Roundworms** are universally present in the sea, fresh water, and the soil. They are typically unsegmented, long slender animals, usually with a smooth, glistening surface and tapering at both ends. The free-living species are usually smaller than the parasitic roundworms, which occur in plants and animals, including human beings.

Parasitic plant roundworms number over 1000 species and live in or on a large variety of plants, causing enormous crop loss. Many of them possess an **oral spear** with which they puncture the plants' roots; others enter leaves, eating the contents of the cells. Some worms stay on the surface of the plant, bury their heads into the tissue, and suck out the juice. These parasites sap the plant's vigor, open the way for bacteria and fungi, and injure growing points. The only means of control are crop rotation, soil sterilization, and development of resistant varieties of plants.

A brief description of the animal and human parasite *Ascaris*, which has a relatively simple life cycle compared to other, more complex roundworms, and an account of other medically important parasitic species follows.

### *Ascaris lumbricoides*

*Ascaris lumbricoides* (Fig. 10–2) is the common roundworm parasitic in the intestine of human beings and pigs. The sexes are separate: the male has a sharply curved posterior end and is considerably smaller and more slender than the female, which measures up to 25 cm long and 6 mm wide. Both have a dorsal and ventral narrow white line (nerve tracts) running the entire length of the body, and a broader **lateral** line on either side. A tough chitinous **cuticle** is smooth and marked with fine **striations.** The mouth opening is anterior and is surrounded by one **dorsal** and two **lateroventral lips.**

Between the intestine and the body wall is a **pseudocoel,** containing a straight, relatively simple **digestive tract** (consisting of a small **mouth,** and **muscular esophagus** or **pharynx**) and other organs. The esophagus draws fluids from the host into the long nonmuscular **intestine,** and the nutriment is absorbed through the wall. In the female, the posterior portion of the intestine is the **rectum,** which discharges through the **anus.** In the male, the intestine and reproductive system open into a common passageway, the **cloaca;** the opening to the outside should be called a cloacal opening, but is usually termed an anus. Near the cloacal opening in the male extend two chitinous rods, the **penial spicules,** used during copulation. The female has a **vulva** or **genital pore,** located ventrally about one-third the length of the body from the anterior end.

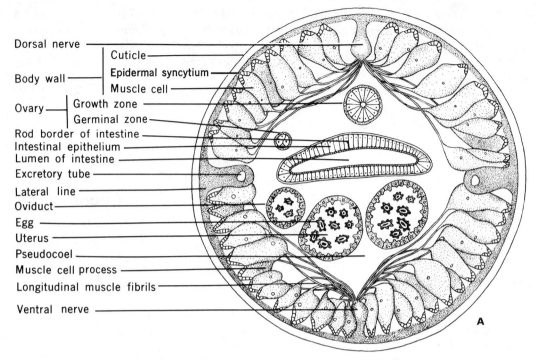

**Figure 10–2** Female Ascaris. (A) Cross-section. (B) Side view. (C) Dissection showing internal structure.

The male **reproductive organ** is a single, coiled threadlike **testis,** from which a **vas deferens** leads to a wider tube, the **seminal vesicle;** this is followed by a short, muscular **ejaculatory duct** opening into the **cloaca.** The female has a Y-shaped **reproductive system,** each branch consisting of a coiled, threadlike **ovary** which is continuous with the **oviduct** and **uterus.** The uteri of the two branches unite into a short muscular tube, the **vagina,** which opens to the outside through the **vulva.**

**Fertilization** takes place in the oviduct. The fertilized **egg** is surrounded by a thick, rough-surfaced shell, and passes out through the vulva. The genital tubules of a female worm may contain as many as 27 million eggs in various stages of development at one time; each mature female lays about 200,000 eggs per day.

The eggs of *Ascaris* are laid inside the intestine of the host and pass out in the feces. They are very resistant; if deposited on the soil they may remain alive for many months. **Embryos** are formed, under favorable conditions, in about 14 days. Infection with *Ascaris* results from ingesting eggs containing embryos, usually carried to the mouth with either food or water or by accidental transfer of soil. They do not regularly hatch in the stomach but pass on to the small intestine, where they begin to hatch within a few hours.

The newly hatched **larvae** burrow into the wall of the small intestine and enter the veins or lymphatic vessels, eventually carried to the right side of the heart. From here they pass to the lungs, through the air passages, trachea, esophagus, and stomach to the small intestine again. This journey through the host takes about 10 days. They become mature worms in the small intestine in about 2.5 months.

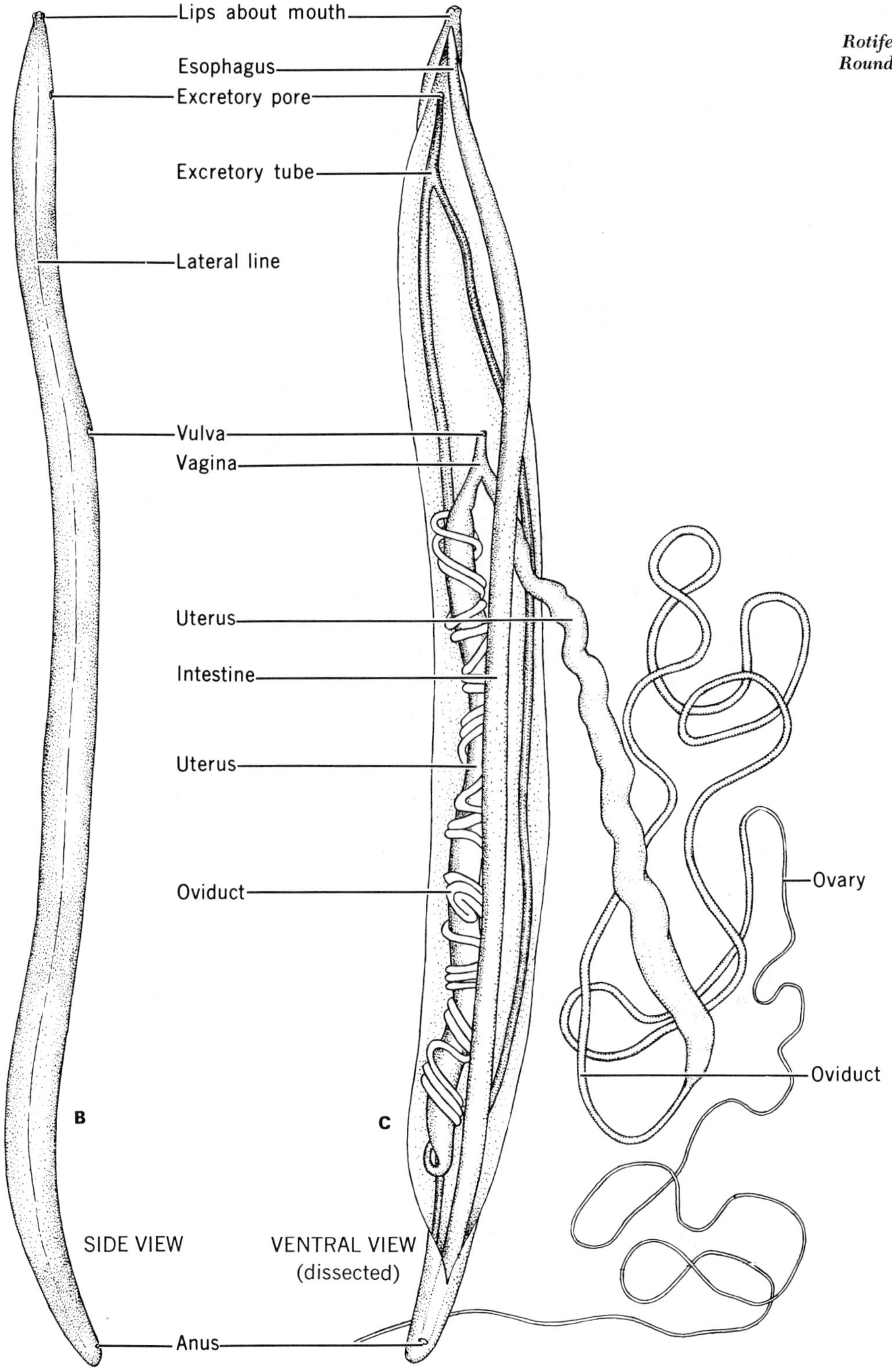

Lips about mouth
Esophagus
Excretory pore
Excretory tube
Lateral line
Vulva
Vagina
Uterus
Intestine
Uterus
Oviduct
Ovary
Oviduct

B  
SIDE VIEW

C  
VENTRAL VIEW  
(dissected)

Anus

*Ascaris* worms found in human beings and pigs are difficult to distinguish, but they differ physiologically in that the human *Ascaris* eggs do not usually produce mature worms in pigs and vice versa. *Ascaris lumbricoides* in human beings is one of the more serious parasites and is widespread where sanitation conditions are poor. It occurs more frequently among children because of their carelessness regarding sanitation.

When large numbers of *Ascaris* larvae pass through the lungs, inflammation begins and generalized pneumonia may result. Adult ascaris may be present in the intestine in such large numbers as to produce fatal intestinal obstruction. One thousand to five thousand worms have been found in a patient, but even 100 worms can cause a fatal blockage. Abdominal pain, headaches, or convulsion may result from the presence and toxic secretions of the worms. Fortunately, several drugs are available which remove the worms.

## HOOKWORMS

The hookworms *Ancylostoma duodenale* and *Necator americanus* are also widespread, and are considered the most injurious parasitic intestinal worms in human beings (Fig. 10–3). The larvae develop in moist earth and usually enter human bodies by boring through the skin of the foot, especially between the toes. They enter a lymph or blood vessel and pass to the heart; from here they reach the lungs, pass through the air passages into the windpipe (trachea), and into the intestine. The adults attach themselves by hooks or cutting plates to the intestinal wall and by suction feed upon the blood and tissue juices. When the intestinal wall is pierced, an anticoagulating substance is poured into the wound, resulting in a considerable loss of blood even after the worm has left the wound. In the case of the dog hookworm (and probably the human hookworm), blood is continuously sucked into the worm and is expelled through the anus in the form of red droplets consisting mostly of red corpuscles. Victims of hookworm are therefore anemic and suffer from malnutrition, subjecting them to other diseases as well.

Hookworm disease is very prevalent in large areas of the tropics, where soil and climate favor these parasites. It is also prevalent in the United States but not to the degree it was 50 years ago. It can be cured by several drugs.

Wearing shoes will help to prevent infection. The most important preventive measure is the disposal of human feces in a manner that avoids contaminating the soil, thus giving parasitic eggs possibly contained therein no opportunity to hatch and develop into infective larvae.

## TRICHINA WORMS

*Trichinella spiralis* (Fig. 10–4) causes trichinosis disease in human beings, pigs, rats, and many other mammals. Human beings are infected by eating inadequately cooked pork containing the parasitic larvae, which soon mature in the human intestine. Each mature female worm deposits from 1500 to 2500 living larvae into the lymph or blood vessels or the intestinal wall. The larvae eventually encyst in muscular tissue in various parts of the body, particularly the tongue and diaphragm. As many as 15,000 encysted parasites have been counted in a single gram of muscle.

Pigs usually acquire the disease by eating restaurant raw meat scraps or slaughterhouse garbage, and less frequently by eating infected rats.

Estimates in 1953 placed the number of persons in the United States infected with trichinosis at several million, undiagnosed cases at several hundred

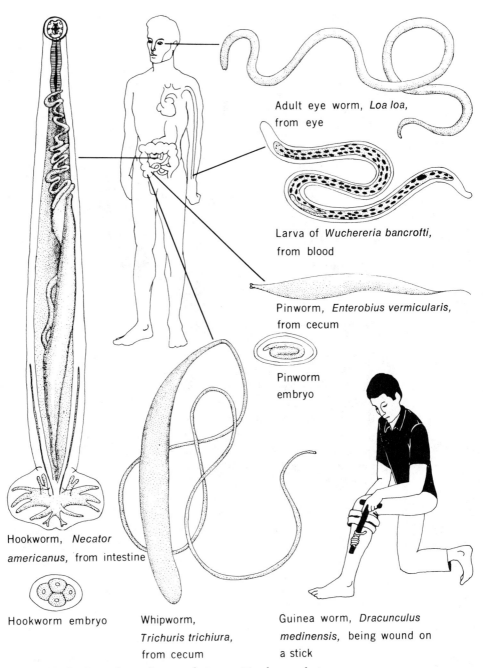

**Figure 10-3** Several roundworms that parasitize human beings.

thousand, and animal deaths at several thousand. A death rate of 5 percent of persons showing symptoms of the disease has been reported; it has, however, been decreasing since 1953, when a law was effected in most states requiring slaughterhouse garbage to be cooked before being fed to pigs.

Although specific treatment is lacking, certain drugs provide relief and may prevent death. Recent experiments have shown that 25,000 rep of gamma radiation will destroy the *Trichinella* larvae in undercooked pork, making it safe for consumption.

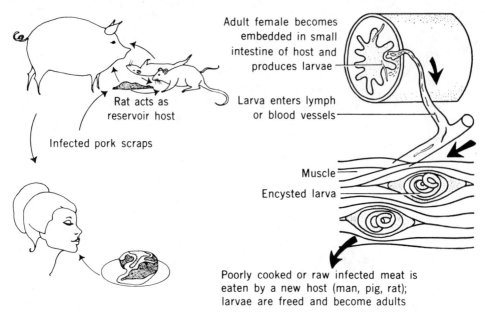

**Figure 10-4** Life cycle of *Trichinella*.

### PINWORMS

*Enterobius vermicularis* causes pinworm infection in human beings. Other, closely related species cause the disease in monkeys and apes. This parasitic worm is 9 to 12 mm long and lives (during the adult stage) in the upper part of the large intestine. The disease is worldwide in distribution, but predominantly infects Caucasians. It is most prevalent in infants; most cases show no symptoms, and many children recover without treatment. In some cases, however, infection persists into adulthood; over 5000 worms have been present in a single adult.

Parasitic roundworms found in other vertebrate animals include the dog ascarid, *Toxocara canis*, especially prevalent in puppies; *Heterakis gallinae*, the cecum worm of chicken; and the horse strongyle (*Strongylus vulgaris*), distributed worldwide and especially prevalent in warm countries.

Other classes of the phylum Aschelminthes are: class **Gastrotricha** (approximately 200 species), inhabiting both fresh and salt water, particularly abundant among algae and debris upon which they feed; class **Kinorhyncha** (approximately 30 species), very small worms living in mud or sand at the bottom of salt waters; and class **Nematomorpha**, or **horsehair worms** (approximately 225 species, 15 occurring in North America), transmitted through drinking water and found in the body cavities of beetles, crickets, and grasshoppers.

## Spiny-head Worms

These parasitic worms are named "spiny-heads" because of their retractile proboscis, armed with rows of recurved hooks that attach to the intestinal walls of the vertebrates they inhabit. They vary in length from less than 2 cm to more than 3 dm. Most are elongated and capable of extension. No digestive tract is present; food is absorbed directly from the host's intestine. The sexes are separate, and the reproductive systems are complex. Species have been found in fish, turtles, birds, rats, mice, pigs, squirrels, dogs, and human beings.

## Entoprocts

These organisms were formerly classified with the next highest phylum, the bryozoans (to be studied in Chapter 11). They are now placed in their own phylum because, unlike the bryozoans, they have a pseudocoelom, and their mouth and anus open inside the tentacles (in bryozoans only the mouth opens inside the tentacles). Entoprocts (Gr. *entos*, inside + *proktos*, anus) reproduce asexually (by budding) and sexually (resulting in free-swimming undulipodiated larvae). Structurally, they are a link between the pseudocoelomates and the coelomates, to be studied in the following chapter.

## Summary

Rotifers and roundworms are pseudocoelomate worms. Rotifers are usually free living. Roundworms are typically parasitic; *Ascaris* is a common representative. Other parasitic roundworms of agricultural and medical importance include hookworms, trichina worms, and pinworms. Spiny-head worms are also parasitic pseudocoelomates, named for their many-hooked retractile proboscis. Entoprocts are pseudocoelomates with both mouth and anal openings within the tentacles; they are the structural link between the pseudocoelomate animals and the coelomates.

# 11

# Bryozoans, Lamp Shells, and Arrow Worms

In Chapters 9 and 10 we saw the transition from organisms lacking a coelom to those possessing an unlined body cavity (pseudocoelomates). This chapter will study the first representative coelomates—animals with a true body cavity (coelom) lined with a mesodermal epithelium (**peritoneum**).

Bryozoans are the first true coelomates. A discussion on these "moss animals" will be followed by descriptions of other minor coelomate groups, including lamp shells and arrow worms.

## Bryozoans (Polyzoans)

**Bryozoans,** or "moss animals," are so named due to their plantlike structure; they are also referred to as **ectoprocts** since their anus is outside their tentacles. Bryozoans are mostly colonial and resemble hydroids in form, but are more advanced in internal structure. The majority are marine, a few are freshwater.

*Plumatella* (Fig. 11-1) is a common colonial freshwater genus, made up of cylindrical branched **tubes** that protect the softer parts of the body. The anterior end consists of a rounded ridge (**lophophore**) which bears a horseshoe-shaped double row of **tentacles,** from 40 to 60 in number, hollow, and undulipodiated. When the tentacles are spread out in water, the undulipodia cause currents that sweep microscopic organisms into the mouth.

Between the digestive tract (which includes a **mouth, esophagus, stomach, cecum, intestine,** and **anus**) and the body wall is a **true coelom,** lined with a peritoneum. There are no excretory, respiratory, or circulatory organs.

Bryozoans are hermaphroditic; certain freshwater species produce disklike buds (**statoblasts**), which secrete a hard chitinous shell. These survive when the animal dies in the fall or during a drought, giving rise to a new colony in the spring or when the wet season returns.

Certain freshwater bryozoans form thick crusts inside pipes, and dead colonies sometimes break loose, become fragmented, and clog small pipes. In prehistoric times there were many more species than today. Since their first appearance during the Cambrian epoch, bryozoans have continuously made substantial contributions to layers of calcareous rock.

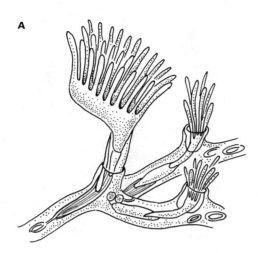
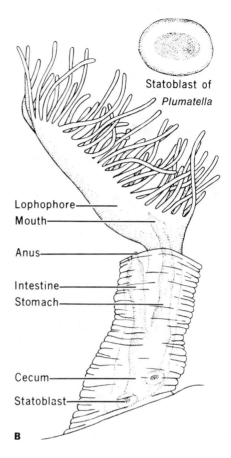

**Figure 11-1** (A) *Plumatella*, a common freshwater colonial bryozoan. (B) Structure of a single individual of this "moss animal" enlarged.

## Phylum Brachiopoda (Lamp Shells)

Brachiopods or **lamp shells** (Fig. 11-2) are marine animals living within a calcareous **bivalve shell** which resembles oil lamps used by the ancient Romans. They are usually attached to some object by a muscular stalk (**peduncle**). Within the shell is a conspicuous structure, the **lophophore,** which consists of two coiled ridges bearing undulipodiated tentacles.

Food is drawn into the **mouth** by the lophophore. A true coelom is present, within which lie the **stomach, digestive gland,** short blind **intestine,** and the "**heart.**" The Brachiopoda show the beginnings of a transition from *protostome* (the first embryonic hole, the blastopore, becomes the mouth) to the deuterostome (the blastopore becomes the anus and a second embryonic hole develops into the mouth).

The brachiopods are extremely old, dating from before Cambrian time; although found in all seas today, brachiopods were formerly more numerous in species and of much greater variety in form. *Lingula* is thought to be the oldest living genus known; it is called a "living fossil," since it apparently has not changed in over 400 million years.

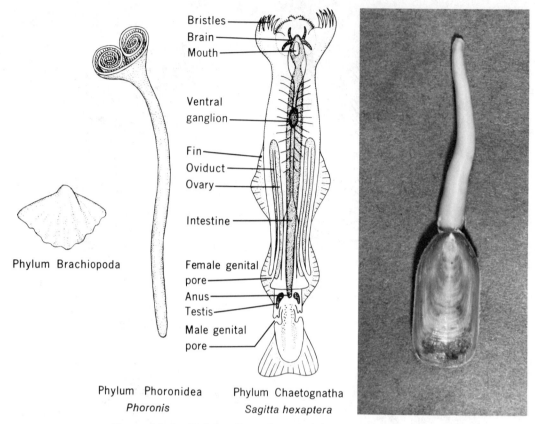

**Figure 11-2** *Left:* brachiopod, ventral view of shell. *Middle left: Phoronis,* removed from its tube. *Middle: Sagitta hexaptera,* an arrow worm. *Right: Lingula* sp.

## Phylum Chaetognatha (Arrow Worms)

The Chaetognatha are marine animals that swim about near the surface of the sea. The best-known genus is *Sagitta,* the arrow worm (Fig. 11-2). The bilaterally symmetrical body consists of three regions: **head, trunk,** and **tail. Lateral** and **caudal fins** are present. There is a distinct **coelom,** a **digestive tract** with **mouth, intestine** and **anus,** a well-developed **nervous system,** two **eyes,** and other **sensory organs.** There are no circulatory, respiratory, or excretory organs. The mouth has a lobe on either side provided with **bristles,** which are used in capturing the minute animals and plants that serve as food. These organisms are **hermaphroditic.** Many species are widely distributed.

Other minor coelomate phyla include: **phoronids** (Fig. 11-2), approximately 15 species of small, wormlike marine animals of sedentary habit that live in tubes; **peanut worms,** approximately 250 species, are large animals living on the seashore in sand and on rocks; **echiuroids,** approximately 50 species (Fig. 11-3), are hot-dog-shaped, unsegmented worms that live in the sea bottom near the shore; and **beard worms,** approximately 40 species, that live at great depths on the sea floor inside secreted tubes.

**Figure 11-3** *Urechis caupo,* an echiuroid common in California.

## Summary

Bryozoans are the first organisms possessing a true coelom with peritoneum lining. They resemble hydroids in their colonial structure but are more advanced physiologically.

Lamp shells date from pre-Cambrian time and are noted for their calcareous bivalve shells.

Arrow worms are bilaterally symmetric, with a body divided into head, trunk and tail regions and possessing lateral and caudal fins.

Other minor coelomate phyla are essentially tubular marine organisms with varied adaptive features that provide zoologists with opportunities for studying adaptation in particular habitats.

# 12
# Segmented Worms

Most annelids are marine (classes Polychaeta and Archiannelida); the rest (class Oligochaeta—earthworms, and class Hirudinea—the leeches), live in fresh water or in the soil. They are composed of a linear series of similar parts, known as **segments, somites,** or **metameres.** Thus, annelids are called **seg-**

**Figure 12–1** Representatives of four classes of segmented worms.

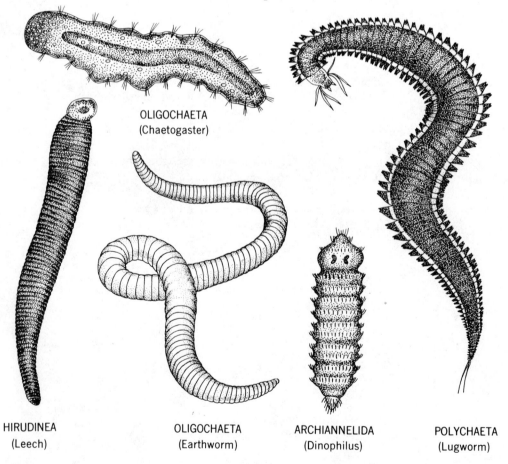

HIRUDINEA (Leech)  OLIGOCHAETA (Earthworm)  ARCHIANNELIDA (Dinophilus)  POLYCHAETA (Lugworm)

mented worms, distinguishing them from unsegmented flatworms and roundworms. The biological principle of segmentation, **metamerism,** is exhibited here for the first time. This type of structure is of considerable interest, since the most successful groups in the animal kingdom, the arthropods and vertebrates, have metamerically arranged parts.

In a popular annelid, the earthworm, metamerism (both external and internal) is conspicuous; the coelom is large and spacious. As organisms become larger, the circulatory and nervous systems must be well developed in order to nourish isolated cells, coordinate the body mass, and monitor the environment.

In this chapter, we will use the common earthworm, *Lumbricus terrestris* (class Oligochaeta), to illustrate the principal characteristics of annelids. Several other annelid species will be briefly discussed, including the sandworm, *Neanthes virens* (class Polychaeta), and the leech, *Hirudo medicinalis* (class Hirudinea); Fig. 12–1).

## Class Oligochaeta

The members of the class Oligochaeta are mostly terrestrial; some inhabit fresh water. No parapodia and few setae are present, and the head has no distinct appendages. They are hermaphroditic, but no trochophore larva develops from the egg.

## Lumbricus terrestris—An Earthworm

Earthworms are found worldwide; they are soft and naked and hence must live in moist earth. They are nocturnal, remaining hidden in their burrows (about 2 feet underground) during the day and coming out to feed at night. Earthworms can move through soft earth, but must eat their way through harder soil. This soil passes through their digestive tract and is deposited on the surface as castings.

### EXTERNAL ANATOMY

The body of *Lumbricus* is cylindrical and varies in length from about 6 inches to 1 foot. The ventral surface is slightly flattened, and lighter than the dorsal surface. There are over 100 **segments,** easy to distinguish by grooves extending around the body. At the anterior end (Fig. 12–2) a fleshy lobe, the **prostomium,** projects over a **mouth** (some authors regard the prostomium as the first true segment). It is customary to number the segments, beginning at the anterior end, since both external and internal structures bear a constant relation to them. Segments 31 or 32 to 37 are swollen in mature worms, forming a saddle-shaped enlargement, the **clitellum,** used during reproduction. Every segment, except the first and last, bears four pairs of chitinous bristles (**setae**); these are moved by retractor and protractor muscles and renewed if lost. In mature worms, the setae on segment 36 are modified for reproductive purposes.

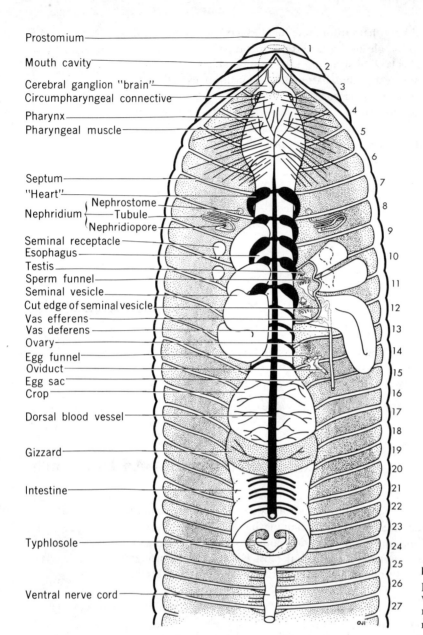

**Figure 12-2** Anterior part of an earthworm, with dorsal body wall removed to show internal structures.

The body is covered and protected by a thin transparent **cuticle** secreted by the **epidermis,** lying just beneath it. The cuticle contains numerous **pores** that allow secretions from unicellular epidermal glands to pass through and is marked with fine **striations,** causing the surface to appear iridescent.

There are a number of **external openings** of various sizes and functions: the **mouth** is a crescentic opening situated in the ventral half of the first segment; an oval **anal opening** lies in the last segment; sperm duct openings or **vasa deferentia,** situated on either side of segment 15, have swollen lips and a slight ridge that extends posteriorly to the clitellum. The **oviduct** openings, through which eggs pass out of the body, are small round pores, one on either side of segment 14. Concealed within the grooves separating segments 9, 10, and 11 are two pairs of minute pores, the **seminal receptacles**. A pair of **nephridiopores**, the external openings of the excretory organs, appear on every segment except the first three and the last. The body **cavity** or **coelom** communicates with the exterior through **dorsal pores**, one located in the middorsal line at the anterior edge of each segment from 8 or 9 to the posterior end of the body.

## INTERNAL ANATOMY AND PHYSIOLOGY

The internal structures may be seen by making a lengthwise incision through the body wall of the earthworm (Fig. 12–3). The body wall contains two layers of **muscles**: an outer **circular muscle tissue** consisting of long, spindle-shaped fibers that contract making the worm smaller in diameter and longer. Beneath it lies a thick **longitudinal layer** of muscle fibers arranged parallel to the length of the worm; when these contract, the diameter of the body increases and the worm is made shorter.

The wall of the coelom is lined with an epithelial tissue known as **peritoneum**. The coelom is divided into segments by partitions called **septa**, which lie beneath and correspond to the earthworms' outer grooves. This cavity is filled with a colorless fluid which flows from one segment to another (through a large opening in each septum) when the body of the worm contracts.

**DIGESTIVE SYSTEM.** The digestive tract passes through the center of the body and is suspended in the coelom by the septa. It consists of a **mouth** (**buccal**)

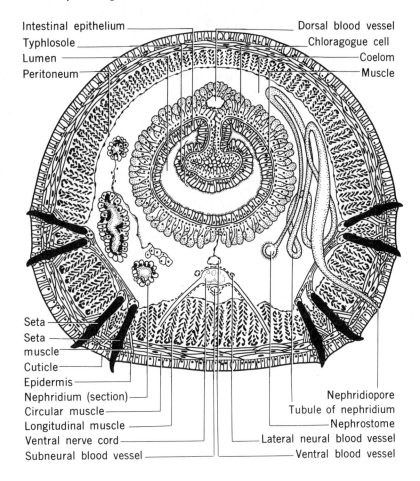

**Figure 12–3** Cross section of an earthworm.

cavity (segments 1 to 3); a thick muscular **pharynx** (segments 4 and 5); a narrow, straight tube, the **esophagus** (segments 6 to 14); a thin-walled enlargement, the **crop** (segments 15 and 16); a thick muscular-walled **gizzard** (segments 17 and 18); and a thin-walled **intestine** (segment 19 to the anal opening). The intestine is not a simple cylindrical tube; its dorsal wall is unfolded, forming an internal longitudinal ridge, the **typhlosole,** that increases the digestive surface. Surrounding the digestive tract and dorsal blood vessel is a layer of **chloragogue** cells, which provide transportation of food to developing eggs in the ova and release of ammonia and urea, thus serving an excretory function as well. Three pairs of **calciferous glands** lie at the sides of the esophagus (segments 10 to 12); the first pair are storage pouches, the second and third are true glands. Their primary function is excretion of calcium carbonate in the regulation of acid-base balance of the body fluids.

Earthworms feed principally on pieces of leaves and other vegetation, particles of animal matter, and soil. At night they crawl out, holding fast to the top of their burrows with their posterior ends to explore the neighborhood. Muscles extending from the body pharynx to the body wall are contracted, enlarging the pharyngeal cavity and producing suction, which draws food particles into the mouth. In the pharynx a secretion is added to the food; it then passes through the esophagus to the crop, where it is stored temporarily. In the gizzard it goes through a grinding process and passes into the intestine, where most of the digestion and absorption take place.

Digestion in the earthworm is very similar to that in higher animals. Enzymes aid in the breakdown of food: **amylase** acts upon carbohydrates, **cellulase** upon cellulose, **pepsin** and **trypsin** upon proteins, and **lipase** upon fats. Digested food is absorbed through the wall of the intestine, assisted by the amoeboid activity of some epithelial cells. Absorbed food is carried to various parts of the body through the bloodstream; it also makes its way into the coelomic cavity and is carried directly to tissues bathed by the coelomic fluid. In one-celled organisms, and simple metazoans such as the hydra, planaria, and ascaris, no circulatory system is necessary, since food is either digested within the cells or comes in direct contact with them; in large, complex animals, a special system of organs must be provided to bring about the proper distribution of digested food.

**CIRCULATORY SYSTEM.** The **blood** of the earthworm is contained in a complicated system of tubes which extend through all parts of the body (Fig. 12–4). Large centrally located tubes branch and rebranch, ending in very thin tubules, the **capillaries.** The blood consists of a great number of colorless amoeboid cells (**corpuscles**) suspended in plasma. The plasma also contains a respiratory pigment, **hemoglobin,** to transport oxygen and which renders the blood red.

The **dorsal vessel** is a pump with valves, thus serving the function of a true heart. Blood is forced forward by wavelike (**peristaltic**) contractions of the dorsal vessel, beginning at the posterior end and quickly traveling anteriorly. **Valves** in the walls of the dorsal vessel prevent the return of the blood from the anterior end. The **aortic arches** (the so-called hearts) actually act as a pressure-regulating mechanism, receiving blood in spurts from the dorsal vessel, and contracting to force it under a steady pressure into the **ventral vessels.** Valves in the heart prevent the backward flow. From the ventral vessel, the blood passes to the body wall, the intestine, and the nephridia. The flow in the **subneural vessel** is toward the posterior end, then dorsally through the **parietal vessels** into the **dorsal vessel.** The anterior region receives blood from the dorsal

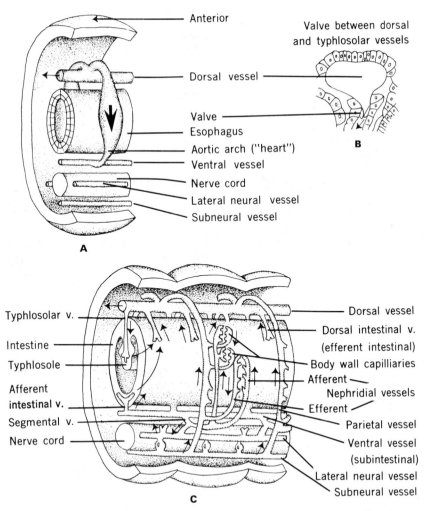

**Figure 12-4** Earthworm circulatory system. (A) One pair of "hearts" and other vessels. (B) Valve structure. (C) Three-dimensional view of circulation.

and ventral vessels. The blood that is carried to the body wall and the skin receives oxygen and eliminates $CO_2$ through the cuticle and is then returned to the dorsal vessel by way of the subneural vessel and the parietal connectives.

The exchange of materials between the blood and the tissue cells takes place in minute tissue spaces. Blood plasma and a few corpuscles, which constitute the tissue fluid, pass from the capillaries into these tissue spaces, where the cells are bathed and the interchange of $O_2$ and $CO_2$ occurs. The tissue fluid also collects waste products of cellular metabolism and makes its way back into the bloodstream.

RESPIRATION. The earthworm possesses no organized respiratory system. It obtains oxygen and releases carbon dioxide through many capillaries lying just beneath the cuticle of the moist skin. Oxygen passes into the blood and combines with the hemoglobin.

EXCRETORY SYSTEM. Most of the excretory waste is carried out of the body by a number of coiled tubes (**nephridia**) occurring in pairs in every segment except the first three and the last. Each nephridium occupies part of two successive segments: an undulipodiated funnel (**nephrostome**) in one segment is connected by a thin undulipodiated tube to the major part of the nephridium, occurring in the segment posterior to it, and consisting of three loops. The undulipodia on the nephrostome and in the nephridium create a current that draws in waste material from the coelomic fluid; other waste is received directly from blood vessels surrounding the nephridium. These excretory products (ammonia, urea, creatine) are eventually carried out through the **nephridiopore.** Chloragogue cells may store excretory matter temporarily before releasing it into the coelomic fluid. Nephridia serve the same function in the earthworm that kidneys do in higher organisms.

NERVOUS SYSTEM. On the dorsal surface of the pharynx is a bilobed mass of nervous tissue, the "**brain**" or **cerebral ganglia** (Fig. 12–5). This is connected by two **circumpharyngeal connectives** to a pair of **subpharyngeal ganglia.** From the latter, a **ventral nerve cord** extends posteriorly, enlarging in each segment into a **ganglion** and (in every segment posterior to segment 4), giving off three pairs of nerves. Near the dorsal surface of the ventral nerve cord are three longitudinal **giant fibers.** These fibers conduct impulses five times faster than regular neurons, allowing the animal to escape quickly from danger. The brain and nerve cord constitute the **central nervous system;** the nerves that pass to and from them represent the **peripheral nervous system.**

The nerves of the peripheral nervous system are either motor or sensory. **Motor fibers** extend to the muscles and other organs; **sensory fibers** originate from the nerve cells in the epidermis and carry impulses to the ventral nerve cord. The functions of nervous tissue are reception, conduction, and stimulation. These are usually performed by nerve cells called **neurons.**

A simple **reflex** in an earthworm begins at a **sensory neuron,** lying at the surface of the body, which sends a sensory fiber into the ventral nerve cord, where it branches out. These branches meet but are not continuous with branches from an **association neuron** lying in the ventral nerve cord. The association neuron is in contact with a motor neuron that sends motor fibers into a reacting organ (in this case, a muscle). The first neuron or **receptor** receives the stimulus and produces the nerve impulse, which is carried on to the association neuron; the association neuron, in turn, transmits the impulse to a motor neuron, which has processes extending to an **effector,** such as a muscle or other organ.

SENSE ORGANS. The sensitiveness of the earthworm to light and other stimuli is due to the presence of a great number of epidermal sensory receptors. The two main types are the light-sensitive **photoreceptor cells** and the **sense organs.** The sense organs are connected with the central nervous system by means of nerve fibers and communicate with the outside through sense hairs which penetrate the cuticle. They occur principally at the anterior and posterior ends of the worm. In addition to these sensory organs, there are free endings of nerve fibers between the cells of the epithelium (Fig. 12–6).

REPRODUCTIVE SYSTEM. Earthworms are hermaphroditic (having both female and male sex organs) and reproduce sexually. Mating takes place at night and requires 2 or 3 hours.

The reproductive organs are in segments 9 to 16. The **female system** consists of a pair of **ovaries** (segment 13); a pair of **oviducts,** which open by a ciliated funnel (in segment 13) and pass to the exterior (in segment 14); an **egg sac,** which is a small pouchlike projection of the septum associated with the funnel; and two pairs of **seminal receptacles** (in segments 9 and 10).

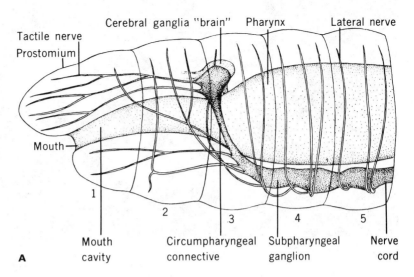

**Figure 12–5** Earthworm nervous system. (A) Side view of anterior end, showing cerebral ganglia and larger nerves. (B) Sensory and motor neurons of the ventral nerve cord connect with the epidermis to form a reflex arc.

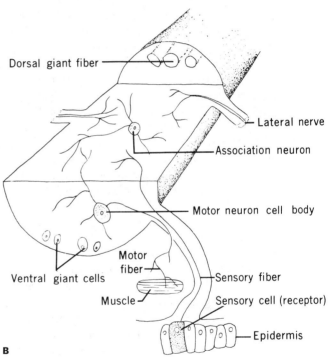

**Figure 12–6** Epidermis of an earthworm, showing sense organs.

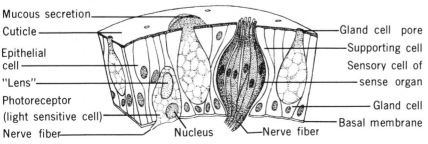

**Figure 12-7**  Pair of earthworms copulating.

The male organs are two pairs of glove-shaped testes (segments 10 and 11) and, in back of each, a ciliated **sperm funnel** which is connected to a tiny duct, the **vas efferens**. The two ducts on each side connect to a **vas deferens**, which leads to the outside. The testes and funnels are contained in the **seminal vesicles**, conspicuous saclike structures which surround the testes and in which sperms mature.

Self-fertilization does not occur. Reciprocal or cross-fertilization takes place with the transfer of spermatozoa from one worm to another during **copulation** (Fig. 12-7). Two worms unite; spermatozoa from the **seminal vesicles** of each worm are discharged, passing along the seminal grooves into the seminal receptacles of the other worm; and the worms separate.

When the time for egg-laying approaches, the glandular clitellum secretes a bandlike mucous tube which is forced forward by movements of the worm. Eggs are discharged from the oviduct openings (segment 14), then sperm are discharged from the two pairs of seminal receptacles (between segments 9–10 and 10–11) into the space between this tube and the body wall. The tube is then forced forward over the anterior end; its ends become closed, and a cocoon, about the size of an apple seed, is thus formed containing fertilized eggs, which develop within the cocoon into minute worms.

The **eggs** of the earthworm are **holoblastic** (divide completely in two), but cleavage is unequal. A hollow blastula is formed, and a gastrula is produced by invagination. The mesoderm develops from two of the blastula cells, called **mesoblastic bands,** which later become the epithelial lining of the coelom. There is no larva stage such as occurs in the marine annelids. The embryo escapes from the cocoon as a small worm in 2 to 3 weeks.

### REGENERATION AND GRAFTING

Earthworms have considerable but limited regenerative ability. No more than five new segments will generate at the anterior end, and no "head" will generate if 15 or more segments have been cut off. An anterior piece regenerates a tail and a posterior piece regenerates a "head" of five segments or, in certain cases, a tail (creating a double-tailed worm, which slowly starves to death). Also, pieces of several worms may be united in various ways.

# BEHAVIOR

The external stimuli that have been most frequently employed in studying the behavior of earthworms are contact, chemicals, and light. Continuous but not overly strong mechanical stimulation, such as would occur in the walls of an earthworm's burrow, elicit a positive reaction. Reactions to sounds are not due to the presence of a sense of hearing, but to the contact stimuli produced by vibrations.

Reactions to chemicals may result in bringing the worm into regions of favorable food conditions or turning it away from unpleasant substances. No positive reactions are produced by chemical stimuli from a distance, whereas negative reactions are produced even when certain unpleasant chemical agents are still some distance from the body. The reactions are quite similar to those caused by contact stimuli.

Light elicits a negative reaction and a partial reaction is given to faint light and red light; the latter may be used to collect the worms at night. A sudden illumination at night will often cause earthworms to quickly retreat into their burrows. Although no definite visual organs (eyes) have been discovered in earthworms, these animals are very sensitive to light. This sensitivity is due to **photoreceptor cells,** found in every segment of the body and concentrated in the epidermis and on nerves at the anterior and posterior ends. By means of these photoreceptor cells, very slight differences in the intensity of the light are distinguished.

Whether or not learning occurs in protists, or in such simple metazoans as sponges and hydras, is uncertain. But at the stage in evolution represented by the earthworm, experiments indicate that this animal is capable of what psychologists call "latent memory," or the storing of impressions until a later time when they may be useful.

In one experiment, worms could escape from a lighted chamber by entering the bottom of a branched passageway constructed of glass tubing in the form of a T. If the worms turned to the right at the top of the T, they entered a dark moist chamber filled with damp earth and moss, a favorable environment for an earthworm. If they turned left, they encountered an electric shock. In the early trials, they turned to the left as often as to the right. At the end of 20 days, they turned to the left only 5 times out of 20, and at the end of 40 days they were turning left only once out of 20 trials.

# Other Oligochaetes

*Tubifex tubifex* is a common bottom-dwelling freshwater annelid, reddish in color and about 4 cm long. It lives in a tube from which the posterior end of the worm projects and waves back and forth in the water. Often large numbers occur in patches on muddy bottoms, looking like masses of tiny reddish hairs.

*Aeolosoma* is only 1 mm long and is spotted with red oil globules. They live among algae, consist of from 7 to 10 segments, and reproduce asexually by transverse fission. The number of segments in oligochaetes varies from 7 in *Aeolosoma* to over 600 in *Rhinodrilus*.

# Class Polychaeta

The polychaetes consist largely of free-living marine annelids in which typical annelid characters occur. The body tends to be long and wormlike and somewhat depressed to a cylindrical shape in cross section. It consists of a prostomial or head region, and a trunk. The outer cuticle is usually soft and moist and is dependent on a wet environment for the prevention of desiccation. The lateral appendages, or parapodia, are usually conspicuous and variously provided with fleshy structures, such as cirri, scales, and gills. The principal characteristics of the Polychaeta are exhibited by the sandworm; however, many variations of this type occur. Many other polychaetes are partly or wholly parasitic, living in water varying in saltiness from briny to fresh; a few are terrestrial.

There is a remarkable structural diversity among the several thousand species of polychaetes, producing a wide range of adaptations to many ocean habitats.

### *Neanthes virens*—THE SANDWORM

*Neanthes virens* (Fig. 12–8) is a common polychaete that lives in burrows in the sand or mud of the seashore at tide level. By day it rests in its burrow, but at night it extends its body in search of food or may leave the burrow entirely.

The body is flattened dorsoventrally and may reach a length of 5 dm or more, with 100 to over 200 internally and externally well-defined segments. The head is well developed. Above the mouth is the **prostomium**, which bears a pair of terminal tentacles, two pairs of **simple eyes**, and, on either side, a thick **palp**. The first segment is the **peristomium**; from each side of this arise four **tentacles**. Small animals are captured by a pair of strong chitinous **jaws**, which are everted with part of the pharynx when *Neanthes* is feeding. Behind the head a number of segments bear fleshy outgrowths on either side, the **parapodia**. On the tips of the parapodia are bundles of setae; they are formed from secretions of specialized cells, and function in locomotion, tube building, food gathering, and in other important ways.

The body wall consists of an outer **cuticle**, secreted by the cells of the epidermis and several muscular layers beneath it. Between the body wall and the intestine is a **coelom**, lined with **peritoneal epithelium**. The **digestive system** consists of a **mouth, pharynx, esophagus** (with an **esophageal gland** on either side opening into it), and a straight **stomach–intestine** extending to the **anus**.

The circulatory system consists of a dorsal vessel and a ventral vessel, with branches to the capillaries in the body wall and the intestine. Almost every segment (except the peristomium and the anal segment) contains a pair of **nephridia**. In the head is a **cerebral ganglion**, the "brain." This is joined by a pair of **subesophageal ganglia** and is followed by a **ventral nerve cord** with a pair of ganglia in each segment.

The sexes are separate. Ova or spermatozoa arise from the wall of the coelom. A **trochophore larva** develops from the fertilized egg.

### OTHER POLYCHAETES

Polychaeta differ from **Oligochaeta** in being largely marine instead of freshwater or terrestrial; parapodia are typically well developed, and the setae

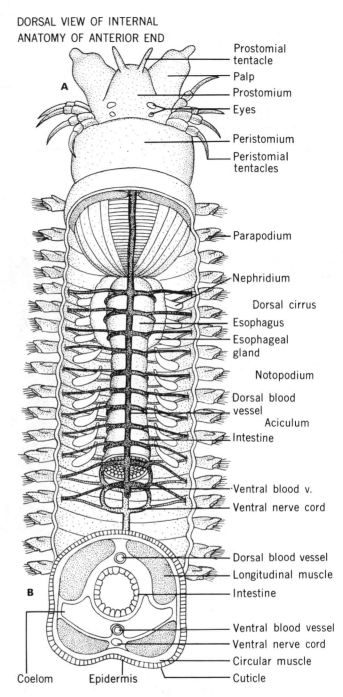

**Figure 12-8** *Neanthes*, the sandworm. (A) Anterior end with dorsal wall removed. (B) Cross section showing internal structure.

are numerous instead of few; the prostomia on some of the first few segments are often highly differentiated from a cephalic region of considerable proportions; sexes are usually separate, with gonads present in a large and variable number of segments. Fertilization of ova is typically external; development is by spiral cleavage and through an ocean-dwelling **trochophore** (Fig. 12-9). In certain species, for example, *Autolytus*, the body, which is only 15 mm long, may produce buds at the posterior ends, forming a linear row of offspring, each of which acquires a head before separating from the parent.

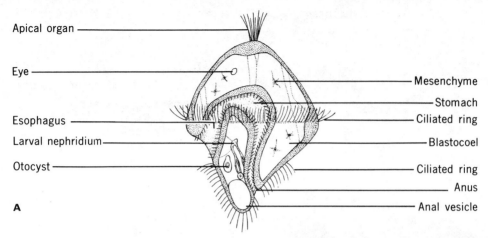

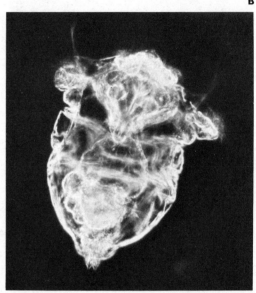

**Figure 12-9** (A) *Eupomatus*, a trochophore polychaete larva. (B) Trochophore larva.

The Pacific palolo worm, *Eunice*, first became known from the Samoan Islands, where it attracted the attention of the missionaries because it was eaten by the natives; also because it appeared periodically in certain localities in enormous quantities for a few hours. The posterior half of the worm breaks off from the parent worm and swims to the surface. The enclosed eggs and sperms are shed into the sea in the early morning, and in some localities in such enormous numbers that the surface of the sea has been likened to a thick noodle soup. The eggs develop into young larvae rapidly, and in 3 days sink to the bottom. Other palolo worms occur in different parts of the world, particularly in warm seas.

## Class Archiannelida

Archiannelida (approximately 45 species) are aberrant marine polychaetes, characterized largely by the persistence of such larval features as ciliary rings and lack of setae, or reduction of organ systems; whether they are primitive or degenerate is not known.

## Class Hirudinea

The class Hirudinea (leeches) are annelids that are usually flattened dorsoventrally. They differ from other annelids in the lack of setae (except in one genus) and in the presence of copulatory organs and genital openings on the ventral side. Leeches are abundant in fresh water but occur also in salt water and on land. Many of them have brilliant and elaborate color patterns. Leeches are commonly thought of as bloodsuckers; large numbers, however, are

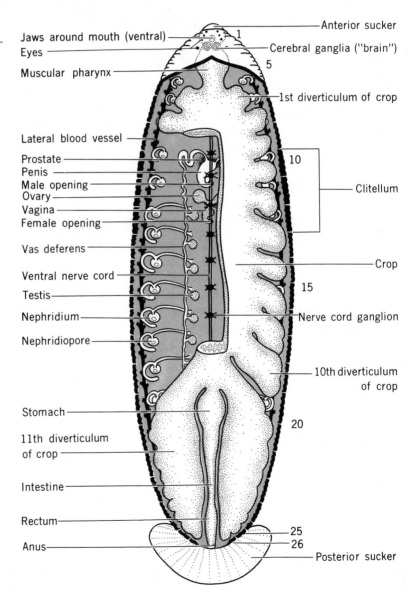

**Figure 12-10** Dorsal view showing segmentation and internal anatomy of the leech.

predaceous (do not act as blood-sucking parasites) but devour other small animals, such as earthworms and mollusks. Leeches are preyed upon by birds, reptiles, flatworms, and other animals.

Their external metamerism is not indicative of the true number of segments; there may be several external rings for every segment, as shown by internal organs.

### *Hirudo medicinalis*—A LEECH

The principal characteristics of hirudineas are exhibited by *Hirudo medicinalis* (Fig. 12–10), which is about 4 inches long, but capable of great contraction and elongation. The **suckers** are used as organs of attachment. The **digestive tract** is fitted for digestion of the blood of vertebrates, which forms the principal food of some leeches. The mouth lies in the anterior sucker and is provided with three **jaws** armed with **chitinous teeth** for biting. Blood is sucked up by the **muscular pharynx.** The short esophagus leads from the pharynx into the **crop,** which has 11 pairs of lateral branches. Here the leech is able to ingest three times its own weight in blood; and since it may take as long as 9 months to digest this amount, meals are few and far between.

**Respiration** is carried on mainly through the surface of the body. Waste products are removed from the blood and coelomic fluid by 17 pairs of **nephridia.** Leeches are hermaphroditic, but the eggs of one animal are fertilized by sperms from another leech. Copulation and formation of a cocoon are similar to those processes in the earthworm. Other leeches carry their eggs on the ventral side, and some deposit them on stones.

## Origin of the Annelida

The archiannelids were once regarded as ancestral annelids; they show some larval features that resemble larval polychaetes, and it is not known whether they are primitive or degenerative.

The polychaetes are by far the oldest, largest, and most diversified of the annelids. The origin of aquatic and terrestrial oligochaetes from an ancestral, generalized polychaete is likely. The leeches, in turn, have many features in common with oligochaetes; their peculiar modifications are the result of parasitism.

## Relations of the Annelida to Man

Of the influences of segmented worms on human welfare, that of the earthworm and leeches is the most obvious. Earthworms are widely used as bait for fishing and have become quite profitable in some resort districts.

Charles Darwin demonstrated, by careful observations, the great ecological importance of earthworms. One acre of soil may contain over 50,000 earthworms, their feces (castings) covering the ground. Darwin estimated that in 1 year, more than 18 tons of earthly castings may be carried to the surface of 1 acre of soil. The **lugworm,** a polychaete, is even more effective than the earthworm; the amount of castings brought up was estimated to be as much as 3700 tons per acre in a single year. The continuous honeycombing of the soil by these worms makes the land more porous and ensures better penetration of air and moisture; the mixing of earth and organic matter in the digestive tract of the worms also helps to increase the humus of the soil.

Earthworms may serve as intermediate hosts of parasitic worms, as in the life cycle of a cestode of chickens, *Amoebotaenia,* and in that of a pig lungworm of the nematode genus *Metastrongylus.*

Oyster pests include polychaete worms of the genus *Polydora;* they cause mud blisters in the nacreous layers of the shells and may weaken or destroy the oyster.

Sedentary polychaetes are among the more conspicuous agents that cause fouling on the bottoms of ships, dikes, and various harbor installations, requiring the periodic drydocking of vessels to clear the hulls of the fouling organisms.

As reef-building agents, some sand- and lime-concreting, tube-building polychaetes are important in some parts of the world, changing shore contours, building up land masses, and transporting vast amounts of inert materials.

The use of the leech in medicine was based on the theory that many illnesses were due to "bad blood"; bloodletting, as the practice is called, was thus considered a cure for many ailments. In modern medical practice, we do this by transfusions of blood *into* the body. So common was leeching in olden times that doctors were often called leeches.

The salivary glands of leeches produce a substance termed *hirudin,* which prevents clotting of blood while the leech is feeding. For this reason a wound made by a leech may bleed for some time after the leech has detached itself. Hirudin is used in modern medicine as an anticoagulant.

# Summary

Annelids possess the basic architecture from which more complex organisms could arise. The body is divided into a linear series of similar segments. Definite organ systems are present to accommodate the increased size and specialization of the organism. Other systems have become more fully developed: The coelom is well developed; the nephridia drain excretory wastes from some blood vessels and the coelom respiratory parapodia anticipate the gill of higher organisms. The nervous system has become centralized, giving rise to a dorsal brain, a pair of ventral nerve cords, usually with a pair of ganglia in each segment. The digestive tract is a straight tube with an anterior mouth and posterior anus. The muscular system consists of an outer circular and an inner longitudinal series. The sperms and eggs are derived from the mesoderm. Cleavage of the egg is spiral unless obscured by excessive yolk. Four classes are recognized:

**Class 1. Oligochaeta.** Terrestrial or freshwater; without parapodia, and a few setae; head not well developed; hermaphroditic; no larva.

Class 2. **Polychaeta.** Marine; parapodia well developed and provided with setae that are variously modified, prostomium and first few segments sometimes highly cephalized; sexes usually separate; larva produced.

Class 3. **Archiannelida.** A small heterogeneous group, most nearly related to Polychaeta; characterized largely by loss of morphological characteristics, such as distinct parapodia or setae, and retention of larval ones, such as ciliary rows. Mainly marine, littoral, and sometimes living in brackish to fresh water. Usually dioecious, sometimes hermaphroditic; or development direct.

Class 4. **Hirudinea.** Parasitic or predaceous; mostly freshwater or terrestrial; without parapodia or setae; body with 33 segments plus prostomium; posterior and often an anterior sucker, hermaphroditic; coelom reduced by encroachment of connective tissue.

# 13

# Joint-footed Animals

The arthropods (Gr. *arthros,* joint + *podos,* foot) include lobsters and crabs (class Crustacea), spiders, scorpions, mites and ticks (class Arachnida), centipedes (class Chilopoda), millipedes (class Diplopoda), and insects (class Insecta) (Fig. 13–1). There are over 1 million species of arthropods, constituting 78 percent of all known species of animals. The diversity of forms of these animals seem infinite, but there is a fundamental body plan displayed by these animals. Their large numbers make them, from an evolutionary perspective, the most successful animals on earth.

**Figure 13–1** Representatives of the major classes of arthropods.

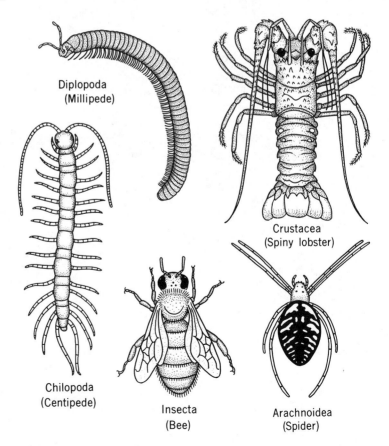

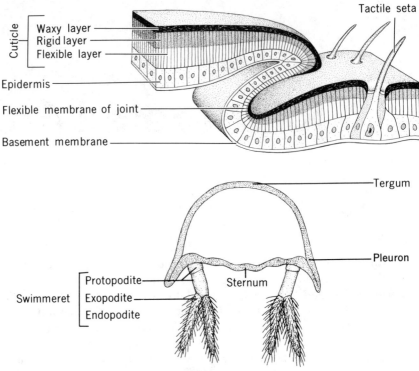

**Figure 13-2** Exoskeleton. *Top:* Arthropod body wall. *Bottom:* Cross section of third abdominal segment of the crayfish.

Their success is due to a number of characteristics, the most important of which is the hard, chitinous outer covering, the **exoskeleton** (Fig. 13-2). Around joints, the exoskeleton is soft and flexible. This exoskeleton serves many functions: as in vertebrates, it provides a place for muscle attachment and structural support for locomotion. In addition, the tough covering provides protection against mechanical, chemical, and disease injuries, and against desiccation, allowing organisms to survive in a terrestrial environment. A hard exoskeleton does limit the size of its possessor (by gravitational pull) and during growth it must be shed (**molt**); however, this small size probably helped arthropods to invade the skies.

The arthropods are segmented, typically having a pair of jointed appendages per segment. In advanced groups, the segments have combined (**tagmosis**) to form body regions (**tagmata**) and the appendages are reduced in number and modified for particular functions. Many of the joint-footed animals are highly social and have complex behavior patterns relating to that way of life.

## Crustaceans

The common crayfish exhibits to excellent advantage the characteristics of the class Crustacea (Fig. 13-3) as well as those of joint-footed animals in general.

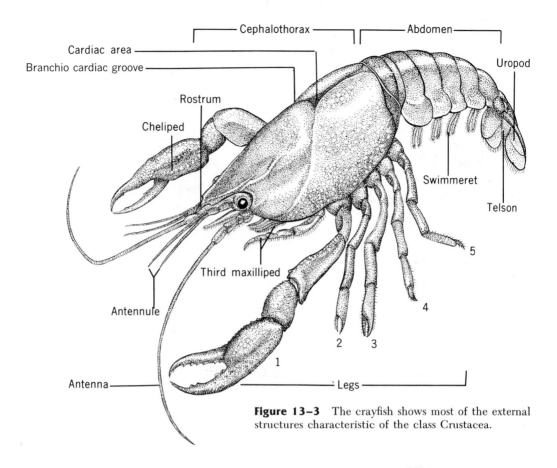

**Figure 13-3** The crayfish shows most of the external structures characteristic of the class Crustacea.

## Cambarus—A Crayfish

The crayfish (or crawfish) is found in freshwater lakes, streams, ponds, and swamps over most of the world. The genus *Cambarus* is common in the central and eastern United States. The lobster, *Homarus americanus*, differs in structure from the crayfish only in minor details.

### EXTERNAL ANATOMY

The outside of the body is covered by the hard exoskeleton, composed of many nitrogenous polysaccharides (including **chitin**), which are secreted by the epidermis and are impregnated with lime salts. At the joints, this exoskeleton is thinner and flexible, allowing for movement.

The entire body is segmented. A typical segment consists of a convex dorsal plate, the **tergum**, a ventral transverse bar, the **sternum**, and plates projecting down at the sides, the **pleura**.

The body consists of two distinct regions: an anterior rigid portion, the **cephalothorax**, and a posterior series of segments, the **abdomen**. The **cephalothorax** consists of segments 1 through 12, which are enclosed dorsally and laterally by a cuticle shield, the **carapace**. The **mouth** is situated on the ventral surface near the posterior end of the head region. It is partly obscured by the neighboring appendages. The carapace of the thorax is separated into three parts by **branchiocardiac grooves**; a median dorsal longitudinal strip, the

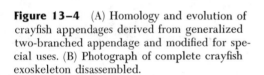

**Figure 13–4** (A) Homology and evolution of crayfish appendages derived from generalized two-branched appendage and modified for special uses. (B) Photograph of complete crayfish exoskeleton disassembled.

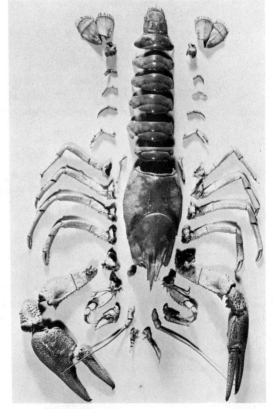

areola; and two large convex flaps, one on either side, the **branchiostegites,** which protect the gills beneath them.

In the **abdomen,** there are six segments and a terminal body extension, the **telson,** bearing on its ventral surface the longitudinal anal opening. Every segment of the body bears a pair of jointed **appendages.** These are all variations of a common type, consisting of a basal region, the **protopodite,** which bears two branches, an inner **endopodite** and an outer **exopodite.** Beginning at the anterior end, the appendages are arranged as follows: the sensory **antennae,** a pair of **mandibles,** the first and second **maxillae;** in the thoracic region are the first, second, and third **maxillipeds,** the **chelipeds** (pincers), and four pairs of **walking legs;** beneath the abdomen are five pairs of **swimmerets** (some are greatly modified). The sixth abdominal segment bears flattened appendages, **uropods.** Different appendages show modifications concerned with different functions, some of which are still in doubt.

Three kinds of appendages can be distinguished in the adult crayfish (Fig. 13-4): **foliaceous,** such as the second maxilla, **biramous,** as are the swimmerets, and **uniramous,** exemplified by the walking legs. These appendages are all probably derived from a biramous-type appendage, indicated by the biramous stage in the embryological development of the foliaceous second maxillae and the uniramous walking legs. Other appendages of the crayfish undergo similar changes in development.

Structures that have a similar fundamental structure, owing to descent from a common ancestor, are said to be **homologous.** The highly specialized chelipeds, walking legs, jaws, and other structures of the crayfish have evidently developed from a fundamental type and have become different in function. When homologous structures are repeated in a series, the result is **serial homology;** the crayfish exhibits a striking example of this condition.

## INTERNAL ANATOMY AND PHYSIOLOGY

The body of the crayfish contains all the important systems of organs characteristic of the higher animals (Fig. 13-5). The coelom is not large and is restricted to cavities enclosing the gonads and the excretory green glands. Certain organs, such as the nervous system, are metamerically arranged; others, such as the excretory organs, are concentrated into a small space.

**DIGESTIVE SYSTEM.** The food of the crayfish is made up principally of living animals such as snails, tadpoles, insect larvae, small fishes, and frogs, but decaying organic matter is also eaten. Crayfishes also prey upon their own kind. They feed at night and are most active at dusk and daybreak.

Food, held by maxillipeds and maxillae, is taken into the **mouth,** where it is torn and crushed by the **mandibles** (jaws). Coarse parts are ejected through the mouth; the food then passes through a short **esophagus** into the stomach. The **stomach** is a large cavity divided by a constriction into an anterior **cardiac chamber** and a small posterior **pyloric chamber.** In the cardiac chamber, food is ground by three hard **teeth** (chitinous ossicles) collectively known as the **gastric mill.** When fine enough, food passes through the **strainer,** which lies between the cardiac and pyloric chambers. The strainer consists of two lateral folds and a median ventral fold, which bear hairlike processes and allow passage only of liquids or very fine particles. **Digestive glands** situated in the thorax and abdomen secrete pancreatic-like enzymes, which pass through the **hepatic ducts** into the short **midgut,** where they mix with the food. From the midgut, some of the dissolved or partially digested food passes into the

**156**
*Animal Diversity: Invertebrates*

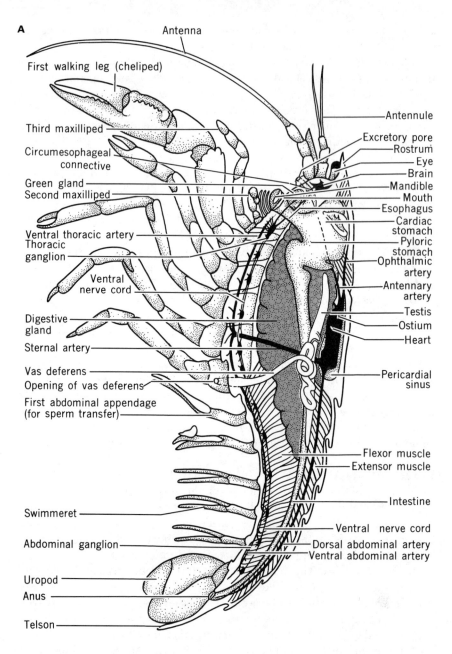

digestive glands, forming digestive enzymes. Undigested particles pass on into the posterior end of a short **intestine,** where they become fecal pellets and pass through the **anus.**

**EXCRETORY SYSTEM.** The excretory organs are a pair of rather large bodies, the **green glands,** situated in the ventral part of the head anterior to the esophagus. Each consists of a glandular portion, which is green, a thin-walled dilation (the bladder), and a duct opening to the exterior through an excretory pore on the basal segment of the antenna.

**NERVOUS SYSTEM.** The general structure of the nervous system of the crayfish is similar to that of the earthworm but is further developed in the head and thorax. The **central nervous system** includes the **brain (supraesophageal gan-**

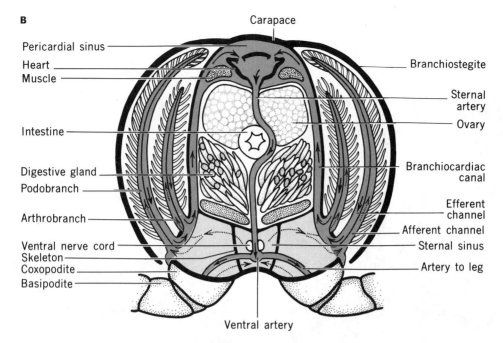

**Figure 13–5** The crayfish. (A) Internal structure of male crayfish. (B) Cross section showing arrangement of gills.

glia), a compact mass, which supplies the eyes, antennules, and antennae with nerves; and two **circumesophageal connectives**, which pass to the **subesophageal ganglion,** the most anterior ganglion of the **ventral nerve cord.** Six ganglia in the **thorax** and six ganglia in the **abdomen** are all joined by the ventral nerve cord.

The **visceral nervous system** consists of an anterior visceral nerve, which arises from the ventral surface of the brain; it is joined by a nerve from each circumesophageal connective, and passing back, branches upon the dorsal wall of the pyloric part of the stomach.

**Sense Organs.** The **statocysts** are organs of equilibrium. In *Cambarus* they are chitin-lined sacs situated on the basal segment of each antennule. In the base of the statocyst is a ridge with many fine sensory hairs which are innervated by a single nerve fiber. Among these hairs are a number of large grains of sand (**statoliths**), placed there by the crayfish.

The **compound eyes** of the crayfish (Fig. 13–6) are situated at the end of movable stalks which extend out from each side of the rostrum. The convex surface of the eye is covered by a modified portion of the transparent cuticle, the **cornea.** This cornea is divided by a large number of fine lines into four-sided areas termed **facets.** Each facet is the external part of a long slender visual rod, the **ommatidium.** These ommatidia lie side by side, separated from each other by dark pigment cells.

Beginning at the outer surface, each ommatidium consists of a **cornea (lens);** two **corneagen cells,** which secrete the cornea; a **crystalline cone** formed by four cone cells; two **retinular cells** surrounding the crystalline cone; several **retinular cells,** which form a central **rhabdome** where they meet; and a number of black **basal pigment cells** around the base of the retinular cells. Fibers from the **optic nerve** enter at the base of the ommatidium and communicate with

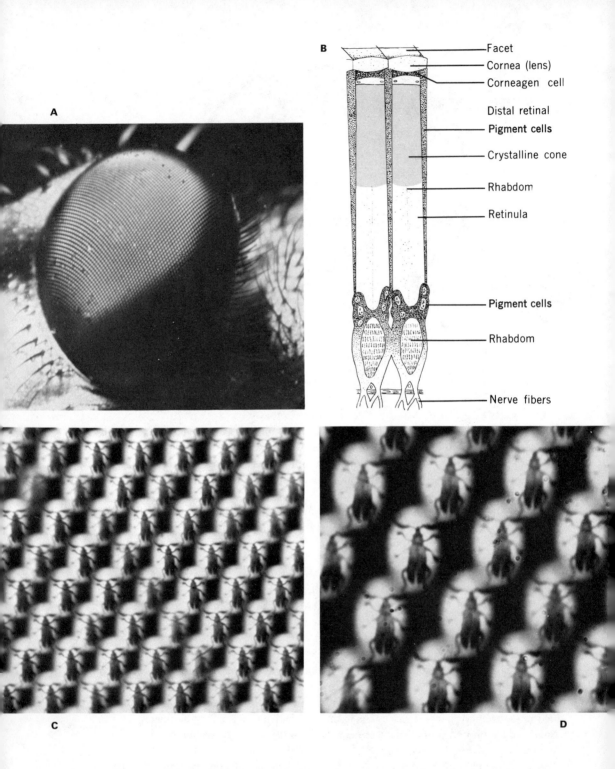

**Figure 13–6** (A) A representative anthropod compound eye, that of an insect. (B) Entire eye in longitudinal section to show general structure. (C, D) View of a beetle through the compound eye. *In vivo*, the neural connections from each ommatidium to the brain integrate the input to form a mosaic image rather than a repeating image as shown here.

the inner ends of the retinular cells. The eyes of the crayfish are presumed to produce a **mosaic** or **apposition image.**

**ENDOCRINE GLANDS.** On each eyestalk is an **x-gland,** which produces hormones that pass along the cell axons to the **sinus gland,** located at the base of the eyestalk. These hormones are distributed by the bloodstream; they appear to control the pigment of the chromatophores in the eyes and body epidermis. To some extent they govern metabolic rate, growth, reproduction, and viability; regulate the frequency of molting (through a hormone produced by the **y-organ,** a small gland at the base of the mandibular muscle); and are necessary for normal disposition of calcium salts in the exoskeleton.

**MUSCULAR SYSTEM.** The complex muscles of the crayfish are all attached to the inner surface of the skeleton; they are not a part of the body wall, as in the coelenterates and annelids, or external to the skeleton, as in vertebrates. The largest muscles are situated in the abdomen and are used to bend that part of the animal forward upon the ventral surface of the thorax, producing backward locomotion in swimming. Muscles of considerable size are situated in the thorax and within the tubular appendages, especially the chelipeds.

**REPRODUCTIVE SYSTEM.** Sexes are normally separate in the crayfish. The **male organs** consist of two white **testes** partially fused into three lobes; on each side is a long, coiled sperm duct, the **vas deferens,** which opens through the base of the fifth walking leg. The sperm remain in the testes and vasa deferentia until copulation takes place. As many as 2 million sperm are contained in the vasa deferentia of a single specimen.

The **female reproductive organs** consist of two **ovaries** and a short **oviduct,** which leads from each ovary to the external opening in the base of the third walking leg. Sperm are received in a cavity called the **seminal receptacle,** located in a fold of cuticle between the fourth and fifth walking legs.

**FERTILIZATION AND DEVELOPMENT.** The sperm are transferred from the male to the seminal receptacle of the female during copulation, which usually takes place between early spring and autumn. The female lays her eggs several weeks to months after copulation; the ova are fertilized by the sperm when laid. The eggs are fastened to the swimmerets with a sticky substance and are aerated by being moved back and forth through the water.

**Cleavage** of the egg is superficial, and the embryo appears first as a thickening on one side. The eggs hatch in 2 to 20 weeks, and the larvae cling to the egg stalk. In about 2 to 7 days they shed (molt) their cuticular covering. Molting is not peculiar to the young, occurring in adult crayfishes and many other arthropods. Molting is necessary before growth can proceed, since the exoskeleton is hard and nonelastic. In the larval crayfish, the cuticle of the first stage becomes loosened and drops off; the epidermal cells have in the meantime secreted a new covering.

The young stay with the mother, attached to her swimmerets until they can care for themselves. They molt at least six times during the first summer. Before the new exoskeleton hardens, an increase in body bulk occurs. It takes several weeks for the new shell to become completely hardened.

## BEHAVIOR

When at rest, the crayfish usually faces outward from its place of concealment and extends its antennae. In this position it may learn the nature of any approaching object without being detected. Activity at this time is reduced to the movements of a few of the appendages and the gills; the gill bailers of the second maxillae move back and forth, bailing water out of the forward end of the gill chambers; the swimmerets are in constant motion, creating a current of water; the maxillipeds are likewise kept moving and the antennules and antennae are in continual motion, exploring the surroundings. Crayfish are more active between dusk and dawn than during the daytime. At night they move about in search of food.

## LOCOMOTION

Locomotion is accomplished by walking or swimming. Crayfishes usually walk forward, but they may walk in any direction. The fourth pair of legs, which are the most effective in walking, bear nearly the entire weight of the animal; the fifth pair serve as props and push the body forward. Swimming is resorted to only when the animal is frightened; the crayfish then extends the abdomen, spreads out the uropods and telson, and, by a sudden contraction of the flexor abdominal muscles, flips the adbomen down and darts backward.

## SENSATION

The crayfish has more highly developed sense organs than the annelids. Their sense of **touch** is perhaps the most unique, aiding them to find food and avoid obstacles, among other things. Touch organs are located in specialized hairlike bristles or **setae** on various parts of the body.

**Vision** is undoubtedly of value to the crayfish in detecting moving objects. Reactions to **sound** have not been observed in crayfishes; the reactions formerly attributed to hearing are probably due to touch reflexes.

In aquatic animals it is so difficult to distinguish between reactions of **taste** and **smell** that both are included in the term **chemical sense.** The end organs of this sense are found in hairs located on the antennules, tips of the antennae, and mouth parts, among other places.

## REACTIONS TO STIMULI

Positive reactions result from stimulation by food substances. If meat juice is placed in the water near an animal, the antennae move slightly and the mouth parts perform vigorous chewing movements, causing general restlessness and movements toward the source of stimulation. The crayfish seems to depend chiefly on touch for accurately locating food.

Positive reactions to contact with solid objects are exhibited to a marked degree by crayfish; the normal position of the crayfish when at rest under a stone brings its side or dorsal surface in contact with the walls of its hiding place. These places are dark, indicating the crayfish's negative reaction to light of various intensities.

## REGENERATION

The crayfish and many other crustaceans have the ability to regenerate lost parts, but to a much more limited extent than the hydra or earthworm. Many species of crayfish of various ages have been used in experiments upon nearly all their appendages. The growth of regenerated tissue is more frequent and rapid in young specimens than in adults.

The new structure is not always like that of the one removed. A remarkable phenomenon is the regeneration of an apparently functional antennalike organ in place of a degenerate eye removed from the blind crayfish, *Cambarus pellucidus testii*. In this case a nonfunctional organ was replaced by a functional one of a different character. The regeneration of a new part that differs from the part removed is termed **heteromorphosis.**

Perhaps the most interesting anatomical structure connected with the regenerative process in *Cambarus* is **autotomy,** the definite breaking off of the walking legs at a point near the base. If a cheliped is grasped or injured, it is broken off by the crayfish at the breaking plane. The other walking legs, if injured, may be thrown off at the free joint between the second and third segments. A new leg as large as the one lost develops from the end of the remaining stump. This phenomenon also occurs in a number of other animals.

## Other Crustaceans

Almost every pond, lake, or stream contains crustaceans of many species (Fig. 13–7); salt water is also inhabited by a large variety of forms; and a few live on land. Only a few of the many interesting species can be mentioned here.

The "modern cyclops," *Cyclops viridis,* is a very successful small crustacean living in fresh water or in the sea. It is usually greenish in color and its single eye is red.

**Figure 13–7** Representative crustaceans.

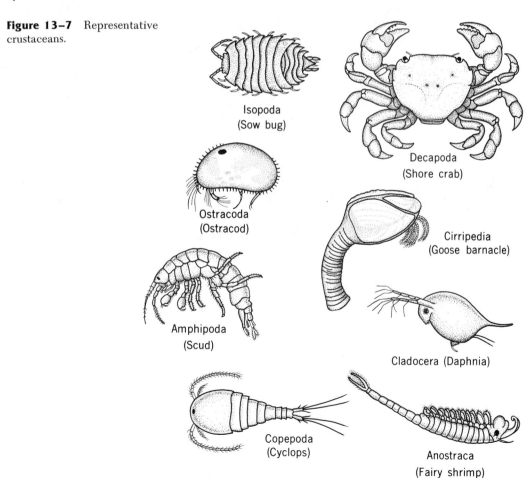

**Barnacles** are small crustaceans which attach themselves to rocks and the bottoms of various larger animals, including whales, and ships. There are many types, including the acorn barnacle (*Balanus*), the goose barnacle, and the parasitic root-headed barnacles.

**Shrimps** and **prawns** are decapods (10-footed animals). The edible shrimp is slender, about 5 cm long, with whiplike antennae. It is very agile, swimming backward with quick jerks of the fin at the posterior end of the body. Prawns are smaller than shrimps. They are both constantly preyed upon by fish, other marine animals, and man.

**Hermit crabs** live in the empty shells of marine snails. As they grow larger, they must move into larger shells. *Pagurus* is a common hermit crab which lives in rock pools and shallow water along the beach.

**Spider crabs** are noted for their long spidery legs. The common species, *Libinia emarginata*, is about 7 cm long and lives on mud flats and oyster beds along the Atlantic Coast and in California. The legs of a Japanese spider crab reach an enormous length; individuals of this species may measure 4 meters across when spread out.

**Fiddler crabs** (named for their custom of waving their large claw back and forth) form colonies which live in burrows that are dug in the mud or sand in salt marshes. In certain species, the males are very pugnacious and fight each other with great vigor.

# Chelicerates

The subphylum chelicerata includes the (terrestrial) arachnids (spiders, harvestmen, scorpions, mites, ticks) and their allies, the sea "arachnids." Among these are the sea spiders, the extinct trilobites, and an ancient fossil group of which the only living member is the horseshoe crab. Although not really arachnids, these classes of arthropods share with the arachnids certain characteristics: they have no antennae or true mandibles (jaws); the body can usually be divided into an anterior cephothorax and a posterior abdomen; and they usually possess six pairs of appendages, the first pair of which are chelicerae.

# Spiders

Since the spider is representative of most arachnids, it will be studied in detail.

In addition to typical arachnid features, spiders show interesting arthropod modifications; especially for breathing, obtaining juices, and spinning webs. Although spiders have for centuries been generally considered to be poisonous, the black widow (Fig. 13–12A) and brown recluse spider are the only two in this country capable of causing death in human beings.

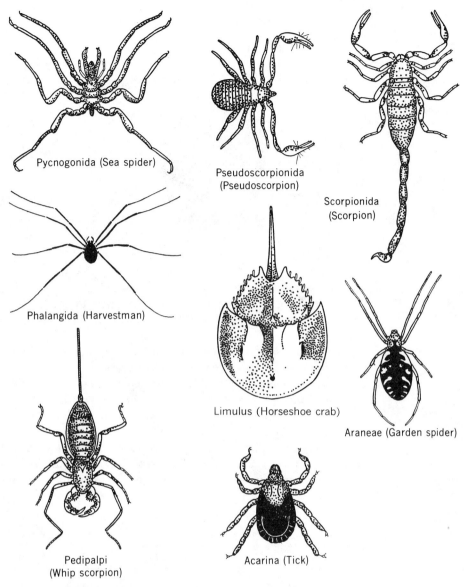

**Figure 13-8** Representatives of the subphylum Chelicerata.

## EXTERNAL ANATOMY

The body of the spider consists of a **cephalothorax** and an unsegmented **abdomen.** There are six pairs of **appendages** attached to the cephalothorax. The first pair are the **chelicerae.** In many species they are composed of two parts: a basal segment and a terminal claw or fang. The second pair of appendages are the **pedipalps;** their basal parts, called **maxillae,** are used as jaws to press or chew food. In mature males, the pedipalps are also used to transfer sperm to females. Four pairs of **walking legs** follow; each consists of seven joints: a coxa, trochanter, femur, patella, tibia, metatarsus, and tarsus. The leg is terminated by two- or three-toothed claws and often a pad of hairs (the claw tuft), which enables the spider to run on ceilings and walls. Antennae are absent and sensory functions are performed in part by the pedipalps and walking legs.

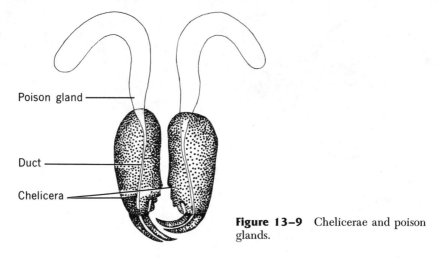

**Figure 13-9** Chelicerae and poison glands.

The **sternum** lies between the legs, and a labium is situated between the maxillae. There are usually eight **eyes,** located on the front of the head. The **mouth** is a minute opening between the bases of the pedipalps; it ingests liquids only.

**Poison glands** are usually located in the cephalothorax (Figs. 13-9; 13-10) the venom is used to kill prey and for defense.

The **abdomen** is connected with the cephalothorax by a slender **peduncle.** Near the anterior end of the abdomen, on the ventral surface, is a **genital opening;** in some females, it is covered by a flat plate, the **epigynum.** On either side of the epigynum are the slitlike openings of the respiratory organs, the **book lungs.** Some spiders also possess **tracheae,** which open to the outside through **spiracles** near the posterior end of the ventral surface. Just back of the tracheal opening are three pairs of tubercles or **spinnerets,** used for spinning threads. The **anus** lies posterior to the spinnerets.

**Figure 13-10** Internal structure of a spider.

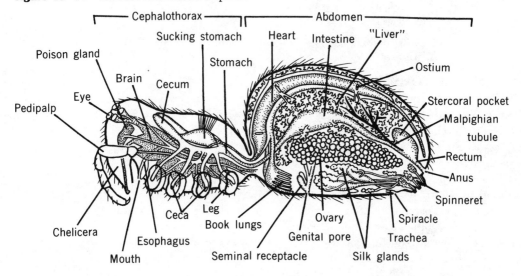

## INTERNAL ANATOMY AND PHYSIOLOGY

The spider feeds mainly on insects; because it can ingest only liquid food, solid parts are liquefied by the action of powerful digestive enzymes contained in a fluid secreted from a large **liver** which surrounds the digestive tract (Fig. 13–10). This fluid pours into the intestine through ducts.

Spiders ingest food in one of two ways: those with weak jaws puncture the body of the insect with their fangs, and then alternate between injecting digestive fluid through this hole and sucking back the liquefied tissues until only an exoskeleton remains; those with strong jaws crush the insect into small pieces between their jaws as the digestive enzymes are regurgitated over them until only a small mass of indigestible material remains.

**DIGESTIVE SYSTEM.** The **digestive system** is made up of a **mouth** and **pharynx**, followed by a horizontal **esophagus** that leads into a **sucking stomach**, which opens into a **true stomach**. The true stomach gives off five pairs of **ceca** or blind tubes in the cephalothorax. The **intestine** passes almost straight through the abdomen; at a point near the posterior end, it forms a sac, the **stercoral pocket,** that connects with the **rectum**, which terminates in the **anus.**

**EXCRETORY SYSTEM.** The **excretory organs** are paired **Malpighian tubes** which open into the intestine, and one or two pairs of **coxal glands** in the floor of the cephalothorax. The coxal glands are sometimes degenerate, and their openings are difficult to find; they are homologous with the green glands of the crayfish.

**CIRCULATORY SYSTEM.** The **circulatory system** consists of a heart, arteries, veins, and a number of sinuses (spaces). The **heart** is situated in the abdomen, dorsal to the digestive tract. It is a muscular contractile tube lying in a sheath, the pericardium. The heart opens into the pericardium, usually by three pairs of openings (ostia). Posteriorly it gives off a **caudal artery**; anteriorly, an **aorta** branches to supply tissues in the cephalothorax; three pairs of **abdominal arteries** are also given off by the heart. The colorless **blood,** containing mostly amoeboid corpuscles, passes from the arteries into sinuses among the tissues, and is carried to the **book lungs,** where it is oxygenated; it then passes to the pericardium by way of the pulmonary veins and reenters the heart through the ostia.

**RESPIRATORY SYSTEM.** **Respiration** is carried on by **tracheae** and **book lungs;** the latter are peculiar to arachnids. There are usually two book lungs; each consists of a blood-filled chamber with an air vestibule posterior to it. This air vestibule communicates with the outside through a slit in the body wall. From the front wall of the vestibule, many narrow parallel air pockets have pushed forward into the blood-filled chamber; these are flattened, fan-shaped tracheae. Some of the more usually shaped tubular tracheae may be present but do not extend to all parts of the body. Air entering through the slit in the body wall circulates through the air pockets where oxygen is taken up by the blood.

**NERVOUS SYSTEM.** The **nervous system** consists of a bilobed ganglion **brain** above the esophagus, a large subesophageal ganglionic mass, and nerves that run to various organs. **Sensory hairs,** over the body and appendages, are the principal sense organs. The eight **simple eyes** form distinct images in only a few families; in others, they function primarily for perception of moving objects (prey) and light intensity. The sense of **smell** is well developed, and an organ of **taste** is located in the pharynx. Very fine, erect, hairlike processes set in sockets may be organs of hearing, but this has not yet been confirmed.

**REPRODUCTIVE SYSTEM.** The **sexes** of spiders are separate, and the **testes** and

**Figure 13-11** Web of orb-weaving spider.

**ovaries** form a network of tubes in the abdomen. The sperm are ejected upon a special "sperm web" and picked up by the pedipalps for transfer to the female **seminal receptacle**. There is courtship activity before mating which varies with the species; the female sometimes kills and eats the male after mating. Eggs are not fertilized until laid; the sperm move from the seminal receptacle to fertilize the eggs as they pass through the "uterus externus." The eggs are laid in a silk cocoon which is attached to the web or to a plant or is carried about by the female. The young leave the cocoon after hatching, as soon as they can run about. Several molts occur before maturity.

The **spinning organs** of spiders are three pairs of appendages, **spinnerets.** They are pierced by hundreds of microscopic tubes through which fluid secreted by abdominal **silk glands** passes externally to form a thread. Threads are used to build snares, spin orb webs (Fig. 13-11), and form cocoons. Many spiders also possess an accessory spinning organ, the **cribellum,** which emits a special kind of silk. The silk threads of spiders are stronger than steel threads of the same size.

## SOME SPIDERS OF SPECIAL INTEREST

More than 30,000 species of spiders are known; about 3000 live in the United States. House spiders are considered a nuisance but are not dangerous. The brightly colored garden spider is one of the more beautiful species.

Crab spiders have the habit of walking sideways. Some are white or yellow and are said to favor flowers of similar color. Jumping spiders have stout front legs for capturing prey; they are famous for their peculiar antics during mating.

Tarantulas (Fig. 13-12B) are the giants of the spider world, reaching a body length of 2.5 inches and a leg spread of 9 or 10 inches; spiders of this size are able to capture small birds. Poison glands in the tarantula are situated in the chelicerae, where ducts open on the fangs to release the venom. Except for the

**Figure 13-12** Arachnids. (A) Black widow spider on web strands. (B) Tarantula, largest of the spiders.

tropical ones, and contrary to popular belief, tarantulas are not harmful to human beings.

## Other Arachnids

**Scorpions** are rapacious and poisonous arachnids measuring from $\frac{1}{2}$ to 8 inches in length. They live in tropical and subtropical regions, hiding during the daytime and active at night. They prey on insects, spiders, and other small

**Figure 13-13** Scorpion capturing its next meal. The prey has been paralyzed by the sharp poisonous sting at the terminal end.

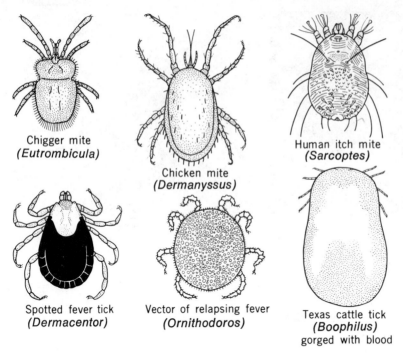

**Figure 13–14** Some parasitic ticks and mites.

animals; they defend themselves against larger animals by a paralyzing sting at the end of their tails (Fig. 13-13). The deadly sculptured scorpion, *Centruroides sculpturatus*, is small and straw-colored; it is the only species considered dangerous to human beings in the United States.

Scorpions are one of the oldest forms of life found on this earth. We see from fossil specimens from many parts of the world that scorpions have remained essentially unchanged for hundreds of millions of years.

**Mites** and **ticks** are found almost everywhere (Fig. 13–14): in fresh and salt water, on the ground, on vegetation, or on the surface, or burrowed into other animals. Animals and plants, either living or dead, serve as their food; the parasitic species live largely on blood and can be disease vectors.

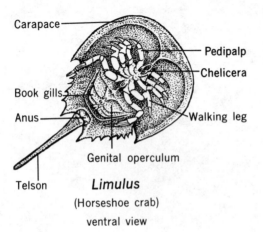

**Figure 13–15** Horseshoe crab, called a living fossil because it has undergone little change during long geological periods.

## HORSESHOE CRABS

Although commonly called horseshoe crabs, referring to their heavy, horseshoe-shaped carapace, these animals are not actually crabs (Fig. 13–15). They are living fossils whose close relatives died millions of years ago and are some of the strangest animals on earth. They occur along the Atlantic Coast, living in shallow water along the shores. Here they shove their way through sand and mud, hunting for worms, bivalves, and other small animals on which they feed.

## Centipedes and Millipedes

**Centipedes** (class Chilopoda) are flattened dorsoventrally and consist of from 15 to 173 segments, depending on the species. Each segment bears **one pair of legs** (Fig. 13–16), except the first and the last two. The antennae are long, consisting of at least 12 segments.

Centipedes are swift-moving creatures, many of which live under the bark of logs or under stones. Their prey consists of insects, worms, mollusks, and other small animals which they kill with **poison claws** (modified legs from the first segment) and chew with their mandibles. Poisonous centipedes reaching a length of 2 dm are found in tropical countries; their bite is painful and dangerous to human beings. The common house centipede, *Scutigera*, has 15 pairs of very long legs and lives in damp areas. It is not only harmless, but actually beneficial to human beings, since it feeds on insects.

**Millipedes** (class Diplopoda) are more or less cylindrical and consist of from

**Figure 13–16** Centipede; each segment bears one pair of legs.

**Figure 13–17** Millipedes are easily distinguished from centipedes by their subcylindrical bodies and two pairs of legs on most segments.

25 to over 100 segments. All segments bear **two pairs of legs** except the thorax, which has only one pair (Fig. 13–17).

The mouth parts are pairs of **mandibles** and **maxillae**. One pair of short **antennae** and clumps of simple eyes are usually present. There are **olfactory hairs** on the antennae and a series of **scent glands** that secrete an objectionable fluid used in defense. One species ejects a highly irritating fluid that causes temporary blindness.

Millipedes move very slowly in spite of their numerous legs. Some are able to roll themselves into a spiral or ball. They live in dark, moist places, and feed principally on decaying vegetable matter and sometimes on living plants.

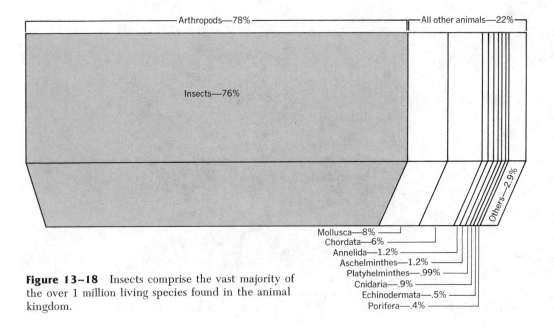

**Figure 13–18** Insects comprise the vast majority of the over 1 million living species found in the animal kingdom.

# Insects

Insects comprise the greatest portion of the phylum Arthropoda. In fact, there are more species of insects than of all other animals combined (Fig. 13–18). Over 850,000 species have been described and it is probable that

**Figure 13–19** Representatives of some orders of insects.

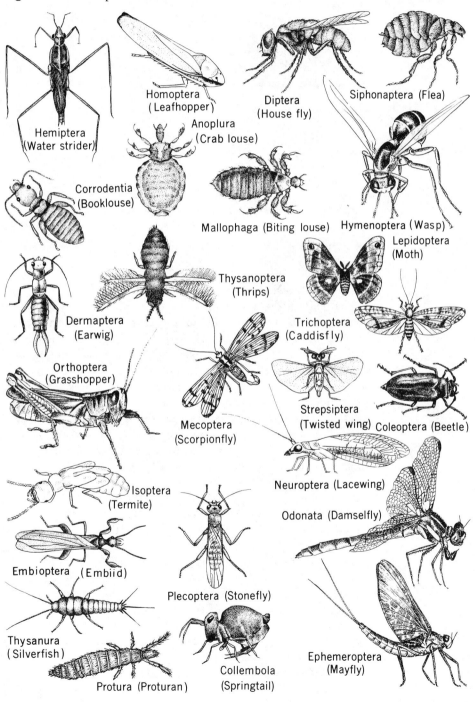

hundreds of thousands remain to be discovered (Fig. 13-19). They live in almost every conceivable type of environment, and their structure, habits, and life cycles are correspondingly modified. Their wide distribution is probably a result of their relative indestructibility and small size, which enables them to fly or be carried by air and water currents. Their extraordinary adaptability, which is due to basic arthropod characteristics (exoskeleton, trachea, musculature, and resistance to desiccation), enables them to remain in new habitats; their resistant eggs may be carried by birds and other animals into new regions.

Because insects differ so widely in habit, physiology, and morphology, and exhibit so many interesting adaptations, only certain general aspects of entomology (the study of insects) can be reviewed here. The grasshopper will be discussed as a representative insect, followed by a discussion of life cycles, coloration, and social behavior.

# The Grasshopper

The grasshopper is a very favorable species for detailed study, exhibiting both internal and external essential features of insects. It is one of the least specialized animals of its class. However, it does exhibit several conspicuous adaptations, such as leathery forewings, enlarged hindlimbs, auditory organs, and structures for producing sound.

### EXTERNAL ANATOMY

The body of the grasshopper and other insects is divided into a head, a thorax, and an abdomen (Fig. 13-20). Like the crayfish, the grasshopper is covered by an exoskeleton, which protects the underlying delicate organ systems. It is soft in certain regions, allowing movements of such structures as abdomen, legs, and antennae. The exoskeleton is divided into a linear row of **segments.** Each segment is composed of separate plates known as **sclerites.** A ridge or **suture** is often seen where sclerites meet.

The body wall is comprised of the **cuticle,** secreted by the **epidermis** directly beneath it, and the **basement membrane,** located beneath the epidermis.

The **head** (Fig. 13-21) is composed of several regions: the front portion is the **frons,** the sides are cheeks or **genae,** and the rectangular sclerite below the frons is the **clypeus.** On either side of the head is a **compound eye;** slightly anterior to these eyes are threadlike **antennae,** composed of many segments. Three **simple eyes (ocelli)** are located on the top of the head, one situated between the antennae and the others near the inner edge of each compound eye.

The food of the grasshopper consists of vegetation, which it bites off and grinds up by means of its chewing **mouthparts.** A **labrum,** or upper lip, is attached to the ventral edge of the clypeus. Beneath this is a membranous tonguelike organ, the **hypopharynx.** On either side is a hard jaw or **mandible,** with a toothed surface fitted for grinding. Behind the mandibles are a pair of **maxillae,** consisting of several parts, and with sensory palps at the sides. The **labium,** or lower lip, has slender palps at the sides. The labrum and labium hold food between the grinding mandibles and maxillae. The **thorax** is separated from the head and abdomen by flexible joints and consists of three segments. A

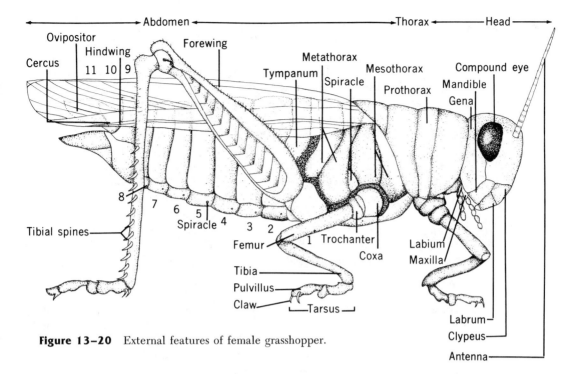

**Figure 13–20**  External features of female grasshopper.

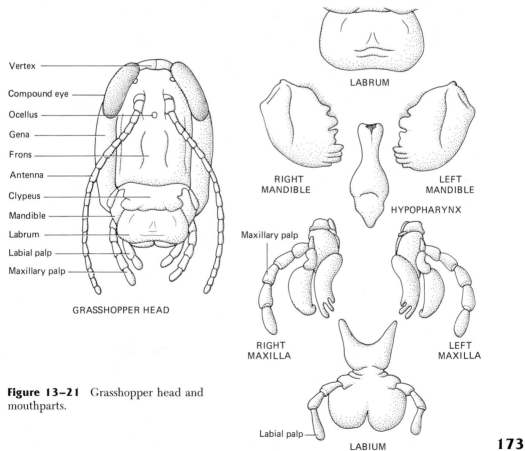

**Figure 13–21**  Grasshopper head and mouthparts.

typical segment consists of a dorsal **tergum** (composed of four fused sclerites), a lateral **pleuron** (three fused sclerites) on each side, and a single ventral sclerite, the **sternum**.

The three segments that comprise the thorax are as follows: The anterior segment, the **prothorax**, has a large dorsal surface (**pronotum**) that extends down on either side; the four sclerites are indicated by transverse grooves and its sternum bears a spine. The middle segment, **mesothorax**, and the posterior segment, **metathorax**, have small terga. The sclerites of the pleuron are distinct and the sternum is large.

The mesothorax and metathorax each bear a pair of **wings**, arising from the region between the tergum and pleuron as a double layer of epidermis which secretes the upper and lower cuticular surfaces. Between these develop longitudinal **veins**, which strengthen the wings. The veins differ in number and arrangement in different species; in certain ones, they are constant enough to be used for classification. In the grasshopper, the wings of the mesothorax are leathery and are not folded; they serve as covers for the thin, folded wings of the metathorax.

Each of the three segments of the thorax bears a pair of legs. Each **leg** consists of a longitudinal series of segments; starting from the segment closest to the body, they are the **coxa, trochanter, femur, tibia,** and **tarsus**. The tarsus consists of three visible segments, the last bearing a pair of **claws**, between which is a fleshy pad or **pulvillus,** used in clinging to surfaces. The femora of the metathoracic legs are enlarged to contain the muscles used in jumping.

The slender **abdomen** consists dorsally of 11 segments; those at the posterior extremity are modified for copulation or egg laying. Along the lower sides of the abdomen are eight pairs of small openings (**spiracles**) through which the animal breathes. In the grasshopper the sternum of segment 1 is fused with the metathorax; on either side of the segment there is an oval **tympanic membrane** covering an **auditory sac**. Segments 2 to 8 are unmodified. In the **male**, the sternum of segment 9 is elongated ventrally, giving an upward twist to the abdomen. The **female** abdomen is more tapered at the end, forming the **ovipositor,** an egg-laying apparatus.

## INTERNAL ANATOMY AND PHYSIOLOGY

The organ systems of the grasshopper (Fig. 13–22) and other insects lie in a **hemocoel,** a body cavity filled with blood (hemolymph); this cavity is not a coelom. All the systems characteristic of higher animals are represented.

MUSCULAR SYSTEM. The muscles are **striated,** very soft and delicate, but strong. In the abdomen, they are segmentally arranged. The most conspicuous muscles are those that move the mandibles, the wings, the metathoracic legs, and the ovipositor.

DIGESTIVE AND EXCRETORY SYSTEMS. The principal parts of the digestive tract are the foregut, midgut, and hindgut. The **foregut** consists of a **mouth,** with **salivary glands** on each side, a tubular **esophagus** enlarging into a crop in the mesothoracic and metathoracic segments, and the **proventriculus,** a grinding organ (or **gizzard**). In the **midgut** is the **ventriculus (stomach),** which extends posteriorly into the abdomen; digestive enzymes are secreted and poured into it by six double cone-shaped pouches, the **gastric ceca**.

The **hindgut** consists of delicate **Malpighian tubules,** which open into an **ileum,** a **colon** that expands into a **rectum,** and an **anus**. Both the foregut and the hindgut are lined with cuticle; hence little absorption takes place in them.

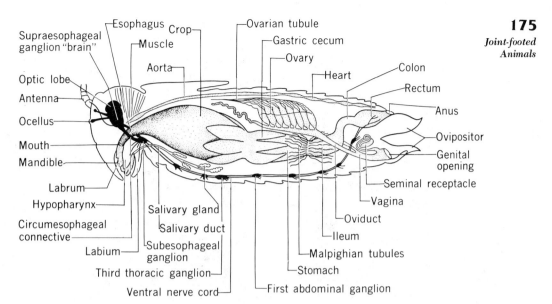

**Figure 13-22** Internal organs of a grasshopper; tracheae not included.

The Malpighian tubules remove metabolic wastes, such as uric acid, which is condensed into crystals; the conservation of water in this process results from its reabsorption by the tubules. The removal of wastes in the dry state is characteristic of small land animals that have only a limited water supply.

CIRCULATORY SYSTEM. As in other arthropods, the blood system is an open one with no capillaries or veins. Hemolymph enters the heart and is forced anteriorly through the aorta into the hemocoel, where it bathes all the organs. The **hemolymph** consists of white corpuscles suspended in a clear **plasma;** its main function is transportation of food and wastes. The white corpuscles act as phagocytes, removing foreign organisms and other substances.

RESPIRATORY SYSTEM. The respiratory system consists of a network of tubes (**tracheae**) that communicate with every part of the body (Fig. 13-23). The tracheae consist of a single layer of cells lined with a layer of cuticula, which is thickened to form **spiral rings** that prevent the tracheae from collapsing. A tracheal branch extends from each **spiracle** (opening) to a longitudinal trunk on each side of the body. **Tracheoles** are the smallest tracheae, connected directly with the tissue, to which they supply oxygen and from which they carry carbon dioxide. The smallest tracheoles contain fluid which dissolves oxygen before it actually reaches the cells; this fluid serves in internal respiration as blood does in other animals. In the grasshopper and certain other animals, some of the tracheae are expanded into thin-walled air sacs. Contraction and expansion of the abdomen enlarges and compresses these air sacs, drawing air into and expelling it out of the tracheal system.

NERVOUS SYSTEM. The nervous system includes a **brain,** consisting of three pairs of ganglia fused together and dorsally located in the head (Fig. 13-24). These ganglia supply the eyes, antennae, and other head organs. The brain joins, by two connectives around the esophagus, to the **subesophageal ganglion.** This ganglion consists of the three anterior parts of ganglia of the ventral nerve chain fused together; it innervates the mouthparts. The ventral nerve chain continues with a pair of large ganglia in each thoracic segment. The ganglia in the metathoracic segment are particularly large and represent the ganglia of

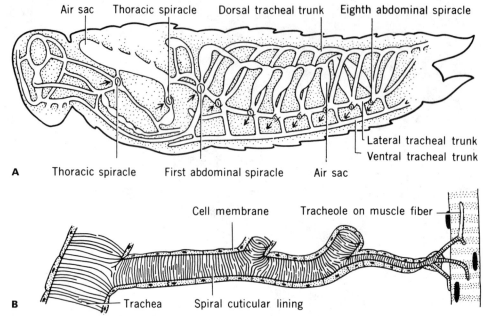

**Figure 13–23** Tracheae of the grasshopper. (A) Arrows indicate inhalation through spiracles and exhalation through abdomen. (Redrawn from E. O. Essig, *College Entomology*, New York: Macmillan Publishing Co., Inc., 1942.) (B) Large tracheal trunk and some of its branches.

this segment and the first abdominal segment. There are five pairs of ganglia present for the other abdominal segments. Connected with the brain are the ganglia of the **sympathetic (autonomic) nervous system,** which controls the "involuntary" movements of the digestive tract, heart, aorta, and reproductive system.

SENSE ORGANS. Grasshoppers possess organs of sight, hearing, touch, taste, and smell. The compound eye is covered by a transparent part of the cuticula, the **cornea,** which is divided into many hexagonal pieces (**facets**). Each facet is the outer end of an **ommatidium** unit, producing mosaic vision as in the crayfish. The ocelli are thought to function primarily in light perception, although it is possible that they may form crude images at a close range. Each ocellus consists of a **retina** (group of visual cells), pigment, and a transparent **lens** (modified from the cuticula).

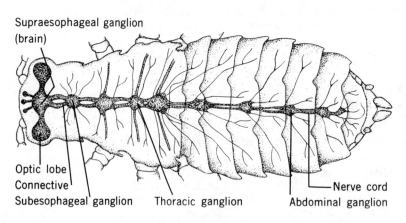

**Figure 13–24** Grasshopper's nervous system in dorsal view. (Redrawn from R. E. Snodgrass, *Principles of Insect Morphology*, New York: McGraw-Hill Book Company, 1935.)

A pair of **auditory organs** are located on the sides of the tergum of the first abdominal segment. Each consists of a tympanic membrane (tympanum) stretched with an almost circular sclerotized ring. Some insects hear sounds beyond the range of the human ear. Grasshoppers produce sound by rubbing the rough-surfaced tibia of the hindleg against a wing vein, causing it to vibrate (stridulation). The **antennae** bear **olfactory pits** and are the principal organs of smell. **Taste organs** are located on the mouthparts. **Tactile hairs** are present on various parts of the body, particularly the antennae.

REPRODUCTIVE SYSTEM. **Female** grasshoppers can easily be distinguished from males by the **ovipositor.** The female also possesses two **ovaries;** each consists of several egg tubules (**ovarioles**), containing **oogonia** and **oocytes** arranged in a linear series, **nurse cells,** and other tissue cells. The oocytes grow as they proceed posteriorly down the ovariole; hence the ovariole becomes gradually larger toward the posterior end. The ovarioles of each ovary are attached posteriorly to an **oviduct,** into which the eggs are discharged. The two oviducts unite to form a short **vagina,** which leads to the **genital opening** between the plates of the ovipositor. A tubular **seminal receptacle (spermatheca)** connects with the dorsal wall of the vagina and receives the **spermatozoa** during copulation, releasing them when the eggs are fertilized.

In the **male,** spermatozoa are developed in the two **testes** and discharged through two **vas deferens,** which unite to form an **ejaculatory duct,** which releases the sperm through a penis. **Accessory glands** are present at the anterior end of the ejaculatory duct.

# Life Cycles of Insects

The most conspicuous differences in the life cycles of various types of insects are associated with the kind of **metamorphosis** involved. There is **no metamorphosis** in certain species, such as the primitive *Campodea staphylinus* (bristletail); the young that hatch from eggs look like miniature adults. As they grow, they molt a number of times, finally reaching sexual maturity and adult functions.

A **gradual metamorphosis** (Fig. 13-25) is present in some species, such as the grasshopper. The young grasshopper that hatches from the egg is called a **nymph.** It resembles its parent but has a large head compared to the rest of the body and lacks wings. As it grows, its body becomes too large for the inflexible exoskeleton and the latter is shed (molted) periodically. Wings are gradually developed and the adult condition is finally assumed.

The metamorphosis of the mayfly is **incomplete** (Fig. 13-25). The changes that take place in the form of the body are greater than in gradual metamorphosis, but less marked than in complete metamorphosis. The young that emerge from eggs are called **naiads.** They live under water, where they breathe through tracheal gills and feed on minute plants. Growth is accompanied by 27 molts and requires from 6 to 9 months. When ready to venture into the air, the naiad swims to the surface; a split appears along the back and a gauzy-winged adult flies out to a nearby object, where it rests for 18 to 24 hours. After molting again, it is ready to fly into the air and find a mate. The adult life of both males and females is but a few hours or days.

**Figure 13–25** Two types of life cycles found in insects: gradual and incomplete metamorphosis.

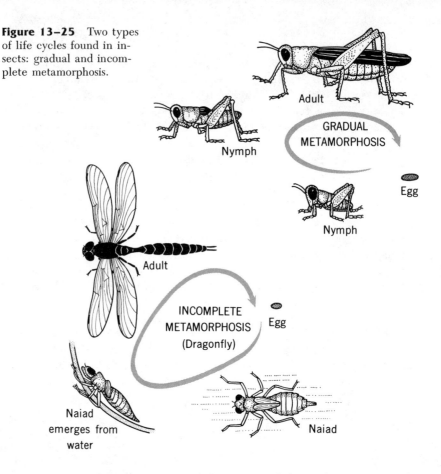

Butterflies exhibit **complete metamorphosis** (Fig. 13–26). Their eggs must be laid on the leaves of certain plants, upon which the **larvae** or caterpillar will feed. The larvas eats its way out of the shell and proceeds to feed on the host plant. When it can grow no larger within its cuticular covering, a split appears and the larvas crawls out. The elasticity of the new exoskeleton enables the larva to expand.

When full grown (about 3 cm long), the caterpillars attach themselves with a silken thread to the underside of a leaf or other object. Butterfly caterpillars do not spin cocoons as many moth caterpillars do. After a time the body becomes thicker and shorter; the skin splits down the back, is pushed off at the posterior end, and a greenish-colored pupa is revealed. The pupa, which does not feed, undergoes a number of major transformations: the digestive system changes from using solid to liquid foods, the muscles are adapted for flight, the nervous system is reconstructed, wings grow out from pads of larval tissue, and the reproductive organs grow to maturity. At the end of 10 days, the pupal skin splits and the adult butterfly emerges, spreads and dries its wings, and flies away.

These are the basic types of metamorphosis; there are many variations occurring in insects of different species.

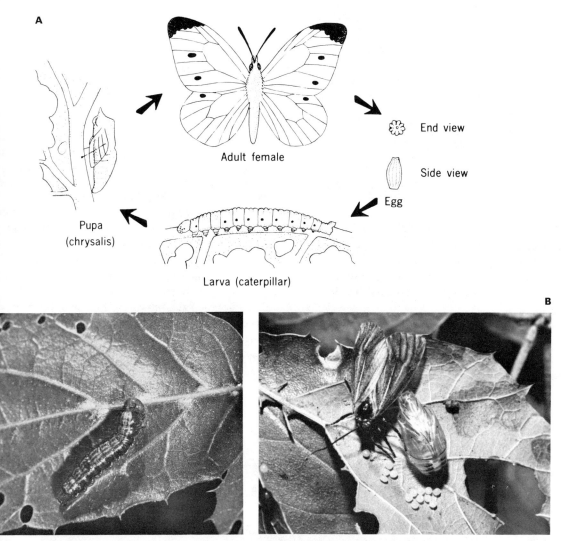

**Figure 13–26** (A) Complete metamorphosis is shown in the life cycle of the cabbage butterfly. (B) Larval, pupal, and adult stages of a moth.

## Coloration

Many insects are brilliantly colored, especially the butterflies, moths, and beetles. Coloration of some insects differs with the season, producing different color patterns in the various broods. Such insects are **seasonally dimorphic** (two types), **trimorphic** (three types), or **polymorphic** (at least three types). Often, males of a species are colored differently from the females; they are **sexually dimorphic**. Insects are often protected from their enemies by their coloration.

In some insects the colors displayed are a result of pigments embedded in the exoskeleton; in others (such as the butterfly), the diffraction of sunlight upon microscopic scales on the exoskeleton produces an iridescent color.

# Social Insects

Many animals exhibit various types of social grouping. The more complex societies involve **division of labor;** the principal types of activity are reproduction, obtaining food, and defending the colony. Wasps, bees, ants, and termites exhibit some of the most interesting behavior of social insects (Fig. 13-27).

## WASPS AND BEES

Wasps and bees may be solitary or social. Solitary wasps and bees dig a hole in the ground, in wood, or construct a nest of mud. Wasps provide their nests with caterpillars or other arthropods that they have paralyzed; bees lay eggs in a nest provided with enough pollen to furnish proteins for the growth of their larvae. After the eggs are laid, they close the nest entrance and cease parental care of their offspring.

BUMBLEBEES AND HONEYBEES. Bumblebees and honeybees are all social. The fertilized queen bumblebee lives through the winter. In the spring she lays a few eggs in a cavity in the ground, from which the infertile female workers develop. They carry on all the activities of the colony except egg laying. At the end of the summer, males (drones) and fertile females (queens) hatch from some of the eggs. These mate and the sperm receptacles of the queens are filled with sperm. The workers and drones die, and the race is maintained by the queens alone.

**Figure 13-27** Insects like these battling red and black ants are noted for their social behavior.

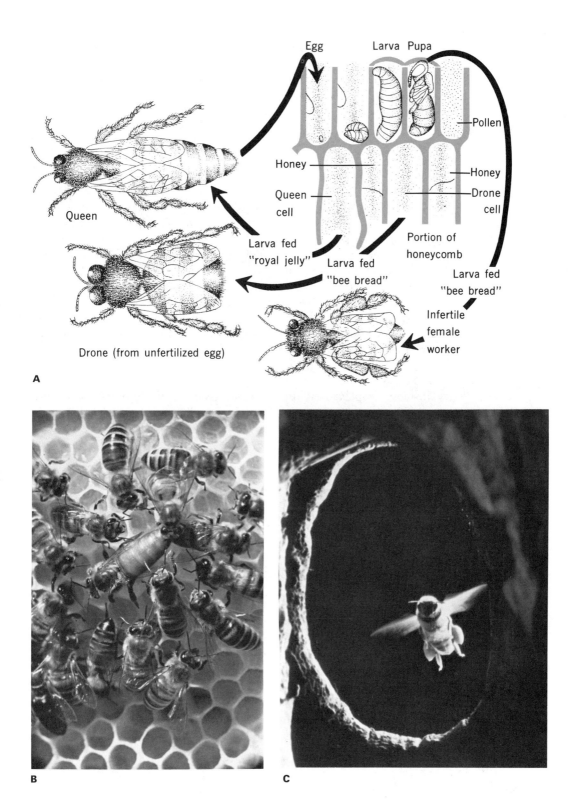

**Figure 13-28** (A) Life cycle of the honeybee, showing growth stages and three adult castes: worker, drone, and queen. (B) Bee hive. The large queen bee is surrounded by the workers. (C) Here a bee returns to the hive laden with pollen.

Honeybees (Fig. 13–28) exhibit an even more complex social organization. Observers of these bees have always realized they had a system of communication. A brilliant Austrian zoologist, Karl von Frisch, made the fascinating discovery that their main method of broadcasting a source of nectar or pollen was by a rhythmic dance and odors. A full discussion of this dance is presented in Chapter 22.

### ANTS

Many ants live a complicated social life. Their colony contains several fertile females (queens) and at certain periods fertile males (drones). Infertile females may be of several types: soldiers to guard the colony, workers to gather food, or workers to care for the eggs and young. These various types are morphologically different. Social ants have developed effective solutions to survival problems, some which closely parallel important advances in human cultures. Agricultural ants maintain crops of fungi, which they use for food; others have domesticated aphids and use their wastes for food. Some complex ant societies migrate great distances, overcoming prey and physical barriers by working together. There are some species in which the queen will invade host ant nests and take over either by force (large vicious ones) or by infiltration (reduced in size and not noticed by the host workers).

### TERMITES

The most complex social life of all insects is that of the termites. The colony contains three principal types of castes (Fig. 13–29): **sexuals** (kings and queens), **workers,** and **soldiers.** In each caste there are males and females. Kings and

**Figure 13–29** Castes and life cycle of the termite.

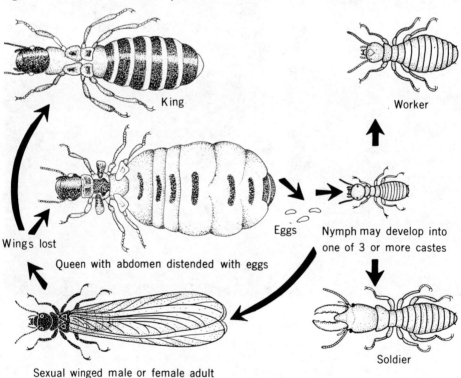

queens may possess functional wings, nonfunctional wings, or no wings at all. Winged kings and queens leave the colony, mate, lose their wings, and form a new colony.

Workers and soldiers have no wings or functional sexual organs. The workers are more numerous than any other caste. They care for the eggs and young, tend the queen, obtain food, excavate tunnels, and construct mounds among other duties.

The soldiers, the most highly specialized, appear to be of two types: one with large bodies, strong heads, and huge mandibles for driving away intruders; the second type carries out chemical warfare by means of a repellent fluid ejected through a pore in the head.

# Origin of the Arthropods

### SUBPHYLUM ONYCHOPHORA—THE MISSING LINK

The onychophorans resemble more closely than any other animals what is believed to have been the ancestral condition of the arthropods. They possess a thin exoskeleton and a continuous muscular body wall, but lack joints.

*Peripatus*, (Fig. 13-30) an onychophoran, possesses a cylindrical body that lacks a distinct head. These animals are especially interesting, exhibiting characteristics of both annelids (paired segmental nephridia; the presence of undulipodia) and arthropods (jaws modified from legs; the presence of hemocoel and tracheae) (Fig. 13-31). *Peripatus* differs from both annelids and arthropods in the possession of a single pair of jaws, scant metamerism, arrangement of the tracheal openings, texture of the skin, and separate nerve cords with no well-developed ganglia. Hence, the onychophorans are not classically like the Phylum Arthropoda and are therefore sometimes placed in a separate phylum.

**Figure 13-30** *Peripatus,* an onychophoran with annelid and arthropod characteristics. (Reproduced by permission from *The Biotic World and Man,* by L. J. and M. J. Milne, p. 48, copyright 1952 by Prentice-Hall, Inc., Englewood Cliffs, N.J.)

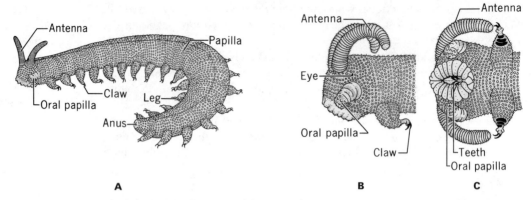

**Figure 13–31** *Peripatus*. (A) External structure. (B) Lateral view of "head." (C) Ventral view showing legs unlike typical arthropods but arthropodlike claws.

The group furnishes an excellent example of *discontinuous distribution*. Species have been reported from Central America, Mexico, the West Indies, and the southern hemisphere. Even in the area where a species occurs, specimens are present in only a few of the many available habitats. This seems to indicate that this group once had a continuous distribution but has disappeared throughout most of its range and is on the road to extinction.

The crustaceans, centipedes, millipedes, and insects appear to have developed along one line, since they have so much in common; the arachnids appear to have developed along another line, since none of their appendages have developed into antennae and none possess mandibles.

## Relations of Arthropods to Man

Crustaceans are of considerable value as food for man, either directly or indirectly (as an important part of the food chain). Commercially, the shrimp is the most important of the crustaceans as human food, followed by crabs, lobsters, and crayfish. Although some crustaceans are parasites of plants and aquatic animals, none is a parasite of man or other land animals.

Domestic animals are infested by a number of species of mites and ticks. The chicken mite is a serious pest of poultry; scab mites attack horses, dogs, and other animals. The sheep scab mite may seriously injure sheep, causing irritation, loss of wool, and decreased vitality. Ticks are carriers of several diseases of human beings and animals. Rocky Mountain spotted fever is due to *Dermacentor*, which transmits rickettsial organisms from rodents and larger mammals to man. This tick is also responsible for tick paralysis, which appears to be due to a poisonous salivary secretion injected by the tick; this disease may prove fatal to children.

Among insects beneficial to human beings are those that produce honey, wax, and silk; those that cross-fertilize (pollenize) flowers (thereby producing fruits such as pears, apples, etc.); and those that destroy harmful insects by either devouring or parasitizing them (Figs. 13–32 and 13–33). Many insects are scavengers; they feed on vast quantities of dead animal and vegetable

**Figure 13-32** Predacious insects that benefit human welfare. (After U.S. Department of Agriculture, *Yearbook*, 1952.)

**Figure 13-33** Some insects that show parasitic–host relationships. (After U.S. Department of Agriculture, *Yearbook*, 1952.)

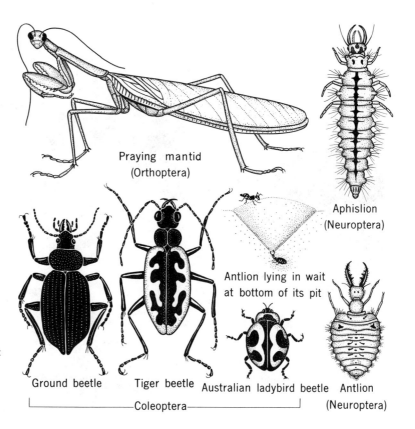
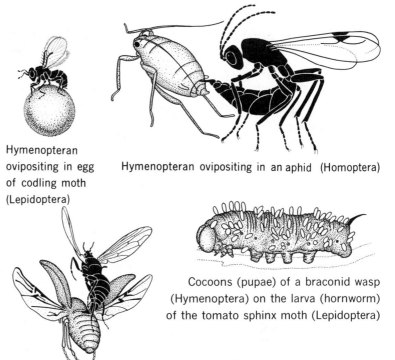

**186**
*Animal Diversity: Invertebrates*

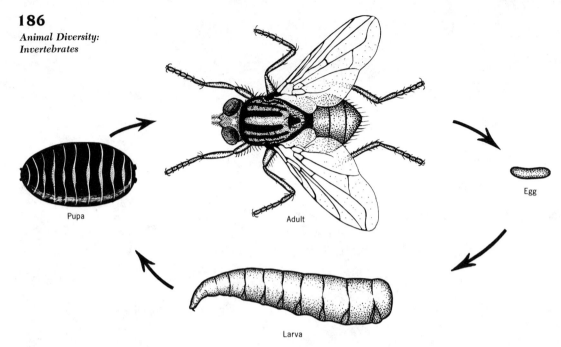

**Figure 13-34** Life cycle of the housefly.

**Figure 13-35** Some carnivorous insects that feed on domestic animals.

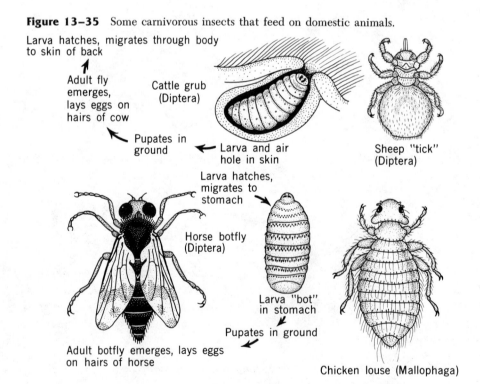

materials. Houseflies (Fig. 13–34) and various types of beetles, including the sacred scarab of the Egyptians, are all scavengers. Disease-carrying insects are one of our greatest enemies (Fig. 13–37); they have a profoundly adverse effect on human welfare. Some species of the following insect groups are among the most harmful: mosquitoes, tsetse flies, fleas, body lice, and houseflies.

Insect pests, according to estimates by the U.S. Department of Agriculture, do about $4 billion damage annually to farm crops, forest, stored foodstuffs, and domestic animals (Fig. 13–35). Our federal and state governments and educational institutions recognize the necessity of controlling injurious insects, and hence economic entomology has become one of the most important activities in the scientific field. Departments of health devote a considerable part of their funds and efforts to insect control, especially of household flies and mosquitoes (Fig. 13–36).

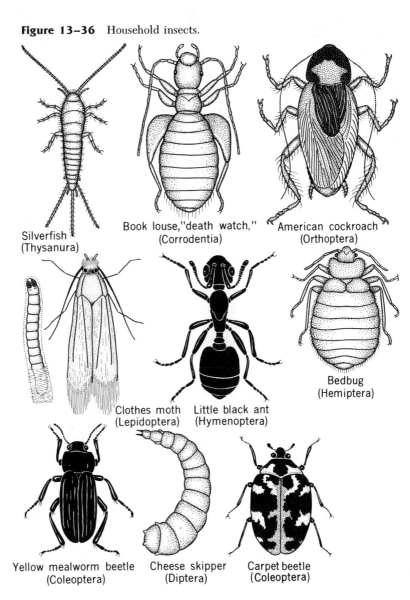

**Figure 13–36** Household insects.

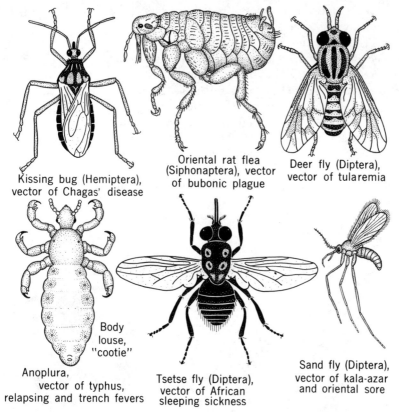

**Figure 13–37** Some insects that transmit human disease (vectors).

## Summary

Arthropods are joint-footed, segmented animals related to the annelids. They have invaded many habitats (sea, land, air) and owe their success largely to their exoskeleton. In advanced groups, tagmosis has reduced the number of segments and the appendages have become modified and less numerous.

Most crustacean arthropods live in water and breathe by means of gills. Their longitudinally segmented bodies are divided into either a cephalothorax and abdomen or a head, thorax, and abdomen. Arachnids possess a cephalothorax and an abdomen, with six pairs of appendages. They include spiders, king crabs, scorpions, harvestmen, mites, and ticks.

There are more species of insects than any other group of animals. Their success is largely due to their exoskeletons, which are protective, supportive, lightweight, and adaptable to new conditions, enabling them to disperse and inhabit new environments easily. All the systems characteristic of higher animals are represented in the insects.

# 14

# Mollusks

Mollusks are soft-bodied animals; members of this phylum include snails, slugs, clams, mussels, oysters, octopi, and squids. All except one class are unsegmented. Most have bilateral symmetry and protective, supportive shells primarily made of calcium carbonate. Although the various types of mollusks are widely different in appearance, they can all be reduced to a basic body plan. Certain structures, such as the **foot,** are found in all mollusks and vary in function from species to species. Snails, for example, use the foot for creeping over surfaces, clams for plowing through mud, and squids for seizing prey.

In Chapter 13 we saw that the arthropods owe much of their success to their exoskeleton; the mollusks are also a highly successful group, but it is not clear what attributes are responsible for this success. The presence of a variety of protective valves (shells) and the fact that shell-less forms have highly specialized alternate defenses suggest that these structures are of primary evolutionary advantage to mollusks.

The mollusks are divided into six classes (Fig. 14–1): **Pelecypods**—clams, oysters, and scallops; **Cephalopods**—squids, octopuses, and nautiluses; **Gastropods**—snails and slugs; **Polyplacophores**—chitons; **Scaphopods**—tooth shells; and **Monoplacophores**—bearing one flat shell. In this chapter we will present some general characteristics of mollusks, followed by a detailed study of a pelecypod, the freshwater clam, and a brief discussion of representatives from other mollusk classes.

## General Characteristics of Mollusks

The bodies of mollusks (Fig. 14–2) are generally covered by a moist integument. They are therefore best suited for aquatic or moist habitats. The **mantle,** which secretes the **shell,** is a fold of the body wall. If there are two lobes, as in the mussel, a bivalve shell is produced. If only one lobe is present, as in snails, a univalve shell is formed. Between the mantle and the body is a **mantle cavity.** A reduced coelom in the form of a pericardial cavity and a cavity for reproductive organs occurs in the adult.

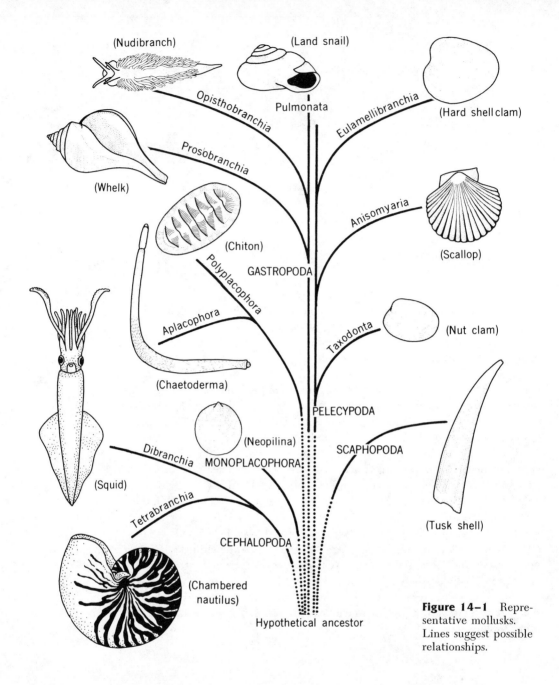

**Figure 14–1** Representative mollusks. Lines suggest possible relationships.

Mollusks may be herbivorous, carnivorous, filter feeders, or both. **Jaws** are present in most gastropods and all cephalopods. A rasping organ (**radula**) is found (usually in the mouth cavity or pharynx) in all mollusks except the bivalves. It consists of rows of chitinous teeth which tear up the food as they are drawn across it.

**Respiration** takes place primarily in the gills and mantle. Most freshwater and land snails (pulmonate gastropods) take air into the vascularized mantle cavity; others breathe cutaneously.

The sexes are usually separate, although certain groups are hermaphroditic. Most mollusks produce a great number of eggs (the oyster produces nearly 500

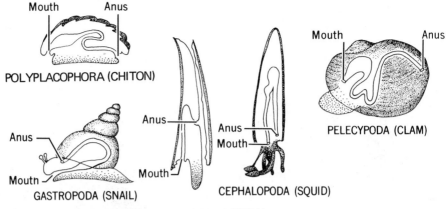

**Figure 14-2** Molluscan body plan and its modifications. Note varying position of shell (heavy lines), foot (stippled), and digestive tract.

million in a single season); these are subject to the dangers of ocean currents and numerous enemies. After hatching, mollusks pass through a metamorphosis which usually includes a **trochophore larval stage** that develops into a **veliger larva,** named for a band of cilia (or **velum**) located in front of the mouth (Fig. 14-3). The velum is an organ of locomotion and is partially responsible for the dispersion of the species.

## Anodonta—A Freshwater Clam

Clams usually lie partly buried in the muddy or sandy bottoms of lakes or streams. They burrow and move from place to place by means of a **foot** that can be extended from the anterior end of the **shell.** Water, loaded with oxygen and food material, is drawn through the **ventral** or **incurrent siphon,** a slitlike opening at the posterior end. Excretory substances and feces, along with deoxygenated water, are carried out through a smaller **dorsal** or **excurrent siphon** (Fig. 14-4).

**Figure 14-3** Two stages in development of a mollusk.

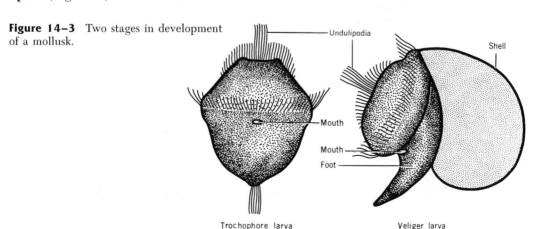

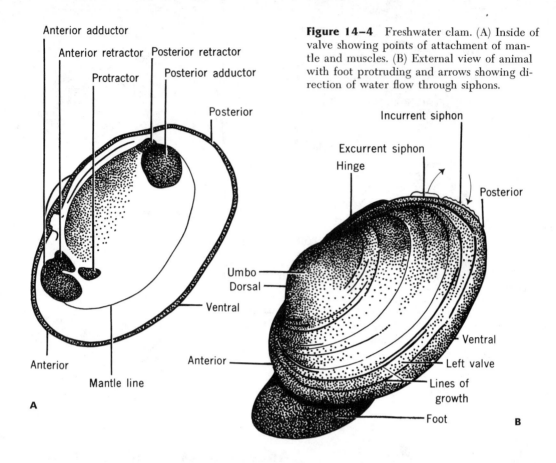

**Figure 14–4** Freshwater clam. (A) Inside of valve showing points of attachment of mantle and muscles. (B) External view of animal with foot protruding and arrows showing direction of water flow through siphons.

The **shell** consists of two halves or **valves.** Concentric ridges (**lines of growth**) appear on the outside of each valve, representing the intervals of rest between successive periods of growth; the annual lines are the most conspicuous. The **umbo** is the first part of the shell to develop and is produced in the late veliger stage; it is usually corroded by carbonic and humic acids in water.

The outer epithelium of the mantle secretes the shell. The shell consists of three layers: an outer, thin, and horny **periostracum,** which protects the underlying layers and gives the exterior part of the shell most of its color; a middle portion of calcium carbonate, the **prismatic layer;** and an inner **nacreous layer** (mother-of-pearl), which is made up of many horizontal layers of calcium carbonate, producing an iridescent sheen (Fig. 14–5).

## ANATOMY AND PHYSIOLOGY

The valves of the shell are held together by two large transverse **muscles,** called **anterior** and **posterior adductors,** and a dorsal elastic tissue hinge. Two folds of the dorsal wall (**mantles**) line the valves; between them is the **mantle cavity,** containing two pairs of gill plates, the foot and the visceral mass.

DIGESTION  Food is brought into the mantle cavity by water circulating through the ventral incurrent siphon. The food consists of minute plants, animals, and debris; these adhere to mucus covering the gills and are carried by undulipodial beating to the **labial palps** (Fig. 14–6), which surround the mouth and serve as a sorting mechanism for foods to be utilized. Foods selected are then carried into the **mouth.** From the mouth, food passes through a short

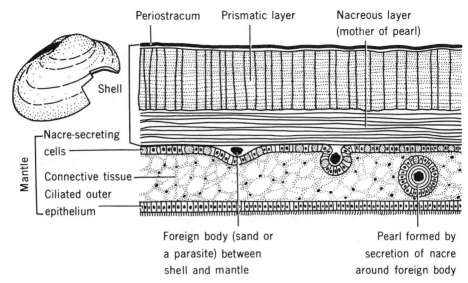

**Figure 14–5** Enlarged cross section of shell and mantle of a freshwater clam.

esophagus into a bulbous **stomach,** which is connected by ducts with a large **digestive gland** or "liver." This gland surrounds the stomach and is the chief source of digestive enzymes. The **intestine** loops through the clam's body, passing near the end through the pericardium and the heart itself, where it becomes the **rectum;** it ends at the **anus,** which opens near the excurrent siphon.

CIRCULATION. The circulatory system consists of a dorsal heart, blood

**Figure 14–6** Internal anatomy of a freshwater clam.

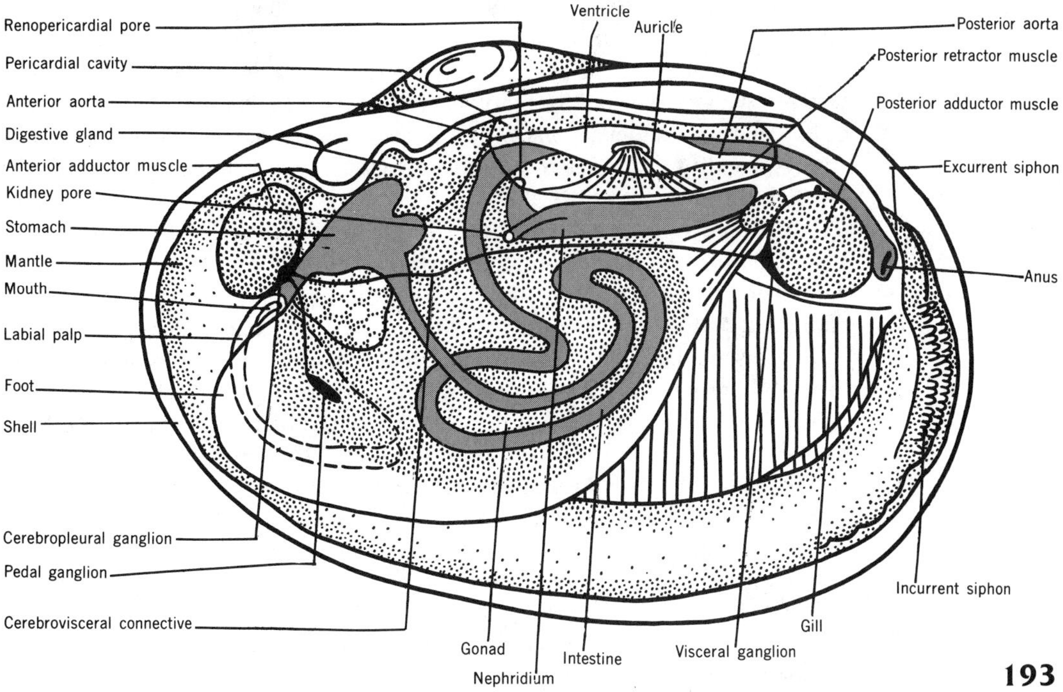

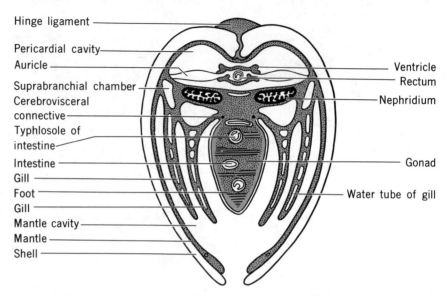

**Figure 14-7** Cross section of a freshwater clam through region of the heart.

vessels, and spaces (or sinuses). The **heart** (Fig. 14-7) lies in the **pericardium.** The **ventricle** drives the blood forward through the **anterior aorta** and backward through the **posterior aorta.** Part of the blood passes into the mantle, where it is oxygenated, and then returns directly to the heart. The rest of the blood circulates through numerous spaces in the body and is collected by a vein just beneath the pericardium. The blood passes from here into the kidneys, the **gills,** and finally through the **auricles** into the ventricle. Nutriment and oxygen are carried by the blood to all parts of the body; carbon dioxide is disposed of in the mantle and gills.

RESPIRATION AND EXCRETION. Although the entire body surface is in contact with water and doubtless functions in respiration, the greater part of the oxygen–carbon dioxide exchange occurs in the gills and mantle. A pair of gills hangs down into the mantle cavity on either side of the foot. Each **gill** consists of two plates, or **lamellae,** made up of a large number of vertical **gill filaments** strengthened by chitinous rods and connected to one another by horizontal bars (Fig. 14-8).

Water enters through the incurrent siphon and flows over the gills, entering many microscopic **water pores** and driven by ciliary action into **water tubes.** The water tubes of each gill join in a **suprabranchial chamber;** from these chambers, water is expelled through the excurrent siphon. Exchange of respiratory gases takes place through the walls of blood spaces located in the **interlamellar partitions,** which separate the water tubes.

The two **kidneys (nephridia)** lie just beneath the pericardial cavity. Each is folded upon itself and differentiated into glandular (dark spongy mass) and bladderlike portions. One end of the nephridium opens into the pericardial cavity and the other into a suprabranchial chamber. Liquid wastes within the pericardial cavity may enter the tubule; metabolic wastes carried by the circulating blood may be removed by the cells in the glandular portion of the nephridium.

NERVOUS SYSTEM AND SENSE ORGANS. Three pairs of ganglia are present: **cerebropleural ganglia, pedal ganglia,** and **visceral ganglia.** These are con-

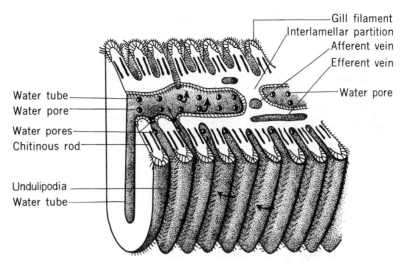

**Figure 14–8** Respiratory system of a freshwater clam. The horizontal section through a gill shows the arrangement of gill filaments, blood vessels, and water tubes. Arrows indicate the direction of water currents.

nected to each other by nervous connectives. The sensory structures include **light receptors** in the siphon margins; an organ of equilibrium (**statocyst**) containing a calcareous concretion (**statolith**) lies just posterior to the pedal ganglia. A thick patch of yellow epithelial cells (**osphradium**) covers each visceral ganglion; these may be useful for detecting foreign materials in water. The edges of the mantle are provided with sensory cells that are probably sensitive to contact and light.

Reproduction. Clams are usually either male or female, a few are hermaphroditic. The reproductive organs are situated in the visceral mass. Many freshwater clams are parasites of fish in a young developmental stage. The eggs develop into larva known as **glochidium,** a modified veliger (Fig. 14–9). In *Anodonta* the eggs are usually fertilized in August and the glochidia that develop from them remain in the gills of the mother all winter. In the following spring they are discharged and seek to attach themselves to fish. The skin of the fish grows around them, forming "blackheads." After a parasitic life within the tissues of the fish (from three to many weeks), the young clam is liberated. As a result of this parasitic habit, clams are widely dispersed by the migrations of fish.

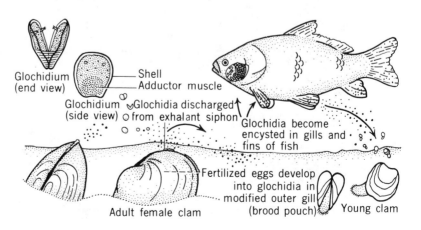

**Figure 14–9** Life cycle of a freshwater clam.

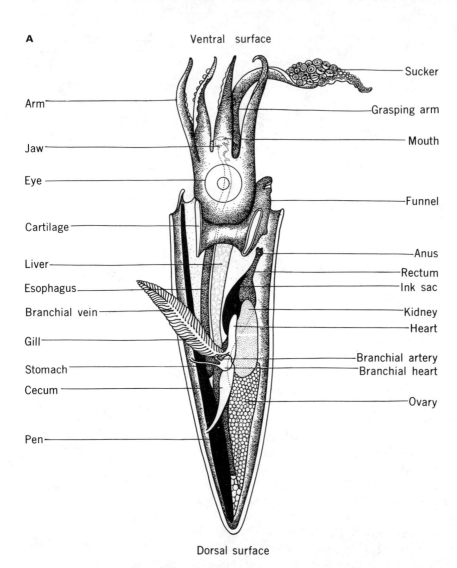

**Figure 14-10** A. Internal structure of a squid. B. *Loligo*, a common North American squid.

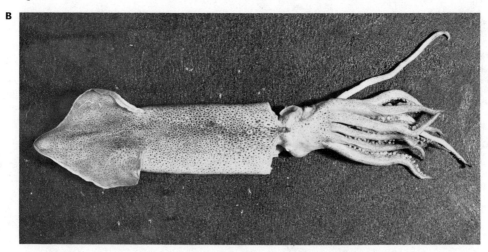

# Other Mollusks

## SQUIDS, OCTOPUSES, AND OTHER CEPHALOPODS

The cephalopods are the most highly developed class of mollusks.

A common squid along the eastern coast of North America is *Loligo pealei*. The foot of this species consists of 10 arms, which bear suckers and are used for capturing prey; and a **funnel (siphon)**, which propels the squid in an opposite direction from its aim by forcing jets of water to pass through it (Fig. 14–10). The mantle in the posterior region is extended into triangular **fins**, which may propel the squid slowly forward or backward or enable it to "hover" by undulatory movements.

The **pen (shell)** is a feather-shaped plate concealed beneath the skin of the back. The true **head** is the short region between the arms and the **mantle collar**; it contains two large eyes. The **digestive system** includes a muscular **pharynx** (buccal mass), **esophagus, salivary glands, stomach, cecum, intestine, rectum, liver,** and a small **pancreas**. There are two powerful horny **jaws** in the pharynx; a **radula** is also present. Above the rectum is the **ink sac**, with a duct that opens near the **anus**. When the squid is attacked, it emits a cloud of inky fluid through the funnel, discharging just enough ink to color a volume of water its own size. An enemy in pursuit often mistakes the ink for the squid; meanwhile the squid escapes.

The blood is contained in a double, closed vascular system. Two **gills** and two **nephridia** are present. The **nervous system** consists of a number of ganglia,

**Figure 14–11** The octopus is a cephalopod without a shell. (Courtesy D. P. Wilson, Plymouth, England.)

mostly in the head. The **sensory organs** are two highly developed eyes, two **statocysts,** and probably two **olfactory organs.**

Squids are especially famous for their color changes. Pigment cells (chromatophores) filled with blue, purple, red, and yellow pigments are present in the skin; when these become larger or smaller, the color changes rapidly.

Near the coast of Newfoundland, **giant squids** are occasionally encountered. These may be 15 meters or more in total length, with arms as large as a man's legs, and are the largest living invertebrate animals known.

**Octopuses** (Fig. 14–11) live in dark crevices and in coral reefs. Most of them are not large enough to harm a human being, but the giant octopus of the Pacific reaches a diameter of about 30 feet and can be dangerous.

## SNAILS AND OTHER GASTROPODS

Gastropods (Fig. 14–12) live in fresh or salt water and on land. A few are parasitic on other animals. Land snails (Fig. 14–13) must protect themselves from drying at certain seasons; they retire into their shells as far as possible and secrete a parchmentlike wall (epiphragm) across the opening to prevent evaporation. The snail moves by wavelike contractions of the foot muscles along a pavement of mucus (secreted by the slime gland) which is laid out ahead of it.

Freshwater snails are numerous in creeks and pools. Snail shells may coil in two directions: **dextral** (clockwise) or **sinistral** (counterclockwise). Some freshwater snails possess gills with which they breathe under water. Others are pulmonates, with a "lung" cavity and must come up to the surface for air when the water is warm; at other times cutaneous respiration may be adequate.

**Land slugs** (pulmonates) are closely related to the land snails, but have no covering. **Whelks** and **periwinkles** are among the commonest of the smaller marine snails. The largest is the **queen conch,** with a shell up to 4 dm long.

**Figure 14–12** The beautiful chambered nautilus is an ancient cephalopod with a coiled shell. The shell is divided by septa into chambers. The density of the fluid within these chambers is controlled by the animal for bouyancy.

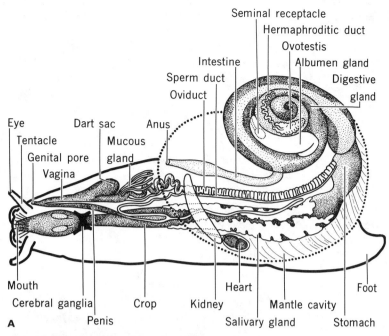

**Figure 14–13** (A) Internal structure of a snail. (B) The common terrestrial garden snail, *Helix*.

Many gastropods—land, freshwater, and marine—serve as intermediate hosts for various trematodes, such as the blood and liver flukes of man.

## THE PRIMITIVE DEEP-SEA MOLLUSK,
### *Neopilina galatheae*—A MONOPLACOPHORAN

The Danish "Galatheae Expedition" dredging off the west coast of Costa Rica found among the many animals collected some extraordinary unidentified deep-sea mollusks. This new limpetlike mollusk, *Neopilina galatheae*, was identified and reported by H. Lemche in February 1957.

*Neopilina* is a living representative of the class Monoplacophora. The class name was originally proposed to cover an extinct group of primitive mollusks dating from the Paleozoic period (270 to 600 million years ago). This recent addition is radically different from all other mollusks in that it is internally segmented, violating one of the basic criteria by which mollusks are characterized (an unsegmented body plan). It therefore occupies a unique and evolutionarily remarkable position—a living "missing link" between the mollusks and the annelids. Further study of *Neopilina* should throw increasing light on our study of molluscan evolution.

### CHITONS

Chitons are of the class Polyplacophora; they are also known as **coat-of-mail shells.** Their shells possess eight transverse plates which are arched above and overlapping; they are not, however, evidence of metamerism. When detached from rocks, to which they cling by means of suction on their foot, chitons roll up like an armadillo, with the soft parts practically covered by the hard shell. Chitons are mostly found in water less than 25 fathoms deep along rocky portions of the seashore.

### TOOTH OR TUSK SHELLS

These animals from the class Scaphopoda partially bury themselves in mud or sand at the bottom of the sea. A pointed foot aids in burrowing. They possess a tubular shell, which is slightly tapered and open at both ends. Their empty shells were once used as currency by Pacific Coast Indians.

## Origin of Mollusks

The Monoplacophora appear to be the most primitive class of mollusks and have changed the least from the ancestral condition. The gastropods have changed to a short creeping type, with the visceral mass and associated structures rotated 180° on their body. The pelecypods probably branched off from the remaining classes at an early date; they became flattened laterally and developed a large bilobed mantle that secretes a shell of two valves, a large mantle cavity containing gills, a burrowing foot in place of the creeping type, and no true head. The cephalopods have become free-swimming animals; the foot is modified into prehensile tentacles and the brain and sense organs are highly developed.

The dominant view of the relation of mollusks to other phyla is based on the presence of the trochophore larvae among both mollusks and annelids, which indicates that they may be derived from the same ancestral type.

## Relations of Mollusks to Man

Freshwater clams are scavengers, continually ingesting organic particles and thus filtering the water in which they live. Some freshwater mussels are used as indicators of toxic contaminants and other environmental conditions of water.

Bivalve mollusks are especially well known as food sources; the Indians used

**Figure 14–14** Pearl in a freshwater clam; arrow points to pearl. (Courtesy F. L. Clark.)

oysters and freshwater clams in great quantities long before white settlers came to America, evidenced by the many piles of shells found around former camping grounds. Other mollusks used as food are: hard and soft shell clams, scallops (only the large adductor muscles are eaten), abalone (the only gastropod popularly used as food), mussels, and snails (both extensively eaten in Europe). Squid, cuttlefish, and octopus are popular in southern Europe and the Orient.

Among the products of value derived from mollusks are pearls and pearl buttons (made from the shells of freshwater mussels). Pearls are primarily found in pearl oysters, *Pinctada*. The Japanese produce cultured pearls by inserting a foreign body into the mantle of the bivalves. Pearls also occur in the common oyster and in clams (Fig. 14–14), but these are not considered valuable.

Overfishing of mussels has seriously depleted them, and stream pollution has greatly damaged their (and our) environment; many of the world's richest freshwater mollusk fauna (of the great Mississippi river system) are now threatened with extinction.

**Figure 14–15** Damage caused by shipworms to wharf pile section, done in less than 18 months. (Courtesy William F. Clapp Laboratories.)

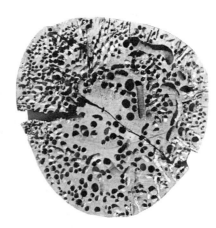

Most mollusks may be considered beneficial to human beings, although some are not. Slugs are sometimes injurious in greenhouses and gardens; shipworms burrow into the bottom of wooden vessels, weakening and destroying them (Fig. 14–15). Some of the large octopi may be dangerous. An important destructive role of mollusks is that of the freshwater snail, which hosts trematode diseases of human beings and their domestic animals.

## Summary

Mollusks are a highly successful group of invertebrates possessing some features unique in the animal kingdom, such as a soft body with a shell-secreting mantle and a cephalized muscular foot. These and other molluscan characteristics have permitted great morphological variability around a common basic "body plan," present in spite of apparently vast differences (such as exist between a snail and an octopus). This adaptability of form to physical and biological requirements has led mollusks to a great variety of habitats, marked by extremely divergent food-getting adaptations, and hence to evolutionary success.

The mollusks do not have jointed appendages and are not segmented except in the class Monoplacophora, in which the presence of segmentation suggests that mollusks are probably derived from an annelidlike ancestor.

# 15
# Echinoderms

Echinoderms are spiny-skinned marine animals found in abundance at seashores. The best known echinoderm is the sea star (starfish, class Asteroidea); the sea urchin (class Echinoidea) and sea cucumber (class Holothuroidea) are also familiar to the seashore visitor. These animals, plus the sea lilies (class Crinoidea) and the brittle star (class Ophiuroidea), comprise the five living classes of the phylum Echinodermata (Fig. 15–1).

Characteristics of echinoderms include a water-vascular system that uses organs known as tube feet, internal skeletons of calcareous plates, and the transition from bilaterally symmetrical larvae to radially symmetrical adults; their eggs are especially suitable for various experimental studies of embryology. Echinoderms exhibit remarkable powers of autotomy and regeneration of lost parts.

## Asterias—A Sea Star

### EXTERNAL ANATOMY

Sea stars provide the best example of the basic body plan and general biology of echinoderms. *Asterias* does not have a head. Five **rays** (each with an **ambulacral groove** on the oral surface) extend from a centrally located mouth. Two or four rows of **tube feet** extend from each groove. Tiny pincers called pedicellariae are used for protection against larvae which might settle on the body surface.

### INTERNAL ANATOMY AND PHYSIOLOGY

**WATER-VASCULAR SYSTEM.** The water-vascular system is a division of the coelom peculiar to echinoderms (Fig. 15–2). Beginning at the entrance with the **madreporite**, the following structures are encountered: the **stone canal**, running downward, enters the **ring canal**, which encircles the mouth. From this canal, five **radial canals**, one in each ray, pass outward just above the **ambulacral grooves**. The radial canals give off side branches from which arise the **tube feet** and **ampullae**. There are nine small spherical swellings on the inner wall of the ring canal, the **Tiedemann's bodies**. Although their function has not been definitely established, evidence points to involvement in amoebocyte production.

The water-vascular system is a hydraulic pressure mechanism for locomo-

**Figure 15–1** (A) Representatives of the living classes of echinoderms. (B) Dredge haul of echinoderms.

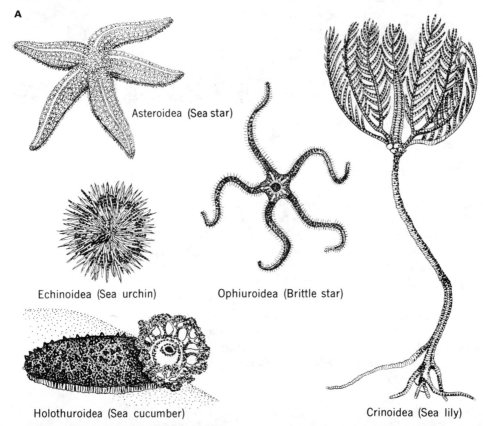

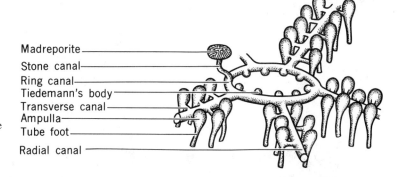

**Figure 15–2** Sea star water vascular system; one of the radial canals is cut off at the base, the other four near the base.

tion. The total effect of the forward thrust of the many tube feet produces movement in the sea star. All the tube feet act in a coordinated way by extending in the same direction, although not all at the same time. The sea star advances slowly, only about 6 inches per minute.

Tube feet are also used for clinging to rocks, capturing and handling food, excretion of urine, and for respiration.

**DIGESTION AND EXCRETION.** The digestive tract is short and greatly modified. A centrally located **mouth** (Fig. 15–3) opens into a very short **esophagus,** which leads into a thin-walled sac, the **stomach.** The stomach consists of two parts—a larger oral or **cardiac** portion and a small aboral or **pyloric** portion. From the pyloric portion a tube passes into each **ray** and divides into two green branches, the **pyloric** or **hepatic ceca,** which have large numbers of lateral pouches. Above the stomach is the slender **intestine,** which opens to the outside through an **anus.** Two branched brown pouches, the **intestinal ceca,** arise from the intestine.

The food of sea stars consists primarily of fish, oysters, mussels, clams, snails, and worms. Small food may be passed to the mouth by epidermal undulipodia, or by tube feet. Larger bivalves are straddled and opened by the sea star, after which part of the stomach is everted. Digestive enzymes from the pyloric ceca begin extracellular digestion of the prey, and the stomach is retracted with its contents. Undigested matter is ejected through the mouth (Fig. 15–4).

Excretion of solid wastes is accomplished by **amoebocytes** in the coelomic fluid, which pass to the outside of the body through the walls of the dermal branchiae.

**RESPIRATION.** Respiration is carried on by the **dermal branchiae (papulae)** on the aboral surface of the rays. Their soft and furry appearance is caused by outpouchings of the thin lining of the body cavity through minute openings in the skeleton. These dermal branchiae are covered with cilia on both the inside and the outside. The external undulipodia keep a current of oxygenated water passing over the branchiae on the outside, and the internal undulipodia cause the body fluid to flow out into the branchiae. While the body fluid is in the branchiae, an exchange of oxygen and carbon dioxide takes place exactly as it does in our own lungs when the blood flows past the tiny air sacs in them. Much respiration also occurs through the tube feet.

**NERVOUS SYSTEM AND SENSE ORGANS.** The sea star lacks a brain; much of the animal's activity is under local control and is coordinated for activities such as locomotion and feeding. Many nerve cells lie among the epidermal cells; in addition, ridges of nervous tissue (radial nerve cords) run along the ambulacral

**Figure 15–3** General structure of the sea star. The bottom portion shows one ray in cross section without gonads.

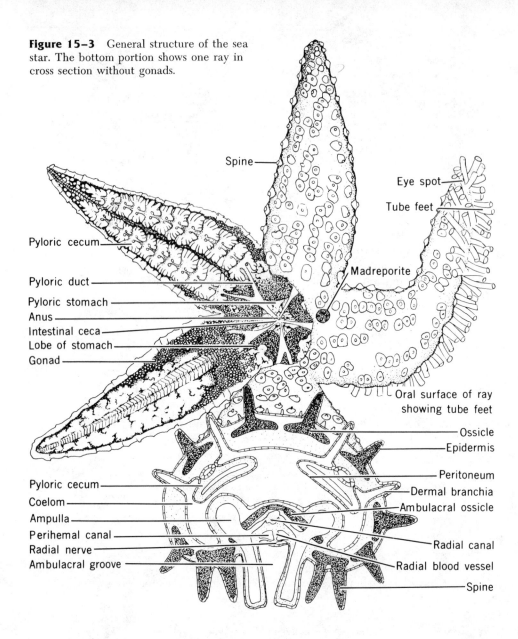

grooves and unite with an **oral nerve (circumoral) ring** encircling the mouth. Each ray possesses a radial nerve cord, a pair of nerves, and a nerve cord in the aboral peritoneum. The **tube feet** are the principal sense organs, receiving fibers from the radial nerves. At the end of each ray is a small, soft tactile tentacle and a light-sensitive eyespot. The dermal branchiae probably also have a sensory function.

REPRODUCTION. The sexes of sea stars are separate. The reproductive organs are branched structures, two in the base of each arm. The female has been known to release as many as 2.5 million eggs in 2 hours, and 200 million eggs may be liberated in a single breeding season. A male produces many times that number of sperm. The eggs of many starfishes are fertilized in the water and develop into larvae called **bipinnarias,** which have bilateral symmetry before they metamorphose into radially symmetrical young sea stars (Fig. 15–5).

**Figure 15-4** Sea star eating a mussel. The sea star attaches its tube feet to the two shells of the mussel and, by a continuous pull, eventually opens it enough to allow the stomach of the sea star to evert through its own mouth, coming in contact with the softer parts of the mussel.

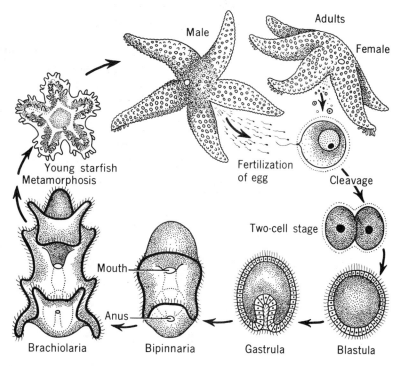

**Figure 15-5** Life cycle of the sea star. Note that in later larval stages the animal has bilateral symmetry before attaining radial symmetry.

## REGENERATION

The sea star has remarkable power to regenerate lost parts. A single arm with part of the disk will regenerate an entire body. In all species studied, arms cut off at any level are regenerated, although at a slow rate. If an arm is injured, it is usually cast off near the base at the fourth or fifth ambulacral ossicle. This autotomous ability allows sea stars to fend off predators by leaving them an arm (which might appease the predator's appetite) while the rest of the sea star escapes.

## Other Asteroidea

Three orders and about 20 families of sea stars are recognized; about 1200 species are known. Usually five or multiples of five rays are present. These rays may be long or short. Sea stars are common marine animals all over the world and are present in both shallow and deep water.

## Other Echinodermata

**Brittle stars** and **basket stars** (class Ophiuroidea) have slender or branched flexible rays which bend like whips, enabling certain of these animals to "run," cling, and swim. Locomotion is thus comparatively rapid. Tube feet have lost most of their locomotor function and serve as sensory and respiratory organs. Food consists of minute organisms and decaying organic matter lying on the sea bottom. The term **brittle star** is derived from the autotomous breaking off of arms when these animals are injured.

A common type of **sea urchin** (class Echinoidea) is the purple *Strongylocentrotus purpuratus,* occurring in both shallow and deep water along the Pacific Coast. It is somewhat globular in shape (Fig. 15-6). The **test** (shell) is made up of calcareous plates which bear movable spines about 25 mm long. Their food consists of plant and animal matter and is ingested by means of a complicated structure known as **Aristotle's lantern.**

The **sea cucumber** (class Holothuroidea) is different from other echinoderms, having an elongated body and lying on its side (Fig. 15-7). Brown, yellow, red, white, black, pink and purple are among their varied color tones. Sea cucumbers live sluggishly on the sea bottom or burrow in the surface mud or sand with only the ends exposed. Instead of a test or skeleton of spine-bearing plates, the soft but muscular body wall contains numerous microscopic calcareous plates. The food of most sea cucumbers consists of organic particles extracted from the sand or mud by 10 to 30 tentacles (modified from the tube feet) which surround the mouth.

There are about 630 living species of sea lilies (class Crinoidea) living in shallow and deep waters. Some are attached to the sea bottom by a long jointed **stalk.** Their rays (five or 10) are often branched near the base and bear smaller branches (**pinnules**) along their sides, giving them a feathery appearance. Fossil remains of crinoids are very abundant in some limestone formations.

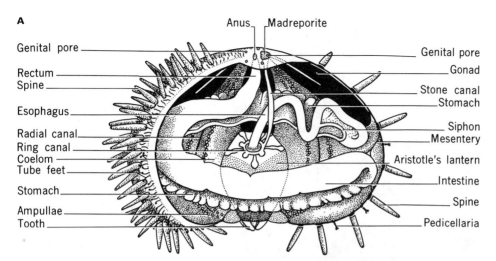

**Figure 15-6** (A) Sea urchin, *Arbacia*, showing both internal and external structure. (B) Sea urchin test.

## Origins of Echinodermata

The oldest echinoderm fossils are crinoidlike filter feeders attached to the substrate with the oral surface up. The tube feet used for feeding by these sessile animals became modified for locomotion as the echinoderms became unanchored and inverted.

Echinoderms show several features common to the chordates and are considered their closest invertebrate relatives. These include the presence of an ectodermal neural system, an endoskeleton, and some embryological similarities.

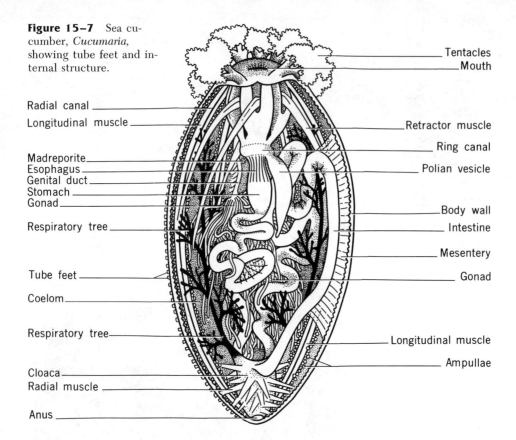

**Figure 15-7** Sea cucumber, *Cucumaria*, showing tube feet and internal structure.

## Relations of Echinodermata to Man

Echinoderms are of considerable importance to man in a variety of ways. In the Orient, sea cucumbers are dried and sold as foods and are especially used in soups. The gonads of sea urchins and the eggs of sea stars are also eaten in certain tropical regions, and recently in Pacific Coast port cities.

The dried skeletons of echinoderms have been crushed and used as fertilizer because of their high calcium and nitrogen content.

Sea stars are very destructive to oyster beds; a sea star has been observed to eat 10 oysters or clams in one day. Control measures to eliminate these harmful animals are in general use at the present time.

A coral-eating sea star (*Acanthaster planci*) has caused much excitement because of its recent spread in the tropics. One theory holds that the phenomenon is a natural periodic occurrence (every 100 years), while another states that it was precipitated by the building of harbors; neither of these theories have been definitely confirmed.

## Summary

Echinoderms are radially symmetrical as adults but bilaterally symmetrical as larvae. There is no segmentation. The body wall usually contains calcareous plates that form an endoskeleton. The nervous system is close to the ectoderm and has an oral nerve ring and radial nerves. Sexes are usually separate. A unique water-vascular system, including tube feet, is generally present.

# PART THREE

# Animal Diversity: Vertebrates

# 16

# Chordates

The chordates are a very large group, comprised of various forms of marine animals, fishes, amphibians, reptiles, birds, and mammals. At some stage in their development, all these animals possess (Fig. 16-1) the following body parts:

1. A **notochord** or skeletal axis, replaced in higher chordates by a **vertebral column.** The skeletal elements are alive and grow internally, thus overcoming the necessity for molting and accommodating animals of a much larger size.
2. A pair of **gill slits** that connect the pharynx with the exterior. All chordates up to and including the fishes carry on respiration by means of gills throughout life. In the higher vertebrates, traces of gill slits are usually only present in embryonic or larval stages. Elements of the gills are modified into respiratory systems.
3. A **hollow nerve chord** dorsal to the digestive tract.

These characteristics may persist, change, or disappear in the adult chordate. Figure 16-1 shows some of the fundamental differences in the body plan of an achordate and a chordate.

Some zoologists think that the chordates arose from a segmented group such as the annelids or arthropods, since chordates are also segmented. There appears however, to be more similarities between the chordates and the echinoderms. Echinoderms possess an endoskeleton, bilateral larva, and many embryological similarities to the chordates.

**Figure 16-1** Body plan of an achordate invertebrate (top) and a chordate (bottom), showing fundamental differences in location of digestive system, heart, and nervous system. Arrows indicate direction of blood flow.

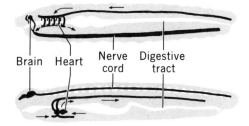

213

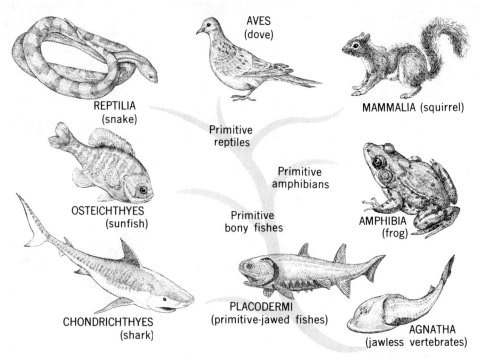

**Figure 16–2** Simplified family tree of vertebrates.

## Divisions Within the Chordate Group

The chordates differ widely from each other, and it is customary to separate them into subphyla, and then into classes (Fig. 16–2).

## Hemichordates

The **hemichordates** include two classes of wormlike animals: the **enteropneusts** or **acorn worms** (also known as tongue worms) (Fig. 16–3) which have many gill slits; and the **Pterobranchia**, very small chordates with one or no paired gills.

The hemichordates may be considered modern representatives of an ancient "link" between the echinoderms and the chordates. They possess features common to both groups. Their echinodermlike features include a partially subectodermal nervous system, similar larval stages, water-vascular-like coelomic cavities, a similar marine habitat, and similar patterns of filter feeding. Chordatelike features include a partial dorsal hollow nerve chord, gill pouches (unlike chordates, these are used primarily for feeding), and a structure that may or may not be a notochord (Fig. 16–4). Many zoologists question the presence of a true notochord in the hemichordates; they are sometimes placed in a separate phylum.

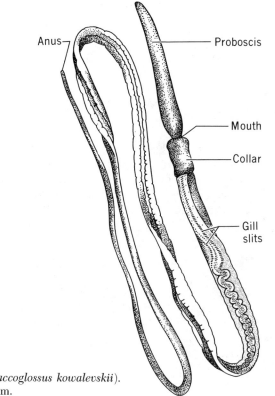

**Figure 16–3** Acorn worm (*Saccoglossus kowalevskii*). Its natural length is about 17 cm.

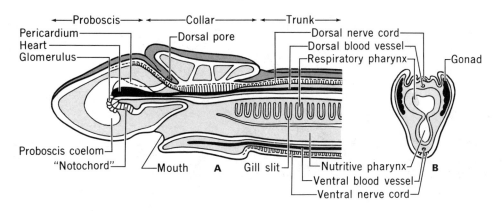

**Figure 16–4** Acorn worm (*Balanoglossus*). (A) Midsaggital section, showing some internal structures. (B) Cross section through trunk region.

# Protochordates

Commonly accepted as chordate subphyla are the **Urochordata** (Tunicates), including sea squirts and a number of other marine forms; and **Cephalochordata,** two families of fishlike animals called lancelets. These subphyla are often referred to as **protochordates;** they are relatively primitive and may be the animals from which the final and largest chordate subphylum, the **vertebrates,** evolved. The vertebrates all possess a **cranium** and **vertebrae**. They include fish, amphibians, reptiles, birds, and mammals.

A short discussion on the protochordate subphyla will be followed by an introduction to the vertebrates. Vertebrates will be discussed by class in Chapters 17 through 21.

## Urochordata (Tunicates)

The term **Tunicata** was formerly applied to members of this group because of their leathery cuticular outer covering, known as a **tunic** or **test**. The tunicates are entirely marine, ranging in size from 1 mm to over 4 dm in diameter. They are either free-swimming as larvae or attached as adults and are widely distributed in all levels of oceans. Some species are brilliantly colored.

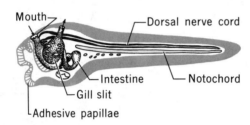

**Figure 16–5** Free-swimming larva of a tunicate, showing notochord, dorsal nerve cord, and gill slits.

**Figure 16–6** Internal structure of a tunicate (*Molgula*). The tunic, mantle, and pharynx are removed from the left side. Arrows indicate the flow of water currents.

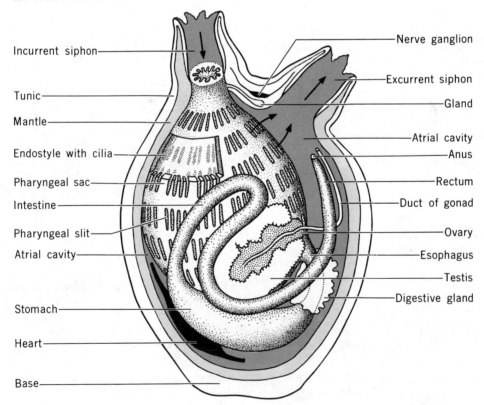

The chordate characteristics of tunicates were not recognized until the development of the egg and metamorphosis of the larva were fully investigated. It was then discovered that the typical larva, which is about $\frac{1}{4}$ inch long and resembles a frog tadpole, possesses a distinct **notochord**; a dorsal **neural tube** in the tail, which is considered to be the forerunner of the brain of the vertebrates; and a **pharyngeal sac**, which opens to the exterior by innumerable undulipodiated **gill slits** (Fig. 16-5). The larva undergoes a **retrogressive metamorphosis**, during which the tail with the notochord disappears and the nervous system is reduced to a ganglion. The typical adult tunicate is attached by a **base** or **stalk** and surrounded by the thick, elastic tunic; the internal anatomy of this animal is shown in Fig. 16-6.

## Cephalochordata—Amphioxus

The Cephalochordata comprise about 30 species of marine animals. *Branchiostoma lanceolatus,* commonly known as amphioxus, lives in waters of tropical and temperate seacoasts. Amphioxus is of special interest because it exhibits characteristics (notochord, gill slits, and a dorsal tubular nerve cord) of the chordates in a simple condition. It may also be similar to some ancient ancestor of the vertebrates.

### EXTERNAL ANATOMY

Amphioxus measures about 5 cm in length. The semitransparent body is pointed at both ends and laterally compressed, with no lateral fins or distinct head (Fig. 16-7). Along the middorsal line is a low **dorsal fin** which extends the entire length of the body and widens at the posterior end into a **caudal fin.** The caudal fin extends forward on the ventral surface to form the short **ventral fin.** Both dorsal and ventral fins are strengthened by rods of connective tissue called **fin rays.** In front of the ventral fin, the lower surface of the body is flattened, and on each side is an expansion of the integument called the **metapleural fold.**

The **body wall** is divided into V-shaped muscle segments, the **myotomes**, arranged alternately on either side of the body and separated by septa of connective tissue, thus producing the lateral body movements used in swimming.

**Figure 16-7** Amphioxus illustrates the three fundamental chordate characteristics.

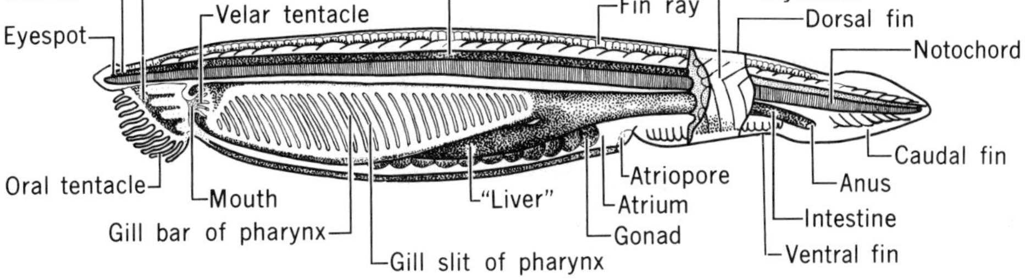

## INTERNAL ANATOMY AND PHYSIOLOGY

**SKELETON.** Amphioxus has a well-developed **notochord,** which is the main support of the body. This rod of connective tissue lies near the dorsal surface and extends almost the entire length of the body. Other skeletal structures are the connective tissue rods that form the fin rays, and similar structures that support the **oral tentacles** (**cirri**) of the **oral hood** and gill bars.

**DIGESTIVE SYSTEM.** Food consists of minute organisms carried into the mouth with a current of water produced by undulipodia on the gills. The **mouth** is an opening in the **velum** (a membrane) and may be closed by circular muscle fibers surrounding it. Twelve **velar tentacles** protect the mouth and fold across it, acting as a strainer to prevent the entrance of coarse, solid objects. The mouth opens into a large, laterally compressed **pharynx** (Fig. 16–8). Food particles are collected in strings of mucus secreted by glands on either side of the **endostyle** or **hypobranchial groove.** Undulipodia in the median portion of the endostyle drive the mucus forward and upward into the **hyperbranchial groove** and posteriorly into the **intestine.** A ventral fingerlike outpocketing of the intestine known as the "**liver**" is an outgrowth of the **midgut.** Although this structure secretes digestive enzymes, it is not known to function as does the vertebrate liver. The intestine leads directly to the **anus.**

In the adult form, the coelom is reduced to coelomic cavities in the pharyngeal region of the digestive tract.

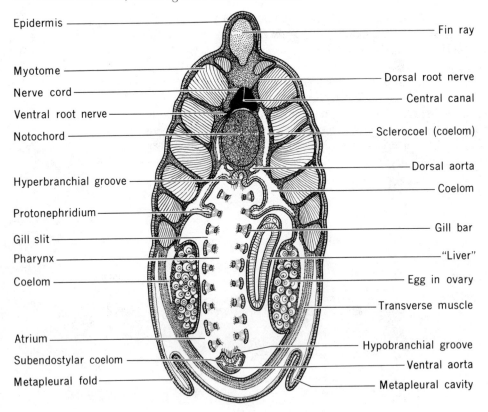

**Figure 16–8** A cross section of an amphioxus through the pharyngeal region shows various internal structures, including some coelomic cavities.

**Respiratory System.** The **pharynx** is attached dorsally and hangs down into a cavity called the **atrium.** The atrium is not a coelom, but is lined with an ectodermal epithelium and is really external to the body. Respiration takes place as water driven by undulipodiated **gill bars** flows through the **gill slits** into the atrium and out of the **atriopore.**

**Circulatory System.** The circulatory system is similar to that of fishes, but lacks a heart. It consists of definite blood vessels and tissue spaces into which the colorless blood escapes. The **subintestinal vein** carries blood loaded with nutriment from the intestine forward to the **hepatic portal vein** and into the liver. The **hepatic vein** leads from the liver to the **ventral aorta.** Blood is forced by the rhythmic contractions of the ventral aorta into the **afferent branchial arteries,** where it is oxygenated as it passes through the gill slits. It then passes through the **efferent branchial arteries** into paired **dorsal aortae** and back into the median dorsal aorta. **Intestinal capillaries** carry it back into the **subintestinal vein.**

The direction of the blood flow, backward in the dorsal and forward in the ventral vessel, is the same as that of the vertebrates.

**Excretory System.** The excretory organs are nephridia (**protonephridia**) situated near the dorsal region of the pharynx. Each nephridium bears several clusters of undulipodiated solenocytes (flame cells) that extend out of a tube in which the undulipodia move. Approximately 100 pairs of protonephridia connect the dorsal coelom with the atrial cavity.

**Nervous System and Sense Organs.** Amphioxus possesses a central **nerve cord** which rests on the notochord and is almost as long. (This is in contrast to the ventral nerve cord of annelids and arthropods.) A minute **central canal** transverses its entire length and widens at the anterior end to form a **brain** or **cerebral vesicle.** In young specimens, an **olfactory pit** opens into this vesicle. Two pairs of sensory nerves arise from the cerebral vesicle and supply the anterior region of the body. Two types of alternating nerves extend from either side of the rest of the nerve cord: **dorsal nerves** have a sensory function and pass to the skin; **ventral nerves** have a motor function and enter the myotomes.

The sense organs include the olfactory pit; 12 oral **velar tentacles,** and 22 **undulipodiated tentacles** (provided with **sensory cells**); "eyes," consisting of numerous light-sensory cells, and a black pigmented "**eyespot,**" which is not sensitive to light and is of questionable function.

**Reproduction.** In amphioxus the sexes are separate. The paired gonads project into the atrium. Eggs and sperms are discharged into the atrial cavity and reach the exterior through the atriopore. Fertilization takes place externally in the water.

# Vertebrata

The vertebrates are chordates with a segmented backbone or vertebral column. A notochord, present in a developmental stage in all vertebrates, persists in some of the lower forms, modified by a buildup of cartilage, which becomes segmented and forms the vertebral column (Fig. 16–9). In the higher vertebrates, the vertebral column is made up of a series of bones (vertebrae) and the notochord disappears before the adult stage is reached. Nine classes of vertebrates are recognized, including two that are extinct.

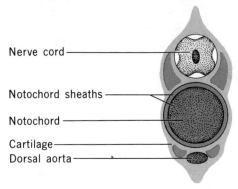

**Figure 16–9** Notochord of young dogfish.

## BASIC BODY STRUCTURE

Vertebrates are similar to other chordates in exhibiting metamerism and bilateral symmetry and in possessing the basic chordate structures (notochord, gill slits, and a dorsal, hollow nerve cord). They differ from other chordates in possessing cartilaginous or bony **vertebrae**, usually two pairs of **appendages**, an internal and jointed **skeleton**, a **ventrally situated heart** (with at least two chambers), and **red blood corpuscles**.

The body of a vertebrate (Fig. 16–10) may be divided into a **head** and a **trunk**, usually separated by a **neck**. In many species there is a posterior extension, the **tail**. Two pairs of **lateral appendages** are generally present, the **thoracic** (pectoral fins, forelimbs, wings, or arms) and the **pelvic** (pelvic fins, hindlimbs, or legs). The limbs support the body, serve in locomotion, and usually have other special functions.

The nerve cord is dorsal and extends anteriorly past the notochord, where it enlarges into a **brain**. The **notochord** becomes invested by the vertebrae. The **coelom** is large; the **digestive canal** is a more-or-less convoluted tube lying within the coelom. The **liver, pancreas,** and **spleen** are situated near the digestive canal. In the anterior trunk region are the **lungs** and **heart**. The **kidneys** and **gonads** lie above the digestive canal.

The vertebrates may be grouped in a number of ways. The **Agnatha** lack

**Figure 16–10** Longitudinal section of generalized vertebrate, showing plan of structure.

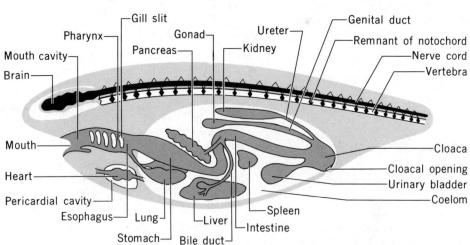

**Table 16-1** General Organization of the Vertebrates

| Infraphylum | Superclass | Class |
|---|---|---|
| I. Agnatha | | 1. Ostracodermi—extinct armored fishes |
| | | 2. Cyclostomata—lampreys and hagfishes |
| II. Gnathostomata | (a) Pisces | 3. Placodermi—extinct fishes with primitive jaws |
| | | 4. Chondrichthyes—sharks, rays, and skates |
| | | 5. Osteichthyes—bony fishes |
| | (b) Tetrapoda | 6. Amphibia—frogs, toads, salamanders |
| | | 7. Reptilia—turtles, lizards, and snakes |
| | | 8. Aves—birds |
| | | 9. Mammalia—mice, whales, and human beings |

jaws; the **Gnathostomata** have jaws. The Gnathostomes are divided into **Pisces** (fishes) and **Tetrapods** (four-legged land vertebrates). Table 16-1 shows the general organization of vertebrates.

As we will see in the following chapters, the vertebrates are a highly diverse group that exhibit many interesting features.

# Summary

The chordates have several features in common, including gill pouches, with their associated gill slits (which do not form in mammals); a dorsal hollow nerve cord; and an endoskeleton consisting of an axial notochord, which in the higher chordates is replaced by vertebrae. The chordates can be pictured as "upside-down" invertebrates, owing to the position of their nervous and circulatory systems. Perhaps the "flip" took place in the echinoderms, to which the chordates seem to be linked, as evidenced by embryological similarities.

# 17

# Lampreys, Sharks, and Bony Fishes

In Chapter 16, we saw that the chordates are divided into Agnatha (jawless primitive fish), and the Pisces (fish) and Tetrapoda (four-legged land animals: amphibians, reptiles, birds, and mammals) groups of the Gnathostomata (jawed animals). Fishes are represented by the Agnatha and Pisces groups; these lower aquatic vertebrates range from the sea lamprey (class Cyclostomata), to the dogfish shark (class Chondricthyes) and finally to the higher bony fish (class Osteichthyes), represented here by the yellow perch. Representatives from these three classes, plus descriptions of two extinct classes, will follow a description of some general characteristics of fishes.

## General Characteristics of Fishes

The bodies of the majority of fish are spindle-shaped and laterally compressed, a form that offers least resistance to progress through water. Variations in form are correlated with the habits of the fish: flounders have flat bodies, adapted for life on the sea bottom, and eels have long, cylindrical bodies, enabling them to enter holes and crevices.

Many fish, especially those of tropical waters, have exceedingly brilliant coloring. Their colors are red, orange, yellow, and black (and combinations of these, producing still other colors). Coloration is due to pigments within chromatophores or to reflection and iridescence resulting from the physical structure of the scales. Usually, the colors are arranged in a definite pattern consisting of transverse or longitudinal stripes and spots of various sizes. The dispersion or concentration of pigment in the chromatophores of certain fishes results in changes in coloration; these are usually protective adaptations which camouflage the animal in its surroundings.

A surprisingly large number of fishes can produce sounds audible to human beings; these noises are used either to bring the sexes together or to warn or startle enemies.

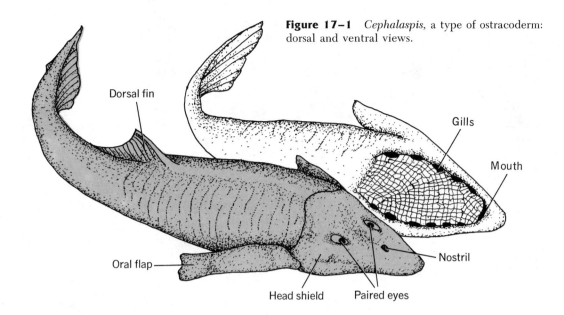

**Figure 17–1** *Cephalaspis*, a type of ostracoderm: dorsal and ventral views.

## Ostracoderms and Cyclostomes

The **ostracoderms** (Fig. 17–1) are an extinct class of jawless primitive fish; they are the oldest known vertebrates, with fossils dating back approximately 0.5 billion years. These primitive animals were small (about 3 dm) and lived in freshwater streams. Bony plates surrounded the body and the head was encased in a bony shield which had openings on the dorsal side for the eyes and a single nostril and on the ventral side for the gills and a small mouth. Ostracoderms were filter feeders.

The **cyclostomes** ("round mouths") represent a primitive level of vertebrate development. They are eel-shaped, without jaws or paired fins, and with only one olfactory pit. These animals include hagfishes, slime eels, and lampreys. *Petromyzon marinus*, the sea lamprey, is a modified survivor of some of the first vertebrates that lived on the earth and will be studied as a representative fish from the Agnathan group.

## Sea Lamprey

The sea lamprey swims about near the bottom of salt or fresh water by undulations of its body. When in a strong current, it progresses by darting suddenly forward and attaching itself to a rock by means of a suctorial mouth. In the spring, all adult lampreys ascend rivers to spawn.

### EXTERNAL ANATOMY

The marine lamprey reaches a length of about 9 dm. Land-locked populations such as those in the Great Lakes attain a maximum size of only 6 dm. The body of the lamprey (Fig. 17–2) is nearly cylindrical, except at the posterior

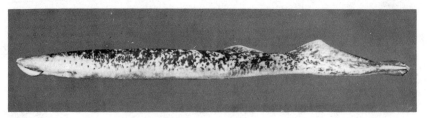

**Figure 17–2** Female sea lamprey, showing characteristically mottled back of sexually mature adult. (Courtesy Institute for Fisheries Research, Michigan Department of Natural Resources.)

end, where it is laterally compressed. The skin, a mottled greenish brown, is soft and made slimy by secretions from **epidermal glands.** A row of segmental sense pits, the **lateral line,** is located on each side of the body and on the head. The **mouth** lies at the bottom of a suctorial disk, the **oral (buccal) funnel** (Fig. 17–3), and is held open by a ring of cartilage.

### INTERNAL ANATOMY AND PHYSIOLOGY

**SKELETAL SYSTEM.** The skeletal system includes a well-developed **notochord** which persists in the adult. In the trunk region it is supplemented by small

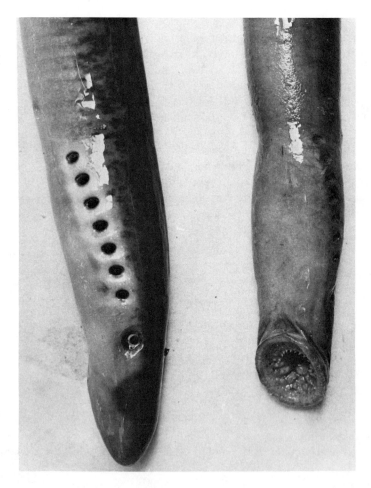

**Figure 17–3** This view of the head of the sea lamprey shows the oral funnel, which serves as a suction cup that attaches this animal to its prey. Sharply pointed teeth and a rasplike tongue allow the lamprey to penetrate through scales and flesh of prey.

cartilaginous neural arches. The organs in the head are supported by a cartilaginous cranium and a cartilaginous branchial basket.

**Muscular System.** The muscles in the walls of the trunk and tail are segmental, in a Σ-shaped arrangement, similar to that of fishes. The rasplike tongue is moved by large retractor and small protractor muscles. The buccal funnel is operated by a number of radiating muscles.

**Digestive System.** The adult *Petromyzon* (Fig. 17–4) feeds chiefly on the blood of fishes. The expansion of the oral funnel causes the mouth to act like a sucker and enables the animal to cling to stones or fasten itself to fishes.

The **mouth** cavity opens at its posterior end into two tubes: an **esophagus** and a **pharynx**. The **velum** is a fold at the anterior end of the pharynx, which prevents the passage of food into the respiratory system. There is no distinct stomach. The posterior end of the esophagus is separated from the straight

**Figure 17–4** Anterior portion of *Petromyzon*, an adult lamprey. *Right:* internal structures related to feeding. *Left:* cross section through gill pouches. Arrows indicate the direction of water current.

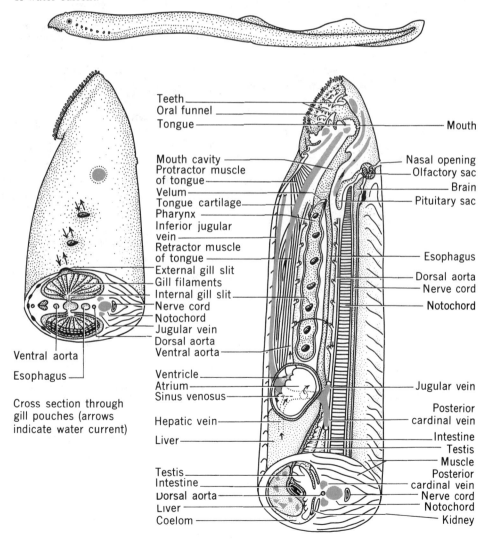

intestine by a valve. The digestive tract ends at the **anus**. A **liver** is present, but there is usually no bile duct in the adult; it is not definitely known if there is a pancreas.

**CIRCULATORY SYSTEM.** *Petromyzon* possesses a two chambered **heart,** a number of **veins** and **arteries,** and many lymphatic **sinuses**. The heart lies in a pericardial cavity and consists of a **ventricle,** which forces blood into the arteries, and an **atrium,** which receives the blood from the veins. A lymphatic system is also present, but a renal portal system is absent.

**RESPIRATORY SYSTEM.** Respiration is carried on by means of seven pairs of **gill pouches,** which open to the outside by **gill slits** and internally to the pharynx (Fig. 17–4). In the adult, water is taken into the gill pouches through the gill slits and discharged through the same openings; this method is necessary because the lamprey cannot take water through the mouth, which is attached to its food source by the oral funnel. In larval lamprey, the water used in respiration enters through the mouth and passes out through the gill slits, as in true fishes.

**NERVOUS SYSTEM AND SENSE ORGANS.** The **brain** of the adult lamprey is very primitive. The forebrain consists of a large pair of **olfactory lobes;** behind these are the small **cerebral hemispheres** attached to the **diencephalon;** ventral to the latter is a broad **infundibulum,** and above it is a **pineal structure.** On the midbrain is a pair of large **optic lobes.** The hindbrain includes a small rudimentary **cerebellum** and a fairly well-developed **medulla oblongata.** There are 10 pairs of **cranial nerves;** the **nerve cord** is flat and lies on the floor of the neural canal. There are no sympathetic ganglia, and the autonomic system consists only of an intestinal plexus linked with the brain.

Organs of taste, smell, equilibrium, and sight are present in the lamprey. The "ears" (balancing organs) lie in the **auditory canal;** although *Petromyzon* has only two, three semicircular canals are usually present in each ear. The **eyes** of the adult lamprey, although primitive, are excellent visual organs. Besides the paired eyes just behind the nasal opening, there is a well-developed median **pineal** eye with a clear lens and pigmented retina. In salt water, lampreys can produce electric fields used for detecting prey.

**ENDOCRINE GLANDS.** The endocrine system includes a **pituitary gland,** formed from numerous small follicles near the connecting point between the infundibulum of the brain and a tubular **pituitary sac.** The endostyle of the larva is the forerunner of the thyroid gland in the adult lamprey.

**UROGENITAL SYSTEM.** The **excretory** and **reproductive** systems are so closely united in the lamprey that it is customary to treat them together as the **urogenital** system. The **kidneys** lie along the dorsal wall of the body cavity, and each pours its secretions by means of a **urinary duct** into the **urogenital sinus** and then to the outside through the **progenital opening.** The immature gonad is hermaphroditic but matures into a male or female; the sexes are therefore separate in the adult. The single **gonad** fills most of the abdominal cavity at the time of sexual maturity. There is no genital duct; eggs or sperms break out into the coelom, make their way through two **genital pores** into the urogenital sinus and pass out through the urogenital opening into the water, where fertilization occurs.

**REPRODUCTION.** Sea lampreys become sexually mature in May or early June; then both sexes migrate into streams. The female fastens to a stone in a nest created by the lampreys and the male attaches to the female by using their oral funnels. Sperms and eggs are discharged into the water where the eggs are fertilized. The average female lays 62,500 eggs and dies after spawning. The eggs hatch in 20 to 21 days, when the blind larvae emerge and drift downstream to quiet water with a mud or silt bottom, into which they burrow. The larval period spent in burrows is from 3 to 12 or more years, depending on the amount of food available and water temperature. In the winter several years after burrowing, the larval lamprey undergoes a metamorphosis and migrates to the sea or one of the Great Lakes, where growth to sexual maturity takes place. At this stage the parasitic species begins to feed on fish.

The distinct head, cranium, better-developed brain, and cartilaginous neural arches are obviously advanced characteristics of the Agnatha over the more primitive amphioxus. The cyclostomes are less advanced than fishes, as is indicated by the absence of hinged jaws, paired limbs, true teeth, and complete vertebrae.

## Placoderms and Chondrichthyes

**Placoderms,** the first of the gnathostomes in the fossil record, are an extinct group (Fig. 17–5). Like the ostracoderms, they were heavily armored and many lived in fresh water, although later placoderms invaded the sea. There were various species, all possessing a primitive **jaw.** In some cases the jaw was composed of bony plates; in others it was derived from the first gill bar of their agnathan ancestors. The jaw made it possible for placoderms to eat new foods and thus to become predaceous. This mode of feeding undoubtedly made speed and agility advantageous traits that were realized through increased musculature, stabilizers (fins; Fig. 17–6), and refinement of senses and nervous system.

It is likely that the placoderms gave rise first to Osteichthyes (bony fish) and then to Chondrichthyes (cartilaginous fish). It is possible that with the rise of the cartilaginous fishes, the placoderms were "phased out." The cartilaginous fish did not have the heavy armor of the placoderms and perhaps "outran" them for food. The placoderms became completely extinct by the beginning of the Mesozoic period (about 230 million years ago).

The **Chondrichthyes** (**cartilaginous fishes,** also called elasmobranchs) are the

**Figure 17–5** Fossil placoderm. The primitive jaw enabled these animals to become carnivorous chordates—a decisive evolutionary step.

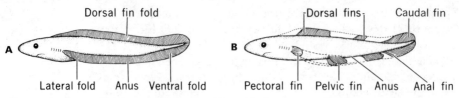

**Figure 17-6** Finfold theory of origin of fins. (A) Continuous folds of paired and unpaired fins in embryo. (B) Parts of the continuous folds disappear to form permanent fins.

sharks, rays, skates, and chimeras. These are all that remain from a group that once dominated the ancient seas. They are the most generalized of the living vertebrates that have complete vertebrae, movable jaws, and paired appendages. The sharks are the largest fishes and the second largest vertebrates alive today (whales are the first).

## Squalus acanthias—The Dogfish Shark

The common dogfish shark is one of an ancient group of fishes; it first appeared in the late Devonian period, about 360 million years ago.

**EXTERNAL ANATOMY**
The body is spindle-shaped and about 7.5 dm long. There are two **dorsal fins,** each with a spine at the anterior end, two **pectoral fins** and two **pelvic fins** (which in the male possess cartilaginous appendages known as **claspers**). The tail (caudal fin) is **heterocercal** (internally and externally asymmetric). The gray skin is covered with **placoid scales.** Over the jaws the placoid scales are modified into teeth, with their points directed backward; these are used for tearing and holding prey. Each placoid scale is composed of a bony basal plate, with a spine of **dentine** in the center and covered with a hard enamel-like **vitrodentine.**

**INTERNAL ANATOMY**
SKELETAL SYSTEM. The skeleton is composed entirely of **cartilage.** The cartilaginous skeleton in the elasmobranchs is, in all probability, a degenerate characteristic. There are two main subdivisions of the skeleton: the **axial** and the **appendicular.**
The **axial skeleton** consists of the vertebral column and the skull. The **vertebrae** are hourglass-shaped (amphiocoelous), and the **notochord** persists in the lens-shaped spaces between them. The highly developed **skull** is composed of a **cranium** or brain case; two large anterior **nasal capsules** and two posterior **auditory capsules;** and the visceral skeleton, made up of the **jaws, hyoid arch,** and five **branchial arches** supporting the gill region.
The **appendicular skeleton** consists of the cartilages of the fins and those of the **pectoral** and **pelvic girdles** that support them.
DIGESTIVE SYSTEM. The **digestive tract** (Fig. 17-8) is longer than the body. Following the **mouth** is a large **pharynx** into which open spiracles and gill slits.

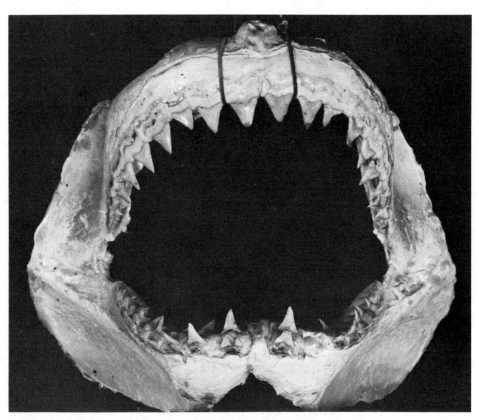

**Figure 17–7** Jaw and teeth of a shark, *Carcharodon carcharias*.

The pharynx leads into the short, wide **esophagus,** which opens into the U-shaped **stomach.** The posterior end of the stomach ends at a circular sphincter muscle, the **pyloric valve.** The **intestine** follows and terminates in the **cloaca** and **cloacal opening.** A slender fingerlike **rectal gland,** which secretes salt, is attached dorsally near the point where the large and small intestines meet. Within the spiral fold of the intestine is a mucous membrane (**spiral valve**), which prevents too-rapid passage of food, increases the surface area, and thus provides for increased absorption (Fig. 17–9). The **liver** is large and consists of two long lobes; its secretion, bile, is stored in a **gallbladder** and empties through the bile duct into the intestine. A **pancreas** and **spleen** are also present.

CIRCULATORY SYSTEM. As in the cyclostomes and most of the true fishes, the heart contains venous blood only. It is pumped through the **ventral aorta** and thence into the **afferent branchial arteries,** becoming oxygenated in the capillaries of the gills. It passes into the **efferent branchial arteries,** to the **dorsal aorta,** and is carried to various parts of the body. Veins carry the blood back into the heart through the **sinus venosus.** The **hepatic portal system** of veins transport the blood from the digestive canal, pancreas, and spleen to the liver; from there the **hepatic sinuses** return it to the sinus venosus. The **renal portal system** conveys the blood from the posterior part of the body to the kidneys; blood from the kidneys is carried by the **renal veins,** which empty into the **posterior cardinal sinuses,** which return the blood to the sinus venosus.

**Figure 17-8** Internal organs of the dogfish shark.

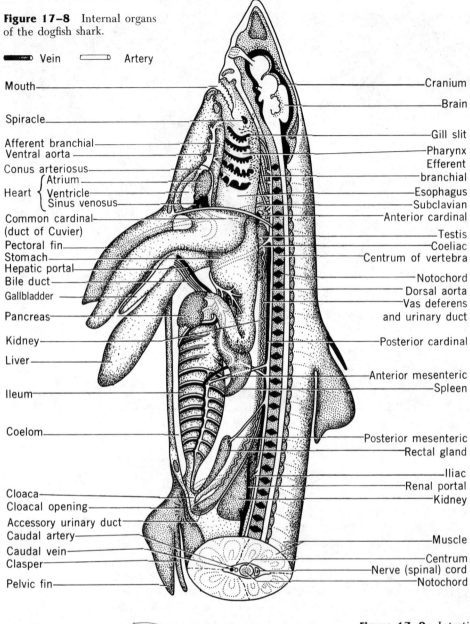

**Figure 17-9** Intestine of a dogfish shark cut open to reveal spiral valve. Food moves in the direction indicated by the arrow, encountering a large surface area where digestion takes place.

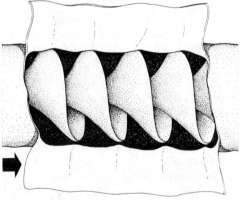

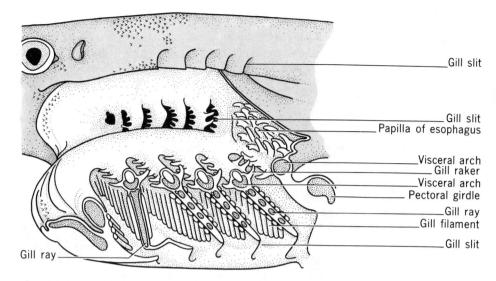

**Figure 17–10** Respiratory structures in the dogfish shark.

RESPIRATORY SYSTEM. Respiration is carried on by means of **gills** (Fig. 17–10). Water entering the mouth passes between the **branchial arches** and out through the **gill slits** and **spiracles**, thus bathing the gills and supplying oxygen to the branchial blood vessels.

NERVOUS SYSTEM AND SENSE ORGANS. The brain is more highly developed in

**Figure 17–11** Dorsal view of brain and cranial nerves (signified by Roman numerals) of the dogfish shark.

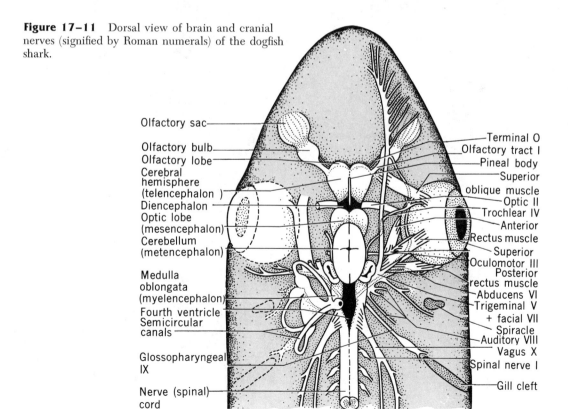

these animals than in the cyclostomes. It possesses two remarkably large **olfactory lobes,** a **cerebrum** of two hemispheres, a pair of **optic lobes,** and a **cerebellum,** which projects backward over the **medulla oblongata** (Fig. 17–11). There are 11 pairs of **cranial nerves.** The **nerve (spinal) cord** is a dorsoventrally flattened tube with a narrow central canal; it is protected by the vertebral column and **spinal nerves** arise from its sides in pairs.

The two **olfactory sacs** are characteristically large in elasmobranchs. The **ears** lie within the **auditory canals;** they are membranous sacs, each with three semicircular canals. The **eyes** are well developed. Along each side of the head and body is a longitudinal groove called the **lateral line;** it contains a canal with numerous openings to the surface. Inside the canal are **sensory hair cells** connected to a branch of the tenth cranial nerve. The lateral line system seems to be a pressure-regulator system able to detect currents in water. Also on the surface of the head are **sensory canals,** which open into pores containing pit organs with sensory hairs. It was once thought that this system was for heat reception; recent findings show that it is actually used for electroreception. At close range it possibly detects the electric activity of the prey's nerves and muscles.

UROGENITAL SYSTEM. The dogfish shark possesses two ribbonlike **kidneys,**

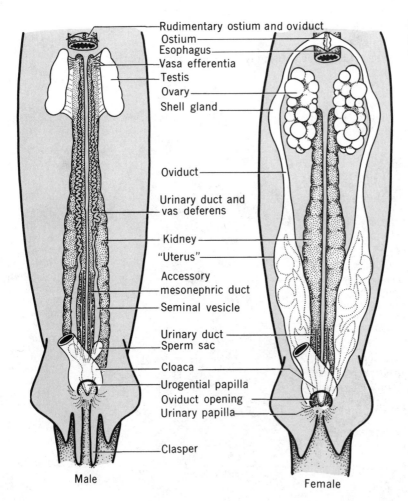

**Figure 17–12** Urogenital system of the dogfish shark.

one on either side of the dorsal aorta (Fig. 17–12). Their secretion is carried by small ducts into a large **urinary duct,** which empties into a **urogenital sinus;** it then passes out of the body through the **cloacal opening.** Not much urea is excreted since it is retained for osmoregulation. A series of yellowish bodies (**adrenals**) are located on the medial border of the kidney.

The sexes are separate. The spermatozoa of the male arise in two **testes** and are carried by the **vasa efferentia** to the convoluted **vasa deferentia,** which empty into the urogenital sinus. During **copulation** the spermatozoa are transferred to the **oviducts** of the female with the aid of **claspers.**

The eggs of the female arise in the paired **ovaries,** which are attached to the dorsal wall of the abdominal cavity. They break out into this cavity and enter the funnel-like opening, the **ostium,** common to both oviducts. An expanded anterior portion of each oviduct is the **shell gland,** and a posterior part is enlarged to form a "**uterus,**" in which the young develop. The oviducts have separate openings into the cloaca.

Other Chondrichthyes include the whale shark, *Rhineodon typus;* the great white shark, *Carcharodon carcharias,* a maneater; the hammerhead; and the electric ray, which is highly specialized to produce up to 1000 watts of electricity for more than 1 hour without fatigue to the animal.

## Osteichthyes—Bony Fishes

The bony fishes range from the ordinary ones, such as the yellow perch (used here as a representative of this class), to the unusual lungfish (Fig. 17–13). They are found in all the waters of the world, from surface level to great ocean depths. They have a bony skeleton and are usually covered by scales or bony plates, which furnish protective covering. They swim by means of body undulations and fins. Their bodies are usually streamlined, but some are grotesque in shape, especially in the deeper parts of the ocean.

## Yellow Perch

*Perca flavescens*, the yellow perch, is a typical bony fish. It inhabits freshwater streams and lakes of the eastern United States.

**EXTERNAL ANATOMY.** The streamlined body of the yellow perch is about $\frac{1}{3}$ m long and is divisible into **head, trunk,** and **tail** regions (Fig. 17–14). There are two **dorsal fins,** a **caudal fin,** a single median **anal fin** just posterior to the anus, two **lateral pelvic fins,** and two **lateral pectoral fins.** The **skin** is provided with a number of overlapping **scales.** Mucous glands are abundant in the skin and produce "slime" (mucus), which makes the fish slippery.

The fish is able to remain stationary without much muscular exertion by use of an air bladder, which are out pocketings of the esophagus. The principal locomotor organ is the caudal fin. By alternating contractions of the muscular bands on the sides of the trunk, the caudal fin is lashed from one side to the other, displacing water and enabling the fish to swim. The paired lateral fins (pectoral and pelvic) are used as oars in swimming when the fish is moving slowly. They also aid the caudal fin in steering. The dorsal, anal, and caudal fins increase the vertical surface of the body and help balance out the weight of the fish, which is centered in the back. The paired lateral fins are also organs of **equilibrium,** acting as balancers.

## INTERNAL ANATOMY

**Skeletal System.** The **exoskeleton** of the perch includes scales and fin rays. The **scales** develop in pouches in the dermis. They are arranged in oblique rows

**Figure 17–13** Representatives of bony fishes (not to scale).

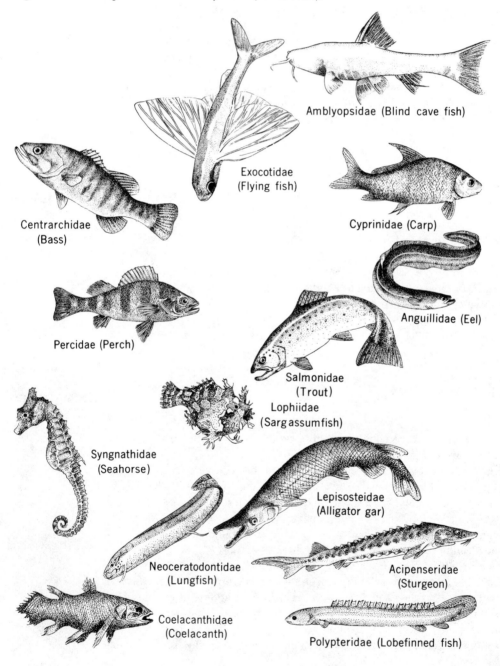

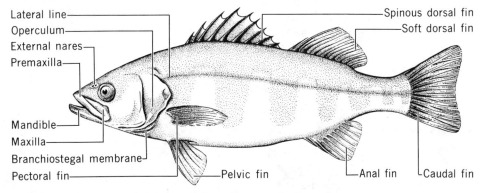

**Figure 17–14** External features of the yellow perch, *Perca flavescens*.

and overlap, forming an efficient protective covering that allows for lateral body movements. The posterior edge of each scale is toothed and rough to the touch (**ctenoid** scales). (Other types of scales are the diamond-shaped **ganoid scales,** occurring in a few primitive fish; and **cycloid scales,** nearly circular with concentric rings and overlapping like the ctenoids) (Fig. 17–16).

The fins are supported by bony cartilaginous **fin rays.** The rays of the spinous dorsal fin and of the anterior edge of the anal and pelvic fins are unjointed and unbranched **spines.** The bones of the **endoskeleton** include the skull, vertebral column, ribs, pectoral girdle, and the interspinous bones, which aid in supporting the unpaired fins (Fig. 17–15). The **vertebrae** are simple and comparatively uniform in structure. Ribs are attached by ligaments to the abdominal vertebra and serve as a protecting framework for the body cavity and its contents. There is no sternum. The **skull** has many cartilaginous and bony parts.

The **visceral skeleton** is composed of seven **paired arches,** more or less modified. The first or **mandibular arch** forms the jaws. The second or **hyoid arch** is modified as a support for the gill covers. Arches 3 to 7 support the gills and are known as **gill arches.** The first four bear spinelike ossifications, **gill rakers,** which intercept solid particles and keep them away from the gills.

**Figure 17–15** Skeleton of a perch. The facial features are artifact.

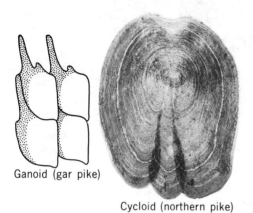
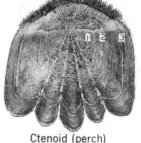

Ganoid (gar pike)

Cycloid (northern pike)

Ctenoid (perch)

**Figure 17–16** Different types of scales on bony fishes. (Cycloid and ctenoid scales courtesy Institute for Fisheries Research, Michigan Department of Natural Resources.)

The **appendicular skeleton** is comprised of the pectoral and pelvic girdles and fins associated with them, and by median fins. The pelvic girdle of the yellow perch is not very typical in form, being degenerate or possibly primitive. The body of the fish is to a considerable extent supported by the surrounding water; consequently, the bones do not need to be as strong as those of land animals, such as birds and mammals, which support the entire body.

MUSCULAR SYSTEM. The principal muscles are those used in locomotion, respiration, and obtaining food. The movements of the body when swimming are produced by four longitudinal bands of muscles, one heavy band on each side along the back and a thinner band on each side of both trunk and tail. These are arranged segmentally as **myomeres.** Weaker muscles move the gill arches, operculum, hyoid, and jaws.

DIGESTIVE SYSTEM. Perch feed chiefly on insects, mollusks, and small fishes. These are captured by the perch's jaws and held by many conical **teeth,** on the mandibles (jaws), premaxillae, and on the roof of the mouth. The teeth are not used for chewing food, only for holding it. A rudimentary **tongue** (Fig. 17-17) projects from the floor of the mouth cavity; it is not capable of independent movement but functions as a tactile organ. The **mouth cavity** is followed by the **pharynx,** on either side of which are four **gill slits.** Food passes directly to the stomach through a short **esophagus.** The **stomach, intestine, liver, gallbladder, pyloric cecum,** and **anus** are shown in Fig. 17-17. The **pancreas,** located in the first loop of the intestine, is so diffuse that it is usually not seen in gross dissection.

CIRCULATORY SYSTEM. The blood of the perch contains oval nucleated **red corpuscles** and **amoeboid white corpuscles.** The **heart** lies in the **pericardial cavity** (a portion of the coelom) (Fig. 17–18). Circulation is much slower in fishes than in the higher vertebrates.

RESPIRATORY SYSTEM. The perch breathes by way of four pairs of gills supported by the first four gill arches. Each gill bears a double row of **gill filaments,** which are abundantly supplied with capillaries. The **afferent branchial artery** brings the blood from the heart to the gill filaments; here an exchange of gases takes place. Carbon dioxide, with which the blood is loaded, passes out of the gill and a supply of oxygen is taken in from the continuous stream of water which enters the pharynx through the mouth and bathes the gills on its way out through the **gill slits.** The oxygenated blood is collected into the **efferent branchial artery** and is carried about the body.

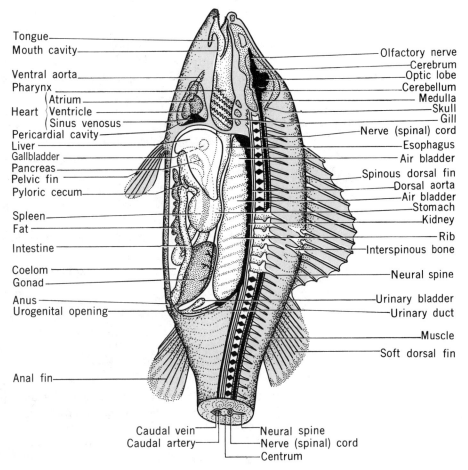

**Figure 17–17** Internal anatomy of the yellow perch.

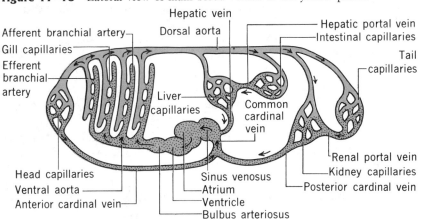

**Figure 17–18** Lateral view of main blood vessels of the yellow perch.

237

**EXCRETORY SYSTEM.** The **kidneys** lie just beneath the vertebral column in the abdominal cavity. They extract urea and other waste products from the blood. Two thin tubes, the **urinary (mesonephric) ducts** or "ureters," carry the excretory matter into a **urinary bladder,** where it is stored for a time and then expelled through the **urogenital opening,** located just posterior to the anus.

**NERVOUS SYSTEM AND SENSE ORGANS.** The brain of the perch is more highly developed than that of the cyclostome or shark. The four main divisions are well marked: the **cerebrum, optic lobes, cerebellum,** and **medulla.** The brain gives off **cranial nerves** to the sense organs and other parts of the anterior portion of the body. The **nerve (spinal) cord** lies above the center of the vertebral column and passes through the neural arches of the vertebrae. **Spinal nerves** arise laterally from the spinal cord.

The **ear** consists of the membranous labyrinth only. Three **semicircular canals** are present, and the **sacculus** contains concretions of calcium carbonate called ear stones or **otoliths.** The ear serves as an organ of hearing and equilibrium. The **eye** of the perch is different in several ways from that of terrestrial vertebrates. The **eyelids** are absent in bony fishes, since the water keeps the eyeball moist. The **cornea** is flattened and of about the same refractive index as the water. The **lens** is almost spherical; the **pupil** is usually larger than in other vertebrates and allows more light rays to enter, a necessary feature since semidarkness prevails at moderate depths. Two **olfactory sacs** with the ability to detect odors (actually a form of taste in aquatic animals) lie in the anterior part of the skull and communicate with the outside by a pair of openings in front of each eye. The integument, especially of the lips, serves as an organ of touch. Some fishes have **dermal barbels,** which function as sensory organs for locating food. The **lateral line** contains sensory cells, which detect vibrations in water and pressure stimuli. There are also numerous other cutaneous sense organs, as specified in Chapter 26.

**REPRODUCTION.** The sexes are separate in bony fishes. The single **ovary** in the female is probably the result of a fusion of two ovaries in the embryo. The ovary or, in the male, **testes,** lie in the body cavity. Eggs and sperm pass through the reproductive ducts and out of the urogenital opening.

Perch migrate in the spring from the deep waters of lakes and ponds to shallow waters near the shore. The female lays many thousands of eggs in a long ribbonlike mass. The male fertilizes the eggs by depositing sperm (milt) over them. Very few eggs develop, because of the numerous predators that feed on them.

The embryological development of the goldfish (Fig. 17–19) is similar to and better known than the perch and is therefore substituted for this portion of the discussion. The young goldfish hatches from the egg in about 3 to 14 days, depending upon the temperature of the water. The egg passes through the stages shown in the figure. A large part of the egg consists of yolk. A cytoplasmic accumulation which forms a slight projection at one end is the **germinal disk. Cleavage** of the germinal disk takes place and the **blastoderm** produced gradually grows around the yolk. The **embryo** appears as a thickening of the edge of the blastoderm; this grows in size at the expense of the yolk. After a time the head and tail become free of the yolk, and the young fish breaks out of the egg membranes. The young fish lives at first upon the yolk in the yolk sac but is soon able to obtain food from the water.

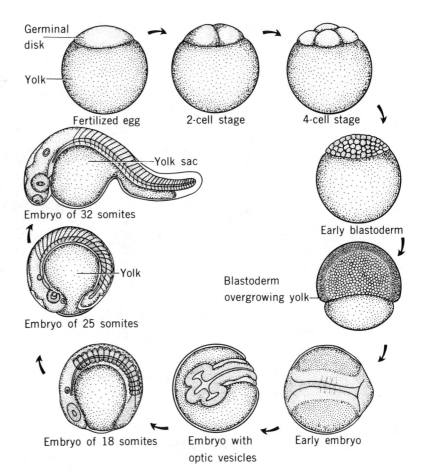

**Figure 17-19** Basic embryology of a goldfish. (After Helen L. Battle, *Ohio J. Sci.* **40:** 85, 1940.)

## Other Bony Fishes

Other bony fish include such a wide range that only a few of the more interesting ones can be mentioned. Sixty-five or more species of **flying fish** live in warm seas. Some are able to leave the water and, rising in the air a few yards, "fly" a distance of from a few rods to more than an eighth of a mile.

The **blind cavefish** (*Amblyopsis*), include six species; some have rudimentary eyes covered with a thick skin. **Eels** are long and slender with inconspicuous scales and continuous dorsal, caudal, and anal fins. They should not be confused with the lamprey "eels," which are really cyclostomes.

**Seahorses** are small and do not resemble other fishes; the head resembles that of a horse. They swim slowly and often cling to objects with their prehensile tails. Seahorses represent a more specialized stage in evolution than the streamlined fishes. Many families of fish contain **deep sea** species, which are often curiously modified. Some have very large eyes, enabling them to catch as many rays of light as possible; others are blind, with small rudimentary eyes, and depend upon organs of touch. Many deep-sea fishes have large mouths with long, sharp teeth, and enormous stomachs. Luminescent organs are variously distributed over the body, secreting a substance that sometimes acts as a lure and may also enable some fish to see in the dark abyss of the ocean.

The ability of the **lungfish** to breathe air suggests an intermediate step between fishes and amphibians. Yet the overall evidence shows clearly that these vertebrates have never been in the direct line of evolution leading from fish to the first land-living vertebrates. They are now regarded as an ancient group that has changed little through recent geologic ages.

## Relation of Lampreys, Sharks, and Bony Fishes to Man

Lampreys are popular as food in many parts of Europe; they are not used extensively for food or other commercial means in the United States. Lampreys have an extremely adverse effect on lake trout. In 1946 commercial fisherman caught approximately 5,500,000 pounds of trout from the upper Great Lakes and Lake Ontario. Lampreys that entered through man-made canals (constructed without ecological-impact studies) reduced the catch to 482 pounds by 1953.

Sharks feed on crustaceans, squids, fish, and other aquatic animals—not, as a rule, on human beings, contrary to popular notion and certain motion pictures. Sharks have, however, injured swimmers and divers in temperate waters, mostly near beaches or reefs.

Sharks (Fig. 17–20) compete with human beings for food, since they eat lobsters, crabs, and food fishes. In the Orient and other parts of the world, they are used as food. Their leather is used for shoes and handbags. The cubshark yields a considerable quantity of shark-liver oil, a source of vitamin A.

Although a few bony fishes are injurious because they destroy valuable food fishes and other useful aquatic animals, many are beneficial to human beings, serving as food or as a means of recreation. Among the game fishes are the yellow perch, bass, various species of trout, and tuna. Marine food fish are of great value; these include herring, anchovies, salmon, mackerel, and the

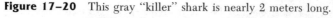

**Figure 17–20** This gray "killer" shark is nearly 2 meters long.

flounder family (halibuts, soles, plaice, and turbots). Codfish are especially valuable; the average annual catch is over 2 billion pounds. Cod-liver oil is a principal source of vitamins A and D. The use of fishmeal has become important in the fertilizer and pet-food industries. Enormous sums are spent by the federal and state governments in rearing fish and in stream and lake improvements.

The topminnows, *Gambusia*, feed on mosquito larvae; they are used to prevent the breeding of certain mosquitoes that transmit malaria and yellow fever.

In recent years the use of fish in experimental studies has increased notably. Fishes, being the largest groups of vertebrates, show enormous variation and remarkable adaptations of great biological and evolutionary importance.

# Summary

The first known vertebrates were the ostracoderms, heavily armored freshwater feeders. Modern agnathans include the lampreys, of the class Cyclostoma. Most cyclostomes are parasitic (hagfish and lampreys), with specialized features for that way of life.

Placoderms were the first vertebrates with jaws (gnathostomes). They may have lived in freshwater streams and eventually invaded the seas, probably giving rise to the cartilaginous fishes. Predation became possible with the jaw and the new breed became modified for this method of feeding—with the development of paired fins, reduced bones, and improved senses, among other modifications.

Bony fishes are the highest group of gnathostomes. Dorsal, caudal, anal, pelvic, and pectoral fins are characteristic, although some variations exist. Their lateral line of sense organs is unique to fishes. Most fishes produce some kind of sound, which arises from the air bladder. Various fish were briefly discussed to give the reader some idea of the wide range of species found in the class Osteichthyes.

Fish are the largest group of vertebrates and exhibit many interesting adaptations, making them an extremely valuable group from both a biological and an evolutionary viewpoint.

# 18

# Frogs and Other Amphibians

Frogs, toads, and salamanders are part of the class Amphibia—the first of the Tetrapoda (four-legged land animals) (Fig. 18–1). All amphibians require moisture and are found in or near fresh water, under logs and stones in damp woods, and in other moist places.

Ten orders of the class Amphibia are now extinct. The three living orders of this class are:

1. **Apoda,** commonly called *caecilians*, are nearly blind, wormlike amphibians that inhabit tropical and subtropical regions.
2. **Caudata** are amphibians with tails; they include mud puppies, sirens, and salamanders.
3. **Salientia** are frogs and toads, which are tail-less in the adult stage.

Frogs and toads are often confused. Toads are generally bulkier with thicker skin and warts. Frogs (Fig. 18–2) are streamlined and good jumpers. Their skin is relatively thin; thus frogs are usually found in aquatic habitats, whereas toads (Fig. 18–3) are found in a variety of places. Frogs possess maxillary teeth on the upper jaw, whereas toads do not. Toads have a **bidder's organ** (thought to be a potential ovary), while frogs do not. Another way of telling them apart is by their eggs: frogs generally lay them in clusters, while toads usually produce a string of eggs.

The frog is a representative amphibian and a typical vertebrate. Since they are relatively easy to work with and abundantly available, frogs are frequently used for laboratory study. A knowledge of the frog's morphology and physiology will give you an understanding of vertebrates in general and serve as a background for the study of more complex vertebrates, such as man. As we suggested in our introduction, Chapters 20 to 25 of this text deal with the various organ systems and an account of animal behavior; reference to this section might be of particular value during our discussion of the frog.

## The Frog—A Representative Amphibian and Typical Vertebrate

*Rana pipiens*, the leopard frog (Fig. 18–2), lives in or near freshwater lakes, ponds, and streams; it walks or leaps on land and swims in water. The hindlimbs are large and powerful; when the frog is on land, they are folded, but for

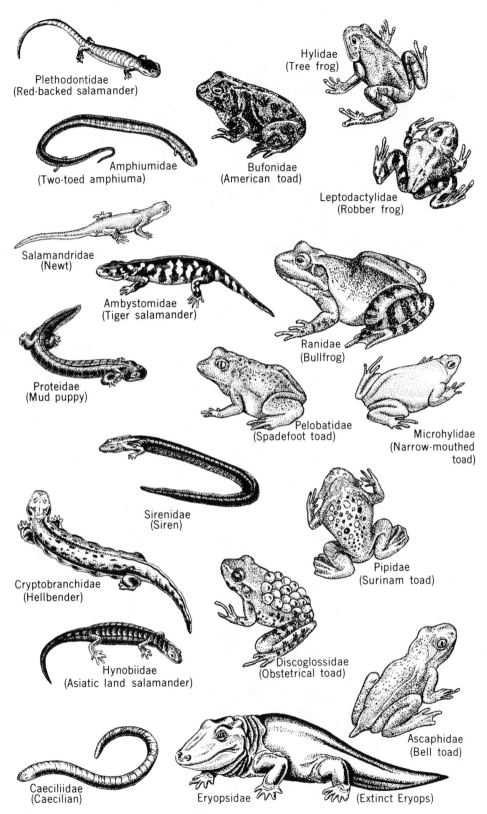

**Figure 18-1** Representative amphibians.

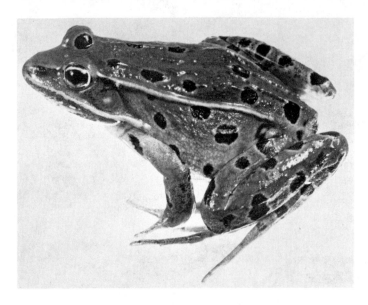

**Figure 18-2** *Rana pipiens,* the leopard frog.

leaping and swimming they are used extensively. The leopard frog is the most common of the 16 species found in the United States. The following account of structure and physiology applies specifically to it and to *Rana catesbeiana,* the bullfrog, and to most frogs in general.

### EXTERNAL ANATOMY

The body of the frog may be divided into a head and trunk; there is no neck region. The **eyes** usually protrude from the head, but they are drawn into their orbits when closed. The frog, along with many other amphibia and reptiles, has a **nictitating membrane** which is lucid and can move laterally across the eye for additional protection (especially underwater). Behind and slightly below each eye is a flat **tympanic membrane (eardrum)**. A pair of **external nares (nostrils)** is situated on the dorsal surface near the end of the snout, and internal nares are found on the hard palate within the mouth. The **mouth** is relatively wide,

**Figure 18-3** The common American toad, *Bufo americanus.* (Courtesy American Museum of Natural History.)

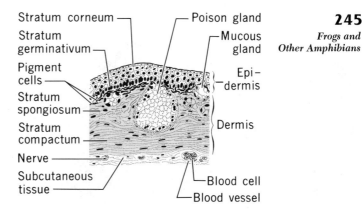

**Figure 18–4** Vertical section of frog skin, showing microscopic structure.

extending from one side of the head to the other. The **cloacal opening** is situated at the caudal end of the body.

Two short **forelimbs** support the anterior portion of the body. The **hand** possesses four digits and a rudiment of the fifth (thumb). The **hindlimbs** are folded together when the frog is at rest; the five toes are connected by a web. The toes are spread during swimming to increase the webbed surface, thus adding more resistance to the water.

The **integument** (**skin**) is usually smooth and loosely attached to the body (Fig. 18–4). Along either side of the body, behind the eyes, is a ridge formed by a thickening of the skin known as the **dermal plica** (**dorsolateral fold**); this may be of a different color than the rest of the skin. The skin is colored by scattered **pigment granules** in the epidermis and by **chromatophores** and **interference cells** in the dermis. The chromatophores are of several types, the most important being those that contain yellow or black pigment. The interference cells contain whitish crystals. The frog's skin lacks green pigment cells; the green color results from a combination of light reflected from the granules and interference cells and the yellow pigment through which the light passes.

Most amphibians, including frogs, can change color. This occurs through a change in the concentration of the black and yellow pigment cells (Fig. 18–5). These color changes are usually protective, making the animal less conspicuous by camouflaging it in its surroundings.

The skin is richly provided with glands, which open to the outside by way of the ducts. These are of two principal types: **mucous glands,** which are numerous and small; and **poison glands,** which are larger and less common. Each mucous gland consists of an epithelial layer of secreting cells, surrounded by muscle fibers and connective tissue. Mucus formed by the secreting epithelium

**Figure 18–5** Stages in changes of pigment-bearing cells (melanophores) in frog's skin.

Melanophore with pigment dispersed throughout cell

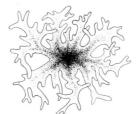

Pigment beginning to concentrate

Pigment concentrated in body of cell

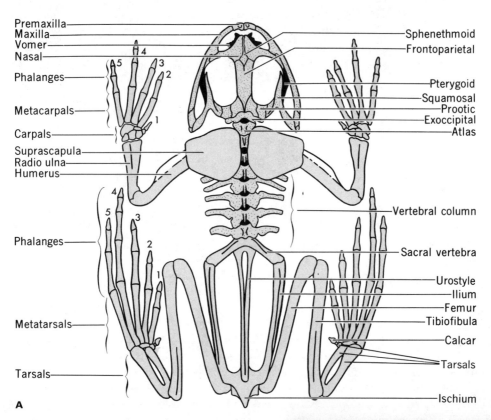
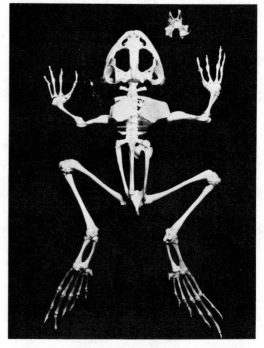

**Figure 18–6** (A) and (B). Skeleton of the frog (dorsal view). Note axial and appendicular skeletons.

is discharged into the gland and forced through the skin by the muscle cells. The mucus covering makes the frog slippery, often allowing it to escape from the grasp of enemies. The poison glands secrete a whitish fluid with a burning taste, which also serves as protection against enemies.

## ANATOMY OF SKELETAL AND MUSCULAR SYSTEMS

**SKELETAL SYSTEM.** The **endoskeleton** of the frog consists largely of cartilage and bone. It supports the soft parts of the body, furnishes points of attachment for muscles, and protects delicate organs such as the brain, spinal cord, and eyes.

There are two main subdivisions of the skeleton: the **axial skeleton** is comprised of the **skull** (divided into a cranium and visceral skeleton), **vertebral column**, and **sternum**. The **appendicular skeleton** consists of the **pectoral** and **pelvic girdles** and the bones of the limbs they support (Fig. 18-6).

The **cranium** contains the **brain case** and **auditory** and **olfactory capsules** (Fig. 18-7). A large part of the cranium consists of cartilage. The bones are either ossifications that replace cartilage or have developed into connective tissue without passing through a cartilage stage. The spinal cord passes through a large opening, the **foramen magnum**, in the posterior end of the cranium. On either side of this opening the exoccipital bones are convex, forming the two **occipital condyles**, which articulate with two concave areas of the first vertebra. This enables the frog to move its head.

The **jaws, hyoid,** and **cartilages of the larynx** constitute the **visceral skeleton** (Fig. 18-8); they are preformed in cartilage and then strengthened by ossifications. The upper jaw (**maxilla**) consists of paired **premaxillae, maxillae** (both bearing teeth), and **quadratojugals**. The lower jaw is the **mandible**. The associated **visceral arches** are represented in the adult by the **hyoid** and its processes.

The **vertebral column** or backbone consists of nine **vertebrae** and a bladelike posterior extension, the **urostyle**. The vertebrae are held together by ligaments. The column thus serves as a firm axial support which allows bending of the body anterior to the rostral edge of the urostyle. Frogs do not have ribs.

**Figure 18-7** Bullfrog skull: dorsal and ventral views.

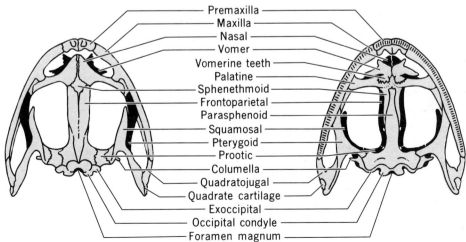

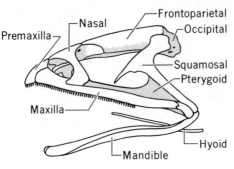
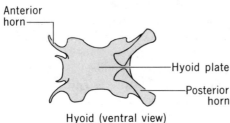

**Figure 18–8** Bullfrog skull and hyoid.

The **pectoral girdle** and **sternum** (Fig. 18–9) are composed of bone and cartilage. They support the forelimbs, serve as attachments for the muscles that move the forelimbs, and protect the organs lying within the anterior portion of the trunk. The **suprascapulae** lie above the vertebral column; the remainder of the girdle passes downward on either side and unites with the sternum.

The **pelvic girdle** (Fig. 18–9) supports the hindlimbs. It consists of two sets of three parts each: the **ilium,** the **ischium,** and the **pubis** (which is cartilaginous). The parts of each half of the pelvic girdle unite, forming a concave space (the **acetabulum**) that receives the head of the long leg bone, the femur.

The forelimbs and hindlimbs differ in size but have similar component parts. The bones that comprise them are listed in Table 18–1.

**Muscular System.** Muscles of the frog (Figs. 18–10 and 18–11) and other vertebrates are of three principal types: **smooth** (**visceral**), **cardiac,** and **striated.**

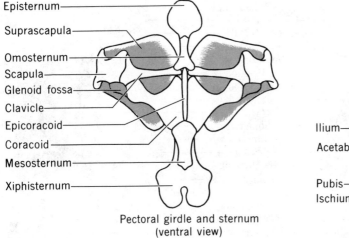
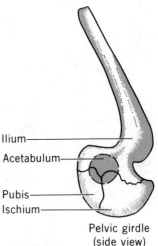

**Figure 18–9** Girdles of the frog.

**Table 18-1** Major Bones of the Frog

| Forelimb (Arm) | Number of Bones | Hindlimb (Leg) | Number of Bones |
|---|---|---|---|
| Humerus (upper arm) | 1 | Femur (thighbone) | 1 |
| Radioulna (forearm) | 2 fused | Tibiofibula (lower leg bone or shank) | 2 fused |
| Carpals (wrist) | 6 | Tarsals (ankle) | 4 |
| Metacarpals (palm of hand) | 5 | Metatarsals (sole of foot) | 5 |
| Phalanges (fingers) | 10 | Phalanges (toes) | 14 + prehallux |

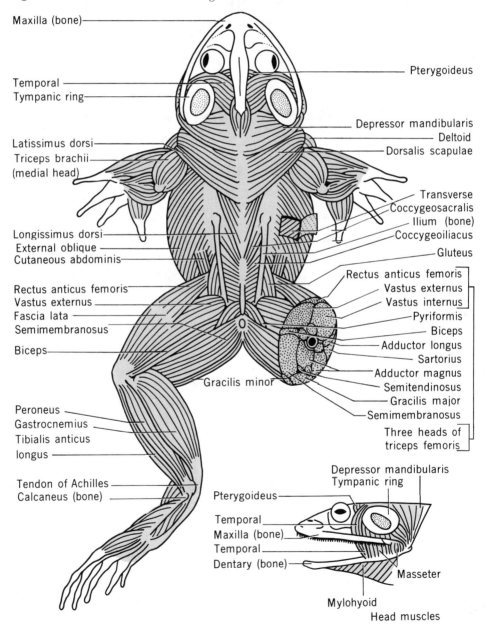

**Figure 18-10** Muscles of the bullfrog: dorsal view.

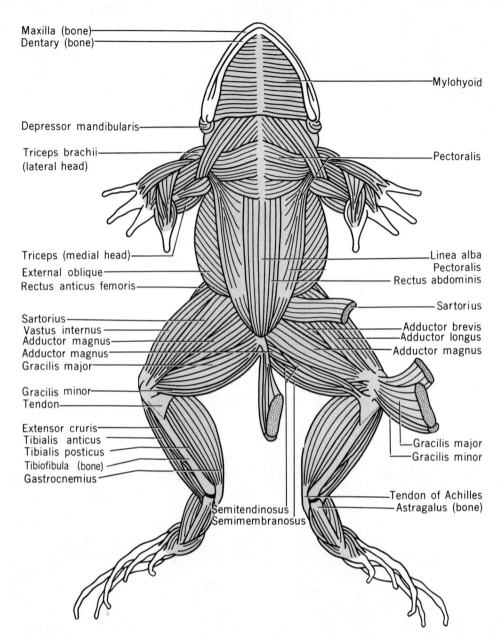

**Figure 18-11** Muscles of the bullfrog: ventral view.

Smooth muscle is always **involuntary** muscle; cardiac muscle is **involuntary striated** muscle that occurs only in the vertebrate heart. Striated muscle of the external muscular system is **voluntary muscle**.

Muscles are named for the specific type of motion they produce in the body. **Flexor** muscles bend one part upon another, **extensors** straighten out a part, **adductors** draw a part toward the midline of the body, **abductors** move a part away from the midline, **depressors** lower a part of the body, **levators** raise a part, and **rotators** rotate one part on another. The movements of an organ depend on the **origin** (less movable end) and **insertion** (more movable end) of

**Table 18-2** Name, Origin, Insertion, and Action of Some of the Muscles of the Hindlimb of the Frog

| Name | Origin | Insertion | Action |
|---|---|---|---|
| Sartorius | Ilium, just in front of pubis | Just below head of tibia | Flexes leg; and adducts thigh |
| Adductor magnus | Ischium and pubis | Distal end of femur | Adducts thigh and leg |
| Adductor longus | Ventral part of ilium | Joins adductor magnus to attach on femur | Adducts thigh and leg |
| Triceps femoris | From three heads: one from acetabulum, and two heads from ilium | Upper end of tibiofibula | Abducts thigh and extends leg |
| Gracilis major | Posterior margin of ischium | Proximal end of tibiofibula | Adducts thigh; flexes leg |
| Gracilis minor | Tendon behind ischium | Joins tendon of gracilis major and tibiofibula | Adducts thigh; flexes leg |
| Semimembranosus | Dorsal half of ischium | Proximal end of tibiofibula | Adducts thigh; flexes leg |
| Biceps (iliofibularis) | Dorsal side of ilium | Tibiofibula | Adducts thigh; flexes leg |
| Gastrocnemius | Distal end of femur; tendon of triceps | Tendon of Achilles | Flexes leg; extends foot |
| Tibialis posticus | Posterior side of tibiofibula | Proximal end of astragalus | When foot is flexed, acts as an extensor |
| Tibialis anticus | Distal end of femur longus | Proximal end of astragalus and calcaneus (ankle bones) | Extends leg; flexes foot |
| Peroneus | Distal end of femur | Distal end of tibiofibula; head of calcaneus | Extends leg |
| Extensor cruris | Distal end of femur | Anterior surface of tibiofibula | Extends leg |

the muscles and the nature of the articulations of its bones with each other and with other parts of the body.

Table 18–2 gives the name, origin, insertion, and action of some of the muscles of the frog's hindlimb.

## GROSS ANATOMY OF INTERNAL ORGAN SYSTEMS

If the body wall of the frog is opened in the ventral midline, from the posterior end to the angle of the jaw, the organs in the body cavity or **coelom** will be exposed (Fig. 18–12).

The **heart** lies within the saclike pericardium; it is partially surrounded by the three lobes of the reddish-brown **liver**. The two **lungs** lie, one on either side, near the anterior end of the abdominal cavity. Coiled about within the body cavity are the **stomach** and **intestine**. The **kidneys** are flat, reddish bodies attached to the dorsal body wall; they lie outside the coelom just behind a thin membrane, the **peritoneum** (referred to as retroperitoneal). The two **testes** of the male are small ovoid organs suspended by membranes and lying at the sides of the digestive tract. The egg-filled **ovaries** and **oviducts** of the female occupy a large part of the body cavity during the breeding season. The reproductive organs and digestive tract are suspended by double layers of peritoneum called **mesenteries**.

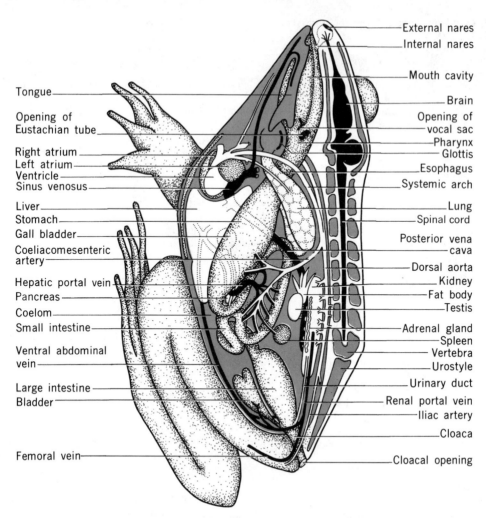

**Figure 18-12** Gross internal anatomy of the frog.

We will now look at each organ system of the frog to illustrate the physiology of both amphibians and vertebrates in general.

**DIGESTION AND EXCRETION.** The food of frogs consists mainly of living worms and insects which are captured by a specialized **extensile tongue**. The food adheres by a sticky secretion that covers the tongue and is drawn into the mouth. The tongue will not extend unless a small spheroidal object moves across its field of sight at a minimum velocity; immobile objects are not seen by the frog. Larger insects are pushed into the mouth with the forefeet. If the object swallowed proves to be unpalatable, it is regurgitated through the mouth.

The **mouth cavity** is large. The forked tongue lies on the floor of the cavity attached at the anterior end to the jaw. The **teeth** are conical in shape and used only for holding food. New teeth will continually replace those that may be lost.

The **esophagus** opens into the mouth cavity by a horizontal slit. Its inner surface bears longitudinal folds, allowing it to distend to swallow large animals. The **stomach** (Fig. 18-13) is crescent shaped and lies primarily on the left side of the body. The anterior or **cardiac** end is larger than the esophagus. It decreases in size toward the posterior or **pyloric** end, where it joins with the small intestine.

The two largest **digestive glands** are the pancreas and liver. The **pancreas** lies between the duodenum (first part of the small intestine) and the stomach. It is a much-branched tubular gland which secretes an alkaline digestive fluid that it empties into a common bile duct. The **liver** is a large, three-lobed reddish organ which serves many functions; most notably it filters toxic (poisonous) substances from the blood, collects dead erythrocytes (red blood cells), and produces **bile,** an alkaline secretion associated with the emulsification of fatty material. The bile is stored in the **gall bladder** until food enters the intestine; it is then released through the **common bile duct** into the duodenum.

Food leaving the stomach (through the **pyloric sphincter muscle**) enters the **duodenum** and then passes to the much-coiled **ileum** (small intestine), which widens abruptly into the **colon** (large intestine).

A certain amount of waste substance resulting from the breakdown of living matter is excreted by the skin, lungs, liver, and intestinal walls, but most of it is removed from the blood in the **kidney.** From the kidney, wastes pass through the **urinary** (mesonephric) **duct** and into the **cloaca;** the digestive canal, urinary bladder, and reproductive ducts all open into this common cavity. Urine may be stored temporarily in the **bladder** or voided at once through the **cloacal opening.**

For a detailed discussion of digestion and excretion in vertebrates, see Chapter 24.

RESPIRATION. Respiration in the frog and other terrestrial vertebrates is carried out largely by the lungs. Much gas exchange takes place through the skin and through the mucous membrane lining the mouth. Air passes through the nostrils or external nares into the nasal cavity and then through the internal nares into the mouth cavity. The external nares then close, the floor of the mouth is raised, and the air is forced through the **glottis.** The air then enters a short tube, the **larynx,** passes into a very short tube, the **bronchus,** and thence into the **lungs.** Air is expelled from the lungs into the mouth cavity by the contraction of the muscles of the body wall.

The lungs are ovoid sacs with thin elastic walls. The inner surface of the lungs is divided into many minute chambers called **alveoli.** The alveoli are the

**Figure 18-13**  Liver, gallbladder, pancreas, stomach, and portion of intestine of the frog.

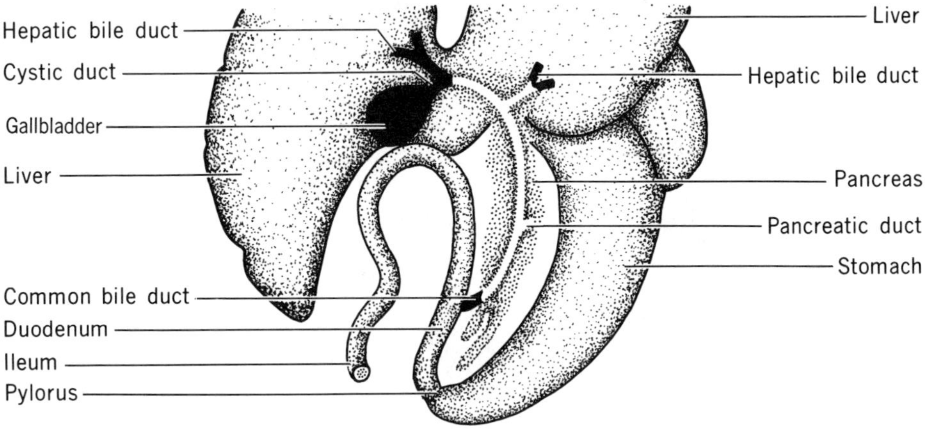

functional units of respiration, containing numerous blood capillaries in their walls, allowing oxygen to diffuse into the blood and carbon dioxide to be released into the lung.

In aquatic vertebrates, respiration is carried on by **gills.** Associated with respiration is the production of sound. Croaking is produced by the vibrations of **vocal cords,** which are two elastic bands stretched across the **larynx** (voice box). The inner edges of these vocal cords vibrate when air is taken into or expelled from the lungs. The laryngeal muscles regulate the tension of the cords, and hence the pitch of the sound. Many male frogs have **vocal sacs,** which open into the mouth cavity and increase the intensity of the sound.

Frogs croak mostly during the breeding season. Croaking may occur underwater; in this case, air is forced into the lungs, past the vocal cords into the mouth cavity, and back again.

Respiration in vertebrates is discussed more fully in Chapter 25.

CIRCULATION. The circulatory system of the frog consists of a heart, arteries, veins, capillaries, and lymph spaces. The liquid portion of the blood, the **plasma,** carries food and waste matter in solution. It contains three kinds of corpuscles (Fig. 18–14): red corpuscles (erythrocytes), white corpuscles (leukocytes), and spindle cells (thrombocytes).

The **red corpuscles (erythrocytes)** are elliptical, flattened cells containing a respiratory pigment called *hemoglobin.* The **white corpuscles (leukocytes)** are of several types, varying in size; most of them are capable of amoeboid movement. A particular type of white corpuscle (**phagocytes**) is of great value to the animal, engulfing foreign bodies such as bacteria and thus helping to overcome infectious diseases. White corpuscles also aid in the removal of broken-down tissue. **Thrombocytes** are usually spindle-shaped; when brought into contact with foreign substances, they release an enzyme (**thromboplastin**) which through a complex process produces an insoluble substance, **fibrin,** which is necessary for blood clotting.

Blood corpuscles arise principally in the **marrow** of the bone. They also increase in number by division while in the blood vessels. Some white corpuscles are probably formed in the **spleen,** a gland that also destroys worn-out red corpuscles.

The frog's **heart** differs from that of many other vertebrates in having only one **ventricle** (instead of two) (Fig. 18–15). In the frog the muscular ventricle is conical in shape; rostral to it are the two less muscular **atria** (Fig. 18–16). The **conus arteriosus** ("arterial cone," where the aorta originates) arises from the base of the ventricle. The **sinus venosus** is a dorsal, triangular-shaped, thin-

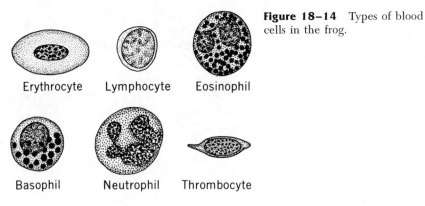

**Figure 18–14** Types of blood cells in the frog.

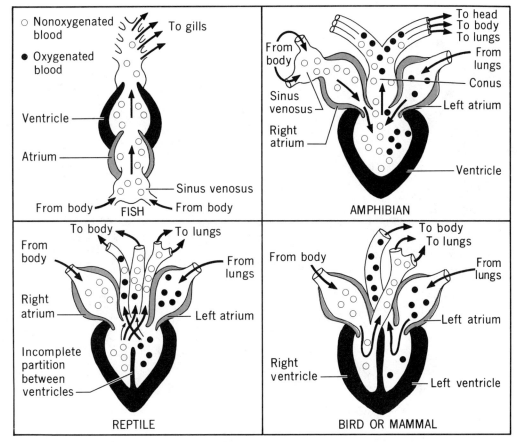

**Figure 18–15** Comparison of heart structure among vertebrates. Arrows indicate the direction of blood flow.

walled chamber which collects the body's deoxygenated blood from the major veins (**venae cavae**) and transmits it to the right atrium. This structure is present in all amphibians, fishes, and reptiles.

**Arteries** carry blood away from the heart (Fig. 18–17). The conus arteriosus divides near the anterior border of the atria into two vessels. Each branch is called a **truncus arteriosus,** and each gives rise to the following three arteries:

1. The **common carotid** divides into the **external carotid (lingual),** which divides to supply the tongue, roof of the mouth, eyes, and brain.
2. The **pulmocutaneous artery** branches, forming the **pulmonary artery,** which passes to the lungs, and the **cutaneous artery,** which splits further, supplying arteries to the skin.
3. The **aortic (systemic) arches,** after passing outward and around the digestive tract, unite to form the **dorsal aorta.** Before the union of the two aortic arches, branches forming the **brachial artery** and other arteries supplying the rest of the body are given off.

**Veins** return blood to the heart. The blood from the lungs is collected in the **pulmonary veins** and poured into the left atrium. Venous blood is carried to the **sinus venosus** by three large trunks; the two **anterior venae cavae** receive

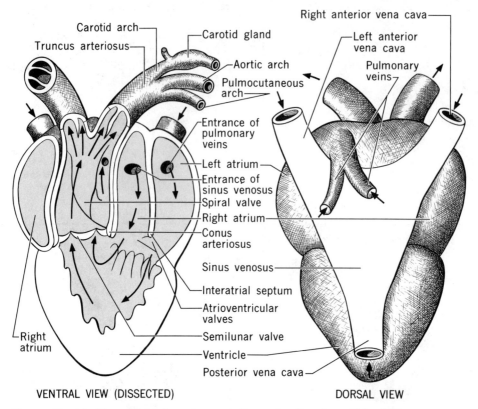

**Figure 18–16** Heart of the frog. Arrows indicate the direction of blood flow.

blood from the tongue, thyroid, head, shoulders, forelimbs, and sides of the body; and the **posterior vena cava** receives blood from the kidneys and liver.

**Circulation** in the frog occurs as follows. The sinus venosus contracts, forcing the deoxygenated venous blood into the right atrium. Oxygenated blood from the lungs passes into the left atrium. Both atria then contract, forcing their contents into the ventricle. Recent experiments have shown that the deoxygenated blood from the right side mixes with the oxygenated blood from the left side. Thus, it is likely that mixed blood is pumped to all parts of the frog's body. The blood is prevented from flowing back into the heart by valves. Respiration through the skin of the frog, both in water and on land, is thought to compensate, at least in part, for failure of all deoxygenated blood to be pumped to the lungs.

The blood that is thus forced through the arteries makes its way into tubular blood vessels, which become smaller and smaller until the extremely narrow **capillaries** are reached. Here food and oxygen are delivered to the tissues. The **renal portal system** carries blood to the kidneys, where urea and similar impurities are removed. The **hepatic portal system** carries blood from the stomach, intestine, spleen, and pancreas to the liver, where bile and glycogen are formed. This passage of venous blood from the intestinal tract through the liver before entering the main circulation makes it possible for the liver to remove food products from, or to add substances to, the blood as physiological needs require. The blood brought to the lungs and skin is oxygenated and then carried back to the heart. The passage of blood through the capillaries can easily be observed in the web of the frog's foot or in the tail of the tadpole.

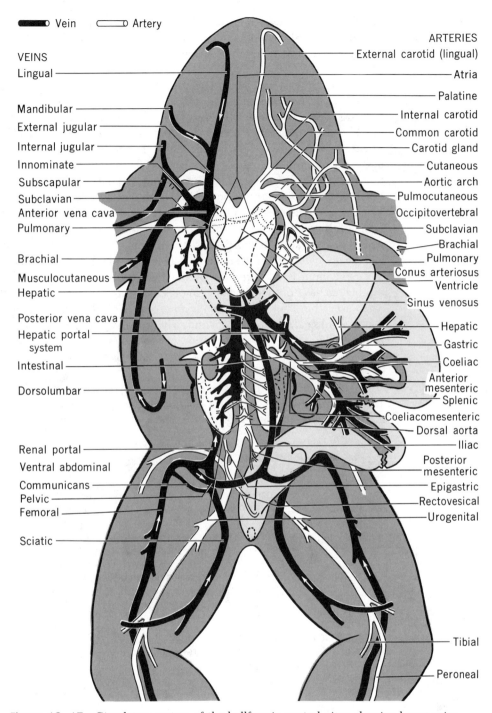

**Figure 18–17** Circulatory system of the bullfrog in ventral view, showing larger veins and arteries in relation to internal organs. Arrows indicate the direction of blood flow.

The **lymphatic system** of the frog includes many lymph vessels of various sizes that form networks coextensive with blood vessels but are difficult to see. Frogs and toads, unlike other vertebrates, have several lymph spaces between the skin and the body. The pulsation of four **lymph hearts,** two near the third

**257**

vertebra and two near the end of the vertebral column, force the lymph into the internal jugular and a branch of the renal portal veins. The watery lymph, which is colorless, contains leukocytes and various constituents of blood plasma.

**Nervous System and Sense Organs.** The nervous system of vertebrates is more complex than that of any other animals. The frog's **brain** (Fig. 18–18) has two large, fused **olfactory lobes,** two large **cerebral hemispheres,** a **diencephalon,** two large **optic lobes,** a very small **cerebellum,** and a **medulla oblongata.** The latter is produced by the broadening of the spinal cord. The **cranial nerves** of the frog are similar to those of higher vertebrates, with the major exception that frogs and other amphibia lack a hypoglossal nerve. The functions of the different parts of the frog's brain have been partially determined by experimentation in which parts of the brain were removed and the effects upon the animals observed. Evidence indicates that the cerebral hemispheres are involved in associative memory, the diencephalon with spontaneous movement, and the optic lobes with the reflex activity of the spinal cord. In human beings, the cerebellum is a center of coordination, but experiments on the frog's cerebellum have produced conflicting results. Many activities are still possible when everything but the medulla is removed. The animal breathes normally, snaps at and swallows food, leaps and swims regularly, and is able to right itself when placed on its back. Extirpation of the posterior region of the medulla results in early death. The brain as a whole controls the actions effected by the nerve centers of the spinal cord.

The frog depends primarily on reflexes, since "conscious" control is limited. Ten pairs of **spinal nerves** arise by **dorsal** and **ventral** roots from the gray matter of the spinal cord (Fig. 18–19). The roots unite to form a trunk, which

**Figure 18–18** Brain of the frog. (A) Dorsal view. (B) Ventral view. (C) Ventricles (cavities) in dorsal view.

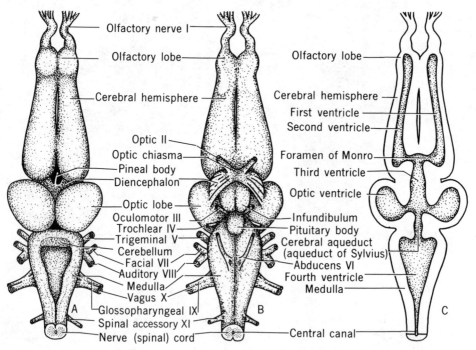

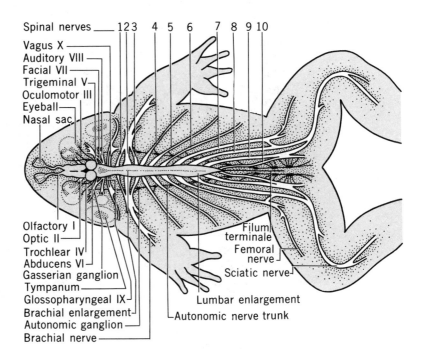

**Figure 18–19** Nervous system of the bullfrog: ventral view.

passes out between the arches of adjacent vertebrae. The two largest pairs of nerves are the **brachials** and the **sciatics**. The brachials are composed of the second pair of spinal nerves and branches from the first and third pairs; they are distributed to the forelimbs and shoulders. The sciatics, which arise from the seventh, eighth, and ninth spinal nerves, are distributed to the hindlimbs.

The **autonomic system** consists of two principal trunks, which begin at the cranium and extend posteriorly, one on each side of the vertebral column. Each trunk is provided with 10 ganglionic enlargements (**ganglia**) at the points where it unites with branches from the **spinal nerves**. The nerves of the autonomic system are distributed to internal organs and regulate many functions that are not under the control of conscious or voluntary action. These include heartbeat, secretions, and movements of the digestive tract, and respiratory, urogenital, and reproductive systems.

The principal **sense organs** are the **eyes, ears,** and **olfactory organs.** There are many smaller structures on the surface of the tongue, and on the floor of the mouth, which probably function as organs of taste. The skin also has many sensory nerve endings which receive contact, chemical, temperature, and light stimuli.

The **olfactory nerves** extend from the olfactory lobe of the brain to the nasal cavities, where they are distributed to the epithelial lining. The importance of the sense of smell in the frog is not known. The frog's **inner ear** lies within the **auditory capsule** and is supplied by branches of the auditory nerve. An external ear is lacking. The middle ear is a cavity that communicates with the mouth cavity through the **eustachian tube** and is closed externally by the **tympanic membrane (eardrum).**

A rod-shaped bone, the **columella,** extends across the cavity of the middle ear from the eardrum to the inner ear. Vibrations of the eardrum, produced by sound waves, are transmitted to the inner ear through the columella. The sensory end organs of the auditory nerve are stimulated by the vibrations, and the impulses carried to the brain give rise to the sensation of sound. The inner ears are also organs of **equilibrium;** without them, frogs cannot maintain an upright position.

The frog's **eyes** resemble human eyes in general structure and function. The **eyeballs** lie in cavities (orbits) in the sides of the head. They are rotated and pulled into the orbits by six muscles. The **upper eyelid** does not move independently. The **lower eyelid** consists of the lower eyelid proper, fused with the transparent third eyelid or **nictitating membrane.** The **lens** is large and almost spherical and is fitted for viewing moving objects at a definite distance. The amount of light that enters the eye can be regulated by the contraction of the **pupil.** The **retina** of the eye is stimulated by the rays of light that pass through the pupil; the impulses, carried through the optic nerve to the brain, produce vision.

ENDOCRINE GLANDS. Physiological processes of the frog, like those of other vertebrates, are regulated by **hormones.** These substances are produced by the endocrine glands and pass directly into the blood. The **pituitary gland,** found at the base of the brain, has three lobes. A **growth-stimulating** hormone is secreted by the anterior lobe in the immature stages. In the adult, the anterior lobe secretes a **gonad-stimulating** (gonadotropic) hormone, which initiates the release of ova or spermatozoa. **Intermedin,** produced by the intermediate lobe, regulates the chromatophores in the skin. The posterior lobe regulates water intake by the skin.

**Thyroxin,** produced by the thyroid gland, regulates general metabolism, while **insulin,** produced by the pancreas, regulates sugar metabolism. **Epinephrine,** secreted by the adrenal gland, has several functions, including the increase of blood pressure and the contraction of chromatophores in the skin, which contain dark pigments.

BEHAVIOR. The activities of the frog enable it to exist within the confines of its habitat. Ordinary movements such as leaping, diving, crawling, burrowing, and maintaining an upright position are due to internal causes, but many are responses to external stimuli. Frogs are sensitive to light; when placed in a light environment, they orient themselves in a direct line to the light sources's rays. Nevertheless, frogs tend to congregate in shady places. Frogs also seem to be stimulated by contact, since they have a tendency to crawl under stones and into crevices.

REPRODUCTIVE SYSTEM. The sexes of frogs are separate. The male can be distinguished from the female by the greater thickness of the first digit of his forelimbs. **Sperm** formed in the **testes** pass through the kidneys through the

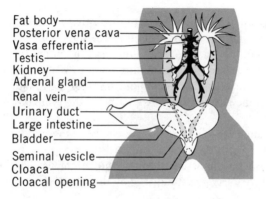

**Figure 18–20** Urogenital system of male frog: ventral view.

Fat body
Posterior vena cava
Vasa efferentia
Testis
Kidney
Adrenal gland
Renal vein
Urinary duct
Large intestine
Bladder
Seminal vesicle
Cloaca
Cloacal opening

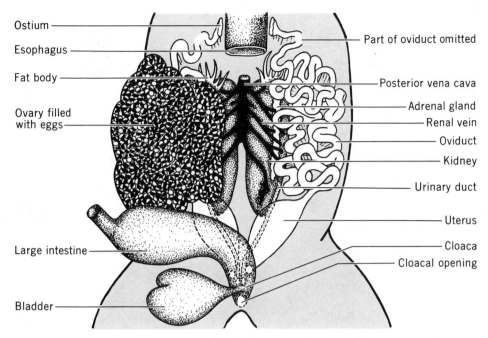

**Figure 18-21** Urogenital system of female frog: ventral view. Left ovary omitted; note large amount of space devoted to ovary.

**vasa deferentia,** then to the **urinary duct** by a route that varies in different species. The urinary duct is dilated at the posterior end in some species to form a **seminal vesicle.** The spermatozoa then pass on from the urinary duct or seminal vesicle, as the case may be, to the outside through the **cloacal opening** (Fig. 18-20).

The **eggs** arise in the **ovaries** of the female; during the breeding season, they break through the ovarian walls into the body cavity (coelom). The eggs then move anteriorly by the beating of undulipodia lining the peritoneum. The undulipodia at the entrance to the oviduct, the **ostium,** create currents that draw the eggs into the convoluted **oviduct;** they are carried down the oviduct into the thin-walled distensible **uterus** by action of the undulipodia in the oviduct (Fig. 18-21). The glandular wall of the oviduct secretes the gelatinous coats of the eggs. A general discussion of reproduction, embryology, and development is given in Chapter 27.

## LIFE CYCLE OF THE FROG

Frogs lay their eggs in water in the early spring. The male, using his forelegs and enlarged thumb pads, firmly clasps the female just behind her forelegs. As the eggs are extruded by the female, they are fertilized by sperms discharged by the male. Each female lays several hundred eggs. The jelly that surrounds and protects the eggs soon swells through absorption of water.

The tadpole breaks out of the membranes, lives for a few days on the yolk in the digestive tract, and then feeds on algae and other vegetable matter (Fig. 18-22). The **external gills** grow out into long branching tufts. A skin fold or **operculum** grows over the external gills; both eventually degenerate and are replaced by **internal gills.**

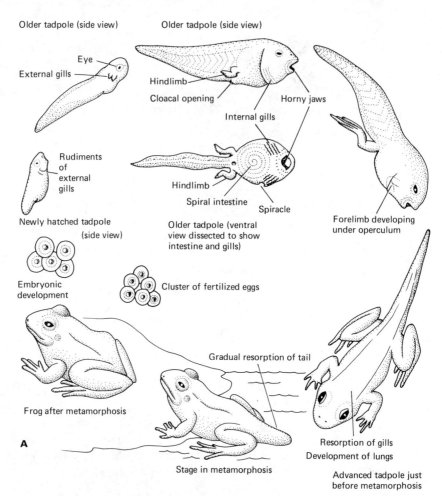

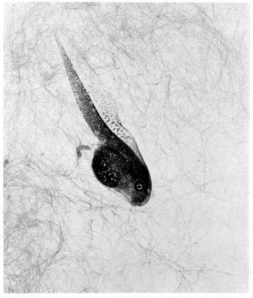

**Figure 18-22** (A) Life cycle of the frog. (B) Older tadpole. Note developing hind legs.

The hindlimbs appear first; later the forelimbs break out. The tail decreases in size because it is gradually resorbed. The gills are then resorbed, and lungs develop. Finally, the adult form is acquired.

## Toads

The family Bufonidae includes well over 100 species of toads, most belonging to the genus *Bufo*. About 15 species of this genus occurs in the United States. The common toad of the northeastern United States, *Bufo terrestris* (formerly *B. americanus*), has a rough, warty skin, but does not cause warts, as was once supposed. Toads secrete a milky poisonous fluid from glands in the skin. The fluid protects them from many enemies. During the day toads remain concealed in dark, damp places. At night they hop about, feeding upon worms, snails, and insects, which are captured with their sticky tongues.

## Salamanders

Newts belong to the salamander family (Fig. 18–23). The crimson-spotted newt, *Diemictylus viridescens*, lives in the water as a larva; but when it is about 2.5 cm long it loses its gills and comes on land. During its terrestrial life (1 to 2 years) it is bright coral red in color and known as the red eft. When it returns to the water, it changes to the adult coloration, yellowish green with black spots on the under surface and a row of black-bordered crimson spots on both sides. The cold, slimy skin of the salamanders gave rise to the medieval belief that salamanders could live in fire and not be injured by it. The skin of the fire salamander of Europe secretes a particularly poisonous substance. This black species has bright yellow spots. Its colors seem to warn other animals that it is dangerous.

**Figure 18–23** Red-bellied newt, *Taricha rivularis*, in water.

# Other Amphibia

### FOSSIL AMPHIBIANS

From fossils, it has been determined that amphibians first appeared during the Devonian period. From that time on they increased so rapidly in numbers that the late Paleozoic or Carboniferous period is spoken of as the "Age of Amphibians." The Paleozoic amphibians, the Stegocephalia, were salamander-like animals about 3 cm long that probably lived in fresh water or on land. Their name refers to their covered or mailed head, roofed over by dermal bones. Primitive reptiles (cotylosaurs) and perhaps mammals stemmed directly from stegocephalians.

### LEGLESS AMPHIBIANS

The family Caeciliidae includes over 50 species of wormlike legless Amphibia. They inhabit the tropical regions of the Americas, Africa, and Asia. They burrow in moist ground with their strong heads and possess small, concealed eyes. A sensory tentacle, which can be protruded from between the eyes and the nose, aids the animal in crawling.

### TREE FROGS

Tree frogs (Fig. 18–24) are usually arboreal amphibians with rough adhesive disks on their toes and fingers that enable them to climb trees. They are provided with large vocal sacs and have a correspondingly loud voice. There are over 180 species. The common tree frog **hyla versicolor** is about 5 cm long. Other common tree frogs are the spring peeper and the cricket frog.

### GIANT SALAMANDERS

The family Cryptobranchidae contains two genera of giant salamanders. The American hellbender, *Cryptobranchus alleganiensis*, occurs only in streams of the eastern United States; it reaches a length of from 45 to 70 cm. The giant salamander of Japan is the largest living amphibian, reaching a length of over 1.5 meters.

### AXOLOTL

The tiger salamander, *Ambystoma tigrinum*, reaches a length of 30 cm. In some parts of its geographic range, it fails to metamorphose and reproduces in

**Figure 18–24** *Hyla andersonii*, a tree frog. (Courtesy N.Y. Zoological Society.)

**Figure 18–25** *Necturus*, commonly called the "mud puppy." Note the featherlike gills of this aquatic animal.

the larval state. Such a larval form is an **axolotl;** its condition of sexual maturity is called **neoteny.** Axolotl had long been considered a separate species that exhibited retarded evolution because of its external gills, which persisted in the adult form. Experiments have since proved this feature to be a secondary specialization for arid regions. Nonmetamorphosing forms arose most probably from stocks that did undergo metamorphosis.

### *Necturus*

*Necturus* or the "mud puppy" (Fig. 18–25) is a strictly aquatic salamander. It is frequently used for study, exhibiting an unusual respiratory process that does not involve the mouth region; its reddish, bushlike gills are just at the caudal extent of the head, and wave about in the water, allowing diffusion to occur.

### POISONOUS AMPHIBIANS

As in the leopard frog, certain salamanders and newts possess poison glands. The warty skin of toads is usually heavily laden with poisonous secretions. As a means of defense the poison is very effective, since an animal that has once felt the effects of an encounter with a poisonous amphibian will not soon repeat the experiment. Some of the most poisonous species, such as *Salamandra salamandra,* may be warningly colored. The poisonous secretions (**bufotoxins**) of *Bufo marinus* are fatal to dogs and cats.

## Relations of Amphibia to Man

Virtually all amphibians are beneficial to man. Many are too rare to be of much value, but frogs and toads are of considerable importance. Frogs are continually being used for laboratory dissections, physiological experiments, human pregnancy tests, pharmacology, and fish bait. The skin of frogs are used for glue and book bindings. Frog legs are eagerly sought as food; more than 3 million pounds are eaten every year in the United States. Mud puppies are also edible. In Japan, the giant salamander is much esteemed for food.

In China, the skin of the toad is used as a medicine; its use may have some therapeutic value, since certain glands contain a digitalislike secretion that increases the blood pressure when injected into humans.

Frogs and toads are widely recognized as enemies of injurious insects. Toads are of special value, since they live in gardens where insects are most injurious. In France, gardeners buy toads to help keep obnoxious insects under control. **Bufo marinus** has proved to be an effective controller of insects in the tropics, especially where sugar cane is grown.

## Summary

The four living types of amphibians are frogs, toads, salamanders, and caecilians.

Although the capacity for regeneration is widespread in animals, it is most pronounced in amphibians. Most amphibians are oviparous, and, usually, fertilization is external. Other notable characteristics of amphibians include the ability to change color, and the production of poisonous secretions from the skin.

The frog, a representative amphibian, exhibits many features typical of all vertebrates, and is frequently used for comparative study. The frog's internal anatomy is a simple version of that found in higher vertebrates. The digestive system is relatively short; the respiratory system includes the skin as an important respiratory organ; the circulatory system incorporates a univentricular heart; the muscular system contains all three types of muscle tissue (smooth, striated, and cardiac); the nervous system is complex, as it is in all vertebrates, and reflexes play a major role in the frog's existence. The frog is capable of audition (although there is no external ear), vision, tactile sensation, olfaction, and, presumably, taste.

# 19

# Reptiles

Reptiles (class Reptilia) constitute one of the most interesting and, in general, one of the least-known classes of the vertebrates. These cold-blooded animals (poikilothermic) are most abundant in warm climates. Contrary to popular notion, very few reptiles are dangerous to man; the majority in the United States are harmless and many are beneficial.

The reptiles living today are a small fraction of the vast hordes that inhabited the earth in prehistoric times. Of the approximately 16 orders recognized (Fig. 19–1), four have living representatives; these are the turtles and tortoises, lizards and snakes, crocodilians, and *Sphenodon punctatus*, a lizardlike reptile confined to New Zealand. We will study each of these animals, and particularly the turtle, more closely following a brief description of some general characteristics of reptiles.

## General Characteristics

Reptiles are better adapted for terrestrial life than are amphibians, as is illustrated by their dry, scaly skin (frequently covered with bony plates), limbs suited for rapid locomotion, ossified skeleton, and shelled eggs suited for development on land (with protective embryonic membranes to prevent drying). Other advances of reptiles over amphibians include the partial or complete separation of the ventricle, resulting in greater separation of oxygenated and deoxygenated blood in the heart, and the existence of a copulatory organ. These animals occupy an important place in vertebrate hierarchy since their anatomy is intermediate between amphibians and birds.

## The Turtle

The turtle (Fig. 19–2) has been selected as our representative reptile. Turtles live on land, in fresh water, or in the sea. The word **tortoise** is mainly applied to the land species, which move slowly on land but can swim quite rapidly; the

**Figure 19-1** Some orders and families of the class Reptilia.

**Figure 19-2** The green turtle, *Chelonia*.

word **turtle** is often applied to the semiaquatic species. The majority of land and freshwater turtles hibernate in the earth during winter; in warmer countries, they estivate during the hotter months.

## EXTERNAL ANATOMY

The turtle is distinguished from all other animals by its **shell,** which is broad and flattened and protects the internal organs. The head, limbs, and tail can be withdrawn into the shell. The **neck** is long and flexible; the **head** is flattened dorsoventrally. The **mouth** is large; instead of teeth, horny plates, used to crush food, form the margin of the jaw. The **external nares** (nostrils) are placed together near the anterior end of the snout; this allows the head, when withdrawn into the shell, to perceive the odors of the environment.

Two **eyes,** situated on either side of the head, are guarded by three **eyelids:** a short, thick, opaque upper lid; a longer, thin lower lid; and a transparent nictitating membrane, which moves over the eyeball from the anterior corner of the eye. Just behind the angle of the jaw on either side is a thin **tympanic membrane.** Each **limb** generally possesses five **digits.** Most digits are armed with large horny claws that are used in crawling, climbing, or digging. The skin is thin and smooth on the head but thick, tough, wrinkled, and scaly over the exposed parts of the body.

## INTERNAL ANATOMY

**SKELETON.** The shell (Fig. 19-3) consists of a convex dorsal armor, the **carapace,** and a flattened ventral armor, the **plastron;** these are strongly bound together on each side by bony bridges varying in width between species. Both carapace and plastron are usually covered by a number of symmetrically arranged horny plates, called **scutes** (shields). The number and shape of the scutes vary according to species, but are usually constant in individuals of the same species. Beneath the scutes are a number of **bony plates** formed by the dermis and closely united by sutures.

The vertebrae and ribs are usually consolidated with the bony carapace; a sternum is lacking.

Reference to Fig. 19-4 will facilitate understanding of the following descriptions of the turtle's organ systems.

**DIGESTIVE SYSTEM.** Turtles are either omnivorous or herbivorous. Their animal prey consists of waterfowl, small mammals, and many species of invertebrates.

The **digestive organs** are simple. The broad, soft **tongue** is attached to the floor of the mouth cavity; it cannot protrude. At the base of the tongue is a longitudinal slit, the **glottis,** and a short distance behind the angles of the jaw are the openings of the **eustachian tubes.** The **pharynx** is thin-walled and can

**Figure 19-3** Turtle shell showing external horny scutes that are produced by the epidermis, which covers the bony plates.

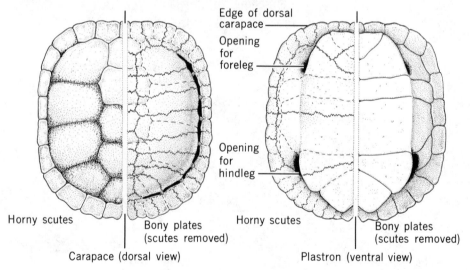

distend greatly; it leads into the thick-walled **esophagus**. The **stomach** opens by a **pyloric valve** into the **ileum (small intestine)** by the **ileocecal valve**. The **liver** discharges **bile** into the intestine through the **bile duct**. Several **pancreatic ducts** lead from the **pancreas** to the intestine. The terminal portion of the digestive canal is the **rectum**, which opens into the **cloaca**.

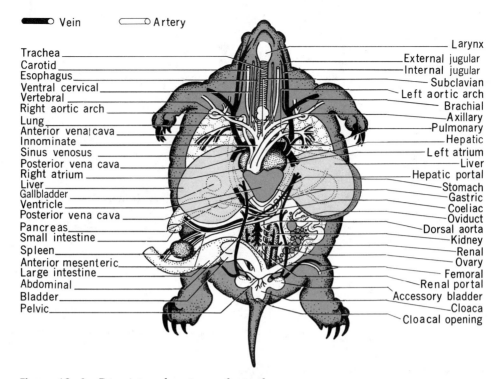

**Figure 19-4** Gross internal anatomy of a turtle.

CIRCULATORY SYSTEM. In the turtle and all reptiles (except crocodilians), the heart consists of two **atria** and a single **ventricle,** which is divided in two by an incomplete septum. (In crocodiles the longitudinal septum is complete, forming a four-chambered heart.) Venous blood from the body is carried by the **posterior vena cava** and the two **anterior venae cavae** into the **sinus venosus** and thence into the **right atrium.** From there it passes into the right side of the **ventricle,** which contracts, forcing the blood out through the pulmonary artery (which sends a branch to each lung), through the left aorta (which conveys blood to the viscera) and into the dorsal aorta.

Blood, oxygenated in the **lungs,** is returned by the pulmonary veins to the left atrium and thence into the left side of the ventricle. This blood is pumped out through the right aortic arch, which merges into the dorsal aorta. Because the septum dividing the two ventricles is incomplete, the blood that enters the right aortic arch is a mixture of oxygenated blood from the left atrium and venous blood from the right atrium. Certain species of turtles have a well-developed renal portal system. The **hepatic portal system** (serving the liver) shows an advance in development over that of the frog.

RESPIRATORY SYSTEM. Turtles and all reptiles breathe by means of **lungs.** Air enters the mouth cavity by way of the nasal passages. The **glottis** opens into the **larynx,** through which the air passes into the trachea or windpipe. The trachea divides, sending one **bronchus** to each lung. The bronchi branch dividing the lung cavity into many spaces and thereby increasing the respiratory surface area. Lungs of reptiles are more complicated and better developed than those of amphibians.

Many aquatic turtles may carry on respiration to some extent by taking water into the cloaca and accessory bladders and then forcing it out through

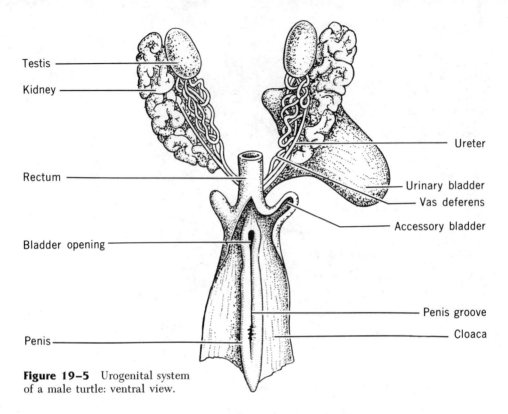

**Figure 19–5** Urogenital system of a male turtle: ventral view.

the cloacal opening; these structures may thus serve as supplementary respiratory organs. It has also been proposed that certain areas of the skin of aquatic forms may be modified for respiration.

**UROGENITAL SYSTEM.** Excretion is carried on by the two **kidneys.** Their secretions pass through the **ureters** into the **cloaca,** are stored in the **urinary bladder,** and exit through the **cloacal opening** (Fig. 19–5).

The sexes are separate. The **male reproductive organs** include two **testes** connected to a pair of **vasa deferentia** and an erectile **penis.** Spermatozoa pass from the vasa deferentia to the **penis,** which is attached to the ventral wall of the cloaca. The **female reproductive organs** are two **ovaries** and a pair of **oviducts;** the latter open into the **cloaca.** The sperms are injected into the female during **copulation,** usually preceded by courtship behavior.

Turtles are **oviparous;** the eggs are round or oval and covered by a hard shell. They are laid in holes dug by the female, in soil or decaying vegetation, where heat aids in incubation.

**NERVOUS SYSTEM AND SENSE ORGANS.** The **brain** is more highly developed than in amphibians. The **cerebral hemispheres** are larger and a distinction can be made between the outer gray area and the central white **medulla.** The **cerebellum** is also larger, indicating an increase in body coordination. There are 12 **cranial nerves,** as opposed to 11 in amphibians.

The **eye** is small, with a round **pupil** and an **iris.** The iris is usually dark in terrestrial forms, but often colored in aquatic turtles. A sense of **hearing** is not well developed, but turtles respond readily to vibrations through the skin and are thus easily frightened by noises. Turtles can distinguish various foods by **smell.** On many parts of the body, their skin is very sensitive to **touch.** We will now direct our attention to other animals in the class Reptilia.

**Figure 19–6** *Anolis*, the American chameleon, changes color from green to dark brown.

## Lizards

Lizards usually have an elongated body and four well-developed limbs for running, clinging, climbing, and digging. The tail is generally long and is easily self-amputated, a process called **autotomy**. Many lizards easily regenerate tails; these new organs lack vertebrae. The skin of lizards is usually covered with small scales.

Geckos are a type of lizard that inhabits all the warmer parts of the globe; they are harmless and usually nocturnal. Many have specialized **lamellae** under the toes, enabling them to climb over trees, rocks, walls, and ceilings with relative ease.

A number of lizards are called chameleons, but the 75 species of true chameleons live in Africa, Madagascar, Arabia, and India; they are most noted for their ability to change color rapidly. The American "chameleon," *Anolis*, also has a great capacity for color change (Fig. 19-6).

"Glass snakes" are lizards without limbs and move as most snakes do, by body undulations. They are distinguished from true snakes by the presence of movable eyelids and ear openings.

The largest of the lizards is the dragon lizard of Komodo, a species of *Varanus*, which lives on some of the small islands of Indonesia. They reach a length of 3 meters and weigh over 115 kg. These ferocious reptiles capture and feed on wild pigs and other large animals. When in captivity, the dragon lizard loses its ferocity and becomes quite tame.

## Snakes

As in other vertebrate groups, snakes are found in almost every kind of habitat. There are 10 families of this order (Serpentes); five of these are found in the United States.

Scales cover the snake's body; those on the head are regular and are used in taxonomy. The outer horny layer of skin is shed several times a year. The eyelids are fused over the eyes, but there is a transparent portion through which the snake sees. Just before being shed, the portion of the skin overlying

**Figure 19-7** Radiograph of a rattlesnake, showing absence of limbs, limb girdles, or sternum. Note the two rattles visible at the caudal end. (Courtesy of Armed Forces Institute of Pathology, Neg. No. 60631.)

the eye (**procornea**) becomes opaque, causing partial blindness. Once the entire skin is shed, normal vision returns.

Appendages are entirely absent (Fig. 19-7), except for a few species like the python and male boa constrictor, which possess short spurlike projections (vestiges of hindlimbs) on each side of the cloacal opening. Snakes move on land by several means; the two principal movements are lateral undulations of the body and the forward shifting of abdominal scutes in alternating sections of the body. Most snakes cannot move forward easily on smooth surfaces. All species can swim; this is, of course, the main method of locomotion for the aquatic species.

The snake has a slender, deeply notched tongue which, because of grooves in the jaws, can be thrust out even when the mouth is closed. The tongue serves as an auxiliary olfactory organ, carrying odorous particles to the paired **organs of Jacobson** (**vomeronasal organs**) on the roof of the mouth.

The sharp teeth curve inward to prevent food from slipping forward once swallowing has commenced. In venomous snakes, certain teeth are grooved or tubular and conduct venom into the bitten prey. Snakes do not chew food; they swallow it whole. A snake can eat animals much larger than itself. Several structural adaptations make this possible. The bones of the palate are movable and the lower jaw is loosely joined with the skull. The lower jaw can thus dislocate from the upper jaw, allowing the mouth to expand several times the diameter of the snake itself. The lower jaw can also spread at the anterior midpoint, allowing for lateral expansion. An extension of the trachea permits the snake to breathe during the long swallowing process.

**Pythons** (Fig. 19–8) and **boa constrictors** (Fig. 19–9) feed almost exclusively on birds and mammals, which they suffocate in their coils. As the prey exhales, the coils tighten until the prey can no longer inhale. Neither the python nor the boa constrictor is venomous; only a few individuals are large enough to be dangerous to human beings.

The common **garter snake** of eastern North America is the most abundant of the harmless snakes (Fig. 19–10). It feeds largely on frogs, toads, fishes, and earthworms. The slender **black snake** reaches a length of 2 meters. West of the Mississippi it gives way to a subspecies called the blue racer, and to another subspecies, the red racer, in Texas.

**King snakes** (Fig. 19–11), so named because they prey on other snakes, are constrictors of various sizes. The scarlet or coral king snake is immune to pit-viper venom, but not to the venom of the coral snake, which it resembles in color.

**Figure 19–8** Indian python in its natural habitat. This is one of the largest snakes, reaching a length of 8 meters. (Courtesy N.Y. Zoological Society.)

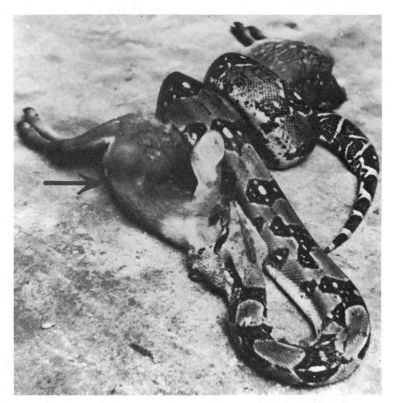

**Figure 19-9** Boa constrictor feeding on captured deer. The arrow indicates the point on the deer to which it had been swallowed when the photographer prepared to take this photograph. As a reaction to the photographer's presence, the snake began to disgorge its prey before the shutter clicked. (Courtesy James M. Keller and Clark Zeek.)

## VENOMOUS SNAKES

The **cobra de capello,** *Naja naja* of India, China, and the Malay Archipelago, is easily aroused. When disturbed, it raises the anterior part of the body from the ground, spreads its hood with a hiss, and threatens to strike. Some are capable of spitting venom accurately (aiming for the victim's eyes) for about 3 meters, and less accurately for greater distances (Fig. 19-12). If cobra venom strikes the human eye, it usually causes only temporary blindness if washed out immediately, but a strong sting will remain for some time.

**Figure 19-10** Garter snakes are frequently used as pets, since they are harmless to human beings.

**Figure 19–11** The king snake is a constrictor which preys upon other snakes, including poisonous species. The color is black and white or yellowish bands.

**Figure 19–12** The ringhals, a South American cobra, is shown with its neck spread, ready to strike and inject venom, which may cause death in a few minutes. (Courtesy of the American Museum of Natural History.)

**Figure 19–13** The copperhead snake *Agkistrodon*. Natural size is about 1 meter. (Courtesy N.Y. Zoological Society.)

Thirty species of poisonous snakes occur in the United States. The **coral "harlequin"** snake and the **Arizona coral snake** have round pupils and are ringed with yellow, black, and red; the red rings border the yellow ones. They occur in the Arizona desert and parts of the southeastern United States. The **copperhead** (Fig. 19–13), **water moccasin** (or **cottonmouth**) (Fig. 19–14), and **rattlesnake** (26 species) (Fig. 19–15), have vertical pupils and are called **pit vipers,** as they all possess a pit between the eye and nostril on each side of the head. The rattlesnake is further distinguished by the rattle at the end of the tail in adults. The **pit organ** (Fig. 19–16) consists of highly vascularized membrane and nerve endings. Experiments have shown it to be a heat-sensitive organ,

**Figure 19–14** Water moccasin (cottonmouth) about to strike. Its bite is occasionally fatal to human beings. (Courtesy of N.Y. Zoological Society.)

**Figure 19-15** *Crotalus horridus*, the rattlesnake. (Courtesy N.Y. Zoological Society.)

enabling the snake to detect a moderately warm body moving a few meters away from its head.

The venom of poisonous snakes is secreted by a pair of glands, located above the jaws, on each side of the head. Muscles surrounding the glands contract, squeezing the venom out through **poison fangs** and into the bitten prey. Several small fangs lie just behind the functional ones. These are held in reserve to replace those that are shed or lost in struggles with prey. Unless the venom is

**Figure 19-16** Rattlesnake, showing pit organ, hollow teeth (fangs), venom gland, venom duct, and muscles used to force poison into victim's flesh. (After drawing, courtesy American Museum of Natural History.)

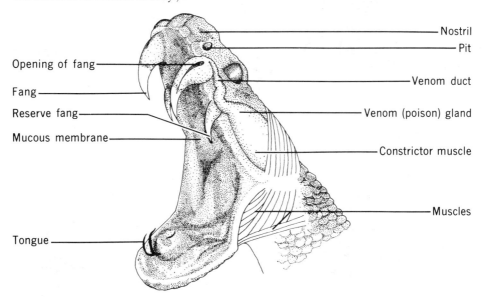

injected directly into a blood vessel, it usually travels slowly. The best **first-aid measures** to take in case of **snakebite** are:

1. Apply a tourniquet *snugly*, but not tight enough to interrupt arterial flow, a few inches above the bite. About every 15 minutes, release for 1 minute (or gangrene will develop).
2. Make an incision about 1.5 cm long and 1 cm deep.
3. Apply suction to the wound by mouth (do *not* do this if your mouth has any cuts or open sores).
4. The victim should lie down and be kept calm.
5. Seek a physician as soon as possible.

A person suffering from snakebite should *not* take liquor, physically exert himself, cauterize the bite with heat or strong acids, or inject potassium permanganate into the wound. The best and only true antidote is specific snake *antivenom*. A commercial polyvalent product effective against all poisonous snakebites can be obtained through Wyeth, Inc., Box 8299, Philadelphia, Pa. 19101.

# Crocodilians

Crocodilians include crocodiles, alligators, gavials, and caimans. They are lizardlike, but their jaws are extended into a long snout. Crocodiles can float on the surface of water with only their nostrils (at the end of the snout) and eyes (which protrude from the head) visible. Their thick, leathery skin is covered with horny epidermal scales and dorsal (and sometimes ventral) bony plates somewhat similar to the shells of turtles. Their nostrils and ears have valves that close when the animal is underwater.

The limbs are well developed with five digits on the forelimbs and four more-or-less webbed digits on the hindlimbs. The tail is a laterally compressed swimming organ. The cloaca opening ("anus") is a longitudinal slit. Two pairs of musk glands are present—one on the throat and the other in the cloaca.

There are 21 living species of crocodilians. The **American crocodile** (Fig. 19–17) inhabits Florida, Mexico, and Central and South America. The **African**

**Figure 19–17** The American crocodile has a pointed snout, webbed hind feet, five toes in front and four behind, and claws on the three inner digits. This is the largest known crocodile, reaching a length of over 7 meters. (Courtesy Science Software Systems, Inc.)

**Figure 19-18** The American alligator. Snout blunt, not pointed as in the crocodile. (Courtesy Science Software Systems, Inc.)

crocodile is one of the few man-eating species. It was once held sacred by the Egyptians, who preserved many specimens as mummies.

There are two species of the genus *Alligator:* the **American alligator** (Fig. 19–18) inhabits the southeastern part of the United States; the **Chinese alligator** is found only in China. Alligators differ from crocodiles in having a broad head with a blunt, rounded mouth and teeth which do not extend past their "lips" at the sides of the jaw. Alligators do not normally attack humans unless provoked.

The jaws of crocodilians are interesting; their musculature for closing is much stronger than for opening. A person of medium strength would have no problem preventing a crocodile from opening its jaws, but the immense power of the jaws when closing could snap a human vertebral column!

# Fossil Reptiles

## DINOSAURS

During the "Age of Reptiles" in the Mesozoic era (about 130 million years ago), enormous dinosaurs roamed the land, ichthyosaurs dominated the sea, and pterosaurs ruled the air.

The dinosaurs lived in swamps and on the uplands; remains have been found in most continents. Some species measured over 23 meters long; both herbivorous and carnivorous forms existed. *Brontosaurus* was herbivorous and about 23 meters long; its remains have been found in Wyoming and Colorado. *Stegosaurus* reached a length of 8 to 9 meters. It was also herbivorous and had huge triangular plates along its back. *Protoceratops* was a small, hornless, herbivorous dinosaur only 2 meters long; fossils were discovered in the deserts of Mongolia.

Many dinosaur fossils have been discovered at the La Brea Tar Pits in Los Angeles, California, including those of *Tyrannosaurus*, a fearsome carnivore.

*Ichthyosaurus* (Fig. 19–19) was a fish-eating aquatic reptile; they have been called the "whales" of the Mesozoic era. *Pterosaurs*, flying reptiles, had forelimbs modified for flight. Their skeletal characteristics somewhat resembled those of birds; *Pteranodon* was the largest form known, with a skull about

**Figure 19-19** *Ichthyosaurus*, a fishlike reptile, had limbs modified as paddles, a remarkable adaptation for swimming. (Courtesy American Museum of Natural History.)

60 cm long and a wing spread of more than 8 meters. These animals lacked teeth and possessed short tails.

Several other types of dinosaurs are known to have existed. It is thought that the decline of dinosaurs was due to large earth movements in the late Cretaceous (last of the Mesozoic periods) and during the Ice Age.

## Other Reptiles

### *Sphenodon*, A LIVING FOSSIL

**Sphenodon punctatus** is the sole surviving species of the reptilian order Rhynchocephalia. Numerous skeletal characteristics are like those of the oldest fossil reptiles, and the ancestors of reptiles were apparently much like this living relic. *Sphenodon* is now restricted to some small islands in the Bay of Plenty in New Zealand and in Cook Straits. Because it is now protected, it is thriving, with an estimated 5000 individuals on Stephan Island alone. One of its most striking peculiarities, shared with many lizards, is the **parietal eye** in the roof of the cranium—a structure with a retina and other characteristics which resemble those of a true eye in juveniles, but are vestigial and scarcely visible in adults.

**Figure 19-20** Common American turtles. The painted turtle is common in ponds. The snapping turtle is less protected by shell than many other turtles.

Among the more interesting types of turtles are the **snapping turtle** (Fig. 19-20), famous for its strong jaws and vicious bite; the **musk turtle,** which emits a disagreeable odor when disturbed; and the **painted turtle** (Fig. 19-20), noted for its brilliant coloring. The plastron of the **box turtle** is hinged so that the shell can be closed completely when the animal is in danger. **Sea turtles** inhabit tropical and semitropical seas and come to land only to lay their eggs on sandy beaches. Their limbs are modified as paddles for swimming. Some of the **giant tortoises,** such as *Testudo,* found on the Galapagos Islands, weigh over 230 kg and are probably over 200 years old. The **leatherback turtle** *Dermochelys* is the largest of all living turtles, sometimes attaining a weight of over 80 kg. It has a leathery covering over the shell instead of horny shields. **Soft-shelled turtles** (Fig. 19-21) also have shells that are leathery and without shields. The **green**

**Figure 19-21** *Amyda,* the soft-shelled turtle. Note the leathery integument, which is not divided into horny scutes. (Courtesy N.Y. Zoological Society.)

**Figure 19–22** Gila monster, the only poisonous lizard in the United States, shown feeding on eggs (a common food). Poison glands are in the lower jaw; grooved teeth carry venom into bitten prey.

turtle *Chelonia* is a marine species that has been extensively hunted for food and may be in danger of extinction.

The poisonous Gila monster or "beaded lizard" (Fig. 19–22) inhabits parts of Arizona, Utah, Nevada, and New Mexico. It is black with conspicuous pink or orange spots. Its bite is fatal to small animals and dangerous to human beings. The mechanism for injecting the poisonous venom is, however, far less efficient than that of the poisonous snakes; as it has to be ground into the victim's skin by nonspecialized teeth.

## Relations of Reptilia to Man

In general, reptiles do little damage by destroying plants and animals for food, since many of the animals preyed upon are insects, pests, and destructive rodents.

Turtles and tortoises rank first as reptilian food for human consumption (Fig. 19–23). The flesh of rattlesnakes is said to have a distinctly agreeable flavor; canned rattlesnake tastes somewhat like chicken.

Tortoise shell, especially from the hawksbill turtle, is used for combs and ornaments of various kinds.

**Figure 19–23** *Malaclemys*, the diamondback terrapin, is one of the most famous of all turtles as food for human beings. (Courtesy Shedd Aquarium, Chicago.)

Skins of lizards, snakes, and crocodilians have been extensively used in manufacturing articles that combine beauty with durability. Alligators, in particular, have been hunted for their valuable hides to the point where they are in danger of total elimination unless they are consistently protected as many now are.

The poisonous snakes of the United States are of very little danger to human beings. According to a World Health Organization study, there are about 300 to 400 deaths per year in North America from snakebite; of these, only about 19 per year occur north of Mexico. In tropical countries, especially India, the death rate due to venomous snakes is higher than that of any other group of animals.

## Summary

Four orders of reptiles exist today, represented by turtles and tortoises, snakes and lizards, crocodiles and alligators, and the living fossil *Sphenodon*. Although its shell sets it apart from all other organisms, the turtle is a simple representative reptile, with a univentricular heart with a partial septum, a relatively simple digestive system, and a nervous system that includes a bilaminar brain.

Lizards are extremely mobile cold-blooded (poikilothermic) creatures. Snakes have special jaws which enable them to prey upon animals much bigger than themselves, which they kill either by suffocation or poisoning. Crocodilians are lizardlike, but generally larger, with powerful jaws. Alligators differ from crocodiles mostly in head structure.

Dinosaurs of various forms dominated the earth, air, and sea during the Mesozoic era. The enormous size of some of these prehistoric species, along with the poisonous nature of certain reptiles and the relation of reptiles to human beings, are some of the most interesting aspects of this class.

# 20
# Birds

Many people consider birds (class Aves) (Fig. 20-1) to be the most interesting of all animals. Their beautiful and varied colors, pleasant songs and call notes, powers of flight, and migration all contribute to our fascination with these animals.

Because birds have a reptilian origin, some zoologists have called them "feathered reptiles." They do have reptilian scales on their legs, and the earliest birds, which we know of only from fossils, had reptilian teeth. Today there are about 10,000 species of birds. Following a discussion of some general characteristics of birds, we will look at our representative, the pigeon, in detail.

## General Characteristics of Birds

### FORM AND FUNCTION

Most birds can fly. The bodies of flying birds are **fusiform** or **tapered** in shape, thus offering little resistance to the air. The attachment of the wings high up on the trunk, the high position of such light organs as the lungs, the low position of the heavy muscles and digestive organs, and the resulting low center of gravity all tend to prevent the body from turning over.

Birds can glide by spreading their **wings** and moving forward by means of acquired velocity. When soaring, birds do not depend solely on acquired velocity, but are aided by favorable air currents. Several different types of flight are recognized. Some aerial birds, such as swallows, gulls, and albatrosses, have long pointed wings that enable them to remain in the air for long periods. Terrestrial birds, such as the bobwhite and song sparrow, have short rounded wings which enable them to fly rapidly for short distances. Besides the wings, other parts of the body of birds, including the tails, feet, and bills, have become adapted to various environments.

During flight, the **tail** acts as a rudder; a long-tailed bird is able to fly in short curves or follow an erratic course without difficulty. The tail is also used as an air brake. It is light and therefore easy to manage. While perching, the tail acts as a balancer. In many birds, such as the peacock, the tail of male birds is more colorful than that of the female.

**Figure 20-1** Representatives of the class Aves.

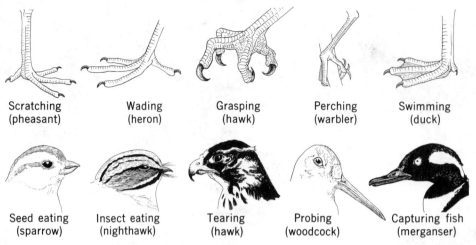

**Figure 20-2** Assorted types of feet and bills of birds.

**Feet** are used for locomotion, obtaining food, building nests, and offensive and defensive purposes. Many ground birds have strong feet adapted for scratching; perching birds have feet adapted for grasping a perch; most swimming birds have webbed feet; wading birds have long legs and long toes, and birds of prey, such as the hawk, have strong feet with long sharp claws for capturing other animals.

The **bills** of birds are not only used to procure food, but are also used to construct nests, preen feathers, and in other duties. Oil, produced by an oil gland at the base of the tail, is used to keep the feathers in good condition and prevents the bill covering from becoming brittle.

Seed-eating birds possess strong short bills for crushing seeds; some birds that eat insects have smaller and weaker bills; birds of prey are provided with strong curved beaks adapted for tearing flesh, and some birds have serrated bills for holding fish (Fig. 20-2).

### COLORS OF BIRDS

Birds are among the most beautifully colored of all animals. Some colors are due to pigments within the feather; these include pigment granules of brown, black, or yellow and red or yellow carotenoids. Green, blue, and iridescent markings, as on some hummingbirds, are due to the peculiar surface structure of particular feathers. The juvenile plumage of birds gives way to the first winter plumage; this is usually dull in color, often resembling the plumage of the adult female. Males and females frequently differ in color (sexual chromatic dimorphism), especially during the breeding season, when the male has a brightly colored coat.

Colors and color patterns of many birds conceal them in their surroundings and are thus of protective value. A striking example of protective coloration is the ptarmigan, which is white in winter when snow is on the ground and mottled brown in other seasons.

### BIRD SONGS AND CALL NOTES

For one who studies birds, a knowledge of bird songs is essential, as many birds are heard rather than seen. **Songs** are usually heard during the breeding season and are generally limited to the males. **Call notes,** on the other hand, are produced throughout the year and correspond in meaning and effect to our conversation. Birds' songs and call notes serve a variety of purposes: to warn of danger, to bring together birds in a gregarious species, for communication between parents and young, to attract mates, and, most frequently, to announce nesting territory.

## MIGRATION

In autumn some birds move southward, returning to their breeding grounds the following spring. Certain species migrate east and west. Birds that breed in the far north may spend part of the winter in the temperate zone. Some birds, such as the great horned owl, do not migrate. Others move southward only in very severe weather.

Most birds migrate on clear nights at an altitude of about 1 km or more; some are daytime migrants. Each species has a more-or-less definite time of migration, and one can predict with some accuracy the date when a species will arrive in a given locality. The speed of migration is rather slow as a rule; the robin averages about 60 km per day during migration.

We will now take a closer look at the structure of birds, using the pigeon as our representative.

# The Common Pigeon

The pigeon (Fig. 20–3) is commonly used for study because of its convenient size, availability, and because it illustrates so well the many adaptations for aerial life.

## EXTERNAL ANATOMY

The body of the pigeon is spindle-shaped and therefore adapted for rapid movement through the air. It is divided into three regions: head, neck, and

**Figure 20–3** The common pigeon.

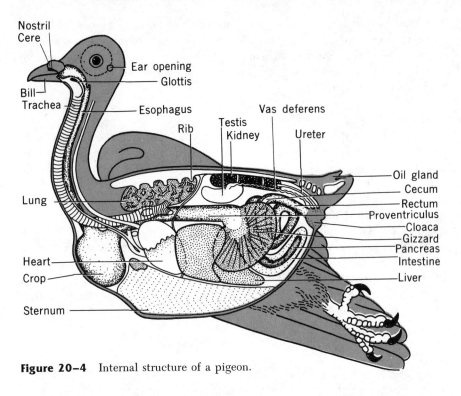

**Figure 20-4** Internal structure of a pigeon.

trunk. The **head** is prolonged in front into a pointed horny bill, at the base of which is a patch of swollen skin, the **cere** (Fig. 20-4). Between the bill and the cere are the two oblique, slitlike **nostrils.** On either side is an **eye,** which is provided with upper and lower **lids** and with a well-developed third eyelid or **nictitating membrane.** Below and behind each eye is an external **ear opening,** which leads to the tympanic cavity.

The **neck** is long and flexible. At the posterior end of the trunk is a projection which bears the tail feathers. The two wings can be folded close to the body or extended in flight. The **feet** are covered with horny epidermal **scales,** and each digit is provided with a horny **claw.**

**FEATHERS.** Feathers are peculiar to birds; they arise, like the scales of reptiles, from dermal papillae (blunt or rounded projections), with a covering of epidermis, and become enveloped in a pit, the feather follicle. A **typical feather** consists of a stiff axial rod, the **shaft,** with a semitransparent, hollow portion proximal to it and a flattish **vane** at the distal end. The vane is composed of a series of parallel **barbs,** each bearing a fringe of small processes, **barbules,** along either side. The barbules on one side of the barb bear **hooklets,** which hold together the adjacent barbs. The whole is thus a pliable, but nevertheless resistant, structure adapted for flight.

There are three principal kinds of feathers. **Contour feathers** (Fig. 20-5) are like those just described. Since these appear on the surface, they determine to a large extent the contour of the body. They include the general body, wing, and tail feathers. **Down feathers** have no shaft; the barbs arise as fluffy tufts from the end of the quill. They provide the first plumage (natal down) of newly hatched birds and provide an insulating covering beneath the contour feathers. **Filoplumes** are the third type; they are hairlike feathers with a few barbs at the tip.

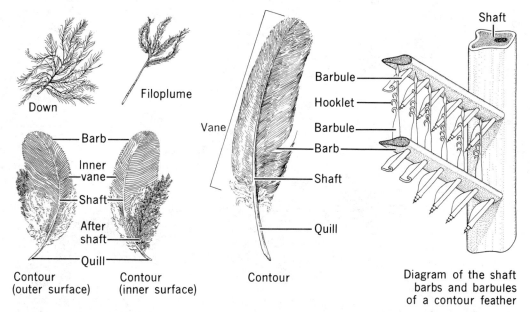

**Figure 20–5** Pigeon feathers. *Left:* types of feathers. *Right:* detail of contour feather.

Only certain portions of the pigeon's body bear feathers; these feather **tracts** are called **pterylae,** and the featherless areas are called **apteria.** Feather tracts differ in different species of birds.

Birds shed (**molt**) their old feathers, usually in late summer, and acquire a complete new set of feathers. There may be a partial or complete molt in the spring, when the bird assumes its **breeding plumage.**

## INTERNAL ANATOMY

SKELETON. Adaptations for flight and other methods of locomotion constitute the principal differences between the skeleton of the pigeon and that of reptiles. The skeleton of the common fowl (Fig. 20–6) is larger and more easily studied than the pigeon and is similar to the latter in most respects. The **skull** is very light, and most of the bones are so fused together that they can be distinguished only in the young bird. The **cranium** is rounded; the **orbits** are large; the **facial bones** extend forward into a bill, and there is a single **occipital condyle** for articulation with the first vertebra. Teeth are lacking.

The **cervical vertebrae** are long and move freely upon one another by saddle-shaped articular surfaces, making the neck very flexible, and enabling the bird to look in all directions. The vertebrae of the trunk are almost completely fused together, forming a rigid skeletal axis that supports the body while in flight. There are four to five free caudal vertebrae which allow for tail movement. These are followed by a terminal **pygostyle,** which supports the large tail feathers and consists of five to six fused vertebrae.

There are two **cervical ribs** and four to five **thoracic ribs** on each side. The second cervical and first four thoracic ribs each bear an **uncinate process,** which overlaps, thus making a firmer framework. The thoracic ribs are connected with the breastbone (**sternum**), which is, in turn, united with the **coracoid** of the pectoral girdle and bears on its ventral surface a median ridge, the **keel,** to which are attached the large muscles that move the wings. The

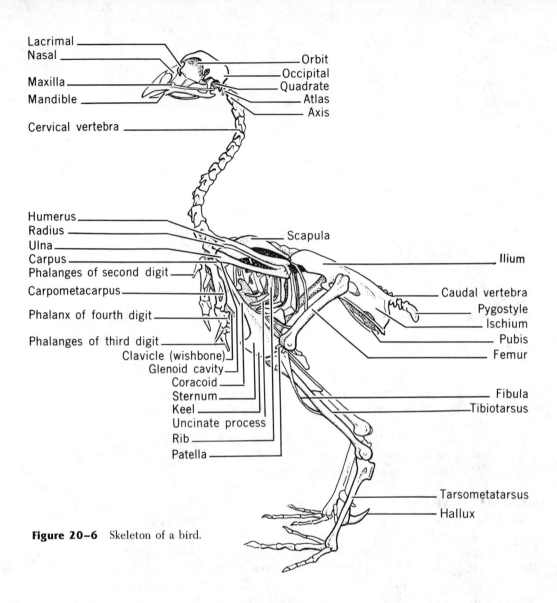

**Figure 20–6** Skeleton of a bird.

**pectoral girdle** consists of a pair of long, narrow **scapulae** (shoulder blades) which lie in the thorax above the ribs, on either side of the vertebral column.

The forelimb or **wing** is modified in many ways from that of primitive vertebrates. Instead of five digits, as was found in the frog, the fowl possesses only the second, third, and fourth digits. The second digit (**alula**) consists of two small bones which support a small tuft of feathers. The middle (third) digit has three **phalanges** and is the most developed. The fourth digit contains a single **phalanx**.

The arm, as in other vertebrates, contains a single bone, the **humerus**, followed by the **radius** and **ulna** in the forearm. The wrist contains two **carpal bones**; the other carpal bones are fused with the **metacarpals**, forming the **carpometacarpus**. The main flight feathers, or **primaries**, are supported by digits 3 and 4, the **secondaries** by the ulna and the tertiaries by the humerus.

The **pelvic girdle** consists of a pair of **ilia**, the **ischia**, and the **pubes**. These bones are firmly fused and united with the posterior part of the vertebral column.

The **hindlimbs** are used for bipedal locomotion. The **femur** is the short, thick thighbone. In each leg there is a slender **fibula** and a long, stout **tibiotarsus** (drumstick); the latter consists of the tibia fused with the proximal row of tarsal bones. The **ankle joint** separates the tibiotarsus from the **tarsometatarsus,** which represents the distal row of tarsal bones and the second, third, fourth, and fifth metatarsals fused. The foot possesses four digits; the first, the **hallux,** is directed backward. Each digit bears a terminal **claw.**

**MUSCULAR SYSTEM.** Muscles of the neck, tail, wings, and legs are especially well developed. The largest muscle is the **pectoralis major,** which produces the downward stroke of the wings; it constitutes about one-fifth of the total body weight. The **pectoralis minor** muscle raises the wing. These muscles comprise what is popularly known as the breast of the bird. Connected with the leg muscles is a perching mechanism, which enables the bird to maintain itself upon a perch even while asleep.

**DIGESTIVE SYSTEM.** Birds have an extremely high metabolism, and thus require large quantities of food, which they digest rapidly. Pigeons feed mostly upon vegetable foods such as seeds. The **mouth cavity** opens into the **esophagus,** which enlarges into a **crop,** where the food is stored and moistened. The **stomach** consists of two parts, an anterior **proventriculus,** which has thick glandular walls that secrete gastric juice; and a thick muscular **gizzard (ventriculus),** which grinds up the food with the aid of grit swallowed by the bird (Fig. 20–4). The slender, many-coiled **intestine** leads to the rectum at a point where two blind pouches, the **ceca,** are given off. The digestive canal leads into the **cloaca,** into which the urinary and genital ducts also open; the cloaca opens to the outside by means of the **cloacal opening** or **vent.** In young birds a thick glandular pouch of lymphatic tissue, the **bursa fabricii,** lies just above the cloaca.

The large **liver** is bilobed and has two **bile ducts.** Although it is present in some birds, there is no gallbladder in the common pigeon. The **pancreas** pours its secretions into the duodenum through three ducts.

**LYMPHATIC SYSTEM.** A **spleen** and, in young pigeons, paired **thymus glands** are present. Lymph nodes are lacking in birds, except in waterfowl.

**EXCRETORY SYSTEM.** The kidneys are a pair of dark-brown three-lobed bodies. Each discharges its semisolid nitrogenous wastes through a duct, the **ureter,** into the **cloaca.** The nitrogenous waste material of birds differs from that of other organisms. It is superconcentrated—the liquid component is almost completely removed just before defecation, a process that aids osmoregulation. There is no urinary bladder, but the semisolid uric acid passes directly out of the cloaca with the feces as a whitish substance.

**CIRCULATORY SYSTEM.** The **heart** of the bird is comparatively large. It is composed of two entirely separated muscular **ventricles** and two thin-walled **atria.** The **right atrium** receives deoxygenated venous blood from the **venae cavae.** This blood passes from the right atrium into the **right ventricle** and is then pumped through the **pulmonary artery,** which divides into right and left pulmonary arteries leading into the right and left lungs (Fig. 20–7).

After being oxygenated in the lungs, the blood returns through four large **pulmonary veins** to the **left atrium.** It passes from the left atrium into the **left ventricle** and is then pumped through the **right aortic arch,** which gives off the **innominate arteries** and continues as the **dorsal aorta.**

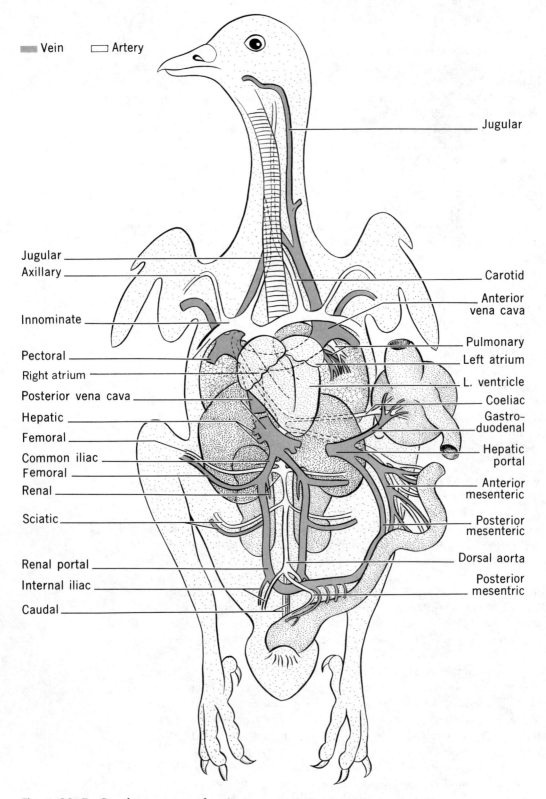

**Figure 20-7** Circulatory system of a pigeon.

Circulation in birds is extremely rapid; the blood is forced through the body by a heart that beats several hundred times per minute when at rest, and up to 1000 or more times per minute under stress.

The **renal portal system** of the pigeon is almost nonexistent since the blood is taken up from the posterior part of the body directly to the heart and not through the renal capillaries as in all lower vertebrates. The **jugular veins** of the pigeon are united just under the head by a cross vein; this special adaptation enables the blood to pass back to the heart from the head whenever the neck becomes twisted and one of the jugular veins is momentarily blocked.

**Respiratory System.** Air enters the mouth cavity through the **nostrils,** as in reptiles, and then passes through the **glottis** into the **trachea** or windpipe, which divides, sending a branch (**bronchus**) to each **lung** (Fig. 20–8). Each bronchus breaks into smaller tubes as it enters the lungs. Several large, thin-walled **air sacs** are attached to the smaller bronchial tubes; these extend out between organs of the body to spaces in the neck region and into the cavities of the larger hollow bones.

Birds have a high respiratory rate; for the pigeon it is 29 times per minute at rest, compared to 14 to 20 times per minute in human beings. This rapid flow of air through the respiratory system explains why the lungs of birds can be so small even though they have, because of their high metabolic rate, the highest oxygen requirements of all animals. Since birds do not have sweat glands, the air sacs also serve as thermal regulators, eliminating excess body heat.

The trachea is held open by partially calcified **cartilaginous rings.** Where the trachea divides into the two bronchi is an enlarged portion that forms the vocal organ or **syrinx,** a structure unique to birds. Note in Fig. 20–8 that the respiratory system is hooked up to the skeletal system.

**Nervous System and Sense Organs.** The **brain** of the pigeon is very short and broad. The **cerebrum** and **cerebellum** are comparatively large, as are the **optic lobes.**

The **bill** and **tongue** are tactile organs; tactile nerves are also present at the base of the feathers, especially those of the wings and tail. Birds have poorly developed senses of taste and smell, but an acute sense of hearing. The **cochlea** of the ear is more complex than that of reptiles. The **eustachian tubes** open by a single aperture on the roof of the pharynx.

The **eyes** of birds are very large. In some birds of prey, **visual acuity** is eight times better than in human beings. Birds also have a wide field of vision and the

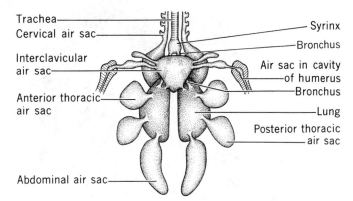

**Figure 20–8** Respiratory organs of a pigeon. Note connection of hollow bones to respiratory system. (After *Thomson's Outlines of Zoology,* 9th ed., revised by James Ritchie, New York: Oxford University Press, 1944.)

night-hunting birds, such as the owl, are adapted for vision in dim light. This ability results from the **tapetum lucidum,** a reflective layer situated behind the visual cells of the retina. Birds have extraordinary powers of eye **accommodation,** changing from farsighted to nearsighted vision in an instant.

**ENDOCRINE GLANDS.** Birds are similar to mammals in possessing typical mammalian endocrine glands. The **pituitary gland** (hypophysis) is at the base of the brain, the **thyroid** in the neck, the **islets of Langerhans** in the **pancreas,** the **adrenals** on the ventral surface of the kidneys, and the endocrine tissues in the **gonads.** The functions of these glands are similar to those of other vertebrates.

**REPRODUCTIVE SYSTEM.** The **male** has a pair of oval **testes.** From each testis the **vas deferens** duct passes posteriorly and opens into the **cloaca;** often it is dilated at its distal end to form a **seminal vesicle.** The sperm pass through the vasa deferentia and are stored in the seminal vesicle. When copulation takes place, they are discharged into the cloaca and transferred to the cloaca of the female. Most birds do not possess copulatory organs; a curved penis, however, arises from the ventral wall of the cloaca in male ducks, geese, swans, and a few other species.

The right ovary of the **female** usually disappears during development so that only the **left ovary** persists in the adult. The ova (eggs) break out of the ovary and enter the **oviduct.** If fertilization occurs, it takes place in the oviduct about 41 hours before the eggs are laid.

## Nests, Eggs, and Development

Some birds, like geese, and eagles, usually mate for life, but the majority live together for a single season. The nesting period varies according to the species.

Many birds conceal their nests or build them in places that are relatively inaccessible. A nest-building bird may construct a flimsy platform of twigs or an intricate hanging basket, but most species build distinctive nests (Fig. 20–9). Some species do not build nests, but lay their one or more eggs directly upon the ground. There are a few birds, such as the European cuckoo, that neither build nests, incubate nor care for their offspring. Their eggs are usually laid in

**Figure 20–9** Nest and eggs of a horned lark. The nest has thick, well-built walls. The coarsely spotted eggs were laid in the nest by a cowbird.

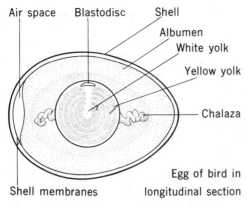 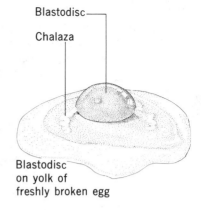

**Figure 20-10** Structure of a bird's egg.

the nests of smaller birds; the young birds are carefully reared by their foster parents and often starve or crowd out the rightful offspring.

The eggs of birds vary in shape, size, color, and number. As a rule, eggs laid in a dark place are white. Many eggs are colored or spotted; these are usually distinguishable by species. The general structure of a bird's egg is shown in Fig. 20-10.

**Figure 20-11** *Top:* young chimney swifts are naked and blind at birth and are cared for in the nest for about 3 weeks (altricial). (Photo by Robert Knickmeyer from National Audubon Society.) *Bottom:* piping plovers, like newly hatched ducks, are covered with down and leave the nest soon after hatching (precocial). (Courtesy Bertha Daubendiek.)

The number of eggs laid in a clutch varies from 1 to 20. The period of incubation is about 12 to 14 days for the smaller passerine birds. The eggs of the ostrich hatch in 45 days and those of the royal albatross, a seabird, have the longest incubation period, approximately 80 days. In some cases only the female incubates; in others, the male and female work in shifts. In a few birds, such as some ostriches and shorebirds, the male performs nearly all of this duty.

Young birds that are able to run about soon after hatching are **precocial birds;** those that remain in the nest for a period of time before they are able to care for themselves are **altricial birds** (Fig. 20–11).

# Other Birds

### FOSSIL AND EXTINCT

The study of fossil birds, although of great interest, is an extremely restricted field, as little is known about this animal group's history. Birds are divided into two subclasses, the Archaeornithes, or ancient birds, and the Neornithes, or recent birds (which includes all living birds).

In the subclass Archaeornithes are two genera, *Archaeornis* and *Archaeopteryx;* they are known from two fairly complete skeletons found in the lithographic slates of Solenhofen, Bavaria, in the Upper Jurassic period. Each was about the size of a crow. They possessed teeth embedded in sockets, forelimbs with three-clawed digits and separate metacarpal bones, and a lizardlike tail with large feathers at both sides (Fig. 20–12). Although reptilian features predominate in the skeleton, the presence of feathers, indicating warm blood, places these curious creatures among the birds.

Other extinct birds include the passenger pigeons, great auks, dodos, Labrador ducks, and heath hens; these have all become extinct in recent times. It is certain that the enormous slaughter of the passenger pigeon as food for man led to its extinction in 1914. The great auk became extinct in 1844; they were destroyed for their feathers, and their eggs were used for food. The North American ivory-billed woodpecker is in extreme danger of extinction (or may already be extinct); in 1970, there were only 20 left. It is obvious that we must take responsibility for our use of these animals and provide protection for threatened species before it is too late.

### WINGLESS AND FLIGHTLESS BIRDS

Five orders of living birds are flightless; they include the **ostriches** (the largest living birds), **rheas** (smaller than true ostriches, but with similar habits), and **cassowaries** and **emus** (ostrichlike birds) (Fig. 20–13). **Kiwis** have rudimentary wings and lack tail feathers; they are about the size of the common fowl.

### AQUATIC BIRDS

The **penquin,** also a flightless bird, is perhaps the most conspicuously adapted of the aquatic birds (Fig. 20–14). The forelimbs are modified as paddles for "flying" under water; the feet are webbed; the cold water can be entirely shaken from the feathers and a layer of fat just beneath the skin serves to keep in the body heat. Other aquatic birds include **loons,** large birds that swim and

**Figure 20-12** *Top:* fossil remains of *Archaeornis* (ancient bird), showing claws on digits of forelimbs, and long tail. *Bottom:* restoration of *Archaeornis*, about 0.5 meter long. (Courtesy American Museum of Natural History.)

**Figure 20-13** Adult emu with chicks that are strikingly striped; this characteristic disappears as the flightless birds mature. (Courtesy Victorian Railways.)

dive with great agility; **albatrosses**, strong fliers with exceptionally long and narrow wings; and **pelicans**, which scoop up small fish and carry them in a huge pouch that distends between the branches of the lower jaw.

Among the common **wading birds** are the **herons** (Fig. 20-15) and **bitterns**, both possessing long legs, broad wings, and short tails; the **flamingos**, gregarious

**Figure 20-14** Penguins are adapted for their antarctic environment, with paddle-like flippers as forelimbs. These flightless birds do not build nests, but have a special brood pocket between their legs for keeping eggs warm.

**Figure 20-15** Great blue heron. Long neck, long legs, a slender body, and a stilettolike bill are characteristics of this water-adapted species. (Courtesy Science Software Systems, Inc.)

birds that congregate in thousands upon mud flats; and the **ducklike birds** (including **swans, geese,** and **ducks**), which have short legs and fully webbed front toes to facilitate swimming. The **marsh birds** are usually wading birds with incompletely webbed front toes. Various birds comprise the **shorebirds,** which spend most of their time at or near fresh or salt waters.

## BIRDS OF PREY

These falconlike birds are noted for their food-hunting ability. Those active during the day (**diurnal**) usually possess powerful wings, a stout hooked bill, and strong toes armed with sharp claws. They include the **bald eagle** (Fig.

**Figure 20-16** Bald eagle. The head and tail are white and the beak is hooked.

**Figure 20–17** The vulture is saprophagous; it feeds on dead animals. (Courtesy Allan D. Cruikshank, National Audubon Society.)

20–16), **Cooper's hawk, sparrow hawk,** and **golden eagle. Vultures** (Fig. 20–17) have weak feet and live on carrion (dead animals); they are valuable scavengers in warm countries where dead bodies would otherwise become a health hazard.

Owls are **nocturnal** birds of prey. They possess large, rounded heads, strong legs, feet armed with sharp claws, strong curved bills, large eyes directed forward and surrounded by a sound-absorbing radiating disk of feathers, and soft, fluffy plumage which renders them almost noiseless in flight. Among the well-known North American species are the **barn owl, screech owl, great horned owl** (Fig. 20–18), and **burrowing owl.**

**Figure 20–18** Great horned owl. Tufts of feathers around large external ears are typical of this species. (Courtesy Science Software Systems, Inc.)

## PERCHING BIRDS

More than half of all living birds belong to the order *Passeriformes*, the perching birds. Their four-toed feet are adapted for grasping. The first toe (the hallux) is directed backward and is on a level with the other three, which are directed forward.

## DOMESTICATED BIRDS

The common hen was probably derived from the red jungle fowl, *Gallus gallus*, of northeastern and central India. The varieties of chickens that have been derived from this species are almost infinite. Many varieties of domesticated pigeons have descended from the wild rock dove, *Columba livia*. Other domesticated birds are the geese, ducks, peacocks, and turkeys.

# Relations of Birds to Man

Birds are of great commercial value; they augment our food supply and furnish feathers for various uses. Before the wearing of wild bird feathers was prohibited by law, vast numbers of birds were killed for their plumes. In certain regions, the excrement of seabirds, particularly the cormorants (Fig. 20–19), is used for fertilizer.

**Figure 20–19** Cormorants are voracious fish-eating birds. Their guano is collected, especially in Peru, and sold for fertilizer. (Courtesy Science Software Systems, Inc.)

One of the beneficial services birds perform for us is the destruction of weed seed. A great number of birds also feed on insects; practically all of the insects preyed upon are destructive to plants and animals, and consequently harmful to man.

A harmful bird is the yellow-bellied sapsucker, which eats the cambium of trees and sucks the sap out, thus disfiguring and devitalizing fruit and ornamental trees, and reducing the market value of lumber. Most birds, however, are beneficial to man.

## Summary

Birds are fusiform, winged animals; the great majority fly. Their feet, bills, wings, and tails are variously adapted to suit their environments. The common pigeon is a standard representative of this class. Its feathers (a feature unique to birds) are composed of a shaft, vane, barbs, barbules, and hooklets. Three major types of feathers exist: contour, down, and filoplumes. The avian skeleton is mostly hollow and modified for flight and bipedal locomotion. The pectoral muscles are responsible for avian flight.

Aquatic birds exhibit adaptations, such as flipperlike wings; predatory birds usually have excellent eyesight and sharp claws. A number of the 10,000 living species of birds were briefly discussed.

Birds are generally extremely beneficial to man, especially as food and in "pest" control.

# 21

# Mammals

**Mammals** are named for their mammary glands, which secrete milk for their developing young. All possess body hair at some time in their lives and are homeothermic—maintaining a constant body temperature. Parental care is generally most highly developed in this group, which includes over 5000 species, living in every major habitat from the tropics to the polar regions and from the oceans to the driest deserts (Fig. 21-1). Most mammals are viviparous, and their young are nourished before birth through placental tissue. The development of mammals reaches its climax in human beings.

Since detailed studies of mammals are ordinarily reserved for special courses in mammalian biology, only a brief account is presented here, using the cat as our representative. Following this is a discussion on some of the more interesting features of mammals and an overview of the diverse organisms found in this class.

## The Domestic Cat

The cat is much like human beings morphologically and physiologically. The domestic cat, *Felis catus*, is a carnivore; it belongs to the same family as do lions, tigers, and wildcats.

### EXTERNAL ANATOMY

The domestic cat is a four-footed (quadruped) animal, a characteristic of higher mammals. It has an external covering of **hair** or **fur**, two **ears**, and separate **genital** and **anal openings**. The mouth is bounded by thin, fleshy **lips**. At the end of the head are two narrow **nostrils**. Two large **eyes** are protected by upper and lower **eyelids** bordered by fine **eyelashes**. A white, hairless third eyelid or **nictitating membrane** may be drawn over the eyeball from the inner corner of the eye. Above and below the eyes and on the upper lip are long hairs or **whiskers (vibrissae)**.

The body of the cat consists of a head, neck, and trunk. The trunk is further separated into an anterior **thorax**, which is supported laterally by the ribs, and

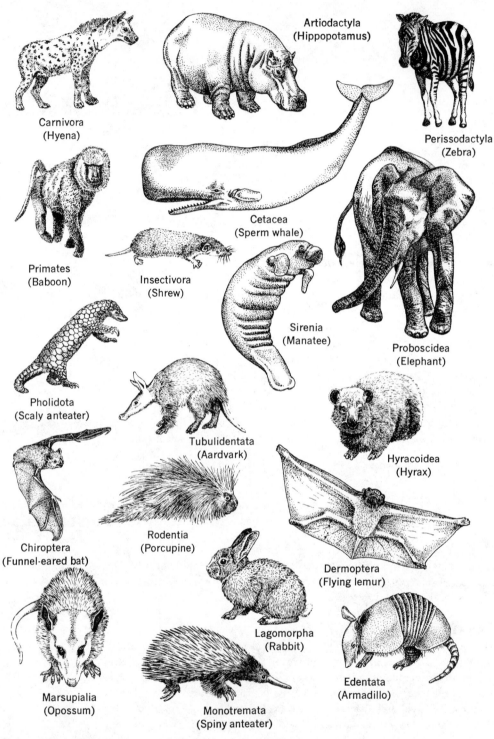

**Figure 21-1** Representatives of 18 orders of mammals (not to scale).

a posterior portion, the **abdomen**. Beneath the base of the **tail** is the **anus**, and just in front of this is the **urogenital opening**. In males, the **scrotum** hangs beneath the anus. There are normally four pairs of small **teats** situated on the ventral surface of the thorax and abdomen. The ducts of the **mammary** or **milk glands** open at the end of the teats. Each **forelimb** possesses five **toes** with fleshy **pads** and retractile **claws**. The **hindlimbs** are stouter and more powerful than the forelimbs and provide the principal power for locomotion. They have four toes; the one corresponding to the big toe in humans is absent. Cats walk on their toes and are therefore said to be **digitigrade**.

## INTERNAL ANATOMY

SKELETON. The cat's skeleton contains bones that closely correspond to those of human beings (Fig. 21–2). It consists principally of bone, but a small amount of cartilage is also present. As in the fishes, amphibians, reptiles, and birds, there are **cartilage bones**, preformed in cartilage, and **membrane bones**, arising by the transformation of connective tissue. A third type, **sesamoid bones**, occurs in the tendons of some of the limb muscles, the action of which they modify.

The **axial skeleton** consists of a **skull, ribs, sternum**, and **vertebral column**. The skull is formed of both cartilage and membrane bones and a small amount of cartilage (Fig. 21–3). The bones of the skull are fused so that their boundaries are usually obliterated in the adult and can be seen only in the embryo. The **vertebral column** of the cat, as in other vertebrates, supports the body and protects the spinal cord. The **vertebrae** move upon one another and are separated by **intervertebral disks** of fibrocartilage (except in the sacrum); they are connected by **intervertebral ligaments**. There are 7 **cervical (neck) vertebrae**; 13 **thoracic (chest) vertebrae**, which bear 13 movably articulated ribs; 7 **lumbar (trunk) vertebrae**; 3 **sacral vertebrae**, which are fused together to support the pelvis; and 16 to 20 **caudal vertebrae**, which form the skeletal axis of the tail.

**Figure 21–2**
Skeleton of the cat.

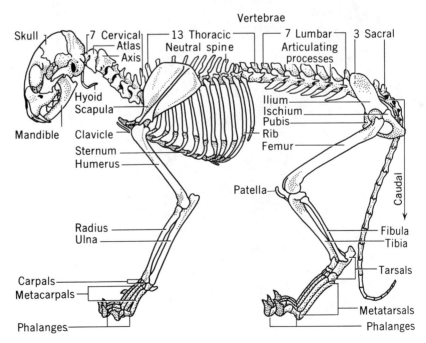

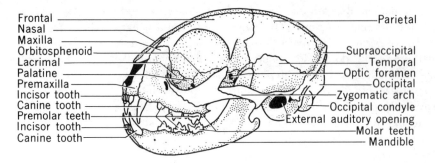

**Figure 21–3** Skull of the cat: lateral view.

The **ribs** and **sternum** constitute the framework of the **thorax;** they protect the vital organs in that region and play an important role in respiration. The **pectoral girdle** consists of 2 **scapulae,** 2 small **clavicles,** and 2 knoblike **coracoid processes.** Each half of the **pelvic girdle** is called the hipbone or **innominate bone** and is made up of the **ilium, ischium,** and **pubis** fused together.

The **forelimb** consists of the **humerus, radius, ulna,** 7 **carpal bones,** 5 **metacarpals,** and the 14 **phalanges** of the toes. The **hindlimb** consists of the **femur, tibia, fibula,** 7 **tarsals** (ankle bones), 4 long **metatarsals** and a rudiment of the first (innermost), and 12 **phalanges.**

**MUSCULAR SYSTEM.** Although many of the muscles are similar to those in lower vertebrates, mammals have a smaller amount of musculature near the vertebrae and ribs and more highly developed muscles in the head, neck, and limbs (Fig. 21–4). A distinctive feature of mammalian musculature is the dome-shaped partition, the **diaphragm,** which separates the coelom into the

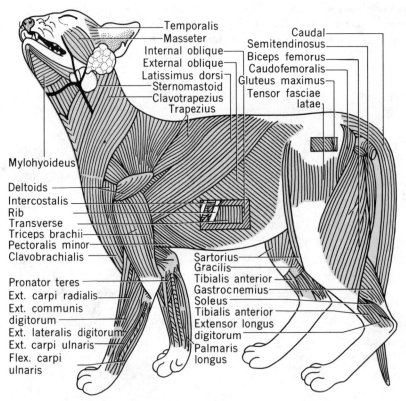

**Figure 21–4** Superficial muscles of the cat. Some of the abdominal muscles are cut away to show the disposition of underlying muscles.

anterior thoracic cavity, containing the heart and lungs, and the posterior abdominal cavity, containing the abdominal viscera. For additional information on the mammalian muscular system and its function, see Chapter 23.

DIGESTIVE SYSTEM. The **mouth** cavity bears a series of transverse ridges on the anterior portion of the roof which help to hold the food; this part of the roof has a bone foundation and is known as the **hard palate.** Posterior to it is a fleshy flap, the **soft palate,** which separates the mouth from the pharynx. At the sides of the posterior part of the soft palate are a pair of small reddish masses of lymphoid tissue called **tonsils.** The **tongue** is attached to the floor of the mouth.

There are four pairs of **salivary glands:** the **parotids,** below the ears; the **infraorbitals,** below the eyes; the **submaxillaries,** behind the lower jaw; and next to these the **sublinguals.** These glands pour watery and mucous secretions into the mouth cavity to moisten and partially digest food.

The posterior continuation of the mouth cavity is the **pharynx** (Fig. 21–5). In the floor of the pharynx is the respiratory opening, the **glottis,** which is covered by a bilobed cartilaginous flap, the **epiglottis,** during the act of swallowing. The pharynx leads downward into the narrow, muscular **esophagus** and is followed by the **stomach.** The U-shaped **duodenum** consists of a **small intestine** several meters long that leads into the **large intestine** (**colon**) and ends in the **rectum.** At the junction of the small and large intestines a short blind sac, the **cecum,** is given off, but there is no appendix as in human beings. Juices from the **pancreas** and **bile** from the **liver** are poured into the duodenum by their respective ducts.

CIRCULATORY SYSTEM. The red blood corpuscles (erythrocytes) of the cat are small and round, a typically mammalian characteristic which distinguishes

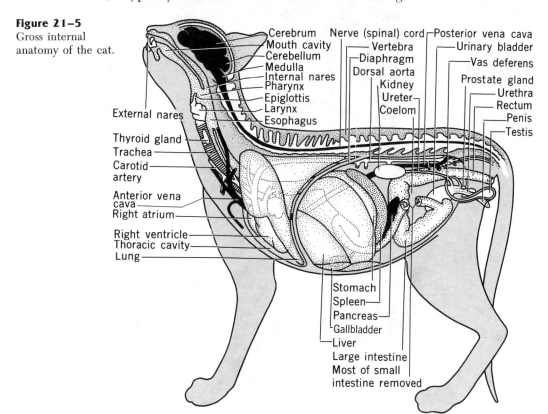

**Figure 21–5**
Gross internal anatomy of the cat.

them from the oval corpuscles of lower vertebrates. The **heart** is completely four-chambered, as in the pigeon, but the main blood vessel, the **aorta**, arising from the left ventricle, has only the left arch. The right arch is represented in the cat by the **innominate artery**, which is the common trunk of the right **carotid** and **subclavian arteries** (Fig. 21–6). A **hepatic portal system** is present, but the renal portal system is lacking. The elongate **spleen,** a dark-reddish organ on the left side behind the stomach, produces lymphocytes (white blood corpuscles) and aids in removing old erythrocytes.

**Figure 21–6** Circulatory system of the cat.

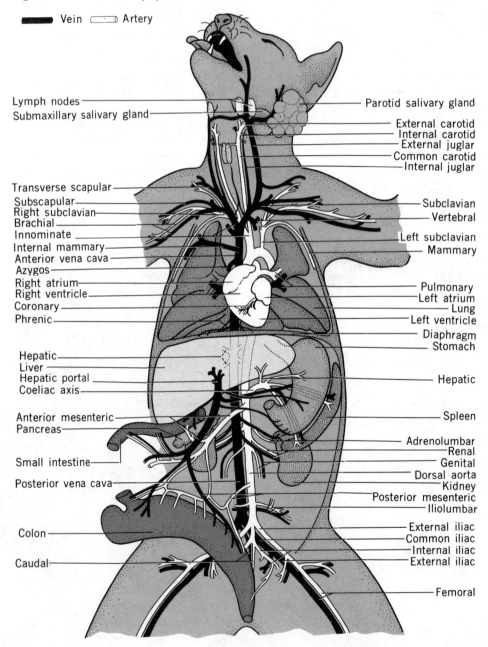

The **lymphatic system** is important in the cat and other mammals. Because of the blood pressure, some of the fluid portion of the blood (along with certain white blood cells) escapes through the walls of the capillaries into the spaces among the tissue and is collected into the lymph vessels. These vessels pass through the **lymph nodes** and empty into the large veins at the base of the neck.

RESPIRATORY SYSTEM. The cat and all other mammals respire by **lungs;** air is drawn into the lungs by the enlargement of the thoracic cavity, accomplished by the extracostal muscles pulling forward and separating the ribs, and by the diaphragm. The diaphragm is normally arched forward; when it contracts, it flattens, enlarging the thoracic cavity.

The glottis opens into the **larynx,** from which the **trachea** or windpipe arises. The trachea is held open by incomplete rings of cartilage; it divides into two **bronchi,** one going to each lung. These distribute the air through smaller and smaller branches that end in microscopic air sacs (**alveoli**) in the lungs. The alveoli are surrounded by a capillary network of blood vessels in which the gaseous exchange of carbon dioxide and oxygen takes place. The larynx is supported by a number of cartilages, and across its cavity extend two elastic cords, the **vocal cords,** which produce sounds.

EXCRETORY SYSTEM. The **urine** excreted by the two **kidneys** is carried by two slender tubes (**ureters**) into a thin-walled muscular sac, the distensible **urinary bladder.** At intervals the muscular walls of the bladder are voluntarily contracted, forcing the urine out through the **urethra.** In the male, the urethra passes through the penis.

NERVOUS SYSTEM AND SENSE ORGANS. The cat possesses a brain, cranial nerves, spinal nerves, and an autonomic nervous system. The **brain,** as in other mammals, differs from that of the lower vertebrates in the large size of the cerebral hemispheres and cerebellum (Fig. 21–7). The **cerebral hemispheres** are slightly marked by depressions or **sulci,** which divide the surface into lobes or **convolutions** not present in the pigeon. The **olfactory lobes** are very large and

**Figure 21–7** Brain of the cat.

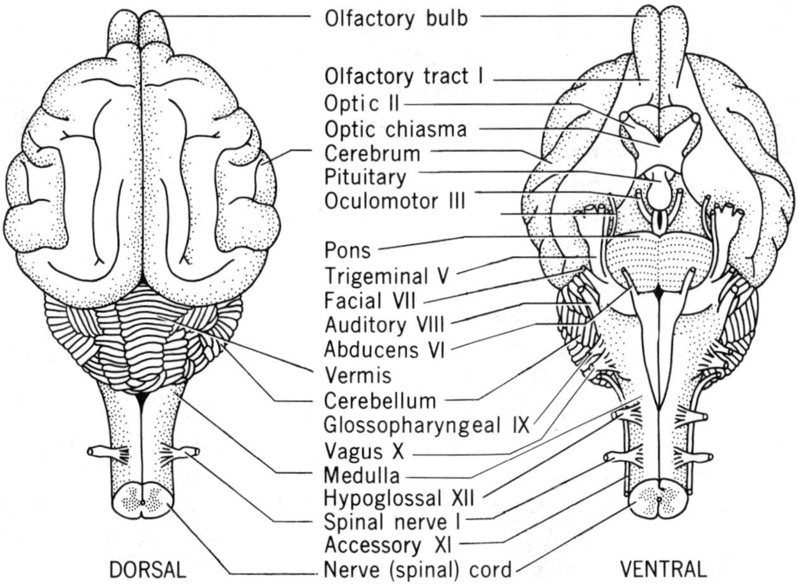

club-shaped. The whole surface of the cerebellum is thrown into numerous folds. There are 12 pairs of **cranial nerves** from the brain; from the nerve cord a pair of **spinal nerves** emerges between successive vertebrae.

The organs of smell, taste, sight, and hearing are very similar to those of man in location and function (see Chapter 26). The **eye** of a cat has a pupil that varies in size and shape, depending on the amount of light striking the iris. The metallic luster which makes cats' eyes shine at night is due to light-reflecting crystals in a part of the eye, the **tapetum lucidum.**

The large **outer ear** collects sound waves which the **middle ear** transmits (as vibrations of the **tympanic membrane** or **eardrum**) by means of three **auditory bones** (malleus, incus, and stapes) to the **inner ear.** The **cochlea** of the inner ear is spirally coiled, not simply curved, as in the pigeon. The **nasal cavities** are large, indicating a highly developed sense of smell, which is also evidenced by the large olfactory bulbs.

REPRODUCTIVE SYSTEM. The two **testes** of the male lie in oval pouches of skin called **scrotal sacs,** one on either side of the copulatory organ, the **penis.** The **sperm** pass from each testis into minute convoluted tubes called the **epididymis**: they then enter the sperm duct, or **vas deferens,** which leads into the abdominal cavity and opens into the **urethra.** During copulation the sperm pass into the urethra and are transferred to the female by the penis. At the base of the urethra is the **prostate gland;** just posterior to the prostate is a pair of **bulbourethral (Cowper's) glands.** The secretions from these glands are added to the spermatozoa, making the seminal mass more fluid and neutralizing the acidity resulting from the passage of urine through the urethra.

The two **ovaries** of the **female** are ovoid bodies exhibiting small, rounded projections on the surface; these are the outlines of the ovarian (**Graafian**) **follicles,** each containing an **egg** or **ovum.** Each ovary lies lateral to the **ostium,** which is the opening to each **oviduct.** The latter is continued posteriorly as a thick-walled **uterus.** The two uteri unite medially to form the body of the uterus, from which the **vagina** extends to the **urogenital opening.** On the ventral wall of the vestibule lies a small rodlike body, the **clitoris,** which corresponds to the penis of the male. The eggs are fertilized while in the oviduct and then pass to the uterus, where development takes place.

The egg is fertilized and undergoes cleavage in the oviduct; it then passes into the uterus, where it receives nourishment and oxygen and disposes of wastes by way of the mother's blood circulation through a structure called the **placenta.** This is formed from the fetal membranes and united with the mucous membrane of the uterine wall. The interval between fertilization and birth, known as the period of **gestation,** is 60 days for cats. For a more detailed discussion of development in mammals, see Chapter 27.

# Structures of Mammals

### HAIR

The hairs that distinguish mammals from all other animals, unlike feathers, which are modifications of horny scales, are new structural elements of the skin. The function of hair and feathers is similar in that both are insulating devices. When hairs are shed, new hairs usually arise to take their place. Secretions from the oil-producing **sebaceous glands** keep the hairs glossy. The two main types are long **guard hairs,** which are strong and may be bristled by the underlying skin muscles, and shorter **wooly hairs,** which constitute the **underfur.**

## CLAWS, HOOFS, AND SO ON

These are modifications of the horny covering on the dorsal surface of the distal ends of the digits. When on the ground, the foot rests partially or entirely upon the digital pads. Dermal papillae occur on the digital pads, often forming concentric ridges such as those that produce fingerprints in man.

## SKIN GLANDS

Mammals possess a greater number of glands than do reptiles or birds; these are for the most part sebaceous and sweat glands, or modifications of them. The **sweat glands** secrete a fluid that evaporates, thereby cooling the skin and regulating the body temperature. In carnivores the sweat glands are generally much reduced in number. Modifications of the sweat glands include the **lacrimal glands,** which keep the eyeball moist; **scent glands,** and the **mammary glands.**

## TEETH

The teeth of mammals, or absence of them, are useful in classification and also indicate food habits. Their embryological development is similar to most vertebrates (Fig. 21–8). The **enamel** is the outer, hard substance; the **dentine**

**Figure 21–8** Teeth of mammals. *Above left:* teeth of a dog. *Upper right and below:* structure of mammalian teeth shown in cross and longitudinal sections.

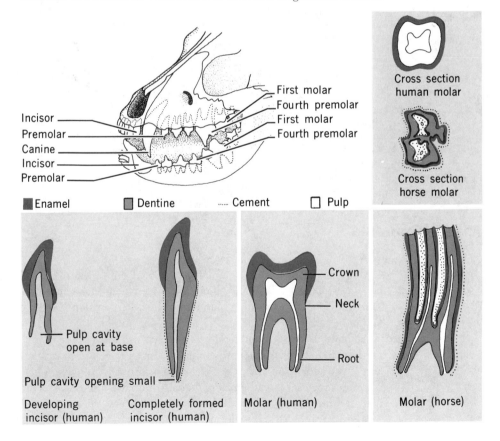

constitutes the largest portion of the tooth, and the **cement** usually covers the part of the tooth embedded in the tissues of the jaw. The central **pulp cavity** of the tooth contains nerves, blood vessels, and connective tissue. A tooth has an open pulp cavity during growth, which in some cases continues throughout life.

There are usually four kinds of teeth in each jaw: **incisors**, the chisel-shaped cutting teeth; **canines**, the conical tearing teeth; **premolars**, the anterior grinding teeth; and **molars**, the posterior grinding teeth. In most mammals the first set of teeth (the **milk** or **deciduous dentition**) is pushed out by the permanent teeth, which last throughout the life of the animal. The milk molars are followed by the premolars, but the permanent molars lack predecessors.

### FOOT POSTURE

Human beings have **plantigrade** foot posture; the entire palm or sole rests on the ground, and neither the wrist nor ankle is raised above the ground. In **digitigrade** posture, as was seen in the cat, the animals walk upon their digits with the bones of the wrist and ankle and upper ends of the palms and soles raised above the ground. **Unguligrade** posture is characteristic of hoofed animals; they walk on modified nails or hoofs, such as those of the horse (Fig. 21–9).

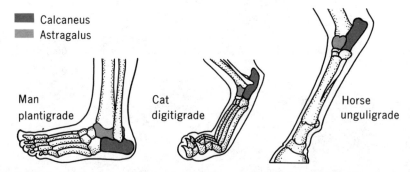

**Figure 21–9** Posture of mammals. *Left:* human beings have a five-digit foot and the entire sole rests on the ground. *Middle:* cats have a four-digit foot and walk with heel raised. *Right:* horses have the most specialized posture, with only one digit (the third), covered by a hoof, upon which the horse walks.

## Fossil Mammals

Many of the orders of mammals are known only from fossils. The earliest known remains are of small species. The Cenozoic era is called the "Age of Mammals" since in this time mammals achieved their dominancy, which they maintain today.

Among the fossils found in North America are the archaic ungulate, *Uintatherium* (Fig. 21–10), which was almost the size of the largest existing elephants and possessed three pairs of conspicuous protuberances on its head; the enormous tortoise armadillo, *Glyptodon*, which was almost 2.8 m in length and had an arched shell of immovable bony plates; and the mastodon of Europe, Asia, and South Africa, which resembled our modern elephants.

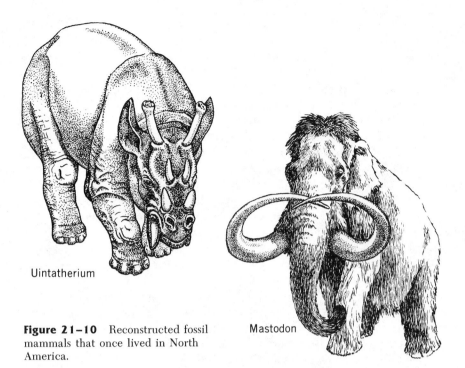

**Figure 21–10** Reconstructed fossil mammals that once lived in North America.

## Other Mammals

Mammals differ widely in their mode of life and in their adaptations to their environments. Only a few of the more interesting groups can be presented here.

**Egg-laying mammals** (Fig. 21–11) are the most primitive group, confined to the Australian region. The young live on the yolk contained in the egg before hatching and are afterward nourished for a time by milk from the mother's mammary glands. The best-known species are the **spiny anteater** or **echidna** and the **duck-billed platypus**.

**Figure 21–11** Spiny anteater, an egg-laying mammal. The leathery-shelled egg is incubated in a pouch, where the newly hatched young are kept for a time. (Courtesy H. C. Reynolds.)

**Pouched mammals** or **marsupials**, such as the **kangaroo** (Fig. 21-12A) and **koalas** (Fig. 21-13), occur mainly in Australia and neighboring islands. The young are born without hair, eyes, or ears, but have good olfactory organs and usually well-developed front teeth. They live for a time in a pouch on the mother's abdomen, feeding on milk from her teats. The young of the **American opossum** (Fig. 21-14) remain in the mother's pouch until they complete their natal development.

**Insect-eating mammals,** such as **moles** (Fig. 21-12B) and **shrews,** are the most primitive of those that nourish their young before birth by means of the placenta. Most of them are terrestrial; a number of these live in burrows; a few are aquatic and some live in trees.

Bats (Fig. 21-12C) are **flying mammals;** they exhibit modifications of their forelimbs for flight. The forearm and fingers are elongated and connected with each other, with the hindfeet, and usually with the tail, by a thin leathery membrane. Because of flight, bats are widely distributed, occurring on small islands devoid of other mammals. Most of the more than 600 species are small and chiefly nocturnal. The **true vampire bats** inhabit tropical America. They live on the blood of horses, cattle, and other warm-blooded animals, and sometimes attack sleeping people. Their front teeth, which they use to cut their victims before sucking their blood, are very sharp. The **little brown bat** is abundant in eastern North America; experiments have shown it to have a well-developed homing instinct.

The **toothless mammals** have either no teeth or a few that are not well developed. These include the **sloths, anteaters** (Fig. 21-12D), and **armadillos** (Fig. 21-12E).

Among the **gnawing mammals** are squirrels, chipmunks, beavers, rats, mice, gophers, porcupines, chinchillas, and golden hamsters. Their front teeth are efficient chisels and grow constantly. **Beavers** (Fig. 21-12F) are adapted for an aquatic mode of life, possessing webbed hindfeet and a broad flat tail and ear and nose valves that close under water, allowing the beaver to gnaw logs below the water surface as well as above.

**Porcupines** possess spines, which normally lie flat on their back but can be elevated by muscles in the skin. The **chinchillas** (Fig. 21-15) of South America are of medium size and are noted for their very soft fur.

The **mustelids** constitute a large family of small, fur-bearing animals, which include the **otter, mink, weasel** and **skunk.**

Among the terrestrial flesh-eating (**carnivorous**) mammals are the **red fox** (most common of all foxes in America), **grizzly bear** (Fig. 21-16), **cat, puma** (also called mountain lion, cougar, or panther) (Fig. 21-17), **leopard, lion, tiger, cheetah,** and **ocelot** (Fig. 21-18).

**Figure 21-12** Some mammals with interesting adaptations. (A) Order Myotis. The bat is a flying mammal with elongated and connected forearms, fingers, hind feet, and tail. (B) Order Insectivora. The mole has forefeet adapted for digging, a sensitive nose, and rudimentary eyes. (C) Order Edentata. The giant anteater has a long snout and sharp foreclaws for digging. (D) Order Edentata. The nine-banded armadillo has a bony shell, scanty hair, and long claws for digging. (E) Order Rodentia. The beaver has chisel-like teeth, webbed hind feet, and a flat, scaly tail used for building dams. (F) Order Marsupialia. Kangaroo, or more specifically, a wallaroo. Note young in pouch, large hindlimbs, small forelimbs, and large tail. (Courtesy N.Y. Zoological Society.)

A

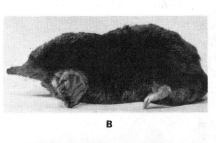

B

C

D

E

F

**318**
*Animal Diversity: Vertebrates*

**Figure 21-13** The koala, the Australian "teddy bear," lives in trees. The females carry their young in a pouch, and later, on their backs. (Courtesy Science Software Systems, Inc.)

**Figure 21-14** (A) Adult female opossum. (B) Young American opossums attached to mammary glands in brood pouches of their mother. The opossum is strikingly undeveloped at birth and is considered the most primitive of America's mammals.

A

B

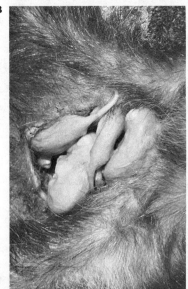
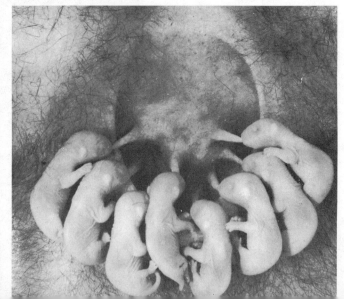

**Figure 21-15** Chinchillas were introduced into the United States in 1923 by Chapman, who obtained 11 animals from Chile. Their fur has a downy-softness, but is not as serviceable as some other furs. (Courtesy N.Y. Zoological Society.)

**Figure 21-16** Grizzly bear. (Courtesy N.Y. Zoological Society.)

**Figure 21-17** The puma (mountain lion) kills calves, deer, and other large animals. (Courtesy N.Y. Zoological Society.)

**Figure 21-18** The ocelot is an inhabitant of Texas and tropical America. (Courtesy N.Y. Zoological Society.)

The **aquatic carnivores** are greatly modified for their water habitat. Either forelimbs, hindlimbs, or both are fully webbed and serve as swimming organs and bodies have acquired a fishlike form suitable for progress through the water. These animals include the **fur seal, California sea lion,** and **walruses** (Fig. 21-19).

**Figure 21–19** (A) Aquatic mammals. California gray-whale. Ever-diminishing populations due to commercial whaling are making this an endangered species.

**Figure 21–19** (B) The killer whale is an example of a toothed whale.

**Figure 21–19** (C) Dolphin. Streamlined, highly intelligent, and well equipped for sonar mapping and auditory communication, these social animals are very well adapted to their lives in the sea.

**Figure 21-19** (D) California sea lions. These social animals are also noted for their complex auditory communication.

D

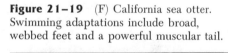

**Figure 21-19** (E) Elephant seals. Note the large trunklike nose appendage used in the defense of territory.

**Figure 21-19** (F) California sea otter. Swimming adaptations include broad, webbed feet and a powerful muscular tail.

E

F

Whales are the largest living animals. The **blue whale,** in fact, is the largest animal that has *ever* lived (including the dinosaurs); this toothless mammal reaches a length of up to 35 meters. Beneath the skin is a thick layer of insulating fat or blubber. The **sperm whale** reaches to 23 meters and is the largest of the toothed whales; its large head contains a cavity filled with up to 900 kg of sperm oil. Other species of toothed whales are the **common dolphin,** the **killer whale,** and the **common porpoise** (Fig. 21–19).

All of the larger species of whales may well disappear in a very few years unless rigidly protected by international law. Modern ships of many nations, particularly the USSR and Japan, hunt these animals with airplanes, radar, whale guns, and other refined apparatus. Whale hunting or importation of whale products is now illegal in the United States.

**Elephants** are the largest land mammals, weighing as much as 6700 kg. The two living species are the **African elephant** and the **Indian** or **Asiatic elephant** (Fig. 21–20). They can be distinguished by their ear size: the African elephant's ears are much larger. There are other differences, both physiological and behavioral, between the species.

**Figure 21–20** *Left:* African elephant. Note the long prehensile trunk with tusks, loose skin, large ears, and hollow back. *Right:* Indian elephant. Note the smaller ears and arched back.

**Hoofed mammals** are divided into those with an odd number of toes, such as **horses, tapirs,** and **rhinoceroses,** and those with even toes, which include most of the "big game animals"; the latter are considered to be the more successful group.

Animals that chew their food (cud), such as cows, are called **ruminants.** They typically possess stomachs consisting of four chambers (Fig. 21–21).

## PRIMATES

Primates are of special interest, since this order includes the human species. Primates live mainly in the warmer regions of the globe. Most are arboreal and can climb among trees because of opposable great toes and thumbs, which adapt the hand and feet for grasping. A few primates lead a solitary life, but most travel in troops. One young is usually produced at birth; it is cared for with great solicitude.

Eleven families of living primates are recognized; some of the more interesting representatives of certain families are the **lemurs,** of small or moderate size, and usually possessing a long nonprehensile tail; **rhesus monkeys,** com-

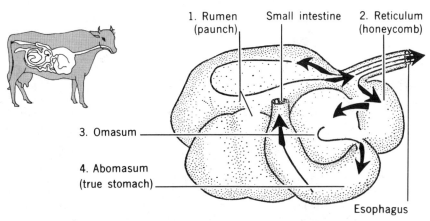

**Figure 21–21** A four-compartment stomach of a ruminant (cow). Arrows show the course of roughage, such as hay. Grains, such as corn, go directly to the reticulum. This type of stomach allows the animal to consume large amounts of grass, which it can rechew afterward. (After *Anatomy of the Domestic Animals*, by S. Sisson and J. D. Grossman, Philadelphia: W. B. Saunders Co., 1938.)

monly used in experiments and noted for its use in identifying the Rh blood factor (see Chapter 23); **South American monkeys,** of small or medium size with opposable thumb and great toe, nails on all digits, and a long prehensile tail that aids in climbing; and the **Old World monkeys,** which usually possess long tails that are never prehensile and include the baboons (Fig. 21–22), macaques, and langurs.

**Figure 21–22** Baboons. These are the largest of the Old World monkeys. They travel in formidable, well-organized troops of as many as a hundred individuals.

The **anthropoid apes** of the family Pongidae are the most nearly related to man. Like man they have a vermiform appendix. The tail is absent; the forelimbs are longer than the legs; locomotion is often bipedal, and when the ape is walking, the feet tend to turn in and the knuckles aid in equilibrium. This family includes the gibbons, orangutan, gorillas, and chimpanzees (Fig. 21–23).

The family Hominidae contains the single living species *Homo sapiens,* or the **human** species. Human beings differ from the other primates in the size of

**Figure 21–23** (A) Gibbons. These arboreal primates can walk upright, unaided by the hands. (B) Gorilla. They are the largest of the great apes, adult males reaching a height of over 6 feet, a weight of 660 pounds, and an arm span of 10 feet. (C) Chimpanzee. These primates have large ears, long lips, and nails on fingers and toes. (D) Orangutan. These herbivorous tree-top dwellers possess brains more nearly akin to man's than any other animal.

**Table 21-1**  Lifespans

| | Years | | Years |
|---|---|---|---|
| *Reptiles* | | *Mammals* | |
| Giant tortoise | 152 | Elephant | 69 |
| Box turtle | 123 | Horse | 50 |
| Alligator | 68 | Hippopotamus | 49 |
| Snapping turtle | 57 | Chimpanzee | 40 |
| Cobra | 28 | Grizzly bear | 32 |
| Cottonmouth | 21 | Bison | 30 |
| | | Lion | 30 |
| *Amphibians* | | Tiger | 25 |
| Giant salamander | 55 | Elk | 22 |
| Toad | 36 | Mountain lion | 20 |
| Bullfrog | 30 | Beaver | 19 |
| Mudpuppy | 23 | Wolf | 16 |
| Green frog | 10 | Squirrel | 16 |
| Newt | 7 | Chipmunk | 12 |
| | | Cottontail | 10 |
| *Fish* | | House mouse | 4 |
| Catfish | 60 | | |
| Eel | 55 | *Birds* | |
| Carp | 47 | Turkey buzzard | 118 |
| Mosquito fish | 2 | Swan | 102 |
| | | Parrot | 80 |
| *Insects* | | Great horned owl | 68 |
| Cicada | 17 | Eagle | 55 |
| Ant (queen) | 15 | English sparrow | 23 |
| | | Canary | 22 |
| | | Hummingbird | 8 |

*The comparative tables of the life-span of animals are reproduced by permission from *Iowa Conservationist*, **16**: 133, 1957, and were prepared by D. H. Thompson and R. Mann.

the brain, which is about twice as large as that of the highest ape, and in erect bipedal locomotion. Hair covering is scarce, and the great toe is not opposable. The mental development has enabled man to accommodate himself in almost every climate and to dominate all other animals.

Table 21-1 contains examples of the life span of several animals, from insects to mammals. The examples were chosen from records of zoological parks and aquaria over the world.

# Summary

Mammals are organisms possessing milk-secreting mammary glands. Parental care is generally most highly developed in this group and reaches its climax in human beings. The cat may serve as a representative of the class Mammalia. The skeleton is made of bone, cartilage, membrane, and sesamoid bones; the muscular system includes a diaphragm for breathing; the digestive system is homologous and analogous to man's; the circulatory system includes the lymphatic system, which retrieves much of the fluid component of the blood; the excretory system includes a distensible bladder; the nervous system incorporates large cerebral hemispheres and cerebellum; the sensory systems are moderately developed (the sense of smell is well developed); and the reproductive system is similar to that in human beings, but the cat possesses two uteri.

Hair, claws, nails, hoofs, skin glands, and teeth are all characteristic of mammals (although lower organisms may also have teeth). Some representatives from the 5000 or more living species of mammals and fossil mammals were briefly discussed.

# PART FOUR

# The Design and Function of Organ Systems

# 22

# Animal Behavior

Throughout this text, great emphasis has been placed on the anatomy and physiology of animals. Knowledge of that sort is necessary for understanding the organisms as a whole, but would be drastically incomplete if not combined with the **behavior** of the organism.

Before we concern ourselves with specific categories of animal behavior, we must gain some background in just what behavior is, how it is studied, and what factors may effect it.

## Background

Decades ago, when nonhuman behavior study was in its infancy, a purely **anthropomorphic** orientation was employed, in which animal behavior was interpreted in terms of human characteristics. In reaction to this relatively unsuccessful method of deciphering and analyzing animal behavior, a new school of thought, based on **mechanistic** considerations, was developed. Animals were conceived of as emotionless automatons, reacting to the immediate environment. This method also proved incomplete and failed to develop a scientifically sound basis for understanding and predicting animal behavior.

The foundations of current animal behavior theory, **ethology**, were laid by Konrad Lorenz and Nikolaas Tinbergen in the 1930s. This method of study is basically a field psychology of animals, involving the quantification and *comparison* of behavioral traits; **patterns** then evolved which reflected the animal's adaptations to the environment.

## General Characteristics of Behavior

The most obvious aspect of behavior is that it is *dynamic*. An organism's actions are constantly modified in various degrees in response to its surroundings. Behavior is not only momentarily variable, but may change over time. As postnatal development continues, behaviors are modified. This type of modification usually increases the specificity of particular actions. A baby, for example, develops a repertoire of noises as it grows. The utterance "mama" gets a different reaction than the sound "waalupo," and soon the baby learns to modify its tonal behavior accordingly.

Behavioral change is, by its very nature, continuous; it has evolved over generations and will continue to do so. Selectional and genetic factors both have long-term effects on behavior, and much of it is also adaptational. In this chapter we will deal with orientational, social, communicational, and other types of behavior that help the organism relate to its world.

# Methods of Behavioral Study

**Ethology** is an interdisciplinary science, collecting data from the fields of anatomy, physiology, embryology, genetics, neurology, astrology, ecology, physics, chemistry, sociology, psychology, and others. The forms of experimentation are as varied as the subjects of the experiments. No portion of the biosphere is unavailable to the ethologist.

Controlled experiments in the laboratory yield data that complement field observations. Study in the natural habitat has great relevance to ethology; ideally, it allows unadulterated observation of the species. A relatively recent innovation in field study is **biotelemetry,** in which miniature mobile electronic transmitting devices are attached to the animal to monitor physiological variables (heart beat, body temperature, etc.).

### ROLE OF HEREDITY

The question of "nature versus nurture" is ancient yet contemporary. Various methods of assessing the role of **heredity** in behavior exist.

In **pleiotropic** genetic studies, the effects on the morphology of a single gene and dominant–recessive gene relationships, which have impressive behavioral implications, are evaluated (see Chapter 28).

In **artificial selection** genetic studies, organisms that possess a particular trait under observation may be bred with each other or with animals which lack that specific trait. In addition, geneticists may study the results of a cross in which the mating individuals have different traits, the combination of which is of interest. The results of such experiments may help to determine whether a particular trait is heritable or even polygenic (Chapter 28).

In a third type of genetic manipulation, **cross matings** are made between individuals of different groups which are reproductively compatible. Matings between different **breeds** or subspecies may provide behavioral information. J. Scott and J. Fuller inbred dogs and concluded that of all behavioral variations between the breeds studied, 25 percent were directly attributable to genetic breed differences.

### SOCIAL METHODS

For all interacting animals, **social** interactions are secondary only to homeostatic behavior (e.g., patterns of feeding). Social studies may have a substantial relevance to the nature versus nurture question.

The role of parental care and indoctrination may be easily and objectively evaluated by statistical comparison of parent-reared and non-parent-reared progeny. An interesting extension of this method is the replacement of the parent by a **surrogate parent.** This has been done with rhesus monkeys in a famous series of experiments by H. Harlow (1961). Infant rhesus monkeys were given cloth-covered and bare-wire "mother" surrogates and their behavior observed. In brief, the monkeys preferred the cloth-covered versions over wire surrogates (for instance, in stressful situations), and showed several abnormal behaviors as adults, notably sexually.

### ENVIRONMENTAL STUDIES

Isolation from a natural habitat is another method for delineating behavioral effects of the environment. Placing an animal in a foreign habitat may have minor, modest, or drastic (including fatal) effects.

In the last century, Carl Hagenbeck, a German who was in the business of catching zoo specimens, pioneered the ecological orientation of animals into zoos. The idea caught on well, although even today there are many zoos with animals simply caged. These can by no means be considered as exhibiting normal or typical behavior of the animal in the wild.

The results of environmental studies may also be applied to conservation efforts. If an animal is shown to be easily adaptable to a foreign environment and if it is an endangered species, it may be transferred to a new (controlled) habitat with reasonable assurance of success.

# Mechanics of Behavior

### STIMULUS–RESPONSE

Behavior may be viewed as a **stimulus–response** relationship. A **stimulus** may be any perceptible variation in the external or internal environment of the organism (conscious or otherwise). What constitutes a stimulus for one animal may not be one for another, even within the same subspecies.

The effectiveness of a stimulus depends in part on its **intensity.** The concern of **psychophysics,** introduced by G. Fechner in the 1860s, is the effect of physiology upon behavior. One of its basic concepts is the **threshold:** a receptor will respond to a stimulus only if that stimulus is of a characteristic minimum value. There are upper *and* lower thresholds. For example, a normal human being will not be able to hear a sound of 5 Hz (hertz; cycles per second) *or* a sound of 30,000 Hz. The normal **range** of human hearing is from 20 to 20,000 Hz, and these values are determined by thresholds.

**External stimuli** are perceived through the five senses: *optic* (visual), *olfactory* (smell), *auditory* (hearing), *tactile* (touch), and *gustatory* (taste) stimuli may affect an organism singly or in complexes. *Orientation* may be perceived through the stimuli of illumination, tactile contact, gravity, wind and water forces, and astronomic cues.

**Internal stimuli** may be either psychogenic or kinesthetic. **Psychogenic** stimuli are extremely diverse. They can be triggered by external stimuli (a particular odor, for instance, may remind one of a previous situation, the memory of which may stimulate further reaction). **Kinesthetic** stimuli are internal cues from receptors in the body's joints, muscles, ligaments, tendons, and various other organs which serve to inform the organism of its state. One's sense of weight, position, balance, motion, temperature, and other states are conveyed, at least in part, kinesthetically.

**Motives** are internal stimuli and may be psychological, physiological, or both. A psychological motive is founded on **hedonistic** bases—an organism capable of purposeful acts actively attempts to preserve or improve its present state. **Biological drives** are motives that are manifested physiologically as **homeostatic mechanisms.** These are processes by which an organism automatically attempts maintenance of a constant internal environment. Complex psychophysiological motives, such as **hunger,** are composed of a psychological component (food tastes good) and a physiological component (high levels of glucose utilization in the CNS and liver).

A **response** is the behavior elicited by one or several stimuli. It may be quite simple or extremely complex; like stimuli, they may also be psychological and/or physiological. The bulk of this chapter is devoted to descriptions of the various responses.

**REFLEXES**

A **reflex** is the simplest form of response; it requires no conscious thought and is not learned. It usually involves one sense organ. Each time an appropriate stimulus is presented, the response will be essentially identical, owing to the fact that most reflexes are neurally built into the organism. The **receptors** (stimulus-sensitive organs) and **effectors** (muscle, gland, etc., which "respond" to the stimulus) are coordinated through the central nervous system. A **spinal reflex** is the simplest of all: although most spinal reflexes include **association neurons** which relay the information to the brain, the neural impulses need not reach the brain (see Chapter 26).

Human beings have relatively few reflexes; the **patellar** (knee jerk) and **sneeze** are examples. Reflexive behavior becomes increasingly prevalent and important in the lower organisms. If a fly lands on sugar, the taste receptors on its pads ("feet") cause the proboscis (feeding tube) to extend automatically. In many invertebrates, copulation is a reflexive activity.

## Learning, Conditioning, and Memory

There are currently several theories attempting to explain, and thus define, **learning.**

The **chemical theories of learning** tend to ignore the purely morphological aspects of the animal and concentrate on molecular conditions. From the viewpoint of these theories, **memory** results from one or a combination of the following factors: a change in the concentration or form of certain molecules, the formation of new molecules, or the varying loci of groups of chemicals.

Experiments with planaria substantially support the chemical learning theories; one experiment showed that regenerated planarians learn tasks taught their "parent" organisms faster than an untrained control group. Experiments using planarians point to RNA (ribonucleic acid) as the chemical of knowledge. One experimenter, J. McConnell, suspecting RNA as the medium of memory transfer, extracted it from trained planarians and injected it into untrained ones. Again, learning was faster. Some experiments indicate that RNA may be *generally* instead of specifically relevant to learning; this question is still being investigated.

The **structural theories of learning** are illustrated by John Eccles's finding that neural pathways function better if they have been used a great deal in the organism's recent past. If a nerve is not used for a considerable period of time, it may at least partially **atrophy** (degenerate); this could explain the act of forgetting. The structural theories of learning are supported by the example of the cortical nerve, which has one or several dendrites covered with thousands of protrusions called **spines.** If input to a cortical dendrite is reduced over a period of time, a great number of spines completely degenerate.

Some scientists believe that learning is the result of newly manufactured cortical connections (synapses). If that is the case, it is possible that spine growth is essential to the new synaptic connections. It would then follow that forgetting, possibly through disuse of particular neurons, would cause degeneration of spines, which would, in turn, destroy various synapses. A logical extension to this is that learning produces new spines and therefore new synapses; experimental verification of this has not yet been brought forth.

No matter what the mechanism of learning, it is generally agreed that it reflects **adaptability.**

## CONDITIONING

An organism learns through processes of **conditioning.** Two major categories are recognized: **classical** and **instrumental (operant) conditioning.**

The Russian psychologist Ivan Pavlov conducted a series of famous experiments which in 1927 established **classical conditioning** as one of the most basic processes of learning. Pavlov placed meat powder in the mouths of hungry dogs and the dogs' responses were salivation, mastication, and swallowing. In classical conditioning terms, the meat powder was the **unconditioned stimulus** (UCS), producing the unlearned or **unconditioned response(s)** (UCR) of salivation, mastication, and swallowing. Pavlov began presenting a buzzer just before he gave the dogs food, and the buzzer became the **conditioned stimulus** (CS). He found that after a few times the buzzer alone would elicit salivation, which thus became the **conditioned response** (CR).

In instrumental or **operant conditioning,** an organism responds, through trial and error until a desirable situation (receipt of food, release from pain, etc.) is reached. The attainment of the desired situation becomes **positive reinforcement** for the response. The frequency of the response increases as more positive reinforcement is experienced.

B. F. Skinner is noted for his experiments in animal conditioning. Skinner placed hungry rats in a box that contained a lever with a bar attached. Gradually, after myriad responses, the rats "realized" that pressing the bar would bring food; they were thus positively reinforced for bar pressing. Consequently, the rats "learned" to bar-press for food, and would go straight for the bar if placed in the chamber at a later date.

Rats have also been used in experiments illustrating **negative reinforcement** (the removal of aversive or competitive stimulation). Whereas hungry rats may press a bar for food pellets (food = positive reinforcement), they may also "learn" to press a bar to stop an electrical shock from crossing their cages (removal of shock = negative reinforcement). In both cases, the bar-pressing response is reinforced and will continue.

**Punishment** can be enforced using operant conditioning responses. If a rat is given an electrical shock whenever it stands on two feet, in a short time the upright behavior will cease.

Conditioning is a highly complex area of animal behavior with virtually unlimited ramifications.

### INSIGHT

Insight is a modified type of trial-and-error learning where an organism may *rapidly* learn a particular response. **Insight** (also called **abridged learning**) is most prevalent among higher organisms and is not usually observed in lower ones. If a piece of food were to be placed behind a barrier yet still be visible, most organisms capable of recognizing food would try indefinitely to break through the barrier, but some of the higher organisms (including dogs, cats, and primates) will travel *away* from the food to circumvent the obstacle after only one failure at traversing the barrier.

### HABITUATION

**Habituation** results from acclimatization to a stimulus. For example, a loud noise will cause a bird to take flight, but if the noise is heard often the bird will soon stop responding to it. Essentially, this represents the erasure of a behavioral response that has become nonadaptive.

### MEMORY

The behavior of **memory** may be defined as the recreation of a prior image from neural traces (neural or biochemical changes in the brain resulting from exposure to new material). The three major categories of memory are psychophysiological, psychological, and recognition.

**Psychophysiological memory** includes the mental recreation of particular sensations (odors, sights, emotions, etc.) and is therefore **experiential** in nature. This type of information retrieval exists in human beings; although it seems reasonable to assume that it also exists in primates and other mammals, no method exists to definitely ascertain this.

**Psychological memory** consists of the reconstruction of thoughts, or thought *patterns*. It is the expression of conditioning and so, by definition, any conditionable animal exhibits psychological memory.

**Recognition** is a subtype of both psychophysiological and psychological memory. It can occur only when an external stimulus is provided—the stimulus activates an existent but previously subconscious memory trace.

# Instinct

**Instinct** is believed to be completely unlearned (or **innate**) behavior, neurally "built in" to the organism so that instinctive actions are stereotyped. A spider will instinctively spin webs, and these will normally be nearly identical to those of the parent spider's, which the spider may never have seen.

In 1931, Lorenz and Tinbergen proposed a three-stage model of instinctive behavior. In the first stage, an organism experiences **appetitive** behavior, becoming restless in preparation for the instinctive act. In the second stage, an **innate releasing mechanism** becomes active in response to a **releasing stimulus**. For example, the male stickleback fish is known to attack other sticklebacks that have red bellies. If an uncolored stickleback model is presented to the male, nothing will happen, but if a crude model with a red belly is displayed, attack will occur. The attack represents the third stage of instinctive behavior, the **final consummatory act,** which relieves tension. Although reasonable in theory, the Lorenz–Tinbergen model for instinctive behavior still lacks firm experimental verification.

# Types of Behavior

## KINESES AND TAXES

Formerly, various orientational and locomotory responses were known as **tropisms.** Several prefixes were used to indicate the *forced movements* of organisms; *phototropism,* for example, meant involuntary movements toward light. Recently, such movements have been divided, for clarity, into **kineses** and **taxes.** Kineses are undirected movement responses to a particular environmental stimulus, such as moving faster in unfavorable conditions and slowing down in a more favorable environment. Taxes are orientational movement responses directed toward or away from a particular environmental stimulus.

## IMPRINTING

Lorenz applied the term **imprinting** to certain behavioral traits of young organisms; these may be elicited by "cues" (stimuli) of the five senses. One of the best examples of imprinting was documented by O. Heinroth in 1910. He found that a chick will follow the first large object it sees just after hatching, and it will do so completely ignoring anything else, including its real mother (Fig. 22-1).

## TOOL USE

**Tool use** is a type of behavior that is almost exclusively indicative of higher forms. Very few subhuman organisms use tools, but some interesting examples do occur. The Egyptian vulture *Neophron percnopterus* drops rocks while flying over ostrich eggs, thereby breaking them open to obtain food. The wasp *Ammophila* seals its burrows by pounding the earth with a pebble. Chimpanzees will insert straws into a termite mound and lick off the termites that attack it.

A form of tool use is the utilization of artificial "homes" for some organisms. W. G. Van der Kloot has described the sea anemone's transfer to a vacant whelk shell in considerable detail. Briefly, the anemone "notices" the shell with its tentacles, which then enter the shell and attach, along with the oral disk, to the inside. The anemone then frees itself from its former attachment. After a time, the pedal disk of the anemone attaches to the shell, the tentacles and oral disk loosen, and the anemone is home (Fig. 22-2).

**Figure 22-1** (A) In nature, ducklings will imprint upon, and follow, the mother duck. (B) In an experimental setting, the duckling, hatched without its real mother, will imprint upon the artificial duck.

## SLEEP

Sleep only has relevance in the higher phyla—an amoeba does not sleep, since its metabolic and attentional activities continue until it dies.

The operational definition of **sleep** is the deactivation of the **reticular formation** in the midbrain (**pons**). If the reticular formation is damaged, organisms seem to be in a constant state of sleep. Basically, sleep is a lack of attention—a psychological insensitivity to the environment.

Most sleep research has been done on human beings; it is characterized by deeper, slower breathing than when awake and a slowing of most metabolic processes. **Paradoxical sleep** occurs four to six times a night; electroencephalogram traces made during this time look as if the organism was awake and rapid eye movements (**REM sleep**) occur. Paradoxical sleep alternates with **slow-wave sleep** (SWS); this refers to the large-amplitude low-frequency EEG tracings noted during this period of sleep (Fig. 22-3).

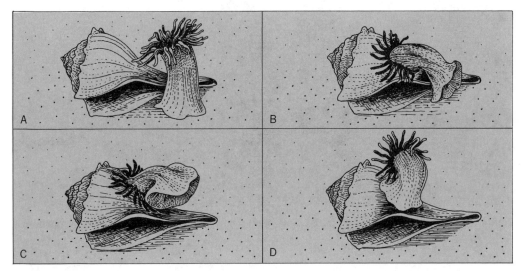

**Figure 22–2** Sea anemone occupying a whelk shell. (A) Anemone attaches to whelk. (B) Tentacles and oral disk attach to shell; pedal disk comes free. (C) Pedal disk swells enormously; tentacles move over to make room for it. (D) Pedal disk attaches; tentacles loosen their grip.

**Figure 22–3** EEG (electroencephalograph) tracings. *Top:* normal EEG of a human being in the waking state. *Bottom:* normal EEG of a sleeping human being. These recordings have proved useful in diagnoses of brain pathology and are used especially in the study of epilepsy. (Courtesy Science Software Systems, Inc.)

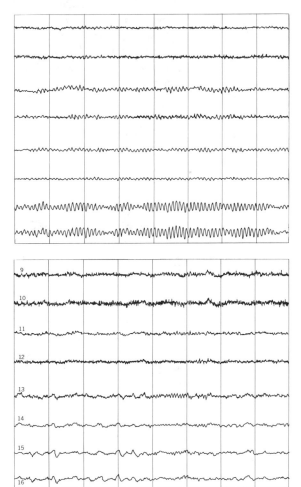

**Figure 22–4** Territoriality, as exhibited by two elephant seals vehemently establishing boundaries during the mating season.

## AGGRESSION

**Physical aggression** and nonphysical or symbolic aggression (such as human "cursing" or a dog's growling) are widespread among animals. Frustration, electric shock, infringement into "personal space" (see "Social Behavior" below), sexuality, and social ascension are just a few of the potential reasons for aggressive behavior.

The territories of the sea lion *Zalophus californianus* are obtained and maintained by aggressive behavior. When two territorial male sea lions see each other, they move rapidly toward each other, barking, and with their vibrassae (whiskers) extended. When they are close, the barking ceases and they fall on their chests, shaking their heads from side to side. They then rear up as high as they can and stare at each other obliquely. This sequence may be repeated or the males will return to their respective territories. Such complex aggressive behavior is not uncommon in Animalia. Figure 22-4 shows two elephant seals in combat.

## COMMUNICATION

**Communicative behavior** may be expressed through visual, tactile, auditory, olfactory, chemoperceptive, or other means. Communication may consist of any one or a combination of these media.

Auditory communication is probably the most prevalent form; its effectiveness is measured by the richness of the specific **vocabulary.** Various vocabularies exist in the animal world; one primate group is known to have nine distinct vocal signals. The human vocabulary seems to be the most extensive, but until communication research has advanced far beyond its present state, the limits of lower organisms' auditory receptive or vocal productive powers cannot be specified.

Bees communicate the location and wealth of food supplies to other bees by a dance performed on a vertical surface of their honeycomb; the intensity of wing beats seems to provide additional information (Fig. 22–5). The sun's position is of supreme importance as a reference point for food location. Because the bee's compound eye is very sensitive to light polarization, the orientation of the dance will conform to the sun's actual movements, even if the sun is blocked from the bee's view. The major direction of the dance

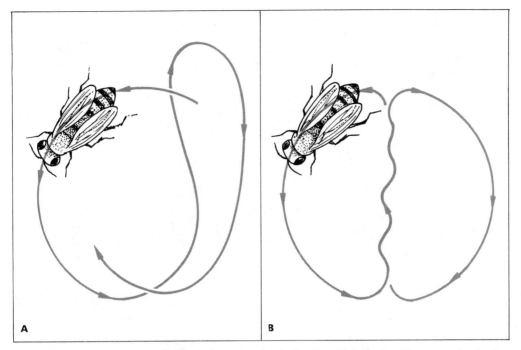

**Figure 22-5** Communication among honeybees. (A) The "round dance" indicates nearby food. (B) The "tail-wagging dance"; the direction of food is indicated by the straight run, and the distance from the hive is communicated through the time spent "wagging the tail."

(up—toward the sun, down—away from the sun) and the number of cycles performed in a given unit of time (the farther away the food, the lower the cyclic frequency) indicate the location of the food supply. Different dance forms are used to indicate different distances. The "round dance" indicates a distance of approximately 85 meters while the "tail wagging dance" is used for further distances. In addition, different species of bees use various "dialects" based on the stereotyped dances.

## MIMICRY

**Mimicry** is a special type of communicative behavior in which an organism **imitates** one or several characteristics of another organism. It is often used by an animal to escape from its normal predators.

An example of **Batesian mimicry** is the viceroy butterfly, which is preyed upon by birds; this organism imitates the monarch butterfly, which is unpalatable to birds.

In **Müllerian mimicry,** an animal may avoid attack by imitating an animal (which it resembles) that has poisonous or retaliatory qualities which the predator may have previously experienced.

An animal that mimics its environment, as the chameleon does by changing its color to match its surroundings, is using the most familiar form of mimicry, **camouflage.**

## COURTSHIP AND COPULATION

Courtship behavior is perhaps the best studied activity. Most often it is cyclic or seasonal, varying with lunar, temperature, or estrous states.

The majority of research on courtship and copulation is concerned with **precopulatory** behavior, the general characteristics of which include:

1. Species-specific amounts of mutual tactile activity (which may further stimulate hormonal release).
2. A soliciting posture of one mate (usually but not exclusively the female), which may facilitate intromission.
3. Special "calls."
4. Protracted precopulatory periods.
5. Selective, as opposed to random, mate choice.

At the beginning of precopulatory activity in the cockroach *Nauphoeta cinerea*, the male *N. cinerea* strokes the female's antennae. The female reciprocates. The male then turns his back to the female and holds his wings at a 90° angle for about 1 minute. This exposes special glands which secrete a substance that is attractive to the female. The female "mounts" the male, he seizes her genitalia, flops over, and copulation begins.

Aggression between the sexes sometimes plays a major role in courtship and precopulatory behavior. In the yellowhammer or flicker, *Emberiza citrinella*, the first meeting between two potential mates frequently results in a fight. Gradually, the male and female become habituated to each other and leave the flock. For several weeks during the courtship period, the male performs mock attacks on the female. In early May, nesting and mating behavior begin.

Many species seem to recognize sex by reaction: If it fights, it is a male; if it is receptive, it is a female.

A rather eccentric copulatory sequence is followed by the mantid. Female mantids may actually decapitate the males, thereby removing a neural inhibition of parts of his (ventral) nerve cord, which allows for continuous abdominal and genital undulations. A similar type of activity has been observed in spiders, where in some instances the male will be eaten by the female.

## SOCIAL BEHAVIOR

Animals that group together (aggregate) due to environmental factors (temperature, presence of water, illumination, etc.) behave very differently than when isolated. Three principles of animal behavior have been proposed:

1. The **dominant-subordinate** hierarchies occur in many societies, including chickens, noted for their society based on **peck order**, where through aggressive behavior the most dominant chicken may peck at all the others, and the next in line may peck at all but the most dominant, and so on.
2. **Territoriality**, well documented in many animals, serves to control the acquisition of nutrients and mates (Fig. 22–6). Birds chirping, cats and dogs marking odors, and the pacing of lions and tigers around their territory are all examples of this behavioral trait.
3. **Leadership** in animal societies is attained differently in almost every species. Wolf packs are lead by one family; red deer by one female (**matriarchal leadership**); fish move and react in packs or schools, which may be asocial with no definite leadership.

Chimpanzees have an extremely complex society which includes much mutual grooming and tactile contact and a most interesting phenomenon, the

**Figure 22–6** Sea gull territoriality. A seemingly placid sea gull changes its temperament when territory is threatened.

"carnival." The animals beat trees, scream, dance, and jump on one another for several hours. Jane Goodall, who did much of the relevant study on chimpanzee society, was permitted to participate in such carnivals, mostly because she learned and joined in the simple dance.

## SYMBIOSIS

When two or more dissimilar organisms form partnerships, they are in a **symbiotic** relationship. Three categories of symbiosis are recognized: mutualism, commensalism, and parasitism.

In **mutualism,** both (or all) organisms concerned benefit from the relationship. Mutualism may occur between plants, plants and animals, or between animals (Fig. 22–7). An interesting example of mutualism is the African honey guide. Herbert Friedman has extensively studied this avian species. The African honey guide eats bees' wax, but does not have the power to break into bees' nests. It circumvents this problem by making a very obvious call ("churring"), which attracts a human being or a honey badger and proceeds to fly in a very conspicuous pattern, leading its "symbiont" to the bees' nest. When the nest has been broken into (and the honey removed), the honey guide moves in and takes its fill of wax! Although Friedman believes the behavior to be instinctual, other scientists attribute it to operant conditioning.

**Commensalism** is a relationship where neither, or only one organism benefits, but neither is disadvantaged (Fig. 22–8). Usually, a larger, unmodified **host** will take on a smaller **guest.** The guest may change structurally to benefit most from the situation. In most cases, dependence is for shelter, locomotion, feeding, or a combination of these. Many bacteria, for example, inhabit the gut tissue of metazoans with no apparent adverse effect on the host.

**Parasitism** is the symbiotic relationship in which one organism benefits at the (sometimes fatal) expense of another. A more detailed discussion of parasitism is found in Chapters 5, 6, and 9.

## MIGRATION AND HOMING

**Migration** is the seasonal shift from one geographical region to another; it occurs in some species of most of the higher groups. The best-documented migrations are those of birds; some fish (such as salmon), reptiles (sea turtles), insects (monarch butterflies), and mammals (bats, whales, and caribou) also migrate.

**Figure 22-7** Mutualism, as shown by an insect pollinating a flower "in return" for a bit of honey.

**Figure 22-8** Commensalism, as shown by a fish residing within the body of a starfish. Neither animal is disadvantaged.

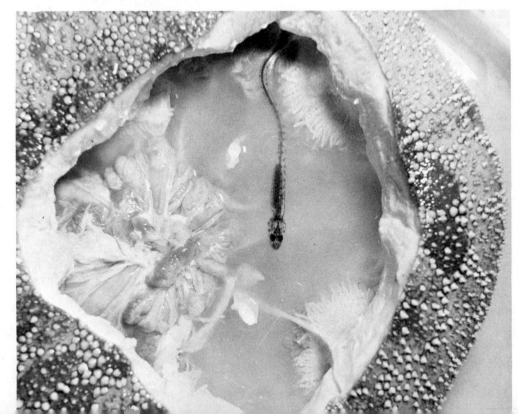

Most bird migrations are **latitudinal,** so called because the north–south movements, which follow *longitudinal routes,* are measured in *latitudinal* increments. This type of migration takes varying lengths of time and spans considerable distances. The warbler may take up to 60 days to perform its summer migration from Central America to Canada, and the arctic tern's migration takes it over a distance of 11,000 miles!

Another type of migration is **altitudinal,** such as the mountain bird, which moves up or down the mountain at a particular season.

The results of migration fall into two major categories. **Climatic migration** brings an animal from a zone of (at least potentially) dysfunctional climate to a zone more climatically suited to daily activity. **Reproductive migration,** as in the salmon, may occur at the beginning of a spawning season. It is thought to be preceded by gonadal changes.

Bird migration is oriented by the sun and sometimes by the stars as well. It has been found that starlings confined during the migratory season tend to face the migratory direction if they are able to see the sun. Birds are capable of compensating for daily solar position changes by using the **solar azimuth** (Fig. 22–9).

The phenomenon of **homing** has been beneficial (e.g., the homing pigeon) and intriguing to mankind for eons. It consists basically of the "return home" of an animal (usually birds) after having been transported to a completely unfamiliar territory.

Two major hypotheses have been presented to explain homing behavior. The first is that birds **extrapolate** position using the sun's arc. This mechanism would be similar to that used in migration. The second is that the task is accomplished through semirandom search. The birds are observed to fly in an ever-increasing spiral, presumably looking for a familiar object. Neither of these theories has been definitely proved yet.

In this chapter, we have tried to show some of the ways in which animals, although extremely diverse in behavioral traits, exhibit patterns through which we can learn more about them (and ourselves) and the ways they relate to each other and their environment.

**Figure 22–9** The solar azimuth is a navigation aid to bird migration. If the sun is too high, a bird wishing to reach home in a northern hemisphere (shown here) will fly away from it. If the sun is too low, a bird will fly toward it. The reverse is true in the southern hemispheres.

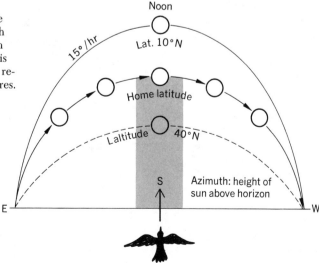

# Summary

Studies of animal behavior are currently based on comparisons of patterns of animal activity (ethology). Behavior is constantly changing, both over time (development) and in repeating segments (patterns, cycles).

Behavior is studied in all environments, inside and outside the laboratory. Much laboratory study is concerned with genetic versus environmental components of behavior (nature versus nurture).

Most behavior is reducible to stimulus–response sets. Stimuli, which are grossly divided into internal and external groups, may be extremely diverse, as are the responses they elicit. Reflexes become more important in the lower organisms.

Chemical theories (which ignore morphological considerations) and structural theories (based on morphological considerations) both attempt to explain the phenomenon of learning, but neither is accepted as yet.

Classical and operant conditioning experiments have told us much about animal behavior. Insight is an abridged form of learning and occurs only in higher organisms. Memory is the recreation of a prior image from neural traces, and may be either psychological, psychophysiological, or recognitional in nature.

Instinct is an often-complex behavior that occurs with no prior learning and has three stages: appetitive, releasing stimulus, and the final consummatory act.

Kineses and taxes are movements. Kineses are undirected locomotor responses, while the more complicated taxes are vectoral reactions to environmental influences.

Other forms of behavior, including sleep, aggression, communication, mimicry, social behavior, symbiosis, and migration were discussed.

An organism expresses itself by its behavior; although behaviors are extremely diverse, major patterns are shared among almost all groups, indicating the underlying unity that exists in the animal world.

# 23

# Integumentary, Skeletal, and Muscular Systems

The integumentary (skin), muscular, and skeletal systems all provide types of support and protection for animals. The representative animals we have studied throughout this text have illustrated the variety of functions these systems may serve.

In this chapter, we will limit our discussion of the integumentary system to vertebrates. This will be followed by an account of skeletal and muscular systems found throughout the animal kingdom.

## Integumentary System

The skin of vertebrates functions mainly in protection and sensation, but may also play a part in respiration, secretion, and excretion. Secretion and excretion take place by means of glands. These glands may be simple, as the **mucous glands** of amphibians and fishes, or complex, as the **sweat, oil,** and **mammary glands.** The skin often produces outgrowths; as we have seen, these modifications include hair, feathers, nails, hoofs, claws, scales, teeth, and bony plates.

### SKIN IN HUMANS

**Human skin** (Fig. 23-1) protects underlying and deeper tissues from drying, from injury, and from invasion of bacteria and other organisms. It contains the end organs of many sensory nerves. It also creates a balance between heat production and heat dissipation: About 87.5 percent of body heat passes out through the skin.

**Sebaceous glands** occur everywhere except on the palms of hands and soles of feet. These are compound alveolar glands with ducts that usually open into a hair follicle. Sebaceous glands secrete **sebum,** a fatty, oily substance that keeps the skin and hair flexible and covers the skin with a layer that prevents too-rapid evaporation of water.

**Sweat glands** are distributed over the surface of the skin, with the exception of the lip margins, under the nails, and on the glans penis. Sweat glands are

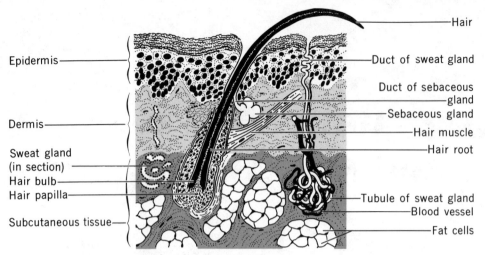

**Figure 23-1** Human skin, showing cell layers, glandular tissue, and other parts: vertical section.

tubular, with the inner portion coiled into a ball. Sweat contains some waste substances, such as urea, but is important mainly in its role in the physics of evaporation. Generally, the heat energy used in evaporating the sweat is lost from the skin, which is then cooled. The average human being produces between 16 and 20 fluid ounces of sweat per day.

We will now turn our attention to the types of skeletal systems found in animals.

## Skeletal Systems

Skeletal systems provide a supportive and sometimes protective framework for the softer parts of the body, and a firm surface for muscle attachment.

Skeletons may be of two types: **exoskeletons** are formed on the outside of the body with muscles attached to their inner surfaces. **Endoskeletons** are formed inside the body; they are surrounded by soft tissues and the muscles are attached to their outer surfaces.

Many skeletal structures, especially among the lower vertebrates, furnish support and protection but consist of more or less isolated parts; the skeletal **spicules** of the sponges are an example.

### PROTECTION AND SUPPORT OF LOWER INVERTEBRATES

Among the lower invertebrates, protective and supportive structures are highly varied. Protists may secrete a shell, such as that of the amoeba *Arcella*, which is made of **chitin**, or they may build a shell of sand held together by a secretion, as in *Difflugia* (Chapter 5). Many other invertebrates, such as the **coral polyp**, secrete internal skeletons of calcium carbonate or silica to support their soft bodies. **Spicules** and **spongin** of sponges effectively prevent the collapsing of a body that would otherwise be nothing but a gelatinous mass. Massive sponges could have probably never evolved without the support of such skeletal structures.

## EXOSKELETONS

Exoskeletons are characteristic of arthropods, and most mollusks. Many lower invertebrates secrete a **cuticle,** which protects the softer parts beneath. In many protists, the **pellicle** also aids in maintaining the shape of the body. The soft body of many coelenterates, such as *Obelia,* is supported and protected by a chitinous tube, the **perisarc.**

The exoskeleton of arthropods is developed in a very characteristic way, including a period of **molting,** which produces an entirely new exoskeleton. The molting process will serve to describe the development of the exoskeleton.

As an insect grows, its body becomes too large for the existing exoskeleton and the animal molts. Prior to the actual molting, the epidermal cells of the skin just beneath the exoskeleton begin to divide and grow. So much new epidermis is formed that it is forced into folds underneath the exoskeleton. During the proliferation, the epidermal cells secrete an **epicuticle,** made of the lipoprotein **cuticulin,** just under the exoskeleton. After sufficient epicuticle has been produced, a fluid is secreted that digests the inner layers of the old cuticle, preparing it to be discarded.

An **exocuticle** is secreted under the new cuticle, containing pigment granules which give the animal its color. The last layer to be secreted is the **endocuticle,** closest to the internal organs. It is composed of protein and **chitin,** a nitrogenous polysaccharide.

The final stage of molting is the splitting and discarding of the old exoskeleton. The new exoskeleton must harden after its first exposure to the air before becoming functional.

The bony **carapace** and **plastron** of the turtle's shell, the bony shell of armadillos, the epidermal **scales** of anteaters, and the calcium carbonate cuticle of the crayfish are examples of exoskeletons in higher phyla. Since an exoskeleton can support only a limited mass, there are not many "large" animals with an exoskeleton.

## ENDOSKELETAL DEVELOPMENT

The origin of **osseous** (bone) tissue may take one of two forms. The **cranium** (skull) and shoulder girdle arise **intermembranously**—converting directly from mesenchyme to bone. Most other osseous tissue has an intermediate **cartilaginous** stage between the mesenchyme origin and ultimate bone formation. This latter process, known as **intercartilaginous** development, is more complex than but similar to **intermembranous** development, a description of the process follows.

During the embryonic development of an animal, migratory mesenchyme cells aggregate in the areas where bone is to be produced. The mesenchyme cells then differentiate into **osteoblasts,** which are active cells secreting the bone's **protein matrix,** on which calcium is deposited. (Note that bone tissue contains about 99 percent of the calcium in the vertebrate body.)

The osteoblasts form **trabeculae,** scattered bars and plates of fibrous material. The matrix also contains **lacunae,** spaces in which the osteoblasts, or osteoblastic **processes** (extensions) reside. The lacunae form an interconnecting network with interbranching canals that allows for communication.

As more tissue is formed, the osteoblasts become **osteocytes,** which play a role in the maintenance of previously formed bone. At this point in development, the lacunae cease to function except as an interconnecting network for the osteocytes.

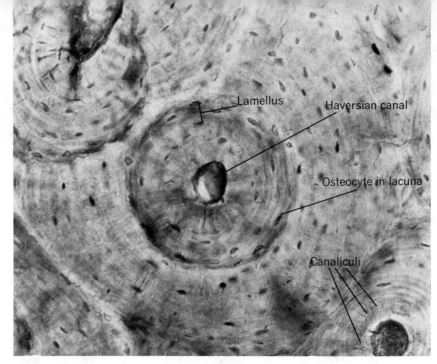

**Figure 23-2** Haversian canal system. 160×.

The newly arriving mesenchyme continues to form osteoblasts, which continue to add to the trabeculae; the latter eventually fuse and become a continuous lattice. The bone is now **spongy** or **cancellous** bone tissue. The areas between the trabeculae are filled with extremely vascular connective tissue, bone **marrow,** which is responsible for the formation of blood cells and in later life for fat storage.

A thin but tough membranous sheath, the **periosteum,** permanently forms over the bone, and dense sheets of material replace the scattered bars and plates, thus resulting in a **compact bone,** which is the final stage of bone development (but not necessarily of *growth*).

The **Haversian system,** which communicates between the periosteum and the marrow, provides for the distribution of blood vessels throughout the osseous tissue (Fig. 23-2).

Stimulation for further bone growth (lengthening) seems to come from the anterior hypophysis (pituitary's) secretion of **growth hormone.** The enzymatic

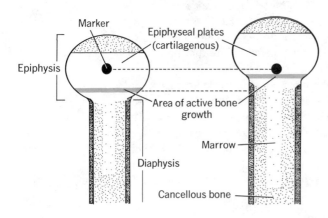

**Figure 23-3** Bone growth. The "marker" cartilage (left) is well within epiphyseal plate area. Later (right), although it has remained in the same horizontal position, the bone-growth area has moved up to it. A later view would show it turned into bone, while the cartilage area continues to produce more cartilage.

action of this hormone speeds up protein synthesis in both the cartilaginous center and bony edge plates of the existing bone. The mechanism for this process is still under debate. It is known that bone structure and growth are related to mechanical stress—the more stress, the more growth.

Whatever the mechanism for elongation may be, the fact remains that some bones *do* grow. On the ends of long bones are **epiphyseal plates** (**epiphyses**); between them lie the main body of the bone, the **diaphysis.** The epiphyseal plates, cartilaginous in nature, contain a thin area of active bone growth. New cartilaginous tissue is formed in the active area, while older, previously active areas are converted to bone (Fig. 23-3).

## MAJOR DIVISIONS OF SKELETAL ELEMENTS

Parts of the skeleton may be divided in various ways. In our discussions of organisms throughout the text, we have divided the skeleton into **visceral** and **somatic** portions.

The **visceral** skeleton occurs in the gill area. It includes such structures as jaw cartilages and ear ossicles.

The remainder of the skeleton is the **somatic** skeleton. It is further divided into axial and appendicular skeletons. The **axial** skeleton includes the **vertebrae, ribs, cranium,** and other related elements of the trunk and tail. The **appendicular** skeleton is composed of the remainder of the skeletal elements: the **limb** girdles and the elements of the free **appendages.**

## NOTOCHORD

The embryos of all vertebrates have at some time in their development both gills and a **notochord,** the latter a precursor of the osseous vertebral column. In some lower chordates, such as amphioxus and the cyclostomes, the notochord is a permanent adult structure.

The semirigid structure stems from the turgid (fluid-filled) thick-walled cells that comprise it, and from the two concentric sheaths of connective tissue surrounding it.

The notochord serves various purposes, including a small amount of support, an axis around which the vertebral column may develop, and an anchoring locus for certain muscles.

## HUMAN SKELETON

The human skeleton generally resembles the frog's in structure and function. The human **skull** (Fig. 23-4) includes the **cranium** (made up of eight bones) and the **face** (containing 14 bones).

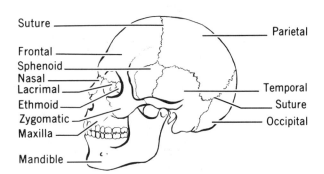

**Figure 23-4** Human skull. Note the distinct sutures between the bones of the cranium; these close and are sometimes completely obliterated in old age.

8 cranium
14 face

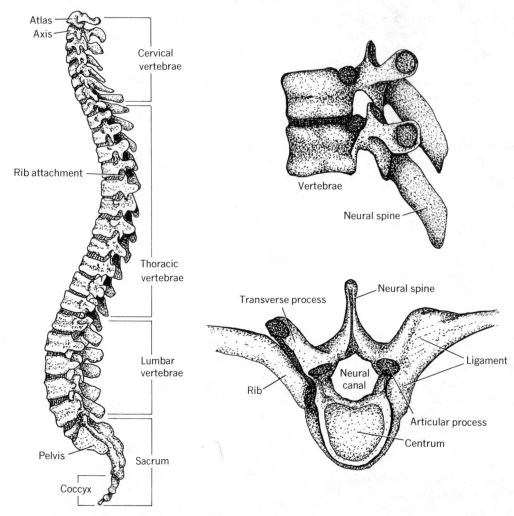

**Figure 23–5** *Left:* vertebral column. The sacrum and coccyx consist of fused vertebrae. *Upper right:* lateral view of two thoracic vertebrae. *Lower right:* superior view of vertebra, drawn to same scale as lateral view above.

Each ear contains three bones: the **malleus, incus,** and **stapes.** The **hyoid bone** lies above the larynx near the base of the tongue.

The **vertebral column,** averaging about 28 inches long in adults, consists of 33 bones in the child, 5 of which are fused in adults to form the **sacrum** (Fig. 23–5). The number of bones in the **coccyx** (the vestigial tail in human beings) varies from 3 to 5. In females, the sacrum and coccyx are less curved and more anteriorly pointed, an adaptation for childbirth that complements the male–female pelvic differences. The other vertebrae are the **lumbar** (5) in the lower back, the **thoracic** (12) in the thorax, and the **cervical** (7) in the neck. The first cervical vertebra is called the **axis** and the second is called the **atlas.** The vertebrae are semiindependent in their movements. The neural arch encloses the spinal cord.

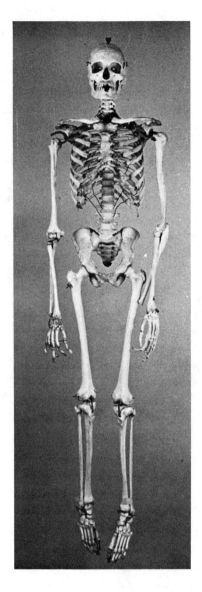
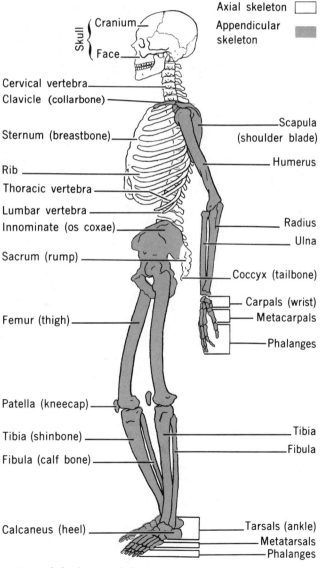

**Figure 23–6** Axial and appendicular portions of the human skeleton.

Twelve pairs of **ribs**, the **sternum** ("breastbone"), and the **costal cartilages** form the bony cage of the **thorax**. The uppermost 7 rib pairs (**vertebrosternal** or **true ribs**) are attached to the sternum by costal cartilages. The next 3 pairs (**vertebrochondral** or **false ribs**) are attached to the local superior costals. The last 2 pairs (**floating ribs**) have no ventral attachment. Some individuals have an extra rib, called a **gorilla rib**, because of its normal presence in gorillas and chimpanzees.

The bones of the appendicular skeleton (Fig. 23–6) are those of the **shoulder (pectoral) girdle** (4), **pelvic girdle** (2), **upper limbs** (60), and **lower limbs** (60). The shoulder girdle consists of 2 **clavicles** ("collarbones") and two **scapulae** (shoulder blades). Bones in each arm are the **humerus**, running from the elbow to the shoulder; the **ulna** (larger bone of the forearm); and the **radius**, which helps the thumb pivot. The 8 **carpals**, 5 **metacarpals**, and 14 **phalanges** make up the hand. Analogously, the leg is made up of the **femur** (longest bone in the body), **patella** ("kneecap"), **tibia**, **fibula**, 7 **tarsals**, 5 **metatarsals**, and 14 **phalanges**. Each side of the **pelvic girdle** is composed of three bones—the **ilium**, **ischium**, and **pubis**—united into a single **innominate** bone or **os coxae**.

The total number of bones in the average adult human skeleton is 206. Primitive vertebrates tend to have a larger number of skull bones than the more recent forms. Some fish have 180 skull bones; amphibians and reptiles may have 50 to 95 bones; and mammals may have 35 or fewer (human beings have 29 bones in the skull). Higher organisms' decreased number of skull bones may be seen as adaptations. Fewer bones result in a stronger and more solid brain case and stronger points of attachment for head and facial muscles.

## JOINTS

Locomotion in vertebrates requires movable joints. Since joints are differentiated on the basis of morphology, three major categories are evident: fibrous joints, cartilaginous joints, and synovial joints.

**Fibrous joints** contain no joint cavity; the bones are opposed directly and held together by thin connective tissue and/or fibrous tissue. A specialized type is the **suture**, which occurs exclusively in the skull; it is the tightest (containing the least interstitial tissue) of all joints.

**Cartilaginous joints** lack a joint cavity; the area between the bones contain either fibrocartilage or hyaline cartilage (see Chapter 3). These joints are only slightly movable, owing to the semirigid character of the cartilages present.

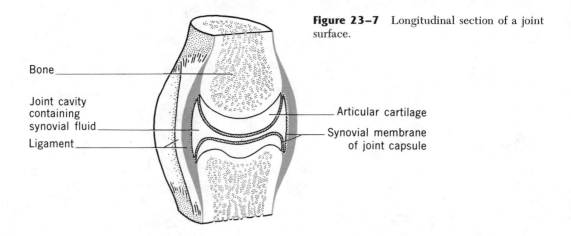

**Figure 23–7** Longitudinal section of a joint surface.

They occur between the ends of growing bone or in special areas such as the **symphysis pubis** within the pelvic girdle.

**Synovial joints** (or **diarthroid joints**) include those that occur in the elbow, shoulder, thumb, and wrist. They have great flexibility in selected directions, owing in part to the presence of synovial fluid in cavities between the bones. Synovial joints have **ligaments** (see Chapter 3) that hold the bones together (Fig. 23–7). Each bone surface, at the joint site, is covered with hyaline cartilage and may additionally possess a cartilaginous pad, which presumably absorbs sharp shocks.

# Muscular System

Muscular tissue in higher animals is usually classified into three types, according to structure and localization: smooth, striated, and cardiac.

## SMOOTH MUSCLE

Smooth muscle is characteristic of many invertebrates and of the visceral organs of vertebrates. It is composed of spindle-shaped cells with a single nucleus located near the center of the cell. They occur in three types of organization: isolated, aggregated in small groups, or in parallel layers of substantial size.

Smooth muscles contract more slowly than the other types, and take from 3 to 180 seconds to perform a single contraction–relaxation sequence. Changes in the shape and size of the visceral organs (as in **peristalsis** in the intestine) are due to the contraction of smooth muscles. These muscles are present in the walls of blood vessels and the bladder, around the openings of ectocrine glands, in the skin (where they may cause hair to rise), in the trachea and bronchi, and in the reproductive ducts. Since smooth muscles are not under voluntary control, they are also referred to as **involuntary muscles.** It is believed that they represent the most primitive form of muscle tissue.

## STRIATED MUSCLE

Striated (voluntary) muscle makes up about 40 percent of the weight of a human being. They are the fastest to react, having a contraction–relaxive time of approximately 0.1 second.

Muscles of arthropods are exclusively striated. In most vertebrates, striated muscle is located in the limbs and body wall; it also appears in the diaphragm, tongue, upper third of the esophagus, pharynx, larynx, and around the eyes.

The fibers of skeletal muscles are spindle-shaped and multinucleated, the nuclei usually lie just beneath the surface membrane. Connective tissue binds the fibers together, thus forming the muscle. Each fiber in the muscle may act separately, producing various gradations in the force of the muscle as a whole.

When viewed through a microscope, each **myofibril** (which run longitudinally through the muscle fibers) appears to consist of alternate light and dark segments or bands, creating the "striated" appearance. The blood supply is rich in capillaries, but is not as great as in cardiac muscle. Since these striated muscles are the type that are attached to the skeleton, they are also called **skeletal muscles.**

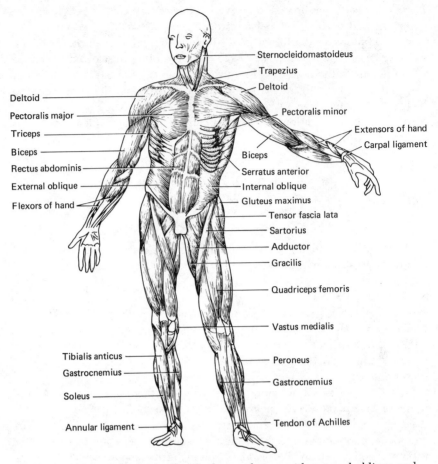

**Figure 23-8** Superficial muscles of a human being, with external oblique and pectoralis major removed from left side to show underlying muscles.

The fibers in striated muscles alternately contract to maintain a constant state of **tonus**. If a muscle is directed to contract for a substantial period of time, fibers again alternate, so that when a muscle is in contraction, not all of its fibers are working simultaneously.

Muscles used in locomotion are usually present in **antagonistic pairs.** For example, when the biceps contracts, it bends (flexes) the arm at the elbow; when it relaxes, it does not "push out" to counter its previous movement, but the antagonistic triceps, on the opposite side of the arm, contracts to bring the arm back to its original (or another) position.

Each striated muscle is attached by tendons to two points on the skeleton. The **origin** is the muscle end attached to the more stationary part of the skeleton, while the **insertion** is the muscle end attached to the skeletal element moved by muscular contraction. Figure 23-8 shows superficial and underlying muscles found in the human body.

## CARDIAC MUSCLE

Cardiac muscle occurs only in the hearts of vertebrates and a few invertebrates. The cardiac muscle in the vertebrate heart consists of cells that are less distinctly striated than those of striated muscle. These fibers are connected with one another so that they form a **syncytium** (mass of cytoplasm that contains scattered nuclei with no apparent cellular membranes). Impulses pass through the ventricles of the heart by way of these connections, and the reaction is that of one large muscle cell.

The contraction–relaxation time of an average cardiac muscle cell is 1 to 5 seconds. After the contraction a relatively long **refractory period** ensues, during which the cells are incapable of further contractions.

## PHYSIOLOGY OF MUSCLE CONTRACTION

Muscle fibers consist of **myofibrils,** which are further composed of **myofilaments,** which may be thin (2.0 $\mu$m long and 5 nm in diameter) or thick (1.5 $\mu$m long and 10 nm in diameter).

A muscle appears, microscopically, as a series of light and dark **bands.** The **A band,** which appears dark, is composed of the protein **myosin** and is made of both thick and thin myofilaments. The A band is flanked by **I bands,** which appear light, are made up of the protein **actin,** and contain exclusively thin myofilaments. In the center of each I band is a dark **Z band,** which contains *no* filaments and separates each muscle unit (**sarcomere**). In the center of the A band lies an area of low density—the **H band**—which represents the area of actin filaments not overlapped by myosin filaments.

The bands shift during contraction, with the following results:

1. The A band width remains constant.
2. The I and H bands become narrower.
3. The Z bands, which separate the sarcomeres, move inward (this is the most obvious microscopic indication of muscle contraction).

The current explanation for these observations is known as the **sliding filament model** of muscle contraction, originally proposed independently by A. F. Huxley and H. E. Huxley in 1954. According to this theory, the filaments do *not* shrink, but slide over each other, thus shortening the muscle fiber as a whole (Fig. 23-9).

A corollary to the sliding filament model states that **interfilament bridges** act as a ratchetlike apparatus, keeping the filaments in proper position. The bridges are seen as permanently attached to the myosin filaments, while the actin filaments contain "active sites" that accept the bridges.

**Nervous stimuli** bring about muscular contraction. The actin and myosin filaments contain different electrical charges. It is thought that the nervous impulse arriving at the muscle reduces the **electrical potential** (difference in charge) among the filaments. The electrical potential is thought to constitute an "obstacle" to contraction, and hence the nerve is seen to remove the inhibition. This theory is still under some discussion.

When a muscle contracts, it converts stored potential chemical energy to mechanical energy and heat. The efficiency of the process is only about 25 to 40 percent; the greater part of the energy is liberated in the form of heat. About four-fifths of all body heat is derived from this source.

During muscle contraction, carbon dioxide is released, glycogen disappears, lactic acid is formed, and changes occur in certain organic phosphates present. There is much evidence that the energy used in muscular movement comes from the breakdown of adenosine triphosphate (ATP).

True **fatigue** is at least partially caused by the effects of waste products

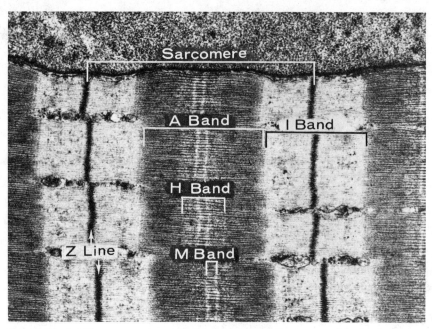

**Figure 23-9** Electron micrograph showing sliding filament model of muscle contraction. 34,000×. (Courtesy W. Bloom and D. W. Fawcett, *A Textbook of Histology*, 9th ed., Philadelphia: W. B. Saunders Co., 1968.)

(carbon dioxide, lactic acid, acid phosphate) that accumulate during exercise; the loss of nutritive substances may also be a factor. Rest is necessary after violent exercise to enable the blood to carry the metabolic waste substances to the excretory organs and the nutritive materials (including oxygen) to the muscles. Exercise stimulates blood circulation and brings about an increase in size, strength, and tone of the muscles.

# Summary

The integumentary (skin) system of vertebrates and some lower animals functions mainly in protection and sensation, but is also active in respiration, secretion, and excretion.

Skeletal structures provide support, protection, and attachment loci for muscles, and may function in osmoregulation. The exoskeleton (prevalent in lower forms, notably the arthropods) appears on the outside of an organism. It is produced, shortly prior to and during molting, by the underlying epidermis.

Endoskeletons (found in most higher forms) appear on the inside of the organism. They are produced initially from mesenchyme cells which differentiate and secrete (or form) trabeculae, which ultimately form a continuous lattice. Marrow is found among the coalesced trabeculae. A periosteum develops around the bone structure, while an interconnecting Haversian system is built up within the bone. Growth hormones stimulate further development, which is also dependent upon stress.

The human skeleton contains approximately 206 bones, more than half of which are present in the limbs.

Three major types of joints occur: fibrous, cartilaginous, and synovial.

The muscular system contains three major muscle forms: smooth, striated, and cardiac. Smooth muscle is under involuntary control and is responsible for such actions as peristalsis. Striated (or skeletal) muscle is under voluntary control and is made up of units called sarcomeres. Each sarcomere is made of many fibers, which are, in turn, composed of myofibrils. Muscle contraction is accomplished by the sliding of actin and myosin filaments. Cardiac muscle appears in heart tissue only, and is in the form of a syncytium.

Muscles are directed to contract through nervous stimulation, and so are effectors. They use energy stored in ATP, and produce lactic acid during contraction.

Skeletal and muscular systems are not necessary for motion or locomotion in the lower invertebrates, but are essential in the rest of the animal kingdom.

# 24
# Digestion and Excretion

In this chapter, we will consider what constitutes the food of various animals we have studied, how this food is captured, ingested, and digested, and how waste products are eliminated. The processes involved constitute **nutrition** and excretion. The term **metabolism** may also be applied to these processes, but is used here in a restricted sense to define the *chemical reactions* that take place within the cells or within the digestive tract of the organism.

## FOOD

All substances taken into the body that are used to build up cytoplasm and to produce energy are **foods**. The principal foods of animals are organic compounds (carbohydrates, fats, and proteins) built up by other organisms, water, inorganic salts, oxygen, and vitamins. Only certain chlorophyll-bearing organisms, such as *Euglena*, and chemosynthetic bacteria can synthesize their food from inorganic substances. All other animals use plants or animals (or both) for their nutritive material.

## NUTRITIONAL MODES

Those protists and plants which photosynthesize are called **holophytic**. Most animals feed on solid material, and nutrition involving ingestion of such matter is **holozoic**. Parasites and other organisms that absorb organic substances in solution through their body surfaces have **saprozoic** nutrition.

Three principal categories of animals may be differentiated based on their diets: **herbivorous** animals feed on vegetation, **carnivorous** on meat, and **omnivorous** on both vegetable and animal matter. **Insectivorous** animals are a subset of carnivores that feed exclusively on insects.

## FOOD CAPTURE

Most every animal capable of locomotion searches about for food. The protists capture food with their undulipodia or pseudopodia. Cnidaria have tentacles with nematocysts used for stinging and/or grasping prey. Many spiders build elaborate webs for catching insects; the frog accomplishes the same thing with its long sticky tongue. Many fish and birds rely on their speed to capture prey; mammals effectively use their claws and teeth.

Civilized man, through **domestication** of animals, has greatly reduced the need for hunting but still retains both canine and grinding teeth for eating—

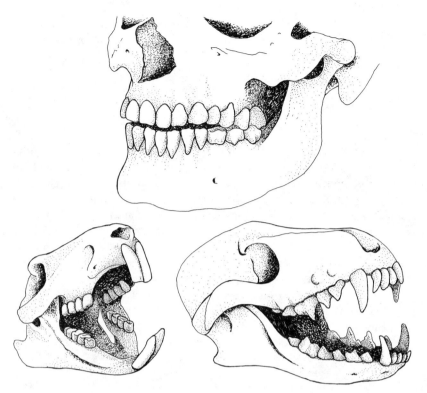

**Figure 24–1** *Top:* human (omnivore) dentition shares characteristics of herbivore dentition (*lower left*) and carnivore dentition (*lower right*).

these tooth types are indicative of carnivorous and herbivorous animals, respectively, and when occurring together indicate an omnivore (Fig. 24–1).

## INGESTION OF FOOD

In many animals, the structures used for obtaining food are also used for ingestion. This is true of the pseudopodia of amoebas; undulipodia of paramecia; tentacles of hydras; muscular pharynx of planarians, earthworms, and certain sucking insects; undulipodia of clams; and tongue of the frog.

In other animals, special methods of ingestion are employed. In sponges, food particles are drawn into a central cavity by long undulipodia and are engulfed by collar cells and amoebocytes. In hydras, the mouth, tentacles, and body force the food into the gastrovascular cavity, but small particles are ingested by the nutritive cells of the inner body wall. In crayfish and many insects, the food is held by specific mouthparts while it is crushed or bitten into small pieces by mandibles. Most fish (and many other vertebrates, including snakes and birds) do not chew their food but hold it with their teeth and swallow it at once. Mammals usually chew their food before swallowing.

Two types of **digestion** occur in animals: **intracellular,** as in protists and sponges, where the food particles are digested inside food vacuoles; and **extracellular,** characteristic of most higher organisms, where food is digested in a gastrovascular cavity.

The remainder of this chapter relates primarily to digestion in higher organisms. The material covered is, in fact, limited in scope, and it should be kept in mind that alternative systems and methods of digestion exist.

## Human Digestive System

More is known about digestion in human beings than in any other animal. The human digestive system resembles that of other vertebrates more closely than invertebrates; but even in the earthworm the parts of the digestive system are given the same names and are analogous (not homologous) to the human organs of digestion. The parts of the human digestive tract (Fig. 24-2) are as follows:

1. **Mouth cavity**, containing the **tongue**, openings of ducts of **salivary glands**, and **teeth**; the 32 permanent teeth of adults include 8 **incisors**, 4 **cuspids** (canines) 8 **bicuspids** (premolars) and 12 **tricuspids** (molars).
2. **Pharynx** or throat cavity, shaped like an inverted cone.
3. **Esophagus**, a muscular tube about 23 cm long.
4. **Stomach**, a saclike dilation of the digestive tube.

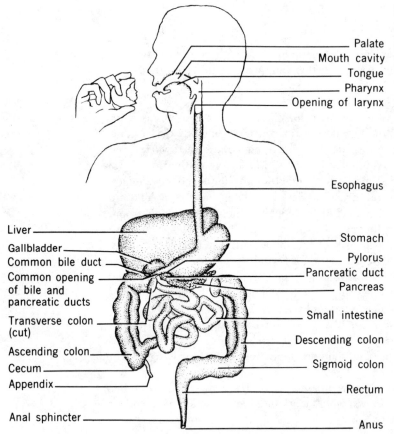

**Figure 24-2** Human digestive system. Not all the coils of the ileum (small intestine) are shown.

5. **Small intestine**, a muscular tube about 7 m long, consisting of a **duodenum**, about .3 m long; a **jejunum**, about 2.5 m long; and an **ileum**, about 4.25 m long.
6. **Large intestine**, a muscular tube about 1.5 m long that is divided consecutively into the **cecum**, a large pouch with a worm-shaped **appendix** appearing at its end; the **colon**, the main part of the large intestine; the **rectum**, about 12.5 cm long; and the **anal canal,** about 3.8 cm long.

## DIGESTIVE PROCESS

The **mouth** cavity receives the food, which is kept among the teeth with the aid of the **tongue** and is **masticated** (chewed). The tongue is also a special sense organ of taste and assists in **swallowing** food. The mouth is lined with a **mucous membrane,** which contains minute oral glands that pour secretions into the mouth cavity. These secretions mix with the secretions from three sets of **salivary glands;** the **saliva** produced contains two enzymes, ptyalin (salivary amylase) and maltase (see below). Approximately 1 liter of saliva is produced each day; its secretion is an involuntary act stimulated by taste, sight, or smell of food, but saliva will be constantly secreted even with no outside stimulus.

The **bolus** (lump of chewed food plus saliva) is passed back to the **pharynx,** which dilates and allows food to pass down the **esophagus.** A series of rhythmic muscular contractions and relaxations (**peristalsis**) continually push the food down the esophagus and through the rest of the digestive system.

At the junction of the esophagus and stomach is a circular muscle, the **cardiac sphincter.** The bolus enters the stomach and is churned for some time (usually about 30 minutes for a light meal). Here the food is acted upon by digestive enzymes secreted by the **cardiac, fundic,** and **pyloric glands** in the stomach wall.

**Digestion** takes place principally in the **small intestine. Bile** from the liver and **pancreatic juice** (containing digestive enzymes) usually enter the duodenum simultaneously to act upon the bolus, now called the **chyme.** Circular folds in the small intestine slow down the peristaltic movement of food, allowing more time for digestion and absorption. Minute fingerlike projections (**villi**) vastly increase the inside surface area throughout the small intestine. Digested food is absorbed into the blood and lymph vessels in these capillary-rich villi (Fig. 24–3).

An **iliocecal valve** allows material to pass from the small intestine to the **large intestine** (but not back again). Most of the water drunk with food or present in the food itself, as well as water secreted in digestive juices, is absorbed in the **colon** of the large intestine. The chyme, now called **feces,** moves up the **ascending colon** from the ileocecal valve, across the **transverse colon** to the **descending colon,** and downward to the **sigmoid colon.** When the feces enter the **rectum** they are semisolid, owing to the absorption of water along the way. Fecal material consists largely of **cellulose** and other indigestible substances, bacteria, and excretions of the colon itself (including excess calcium and iron). The feces remains in the rectum until the organism **defecates.**

## CHEMICAL DIGESTION

Digestion is the **catabolism** (breaking down) of food material to a soluble form that can pass through membranes (**absorption**) and be distributed throughout the body.

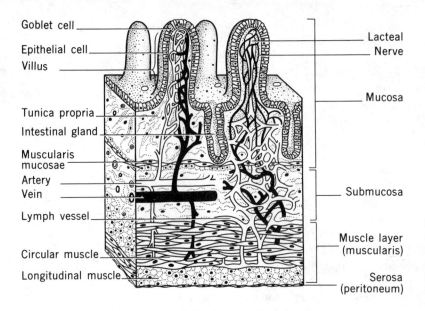

**Figure 24–3** Human small intestine, showing blood vessels (*left*) and (*right*) lacteal vessels and nerves that occur in all villi.

Proteins are broken down into amino acids, fats into glycerol and fatty acids, and carbohydrates into simple sugars. These processes are all accomplished through the use of **digestive enzymes,** complex (usually protein-containing) substances secreted into various parts of the digestive tract. They bring about chemical changes in food substances without undergoing significant change in the process (they are biological **catalysts**). Each enzyme requires a neutral, acid, or alkaline medium. Their action, which may aid in building up as well as breaking down substances, are specific; each enzyme acts upon one type of food.

Chemical digestion (**hydrolysis**—see Chapter 2) is a process whereby a complex molecule combines chemically with water, resulting in a split into two or more simpler molecules.

We will now see what actually happens to food during the digestive process.

**Water.** Most of our tissues are 75 to 90 percent water. Water does not need to be digested—it is used by the body in the same form in which it is ingested. Water is the medium in which digestive changes occur, dissolved food substances are absorbed, and wastes are dissolved and removed by excretion. As a result of these functions, there is hardly a physiological process in which water is not of fundamental importance.

**Mineral Salts.** Inorganic salts make up about 1 percent of the cytoplasm and body fluids, but are necessary in the proper relative concentrations for the body to function normally. The principal mineral salts contained in cytoplasm are listed in Chapter 2; foods that supply us with some of the more common mineral salts are as follows. Milk and leafy vegetables are the best sources of **calcium,** 99 percent of which is contained in our bones and teeth. Milk is also the best source of **phosphorus,** which is necessary for normal bone development.

Beef liver, egg yolk, whole grains, fruits, and green vegetables are rich in **iron,** found principally in erythrocytes (red blood corpuscles).

**Iodine** may be supplied by iodized salt, certain fish, milk, leafy vegetables, and fruits (grown in iodine-rich soil). It is used by the thyroid gland in hormone production (notably **thyroxin**); a deficiency in this mineral could cause an enlargement of the thyroid gland, forming a **goiter.**

**Copper** is found in liver, nuts, legumes, leafy vegetables, and fruits; its role in nutrition, although essential, is not yet fully understood.

Minute quantities of other salts (such as **manganese**) are necessary for normal health.

**CARBOHYDRATES.** Carbohydrates provide energy most abundantly and economically due to the high concentration of carbon, hydrogen, and oxygen. Three kinds are recognized on the basis of the complexity of the molecule: **monosaccharides** (simple sugars) are soluble in water and can therefore be absorbed in the intestine without being "digested"; **disaccharides** (double sugars) are hydrolyzed during digestion into two simple sugars; and **polysaccharides** (complex carbohydrates of three or more sugar units) break up into many simple sugars when hydrolyzed.

The **digestive enzymes** contained in **saliva** which act upon carbohydrates include **ptyalin** (**salivary amylase**), which hydrolyzes starch to dextrin and sugar, and **maltase,** which converts maltose (a disaccharide) to glucose (a simple sugar). In the small intestine, starch that was not acted upon by the saliva is hydrolyzed to maltose by **amylase,** which is produced in the **pancreas.** Other enzymes secreted by the small intestine include **maltase, invertase,** and **lactase.**

**FATS.** Included in this category are true fats (or **lipids**) and compound fats (or **lipoids**). True fats are composed of carbon, hydrogen, and oxygen, with proportionally less oxygen than carbohydrates, making them a more concentrated form of fuel.

The best-known lipoid is **lecithin,** abundant in the yolk of hen's eggs; it contains nitrogen and phosphorus in addition to the carbon, hydrogen, and oxygen. Other types of lipids, including cholesterol, are solid waxy substances called **sterols.**

One molecule of fat can be split into two kinds of smaller molecules: one of glycerol (or glycerin) and three of a fatty acid. Most fat is digested in the small intestine. **Lipase,** an enzyme from the pancreas, splits fats into glycerol and fatty acids that are then absorbed. **Bile** from the liver breaks up the fat droplets into small ones (emulsification), thus hastening the action of the lipase by increasing the surface area of the exposed fats.

The fats we use are derived from plants and animals; some of the most common are those ingested with meat, lard, butter, and olive oil.

**PROTEINS.** Proteins are built up from **amino acids,** which are necessary for growth; they are therefore of primary importance to the cell. Hundreds of thousands of amino acids (of which there are approximately 20 different types) may be found in a single protein molecule. Proteins are often called **nitrogen compounds** because they always contain nitrogen in addition to carbon, hydrogen, and oxygen; sulfur, phosphorus, and iron may also be present.

In the stomach, the enzyme **pepsin** is aided by hydrochloric acid to hydrolyze proteins into proteoses and peptones; these products are further acted upon by the enzyme **erepsin** in the small intestine to form amino acids, which can be absorbed by the cells. Also in the small intestine is the enzyme **trypsin** (from the pancreas), which may break proteins into polypeptides.

Our food contains proteins from such animal sources as meat, milk, fish, and albumin and from such plant sources as gluten from wheat and legumin from peas, beans, and peanuts.

**VITAMINS.** Vitamins are necessary for normal metabolism and growth; deficiency diseases would develop without these substances, which act as coenzymes or influence enzyme systems within cells.

Minute quantities of vitamins are effective; for example, only 0.5 mg of vitamin $B_1$ is required daily for a healthy adult. Plants and fish-liver oils furnish many vitamins; some are also produced synthetically.

Vitamins were originally designated by capital letters, but as their chemical structures became known they were given chemical designations.

Vitamins of demonstrated importance in human nutrition are listed in Table 24–1 and seen in Fig. 24–4.

**Table 24–1** Vitamins Important in Human Nutrition

| Vitamin | Food Source | Deficiency Symptoms or Disease (in human beings unless otherwise noted) |
|---|---|---|
| A | Spinach, asparagus, carrots, sweet potatoes, butter, cream, eggs, liver, fish-liver oils | Xerophthalmia ("dry eye"); night blindness (lack of visual purple in eyes) |
| $B_1$: thiamine | Brewer's yeast, whole grains, peas, nuts, beans, liver | Beri-beri (nervous degeneration, loss of appetite) |
| Niacin | Brewer's yeast, wheat germ, lean meat, milk, green vegetables, beans | Pellagra (rough skin, mouth sores, nervous disturbances, and diahrrea) |
| $B_6$: pyridoxine | Yeast, whole grains, milk, liver | Dermatitis (skin inflammation) in rats |
| Pantothenic acid | Yeast, cane molasses, meat, egg yolks, milk, liver | Dermatitis in chicks; decreased adrenal cortex function in rats; diarrhea and nerve degeneration in swine |
| M: folic acid | Green leaves, soybeans, yeast, egg yolk | Anemia and sprue (a tropical disease) |
| H: biotin | Liver, kidney, yeast (supplied in human beings by intestinal bacteria) | Diarrhea; dermatitis; nervous disorders |
| $B_{12}$: cyanocobalamin | Milk, liver, kidney, lean meat | Pernicious anemia (lack of formation of red blood cells) |
| C: ascorbic acid | Citrus fruits and tomatoes (all animals except primates and guinea pigs produce vitamin C) | Scurvy (rupture of capillaries, resulting in loose teeth, bleeding gums, and fragile bones) |
| D | Fish-liver oils, eggs, enriched flour, vitamin D milk | Rickets (bone softening), leading to deformities; excess vitamin D may result in calcifications in the kidney |
| E | Widely distributed, notably in green leaves and vegetable fats | Necessary for normal reproduction in some animals |
| K | Green leafy vegetables and certain bacteria (including intestinal flora) | Delays blood-clotting time (vitamin K is necessary for the formation of prothrombin) |

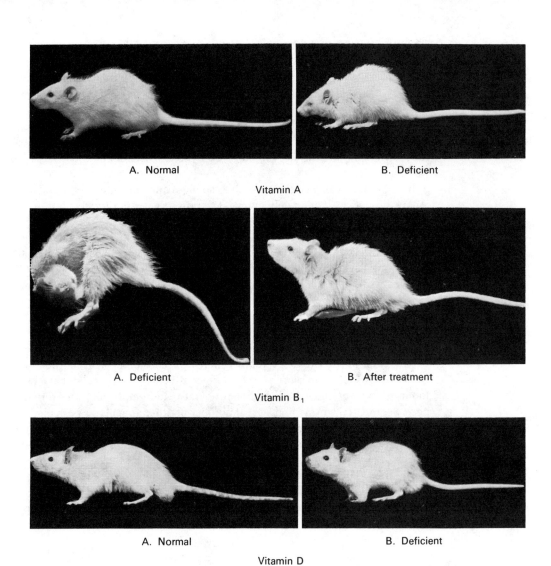

**Figure 24–4** Effects of vitamin deficiencies in rats. (Courtesy U.S. Bureau of Human Nutrition and Home Economics.)

## Excretory Systems and Disposal of Wastes

For cells to carry on the constant process of chemical digestion, metabolic wastes, or excretions, must be continually removed. The removal of these wastes is called **excretion.**

Metabolic wastes include carbon dioxide (see Chapter 25), certain soluble nitrogenous salts (such as urea), soluble inorganic salts (such as sodium chloride) and water, which is one of the chief products of oxidation (see Chapter 2). These substances usually escape from the cells by diffusion, are transported by the circulatory system (when present), and cast out of the body by the excretory system. We will now turn our attention to a few types of excretory apparatus occurring in the animal kingdom.

## EXCRETORY APPARATUS

Most excretory systems have as their primary unit the cell membrane. The true excretory organ of protists (originally thought to be the water expulsion vesicle) is the **plasma membrane;** nitrogenous wastes in the form of ammonia pass easily through this membrane.

Wastes produced by the cells in multicellular animals first diffuse (or are transported) through the membrane into the intercellular fluid, and from there diffuse into the organism's excretory apparatus, or its circulatory system, if present.

In many lower invertebrates, such as sponges and coelenterates, wandering amoeboid cells called **amoebocytes** may transport waste products out of the organism.

Flatworms have **flame cells** (also called **flame bulbs;** Fig. 24–5), which reside in blind tubes throughout the body of the worms. They contain medium-long undulipodia which "flicker," hence the name "flame cell." The degree to which these apparatus actually participate actively in excretion is debated. Since flame cells lie within the tubal excretory system of the worms, one theory suggests that they act in stirring up fluids present, so that more of the liquid is exposed to the surface of the tubes at one time; this would increase the *efficiency* of the excretory apparatus, but not its actual function.

Nematodes have excretory tubes which vary considerably in morphology. **Renettes** are cells with short excretory tubes attached. Larger tubes have been found to extend along the body longitudinally and empty into an **excretory pore.** Segmented worms, and a great number of other animals, have **nephridia.** Three basic parts may be distinguished in a nephridium: the **nephrostome** is the opening into the body cavity; its edge is lined with a ring of short undulipodia that move materials into the cup. The nephrostome empties into a highly coiled **tube.** The waste materials travel down the tube to the bladder, then to the **nephridiopore,** which opens to the outside.

The "kidney" of mollusks is actually a folded group of nephridia; the **green**

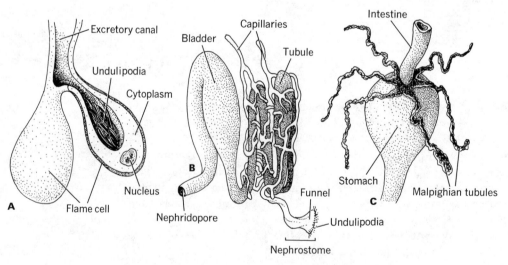

**Figure 24–5** Excretory apparatus of some invertebrates. (A) Flame cells. (B) Nephridia. (C) Malpighian tubules.

glands of crabs and crayfish, which open at the bases of the antennae, are modified nephridia.

**Malpighian tubules,** which appear in terrestrial joint-footed animals, gather liquid from the body cavity and channel it into the intestine through their lumens. They originate at the junction of the stomach and intestine and float out into the hemocoel. Within this fluid environment, the tubules are capable of some movement, which increases the efficiency of their absorption function.

The principal excretory organs of vertebrates are **kidneys.** The excretory process and function of the human kidneys will be discussed following a brief account of other excretory means in vertebrates.

## ACCESSORY EXCRETORY APPARATUS OF VERTEBRATES

A large part of the water in vertebrates is excreted through the **lungs** by evaporation. **Sweat glands,** which are very valuable in regulation of body temperature, also excrete water but are of little importance in the elimination of other excretory material. The **liver** excretes certain substances, such as the decomposition products of hemoglobin, which are carried in the bile to the duodenum and out of the body in the feces. **Feces** are not chiefly excretory products of metabolism; most of their contents have never been a part of the cytoplasm of the body, but are leftover food materials plus large masses of bacteria.

## THE KIDNEY

The **kidneys** excrete **urine,** which is carried by two **ureters** to the bladder and is released from the bladder (during urination) through the **urethra** (Fig. 24-6). The two kidneys in human beings are bean-shaped, tubular glands, about 12 cm long. They consist of an outer **cortex,** an inner striated **medulla** (divided into about 8 to 18 cone-shaped pyramids), and an expanded portion of the ureter known as the **pelvis.**

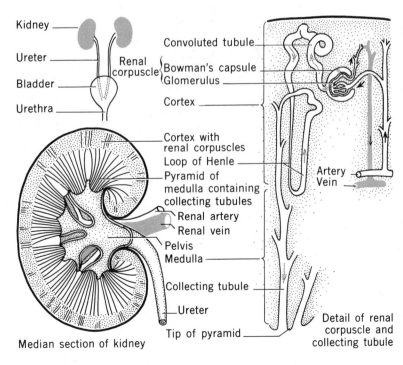

**Figure 24-6** Human excretory system. *Upper left:* position of kidneys, ureters, bladder, and urethra. *Lower left:* cross section of human kidney. *Right:* the nephron, functional unit of the kidney.

Blood entering the kidney brings waste matter which is filtered from it, collected in renal tubules, and discharged into the ureter. The urine travels down into the **bladder,** where it is stored.

## KIDNEY FUNCTION

The formation of urine takes place in the **nephrons,** the functional units of the kidney. Blood from the **renal artery** is channeled through the **afferent arteriole** into a dense capillary network called the **glomerulus.** The glomerulus resides within the **Bowman's capsule,** where the process of **filtration** occurs. Water, salts, sugar, and urea but not blood cells and large protein molecules pass from the blood into the Bowman's capsule; these substances are collectively referred to as the **glomerular filtrate.** The blood cells and various other substances left in the glomerulus travel out through the **efferent arteriole.**

The glomerular filtrate moves from the Bowman's capsule down the **proximal convoluted tubule,** which is continuous with the **loop of Henle** at the **medulla** of the kidney. The loop of Henle bends back to the **cortex,** where it becomes the **distal convoluted tubule.**

During its movement through the convoluted tubules and loop of Henle, the glomerular filtrate is changed considerably. The proximal and distal convoluted tubules **reabsorb** much water and virtually all glucose, amino acids, and other substances which are not waste. These tubules also excrete wastes from the surrounding capillary net into the glomerular filtrate. This function is of limited importance in human beings and higher mammals, but in animals such as the toadfish, which lacks glomeruli and Bowman's capsules, **tubal excretion** is the exclusive mode of waste disposal.

As the glomerular filtrate nears the end of its journey, most material necessary for bodily functioning has been resorbed, while the toxins and other wastes (especially urea) continue along the excretory path. The urine resulting from

**Table 24–2** Human Urine Contents (Approximate Values)

| | |
|---|---|
| $H_2O$ | 96.0% |
| Organic materials | 2.5% |
|     Urea (50% of solids in urine) | |
|     Uric acid | |
|     Urochrome (gives urine its color) | |
| Inorganic materials | 1.5% |
|     Sodium chloride | |
|     Potassium | |
|     Calcium | |
|     Magnesium | |
|     Ammonia | |
|     Ammonium sulfate | |
|     Phosphate | |
|     Bicarbonate | |
| | 100.0% = 1200–1500 ml/day |

Total solids: 60 g
    pH: 4.8–8.0

the filtration–resorption process is much more concentrated than the original glomerular filtrate. The urine travels from the distal convoluted tubule into the **collection duct** or **collecting tubule**. From there it moves to the **renal pelvis**, and further, through the ureters to the bladder, where it is stored until voided.

One of the major functions of the kidney, other than excretion per se, is osmoregulation, accomplished through the resorption of water during the excretory process. The kidney is such an efficient organ that human beings can frequently get along with just one kidney. Each liter of urine represents work on about 125 liters of glomerular filtrate. The urea in the urine is concentrated to about 70 times its original filtrate concentration. In the normal adult person, 1200 to 1500 ml of urine is excreted daily; the contents of urine are presented in Table 24–2.

# Summary

The principal foods of animals are organic compounds built up by other organisms, water, inorganic salts, oxygen, and vitamins. The choice of food for a particular animal is dependent upon its environment, its period of activity, and many other poorly understood biases.

The diverse digestive systems of animals all make ingested substances utilizable to the organism. In human beings, the mouth, pharynx, esophagus, stomach, small intestine, and colon form the digestive tract. Enzymes from the mouth, stomach, liver, pancreas, and intestine act upon the food. Most nutrient absorption and digestion takes place in the small intestine, while the colon (large intestine) functions mostly as a water absorber.

The mineral salts calcium, phosphorus, iron, iodine, copper, and manganese are the most important ions for higher animal nutrition. Carbohydrates are the most energy-rich foods; they are digested by the enzymes ptyalin and maltase while fats and lipids are digested by lipases. Proteins are made of amino acids and are broken down by pepsin and HCl in the stomach and erepsin in the small intestine.

The plasma membrane of protists, amoebocytes of lower invertebrates, flame cells of flatworms and most roundworms, nephridia of segmented worms, and Malpighian tubules of terrestrial joint-footed animals are some of the varied excretory apparatus occurring throughout the animal kingdom.

The human kidney consists basically of an external cortex, an inner medulla, and a pelvis. The functional unit of the kidney, the nephron, is made of a Bowman's capsule, glomerulus, proximal convoluted tubule, loop of Henle, distal convoluted tubule, and collection tube.

A dense system of capillaries surrounds the nephron, through which substances are either lost, gained, or exchanged. Renal filtration by the nephrons results in the concentration of urea, osmoregulation, and the selective filtering out of various blood substances, including toxins. The waste products are finally excreted from the body in the urine.

# 25
# Circulatory and Respiratory Systems

In Chapter 24, we saw how food is digested into nutrients which are used by the body. In order for these nutrients to be used, they must be carried to the cells that need them. The circulatory system provides for the transportation of these nutrients. In addition, it carries oxygen obtained through the respiratory system to all the cells. Both the circulatory and respiratory systems function in removal of metabolic wastes, which if allowed to build up would cause death.

In this chapter, we will review the circulatory and respiratory systems of various animals studied in this text and give emphasis to circulation and respiration in higher mammals, particularly the human species.

## Circulatory Systems and Circulation Development from Invertebrates to Vertebrates

In protists, digestion occurs in the vacuoles and intercellular transportation of nutrients is accomplished by the streaming of the cytoplasm. In sponges, amoebocytes perform the same function. In hydras and other coelenterates, digested food is absorbed by some of the gastrodermal cells, which then distribute the nutrients to neighboring cells. A similar method of distribution occurs in the multibranched intestine of flatworms.

In the earthworm we encounter a complicated system of tubes, the circulatory system, which carries digested food, oxygen, and other substances to all parts of the body. Similar systems are present in most of the higher animals: crustaceans have large body spaces (**sinuses**) in which blood from the arteries accumulates and from which the blood passes into the heart (after flowing through the gills). Insects have a **hemocoel** filled with blood which bathes the tissues and a dorsal heart that pumps the blood (hemolymph). Most of the blood movements, however, are initiated by respiratory, locomotory, and peristaltic movements.

The circulatory system in echinoderms is rather poorly developed; short undulipodia keep the coelomic fluid in motion, thus distributing it throughout the organism.

The lower vertebrates, such as the frog, have a three-chambered heart with a partial septum that allows oxygenated and deoxygenated blood to mix, and includes a complex system of arteries and veins. The crocodile is the first vertebrate to exhibit a fully developed septum which divides the heart into four chambers. Aspects of the circulatory system of the higher vertebrates—the mammals—will be discussed more fully in the following sections.

## Circulation in Higher Vertebrates

### THE HEART AND BLOOD VESSELS

The mammalian **heart** (Fig. 25-1) is **cardiac muscle** and is surrounded by a fibrous sac, the **pericardium.** There are four chambers: the left and right ventricles, and the left and right atria. The **ventricles** are the most muscular parts; their powerful contractions pump blood out of the heart. The **atria** (also called **auricles**) sit atop their respective ventricles and receive blood into the heart.

Communications between the atria and ventricles take place through the **atrioventricular valves.** The **tricuspid valve,** a three-flapped structure, allows blood to move from the right atrium to the right ventricle, but not back. When the blood-filled right ventricle contracts, pressure closes the valve. The structure of the tricuspid valve coupled with the **chordae tendineae** between the valve and the myocardial wall keep the valve shut. **Semilunar valves** perform a similar function between the ventricles and their communicating vessels (the pulmonary artery and aorta).

Near the tricuspid valve is the **sinoatrial node (SA node)** which acts as a "pacemaker" for the entire heart. When an impulse is emitted from the SA node, a wave of contraction spreads over both atria in less than 0.1 second and the atria simultaneously contract.

When the SA node's impulse reaches the **interventricular septum,** the **atrioventricular node (AV node)** becomes activated. The AV node allows 0.1 second for the atria to empty before sending its signal (by means of the **Purkinje fibers**) to the base of both ventricles.

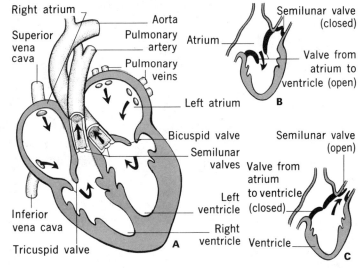

**Figure 25-1** Human heart. (A) Internal structure and direction of blood flow. (B) Blood is flowing from the atrium to the ventricle, and the semilunar valve is closed. (C) The valve between the atrium and the ventricle is closed, forcing blood past the semilunar valve into the artery.

The **sounds** the heart makes while pumping may be heard through a **stethoscope.** The basic sound made is approximated by "lubb-dup"—the "lubb" signifying the closing of the AV valve plus the simultaneous vibrations of the chordae tendineae; the "dup" sound signifies the closing of the semilunar valve and vibrations of the atrial walls and arterial blood column.

The heart is connected to **arteries,** which carry blood (oxygenated) away from the heart, and **veins,** which carry blood (usually deoxygenated) toward the heart. Both arteries and veins narrow until they eventually form **capillaries,** minute tubes with thin walls which allow substances to diffuse in and out of the blood (Fig. 25–2).

The walls of the arteries are especially abundant in elastic and muscle tissues. These regulate the flow of blood forward to the capillaries. The pressure exerted by the blood against the walls of the arteries is known as **blood pressure.** When an artery is cut, blood from it will spurt out, indicating that it is under pressure, while blood from a cut vein will flow out continuously, indicating relatively low pressure.

Blood pressure is in part controlled by the heartbeat. The contraction of the

**Figure 25–2** *Top:* venous valves, which allow blood to flow only toward heart. *Bottom:* histology of blood vessels. Note the thicker muscle layer of arteries.

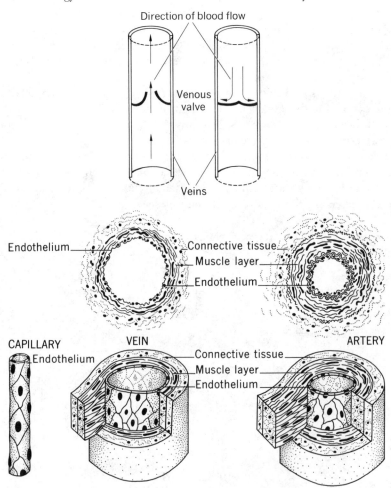

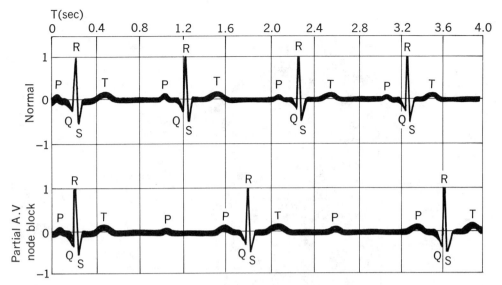

**Figure 25–3** *Above:* normal electrocardiogram (EKG). *Below:* aberrant EKG, representing partial block of AV node. Note absence of QRS and T waves after every other P, indicating that although the atria depolarize, the ventricles do not always get the message.

ventricles (a state known as **systole**) is the major contraction of the heart. The average systolic pressure is 125 mm Hg (measured in millimeters of mercury). During the refractory period between ventricular contraction and atrial contraction, the heart is in **diastole,** the normal pressure of which is 75 mm Hg.

Blood pressure depends, therefore, on volume, the force of the contracting ventricles, the elasticity of the walls of the blood vessels, and the resistance to the flow through the vessels. High blood pressure may be brought about by reduction in the elasticity and interior diameter of the small arteries; this occurs in **arteriosclerosis.**

The **pulse** is created by alternate dilation and elastic recoil of the artery due to the forcing of blood through it by contractions of the heart. The character of the heart's action can be determined by feeling the pulse. Various abnormalities of the heart may be detected with the aid of the **electrocardiogram** (**EKG**), as shown in Fig. 25–3.

**Capillaries** form networks that connect the smallest arteries (**arterioles**) with the smallest veins (**venules**). Capillaries themselves are so small that red blood cells have to squeeze through them one at a time. Capillaries carry out the principal functions of the circulatory system, taking in secretions from ductless glands, transporting oxygen from and carbon dioxide to the lungs, delivering waste products to the kidneys, absorbing food from the digestive system, and transporting food and oxygen to the tissues in exchange for cellular excretions.

We will now see what path the blood takes through the heart and its vessels during circulation.

## CIRCULATORY PATHWAY

The rate of circulation is so rapid as to almost defy imagination. The entire blood supply (approximately 5.3 liters in an average-sized man) passes from the heart, through the body, and back again in about 20 to 40 seconds. The velocity of the blood flow changes, moving faster in the larger arteries, slower in smaller

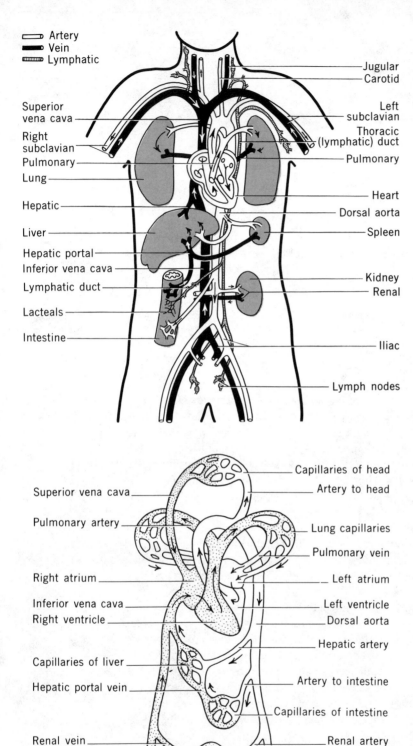

**Figure 25-4** Human circulation. *Top:* the larger blood and lymphatic vessels, and some of the organs they serve. Lymph flows only toward heart. *Bottom:* general scheme for circulation.

ones, and slowest in the veins, where **venous valves** prevent the blood from flowing backward.

Blood exiting the left ventricle flows into the **aorta,** the largest artery in the body (Fig. 25-4). A small portion of this blood is diverted to the heart through the **coronary arteries.** Blood from the **dorsal aorta** travels up the **common carotid artery** toward the head. At about shoulder level, the **subclavian arteries** branch off and run down the arms, becoming the **brachial, ulnar,** and **radial arteries.** At about the level of the jaw, the common carotid splits into the **internal** and **external carotids,** which travel into the head.

On its downward movement, the blood from the **dorsal aorta** splits off to the various major organs (**hepatic** artery—liver; **splenic artery**—spleen; **renal arteries**—kidneys, etc.). About one-fifth of the way into the pelvic girdle, the dorsal aorta splits into two **common iliac arteries,** which become the legs' **femoral, popliteal,** and **tibial arteries.**

The venous system is generally analogous to the arterial system; most of the names of the veins are the same as their corresponding arteries.

The only artery that carries deoxygenated blood is the **pulmonary artery,** going from the right ventricle to the lungs. The blood is oxygenated in the dense capillary beds of the lungs, and is returned by way of the **pulmonary veins** to the left atrium.

This account of the morphology of circulation has covered only the most general aspects; for a more complete reading, consult any specialized textbook.

## LYMPHATIC SYSTEM

Capillaries are under some pressure in the circulatory system, and some fluid thus leaks out of them during the body's normal diffusional activity. If this excess fluid were allowed to accumulate, **edema** (a swelling) would result.

The **lymphatic system** counters this problem; it is composed of thin, one-way vessels that are extremely permeable to almost all tissue fluid substances. These **lymph capillaries** are blind tubes that carry protein and other fluids back to the blood via the lymphatic system.

Other functions of the lymphatic system include the transportation of digested fat from the gastrointestinal tract into the circulatory system and the action of the **lymph nodes.**

The lymph nodes are of great importance, producing **antibodies** that break up **antigens** (foreign substances) present in the body. They also act as lymph "cleansers," removing particulate matter and cells that might have become inflamed. **Leukocytes** and **lymphocytes** are produced by the lymph nodes; these perform a cleansing function (see below).

## THE BLOOD

The chief function of the blood is to **transport** substances throughout the body. Blood carries oxygen from the lungs to the tissues, waste products to the excretory organs, nutrients to the tissues, and hormones and other secretions to wherever they are needed. In addition, blood helps maintain a normal temperature, pH level, and defends the body against infection.

The blood consists of about 55 percent liquid **plasma,** defined as *that part of the blood that is liquid, suspends the corpuscles, and in addition contains dissolved salts and proteins,* and about 45 percent "formed elements" or corpuscles (Fig. 25-5), which are:

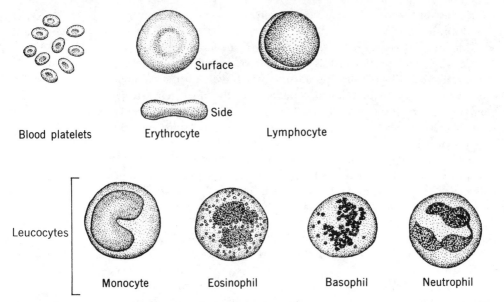

**Figure 25–5** Human blood corpuscles and platelets.

1. **Erythrocytes** (**red blood cells**), which carry oxygen bonded to the hemoglobin molecules within them.
2. **Leukocytes** (**white blood cells**), which are divided into **monocytes, neutrophils, eosinophils,** and **basophils.**
3. **Lymphocytes,** which are sometimes included among the leukocytes.
4. **Platelets,** essential in blood clotting.

**Erythrocytes** are present in great numbers; they are anucleate circular disks which are biconcave in profile. They contain **hemoglobin,** which consists of a protein (**globin**), and a nonprotein, iron-containing pigment (**hematin**). Hemoglobin combines easily with oxygen, becoming **oxyhemoglobin,** and gives up the oxygen to become *reduced* hemoglobin. Erythrocyte formation takes place continuously in the bone marrow; each cell is capable of living about 4 months in the human circulatory system.

Most **leukocytes** act as **phagocytes** (Gr. *phagos,* to eat), cleansing the blood by devouring foreign bodies such as bacteria and lysed tissue and by producing antibodies. The four types of leukocytes are easily differentiated by a light microscope.

**Monocytes** are large phagocytic cells with one nucleus; they comprise 2 to 7 percent of the blood. **Eosinophils** are *non*phagocytic cells with one nucleus and make up 2 to 5 percent of the blood. **Neutrophils** are phagocytic and make up the majority of the blood's leukocytes: 65 to 75 percent of the blood is composed of these cells. Neutrophils are formed in the bone marrow and contain fragmented nuclei. **Basophils** have several nuclei and make up about 0.5 percent of the blood. Their function is unspecified, but current research indicates that they release histamine and may be involved in allergic reactions. Basophils have been observed to proliferate in the body during periods of disease.

**Lymphocytes,** as we have seen, are also phagocytic in nature and are produced by the lymph nodes.

There are several substances within the blood which participate in **blood clotting.** Blood **platelets** are minute discoidal bodies that contain **thrombokinase** (or **thromboplastin**). Thrombokinase plus other plasma factors and substances from injured tissue are necessary for the formation of **thrombin.** Once thrombin is formed, it can react with **fibrinogen** (a soluble form of fibrin) to form insoluble **fibrin.** Fibers made from this fibrin "catch" transient erythrocytes and a **clot** is formed.

In certain individuals, one or more of the preceding factors are lacking in the blood, resulting in such slow clotting that even a minor cut could cause severe bleeding. This condition, known as **hemophilia,** is hereditary and sex-linked: male children are bleeders and females are carriers (see Chapter 28). A drug is now available to prevent bleeding in hemophiliacs.

**HUMAN BLOOD GROUPS.** When blood is given by a donor to a patient (recipient), the red blood cells of the donor's blood must be **compatible** with the serum (the blood minus fibrinogen) of the recipient's blood. The harmful effects of some blood transfusions are due to the clumping (**agglutination**) of the red blood cells. The first such unfavorable blood reactions investigated were due to what is now known as the **ABO group,** discovered by K. Landsteiner in 1900.

Landsteiner found that there were two agglutinogens (**antigens**), designated A and B in the erythrocytes of human beings, and that the serum contained two kinds of agglutinins (**antibodies**), called $a$ and $b$. Whatever antigen one has in his erythrocytes, the corresponding antibody is absent from his blood serum. This makes possible four kinds of ABO blood group, designated A, B, AB, and O:

| Blood Group | Antigen(s) in Erythrocytes | Antibody(ies) in Blood Serum |
|---|---|---|
| A | A | b |
| B | B | a |
| AB | AB | None |
| O | None | a and b |

In transfusions, the primary consideration is what effect the antibodies in the recipient's serum will have on the antigens of the donor's red blood cells. It is possible, for example, to give a type O blood transfusion to a type A person, but giving type A to a type O person would cause severe reaction and possible death.

Many other blood groups, including P, MNS, Kell, Lewis, Lutheran, Kidd, and Duffy, have been discovered in recent years; some of these must be taken into consideration in making blood transfusions, and there is presently much research activity in this field.

The blood of individual anthropoid apes is one of four human types; A and O have been found in chimpanzees, indicating the close chemical relationship between apes and human beings. Blood groups are also known in rabbits, dogs, and cattle, but none are identical with a human type.

**RH FACTOR.** The Rh **blood factor** is another important blood antigen found in human beings and monkeys. About 86 percent of the white population of the United States have this antigen and are thus **Rh-positive.** The remaining 14 percent lack this factor and are therefore **Rh-negative.**

There is no antibody that normally accompanies the Rh antigen as is true of the antigens in the ABO blood group. The Rh antigen will, however, cause formation of antibodies if the blood of an Rh-positive person is transfused into an Rh-negative person. If at a later time another Rh-positive transfusion is made, the Rh antigen will react with the antibodies, causing agglutination, which may result in death. Another serious result of Rh incompatibility may be caused when an Rh-negative woman is pregnant with a Rh-positive child (indicating that the father is Rh-positive). Rh-positive erythrocytes may pass from the bloodstream of the fetus into the mother's blood, where they stimulate the production of antibodies. In a subsequent pregnancy with a Rh-positive fetus, these antibodies of the mother may get into the bloodstream of the fetus and cause agglutination of the fetal erythrocytes. Depending on the concentration of the antibodies present, the effects of this on the fetus may range from mild anemia to severe structural or neural abnormalities, miscarriage, or stillbirth.

The inheritance of some of the other blood groups is discussed in Chapter 28.

We will now turn our attention to the various respiratory systems occurring in some of the animals we have studied and a discussion on human respiration.

## Respiratory Systems and Respiration Development from Invertebrates to Vertebrates

The cells of the body depend upon an uninterrupted flow of energy. This energy is usually obtained through oxidation of foods. The acquisition of oxygen, through the process of **respiration,** is therefore essential to all life systems.

The use of oxygen directly by the cells is called **internal** (**cellular**) respiration. This is distinguished from **external** respiration, which differs greatly among the various groups of animals and involves the exchange of oxygen and carbon dioxide in the lungs, gills, membranes, and so on.

Respiration in the **protists** involves the diffusion of oxygen (dissolved in water) into the body and the release of carbon dioxide through the **plasma membrane.**

In sponges, coelenterates, and many other **aquatic animals,** respiration occurs directly between the cells in the body wall and the water that bathes them.

**Earthworms** and many other animals lack respiratory systems; they take in oxygen and give off carbon dioxide through the moist skin. The gases are transported by capillaries just beneath the skin's surface. The blood transports oxygen to and carbon dioxide from the various organs, where the gases diffuse to and from the blood into intercellular spaces and then into the cells.

**Insects** and other terrestrial arthropods possess respiratory tubes (**tracheae**) that carry air into and out of the body through a number of openings (spiracles) in the sides of the body segments. Gases diffuse from the finest tracheae (tracheoles) into and out of the tissue cells. In these animals, the circulatory system may aid in carbon dioxide transport but is of little importance for transportation of oxygen.

In many aquatic animals with circulatory systems, such as the **crayfish,** respiration takes place with the aid of **gills.** Oxygen contained in water flows through the gill chamber and then diffuses into and through the cells of the gill filaments and into the blood in the efferent gill channels. Carbon dioxide diffuses out of the blood in the afferent gill channels into the gill filaments and from there out of the body. A similar respiratory process occurs in fish.

**Amphibians,** such as the adult frog, carry on respiration through the moist surface of their bodies and by means of **lungs. Reptiles, birds,** and other higher vertebrates respire mostly by lungs, which are more complex than those of the frog. **Mammals** are provided with respiratory systems quite similar to those of human beings, which will be described more fully.

## Respiration in Higher Vertebrates

In human beings, air passes through the nose or mouth (where it is moistened, warmed, and filtered) and into the lungs by way of the larynx, trachea, bronchi, and bronchioles. The **larynx** is a triangular organ that contains the **vocal cords** (see below). The **trachea** or "Adam's Apple" in males has nine pieces of cartilage (**tracheal rings**), which prevent its walls from collapsing. The trachea branches into two **bronchi;** each branches into many smaller **bronchioles.** Each bronchiole terminates in an **alveolar duct,** which leads into the lungs. On the surface of the duct and within the lungs themselves are tiny air sacs (**alveoli**), the functional units of the lungs. There are approximately 750 million alveoli in the human lungs.

The essential feature of the **lungs** (Fig. 25–6) is the moist epithelial lining with oxygen-containing air in the cavities on one side and blood in the capillaries on the other side. Oxygen from the lung cavity diffuses into the epithelial cells and then into the blood; carbon dioxide diffuses from the blood into the cells and then into the lung cavities. The total area of surface within the lungs that is exposed to air is more than 50 times the skin surface of the

**Figure 25–6** Human lungs. The right lung is cut open to show internal structure.

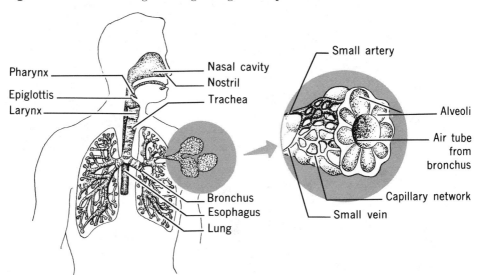

body. Each lung lies within a cavity lined with a serous membrane, the **pleura.** If the pleura become inflamed, **pleurisy** develops.

Air forced out of the lungs may vibrate the **vocal cords** within the larynx; the column of air above the cords gives rise to **sound.** The cavities of the pharynx, mouth, and nose act as resonators. The amplitude of the vibrations, and volume and force of the air current determine the loudness and intensity. The length and tightness of the cords and the frequency of vibration determine the pitch. The vocal cords of women and children are usually short, producing a high-pitched voice. Men's vocal cords are usually longer (brought about by male sex hormones) and the voice is lower.

## BREATHING AND THE RESPIRATORY FUNCTION OF BLOOD

Air must be brought into and expelled from the lungs—if it were allowed to stay in the alveoli, it would come to equilibrium with the gases in the blood. No exchange would occur, fresh oxygen would not be transported to cells, toxins would build up, and death would result.

The lungs, which reside in the **thoracic** cavity, have *no musculature.* Consequently, they do not move of their own accord during the body's **respiratory movements.** They are entirely dependent upon muscular movements of the diaphragm, intercostal, and extracostal muscles.

During **inspiration** (taking in of oxygen in the air), the **external intercostal** muscles attached to the ribs contract, bringing the rib cage slightly up and out, and generally increasing the volume of the thoracic cavity. Their work is greatly augmented by the contraction of the **diaphragm,** which is the muscle dividing the thoracic and abdominal cavities. The combined efforts of these muscles *decrease* the pressure in the cavity. Hence, air outside the body is at a higher pressure and moves into the lungs through the trachea to regain equilibrium.

**Expiration** (the release of carbon dioxide in the air) is the opposite of inspiration. Here, the **internal intercostals,** the muscles on the insides of the ribs contract, while the opposing external intercostals relax. This tends to decrease the volume of the thoracic cavity and is again augmented by the diaphragm, which relaxes (arches up). At this point, the pressure in the lungs is *greater* than the pressure outside them, and air is forced out.

These pressure changes play a significant role in circulation. The **inferior vena cava** passes through this cavity, and when pressure is increased, an action similar to that of the skeletal muscles' contractions pushes blood through the great vein.

The process of diffusion is responsible for the gain in oxygen and the loss of carbon dioxide at the alveolar sites. The changes in quantity of these substances are not great. The normal awake but resting adult breathes between 9 and 20 times per minute:

|  | Inspired Air (%) | Expired Air (%) | Net Gain/Loss (%) |
| --- | --- | --- | --- |
| Oxygen | 21.00 | 16.02 | Gain: 5.80 |
| Carbon dioxide | 0.05 | 3.60 | Loss: 3.55 |
| Nitrogen | 78.30 | 75.00 | Gain: 3.30 |

Most of the oxygen that diffuses into the blood enters into a loose chemical combination with **hemoglobin,** producing **oxyhemoglobin,** which gives the red color to arterial (oxygenated) blood. Oxyhemoglobin at the capillary level gives up its oxygen, which diffuses through the tissue fluid into the cells to take part in **cellular respiration**. The carbon dioxide resulting from the metabolic processes within the cells diffuses into the blood, where it is less concentrated. A small portion (about 20 percent) of it combines with hemoglobin to form **carbaminohemoglobin;** the great bulk is transported in the bloodstream to the lungs in the form of **bicarbonates.** In the lungs, carbon dioxide is freed from these bicarbonates and from the hemoglobin and diffuses through the cells into the air in the lung cavities. From the lung cavities, it is expelled into the environment.

## CONTROL OF RESPIRATION (VENTILATION)

Respiratory rate is both chemically and neurally controlled. The chemical control depends on the pressures of carbon dioxide and oxygen as well as on the hydrogen ion ($H^+$) concentration in the blood. As the pressure of carbon dioxide and the $H^+$ concentration is increased, breathing rate increases. Conversely, if the oxygen level in the blood is high, the body respires more slowly.

H. Houdini, the famed escape artist, used to have himself locked in a trunk, which would then be cast in water. What he did just before being locked into the trunk was to **hyperventilate** (breathe very fast) using pure oxygen. This vastly increased the oxygen content and decreased the carbon dioxide content of his blood which suppressed the stimulus to breathing and allowed him to hold his breath much longer than normally.

One of the controlling factors in respiration is the ease or resistance of airflow. In the disease **asthma,** the smooth muscle in the terminal (smallest, most distal) bronchioles goes into spasm, drastically increasing resistance to airflow, making the bronchioles' diameters smaller and breathing much more difficult.

In the disease **sickle-cell anemia** (which is confined mainly to blacks, especially of West African descent), some of the red blood cells may actually be sickle-shaped. This results in a decreased surface area as well as some trouble in smooth blood flow; thus the blood's capacity for taking up oxygen is reduced.

We have seen in this chapter the interrelations of the circulatory and respiratory systems and their crucial contributions to the living organism.

# Summary

After an organism ingests and digests food, it must pass on the nutritive substances to all parts of its body; this process is accomplished by circulation.

Unicellular organisms pass nutriment around by the streaming of the cytoplasm. Many multicellular organisms disperse nutriment intercellularly with the aid of amoebocytes.

The human circulatory system is comprised of a heart with two ventricles (which pump blood out) and two atria (which receive blood), arteries (carry blood away from the heart) and veins (carry blood toward the heart). The arteries and veins narrow to form capillaries, thin-walled vessels that form a dense network throughout the organism and carry on the principal functions of circulation.

The lymphatic system, extremely permeable to most substances, picks up fluids and proteins that have leaked from the capillaries and returns them to the bloodstream.

Human blood contains erythrocytes (important in oxygen–carbon dioxide exchange), four types of leukocytes (monocytes, neutrophils, eosinophils, and basophils), lymphocytes (which along with leukocytes serve in bodily protection), and platelets (necessary for blood clotting).

The respiratory system is mostly concerned with obtaining oxygen and eliminating carbon dioxide. The respiratory process is diffusional. In higher vertebrates, air, providing oxygen and removing carbon dioxide, is brought into and out of the lungs. This is accomplished through thoracic pressure changes initiated by contractions of the diaphragm and the intercostal and extracostal muscles. Lower organisms obtain life-sustaining oxygen in a variety of ways, including through a moist integument, through holes in the body, and through gills.

# 26

# Nervous System, Sense Organs, and Endocrine System

Coordinated behavior in higher invertebrates and vertebrates is accomplished largely by nervous systems and sense organs and by means of hormones secreted by the endocrine glands. The higher an animal is phyletically, the more complex the coordination of behavior. We have studied the reactions of a number of different animals, their various nervous systems, sense organs, hormones, and their behavior. In this chapter, coordination and behavior are considered as a functional component of biological organization and activity.

## Nervous Systems

### NERVOUS SYSTEMS OF INVERTEBRATES

**Irritability** and **conductivity** were found to be fundamental characteristics of cytoplasm. Amoebas and protists that lack specialized organelles for reception of stimuli nevertheless respond to changes in their environment in a similar manner as animals that possess sense organs and nervous systems. The reactions to stimuli in sponges are hardly more advanced than in protists.

The lowest organisms to exhibit receptor–effector nervous systems are the coelenterates, represented by *Hydra*. In this type of nervous system, sensory cells for reception of stimuli (**receptors**) are in direct continuity with fibers from nerve cells (**conductors**) that send out processes to the contractile fibrils (**effectors**) in the epitheliomuscular cells. All together, these constitute a **nerve net** or **plexus**. Unlike the directed impulses of more complex systems, the conduction of nervous impulses in this system is truly primitive. Nervous conduction is omnidirectional and the organism as a whole respond to the stimulus.

Flatworms have a more advanced nervous system. The planarian exhibits **cephalization** and has long nerve cords connected by transverse nerves, ganglia, and sense organs.

Segmented worms manifest a rather important development: they have ganglia *and* **reflex arcs** which allow them to react to certain stimuli automatically, without neural integration. Joint-footed animals have relatively complex sensory systems, including **antennae** and **eyes**, and **chemosensory apparatus**.

# VERTEBRATE NERVOUS SYSTEM

The nervous system of vertebrates is vastly more complex than that of most invertebrates. Vertebrates have a **dorsal nervous system,** as opposed to the ventral nervous systems of invertebrates, and possess association neurons. Mammals possess what is considered to be the most advanced nervous system. We will use the human nervous system as our example.

The nervous system of human beings (and other vertebrates) has two basic functional divisions: The **central nervous system (CNS)** and the **peripheral nervous system (PNS).** The CNS consists of the brain, its nerves (Table 26–1), and the spinal cord. It is the most massive part of the nervous system and is concerned with interpretation and integration of incoming data. The PNS is comprised of all nerves outside the CNS, and transmits impulses from the CNS to **effectors** (glands, muscles, etc.). The PNS is incapable of the integrational and interpretational functions characteristic of the CNS, and it contains no associational fibers (see below).

Superimposed over the CNS–PNS systems is a division between **somatic** and **autonomic** nervous systems. Voluntary functions, such as most skeletal muscle contractions, come under the control of the somatic nervous system. The autonomic nervous system regulates *in*voluntary functions.

The autonomic nervous system exerts its influence on **visceral organs,** such as the lungs, bladder, heart, and iris of the eye. This system has two parts, with antagonistic functions. The **sympathetic nervous system** is an activating system, used during emergency situations. When this system is activated, several things

**Table 26–1** Number, Name, Origin, Distribution, and Function of the Cranial Nerves of Vertebrates

| Number | Name | Origin | Distribution | Function |
|---|---|---|---|---|
| 0 | Terminal | Forebrain° | Lining of nose | Probably sensory |
| I | Olfactory | Olfactory lobe | Lining of nasal cavities | Sensory |
| II | Optic | Diencephalon | Retina of eye | Sensory |
| III | Oculomotor | Ventral side of midbrain | Four muscles of eye | Motor |
| IV | Trochlear | Dorsal side of midbrain | Superior oblique muscle of eye | Motor |
| V | Trigeminal | Side of medulla | Skin of face, mouth, and tongue, and muscles of jaws | Sensory and motor, mostly sensory |
| VI | Abducens | Ventral side of medulla | External rectus muscle of eye | Motor |
| VII | Facial | Side of medulla | Chiefly to muscles of face | Motor and sensory, mostly motor |
| VIII | Auditory | Side of medulla | Inner ear | Sensory |
| IX | Glossopharyngeal | Side of medulla | Muscles and membranes of pharynx, and tongue | Sensory and motor |
| X | Vagus | Side of medulla | Larynx, lungs, heart, esophagus, stomach, intestines | Sensory and motor |
| XI | Spinal accessory | Side of medulla | Chiefly muscles of shoulder | Sensory and motor |
| XII | Hypoglossal | Ventral side of medulla | Muscles of tongue and neck | Motor |

° According to one authority, it originates from both telencephalon and diencephalon in most species.

happen simultaneously: heart rate increases, blood glucose levels are increased, vasodilation occurs in the skeletal muscles, vasoconstriction occurs in the digestive tract, and other activation phenomena take place. Many of these reactions are initiated by vast amounts of **adrenaline** (epinephrine) secreted in response to activation of the sympathetic nervous system. The **parasympathetic nervous system** returns the body to a normal state following an emergency situation. Its actions (decrease of heart rate and blood glucose, return to normal diameter of blood vessels, etc.) are opposite to those of the sympathetic system.

## THE NEURON

The functional unit of the nervous system is the **neuron** (Fig. 26–1). Neurons are single, nucleated cells which receive impulses through **dendrites** and conduct impulses through rather long processes (extensions) called **axons**. **Nerves** are bundles of neurons that perform specific functions.

Many neurons are **myelinated**. Myelin, a substance made of the membranes of **Schwann cells**, increases the velocity with which neuron impulses are conducted (Fig. 26–2). Schwann cells perform insulatory and nourishing functions for the neurons they encircle. Within the CNS, the myelin-containing cells applied to the neurons are called **oligodendrocytes**, and serve the same function as Schwann cells. Between successive Schwann cells or oligodendrocytes are short areas of nonmyelination, the **nodes of Ranvier.** These also serve to increase the velocity with which an impulse may travel along the neuron.

Four major types of neurons are recognized: **Motor neurons** have radiating

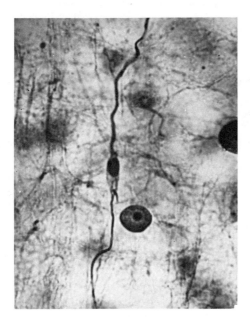

**Figure 26–1** Neurons. *Left:* bipolar neuron. *Right:* multipolar connective neurons. Note the multiple connections. 200×.

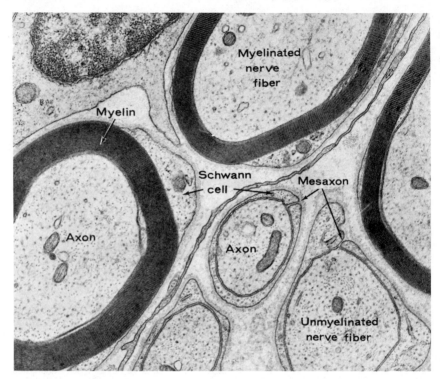

**Figure 26–2** Myelination; note the concentric layers of myelin. (From W. Bloom and D. W. Fawcett, *A Textbook of Histology*, Philadelphia: W. B. Saunders Co., 1968.)

dendrites and long axons. They are usually myelinated and carry impulses from the CNS to the muscles. **Bipolar neurons** may or may not be myelinated and have a threadlike, spiny dendrite that appears in various parts of the nervous system. **Sensory bipolar neurons** are usually myelinated, and conduct impulses from **receptors** to the CNS. **Connective neurons,** or interneurons, appear exclusively in the CNS and are usually nonmyelinated. They have profuse dendritic connections, but do not connect with receptors or effectors.

Neurons connect with other neurons, receptors, and effectors by means of **synapses** (see below).

## MECHANISM OF THE NERVOUS SYSTEM

To understand the flow of an electrical impulse down a neuron, it is first necessary to understand the neuron at rest. When in this state, the neuron is surrounded by positively charged sodium ions ($Na^+$) with a minimal amount of $Na^+$ inside the neuron. Conversely, potassium ions ($K^+$) are in greater concentration inside the neuron. Negatively charged chlorine ($Cl^-$) inside the neuron counterbalances this positive charge, leaving the neuron positive outside and relatively negative inside. This polarization of charge defines the **resting** or **membrane potential.**

When the neuron is activated, a rapid succession of changes occurs. The dendrite's tip becomes **depolarized** at the synapse (see below), and the membrane become very permeable to $Na^+$ ions, which rush into the neuron.

The movement of $Na^+$ ions perpendicular to the longitudinal axis of the neuron creates an electrical current. This current affects the membrane just ahead of the ionic movement by increasing the membrane's permeability to $Na^+$ ions. In the wake of the impulse, the $K^+$ ions within the neuron move out to its surface, thereby **repolarizing** the neuron (by bringing the positive charge back out). Thus, a wave of depolarization (and subsequent repolarization) sweeps down the neuron.

The repolarization is not immediate, and a characteristic **refractory period** ensues during which the neuron cannot repeat the firing process.

If a dendrite is sufficiently depolarized at its tip, the entire neuron will transmit that action potential to the end of the axon. If insufficient stimulus is applied, no action potential will be transmitted. This is known as the **all-or-none law** of nervous function, which applies to all neurons.

THE SYNAPSE. Most nervous impulses are transported chemically across synapses. On the tips of axons and dendrites are **synaptic knobs,** which contain small sacs (**synaptic vesicles**) (Fig. 26–3). Between two synaptic knobs is a space

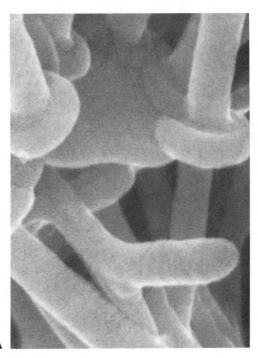
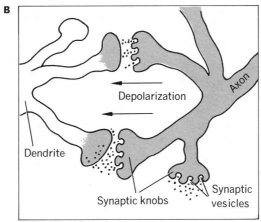

**Figure 26–3** (A) Scanning electron micrograph of nerve endings. (B) Synapse. Here acetylcholine is being released from the synaptic vesicles across the synaptic cleft.

called the **synaptic cleft.** When the wave of depolarization reaches the synaptic knob, synaptic vesicles release their contents. If **acetylcholine** or **norepinephrine (noradrenaline)** is released, the next neuron in line is induced to transmit an impulse. This is known as an **excitatory synapse.** Excitatory synapses occur in the autonomic nervous system and parts of the CNS. Chemical transmitting substances for the remainder of the CNS have not yet been identified.

Synaptic *inhibition* has the reverse effect of synaptic excitation and may occur in two ways. **Presynaptic inhibition** (Fig. 26–4, top) occurs when an axonal synaptic knob chemically contacts an axon behind the synaptic knobs. The resultant depolarization precludes an approaching potential's transmission all the way to the knob. **Postsynaptic inhibition** (Fig. 26–4, bottom) may occur just past or on dendritic synaptic knobs. In this case, the axonal synaptic knobs release a substance quite unlike acetylcholine or norepinephrine. The substance, which current research implies *may* be an amino acid, **hyperpolarizes** the dendrite. This makes the dendrite more stable by further decreasing the permeability of $Na^+$.

**THE SPINAL CORD.** The **spinal cord** is the part of the CNS that extends just under the cerebellum (at the mouth level) to the end of the vertebral column (approximately 45 cm total length). The body of the cord is made exclusively of myelinated and nonmyelinated neurons. The butterfly-shaped core, or **gray matter,** is composed of nonmyelinated neurons. The projections of gray matter are called **horns,** and contain mostly motor neurons and interneurons. Surrounding the core is **white matter,** composed of myelinated tracts of **nerve fibers,** which are divided into **funiculi** (Fig. 26–5).

The **ventral roots** leave the cord bilaterally and ventrally and carry efferent

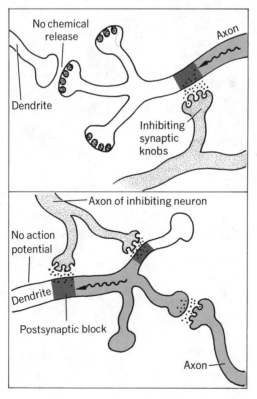

**Figure 26–4** *Top:* presynaptic inhibition. *Bottom:* postsynaptic inhibition. See the text for a description.

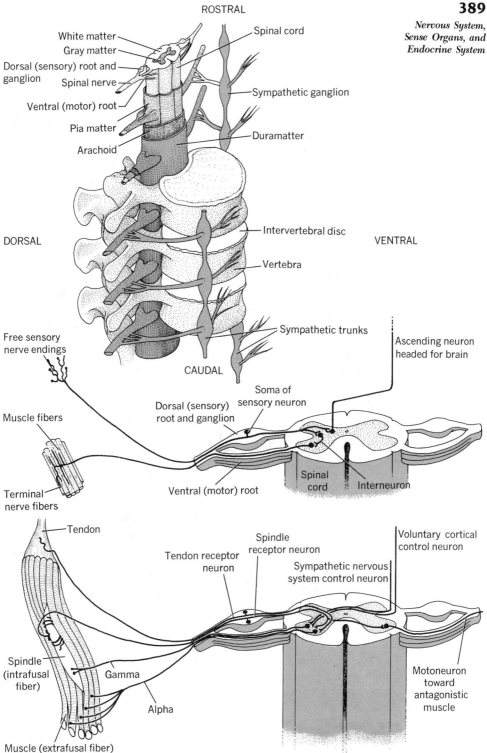

**Figure 26–5** *Top:* human spinal cord and associated structures. *Middle:* spinal reflex arc. *Bottom:* muscle control mechanism of spinal cord.

motor neurons (heading toward the muscle). The **dorsal roots** occur bilaterally and dorsally: they contain afferent sensory neurons (heading toward the cord) and have groups of nerve cells called **ganglia. Spinal nerves** leave the cord at various points.

Surrounding and protecting the spinal cord are **meninges,** which, along with **cerebrospinal fluid,** decrease friction in surrounding vertebrae. The **pia mater, arachnoid,** and **dura mater** successively enclose both the spinal cord and the brain. Running parallel to the cord and outside the vertebrae are **sympathetic trunks** and ganglia.

The spinal cord is a complex instrument, and although it is incapable of *interpretation* of impulses, it can direct some effectors without cerebral influence. This is known as a **simple reflex arc** (Fig. 26-5). Many of these reflexes are of a protective nature, involving rapid withdrawal of a body part from an area of pain or danger.

Although no cerebral control is executed after a simple reflex arc, interneurons may conduct to the brain the sensory neuron's "information" of the reflex and/or pain. Depending on the level at which the reflex occurs, the path lengths to the cord and brain, and the velocity of the impulses along different neurons, one may be "conscious" of the reflex and/or pain before, during, or after the muscles are activated.

We now turn our attention to that part of the CNS responsible for coordination and interpretation of neural impulses, the brain.

THE HUMAN BRAIN. The three major divisions of the brain are the **cerebrum,** the **cerebellum,** and the **brainstem.** Within the brain are **ventricles,** spaces filled with cerebrospinal fluid, that connect with the spinal cord.

The **cerebrum** is the most complex part of the brain (Fig. 26-6). The outside gray matter of the cerebrum is the **cerebral cortex** (covered by pia mater, arachnoid, and dura mater). The cortex is divided into **frontal, temporal, parietal,** and **occipital lobes,** all containing bunches of neurons called **gyri.** The cortex is responsible for "thought" and many essential motor, sensory, and visceral functions.

Below the cerebral cortex are the thalamus and the hypothalamus. The **thalamus** is composed of gray matter and is the center for nonolfactory sensory input, integration, and dispersal. It defines the qualitative aspects of sensation and connects with the sensory cortex for definite localization. The **hypothalamus,** situated below the thalamus, links the "mental" and "somatic" functions of the organism. It coordinates reflexes, and relates especially to homeostasis and the autonomic nervous system.

The **cerebellum,** at the posteroventral portion of the brain, has external gray matter and internal white matter (**arbor vitae**). The cerebellum coordinates muscle groups and is almost exclusively concerned with movement.

The **brainstem** consists of the **medulla** and **pons** (midbrain). The **medulla oblongata** occurs at the point of brain–spinal connection, and is composed mostly of white matter. An interesting facet of brain physiology is the *decussation* (crossing over) of the majority of ascending and descending fibers. This occurs mostly at the level of the medulla.

The **pons Varolii** lies just above the medulla and is also generally formed of white matter. It serves as a routing station for impulses passing between the spinal cord and the cerebrum. The pons plays other roles as well, including that of respiratory system activator–coordinator.

The **reticular formation** spans an area within the medulla and pons. The

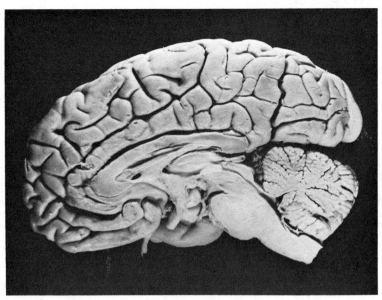
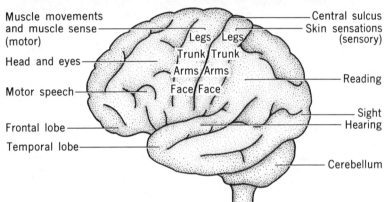

**Figure 26–6** Human brain. *Top:* midsagittal section shows right hemisphere. *Bottom:* left hemisphere, showing main lobes and areas concerned with particular functions. The cerebrum, which reaches its greatest development in human beings, is by far the largest part of the brain. (Courtesy Ernest Gardner, from *Fundamentals of Neurology,* 5th ed. Philadelphia: W. B. Saunders Co., 1968.)

reticular formation is capable of selecting important stimuli and ignoring others—so that, for example, the sound of your name may awaken you from slumber, while the sound of someone else's may not.

On the **ventral surface** of the brain, several important structures are revealed. The **olfactory bulbs,** as the name implies, interpret odors. The **optic chiasm** contains the fibers of the optic nerves. The **hypophysis** or **pituitary gland** extends downward from the brain's ventral surface, and lies just above the roof of the mouth in higher vertebrates.

Twelve **cranial nerves** exit directly from the brain. Most of the cranial nerves originate deep in the brain and pass out near their points of origin (Fig. 26–7).

**COMPARISON OF BRAINS IN VERTEBRATES.** Differences in the brains of vertebrates are probably more striking than in any other comparison of organs. An animal's phyletic level of development is best reflected in the relative sizes of its cerebrum and cerebellum (Fig. 26–8). In **cyclostomes,** both the cerebral lobes and cerebellum are quite small. In **fish,** the size of the cerebellum partially determines a species' swimming ability; the optic lobes and medulla are especially large. In **amphibians,** the olfactory lobes are large and elongated; the optic lobes are smaller; the cerebellum is a mere transverse band; and the

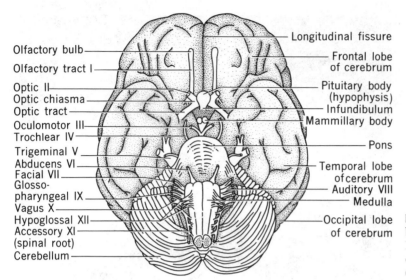

**Figure 26-7** Human brain, ventral surface. Numerals I to XII indicate cranial nerves (see Table 26-1).

**Figure 26-8** Comparison of six vertebrates' brains (dorsal surface). Note the progressive increase in size of the cerebellum and the cerebrum.

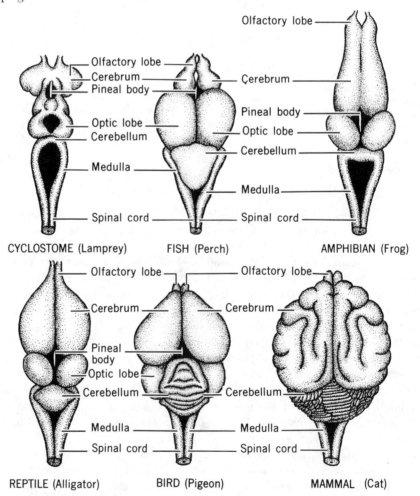

medulla is produced by a broadening of the spinal cord. In **reptiles,** the olfactory lobes are usually not well developed; the cerebral hemispheres are much larger, a small amount of gray matter is present in the cortex; the optic lobes are quite small, the cerebellum is larger; and both pineal and parietal bodies are present. In **birds,** the brain is short and broad; the olfactory lobes are very small; the cerebral hemispheres are large, but the gray matter is limited to the posterior region; and the optic lobes are comparatively large and spread apart by the large cerebellum. In **mammals,** the olfactory lobes are well developed; owing to the growth of grey matter in the cortex, the cerebral hemispheres are very large; four **corpora quadrigemina,** derived from the optic lobes, are covered by the cerebrum; the cerebellum is large and divided into three parts; and the medulla is comparatively short.

Differences in brain structure are correlated with the specific activities of vertebrates and the degree of development of their senses. We will now look at the sense organs found in different animal groups.

# Sense Organs

All five senses depend entirely on neural mechanisms for their existence. Gustation (taste), olfaction (smell), vision, tactility (touch), and audition (hearing) rely on specialized **receptors.** The sensitivity levels (thresholds) for each of these receptors vary greatly throughout the animal kingdom. While human beings have reasonable competence in all senses, lower animals usually have superior capabilities in one or a few senses and are sadly lacking in the remainder.

## GUSTATION

All animals are to some extent capable of discriminating between toxic and nontoxic substances. The higher an animal is phyletically, the greater capacity it has for discriminating between food tastes. Such **food preferences** of higher animals depend on **chemoreceptors,** specialized cells that respond in different degrees to concentrations of selected chemicals (sweetness, saltiness, etc.). Most taste receptors are localized in mouthparts.

In humans, the stimuli of *sweet* (localized mostly on the tip of the tongue), *sour* (sides of the tongue), *bitter* (back and sides of the tongue), and *salty* (tip of the tongue) are responded to by different **taste buds.** Each taste bud is a compact group of chemoreceptive neurons that communicate with the mouth cavity through a **pore.** One function of **saliva** is to partially dissolve ingested foods so that they can "fit" through pores and appropriately stimulate chemoreceptors.

The interpretation of taste stimuli does not depend solely on taste bud response. Food can lose much or all of its flavor when one has a cold. This is greatly influenced by the loss of olfactory sensitivity, which appears to be necessary for adequate cortical integration of taste stimuli.

## OLFACTION

**Olfaction** is generally considered to be of a chemosensory nature.

Several variations of the olfactory system occur in animals. Fish have an olfactory pouch covered with sensory epithelium. Birds and reptiles have rather similar systems which contain an anterior sac, followed by the main olfactory chamber. Many of the higher land animals have a **vomeronasal organ,** which picks up olfactory stimuli directly from the roof of the mouth. The vomeronasal organ acts in conjunction with the nasal portion of the olfactory system.

Sensory elements in the olfactory system of human beings are located high in the nasal cavities, away from the normal respiratory air route. The functional units of the system are embedded in a **mucous membrane.** Substances must be in a **volatile** state to reach the membrane. Whereas other sensory systems use interneurons, the receptor cells of the olfactory system are unified, **bipolar neurons.** These are continuous from their tuftlike sensory ends all the way to the **olfactory bulbs** on the ventral surface of the brain.

## VISION

Vision is perceived in a variety of ways among animals. However varied the visual organs appear, the basic chemical operations involved are the same: a *photosensitive* pigment (often **rhodopsin**) absorbs light and undergoes a change in molecular shape that initiates further activities, culminating in *perception*. The structures constituting the visual organs (**eyes**) are simply devices to refine the image.

The planarian's semicircular ring of opague pigment is only capable of detecting light direction. The vertebrate eye is able to detect light intensity, direction, wavelength (color), and shapes—it is able to furnish the brain with enough information to form images.

There are many interesting variations and modifications of eyes. Insects have what amounts to many small eyes called **ommatidia,** grouped to form a **compound eye,** which is specialized for detecting movement. Nocturnal animals often have pure-rod retinas, or at least the rod density is several thousandfold greater than the cone density (see below). Some of these animals, including the cat, have further specializations, such as relatively large pupils and a reflective structure behind the retina (**tapetum lucidum**) which increases contrast. This specialization limits the cat's visual acuity, since the double passage of light (resulting from the tapetum lucidum's reflection) renders the image less clear.

The human eye, rather typical of vertebrate eyes, has a casing composed of three layers (Fig. 26–9). The outside layer or **sclerotic coat** is the white of the eye, a "bag" of connective tissue with a window that contains and supports other eye parts. The window or **cornea,** admits light, and, as a result of its curvature, helps focus the light. Beneath the sclerotic coat lies the **choroid coat,** a vascular and heavily pigmented layer with a variable hole (the **pupil**) behind the cornea. The vascular portion of the choroid coat nourishes part of the adjacent **retina.** The pupil is centered in a visible portion of the choroid coat called the **iris.** The retina is a light-sensitive layer composed of visual cells and their related neurons.

There are two types of visual cells, **rods** (containing rhodopsin for dim-light vision), and **cones** (containing **iodopsin** and other visual pigments for daylight and color vision). There are approximately 120 million rods and about 7 million cones in a human retina.

Directly behind the pupil, and in the path of light, is the elastic **crystalline lens.** The lens is held in place by fibers (the **zonules of Zinn**) that attach to a

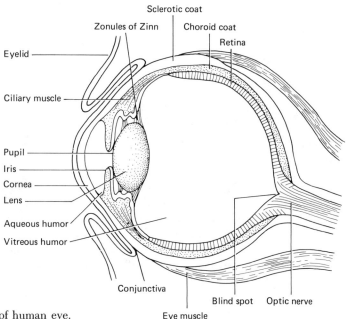

**Figure 26–9** General structure of human eye.

modification of the choroid coat called the **ciliary body.** Normally, the lens is stretched by these fibers so that the curvature is decreased and the eye is *accommodated* for distant vision. For close-up vision, ciliary muscles constrict the eye, the zonules of Zinn exert less pull on the lens, and the lens assumes a more spherical shape, increasing the curvature. As one grows older, the lens becomes stiffer, and accommodation for close-up vision is impaired.

In front of the lens is a fluid called **aqueous humor,** and behind the lens is the **vitreous humor.** These fluids are secreted by the ciliary body, and produce intraocular pressure that inflates the eye. When new humors are produced, old ones leave the eye through the **canal of Schlemm.** If this passage is blocked, increasing intraocular pressure (**glaucoma**) presses blood vessels closed, and the retina may die. In addition, the increased internal pressure may cause the hydration of the cornea and/or lens, rendering them opaque (**cataract**).

## TACTILITY

The senses of touch and pressure are responded to by different types of receptors. **Meissner's corpuscles** and **tactile disks** (mechanoreceptors) lie closer to the epidermis and are the receptors of *light* touch, while *heavier* touch (pressure) affects **Pacinian corpuscles.** These **mechanoreceptors** act when their distal membranes are bent or displaced, changing the permeability of selected ions. The amount of action generated is proportional to the degree of stretch or displacement. Free nerve endings within the dermis may respond to skin damage and send *pain* sensations to the brain.

The brain has the capacity for **localization** of receptor sites; although the brain is the locus for interpretation of nervous impulses, one "feels" the stimulus not in the head but at the sight of stimulation. This is aptly demonstrated in the case of *phantom limb pain*. Frequently, a person who has had an arm or leg amputated may feel pain in the missing fingers or toes. This is due to stimulation of neurons originally innervating these areas that now end near the stump. Since amputation does not alter the morphology of the cortex, and the brain does not "know" that the stimuli arise from a different place than they did originally, the brain interprets the pain sensations as arising from the nonexistent limb.

## AUDITION

**Audition** may be loosely defined as *the response to and interpretation of vibrations*. A primitive example of audition is the "auditory" system of the salamander. Vibrations picked up from the ground through the salamander's legs are carried in a **bone conductor** to the ear bones and interpreted in the head. In fishes and larval amphibians, the **lateral line sense organ** exists bilaterally along the longitudinal borders and in many cases extends to the head. This organ is composed of units called **neuromasts**, which are aggregations of sensory cells, sensory hairs, and **cupulae**. When the cupulae are disturbed by currents, sensory hairs relay the information to sensory cells, and the impulses travel from there to the CNS.

Specialized cells called **ampullae of Lorenzini** resemble hair cells and appear on the snouts of elasmobranch fishes. They are responsive to temperature, pressure, salinity (salt content), and electrical field changes.

The mammalian ear represents a high level of auditory development (Fig. 26–10). Waves of compressed air are "captured" by the **pinnae** (external ears) and channeled through the **meatus** (middle ear). The middle and internal ears are divided by the **tympanum (tympanic membrane** or **eardrum)**, an easily distorted, taut membrane. The **malleus** (hammer) is attached to the upper portion of the tympanum and translates the tympanic membrane's vibrations through the second ear bone, the **incus** (anvil), and the third auditory bone, the **stapes** (stirrup), to the **oval window** of the **cochlea.** Below the oval window is the **round window,** which bulges out when fluid in the cochlea is under pressure. This delicate set of bones and hinges serve to decrease proportionally the vibrational displacement of the tympanum so that when the vibrations reach the cochlea, they are 1.3 to 3.0 times smaller but *more powerful.*

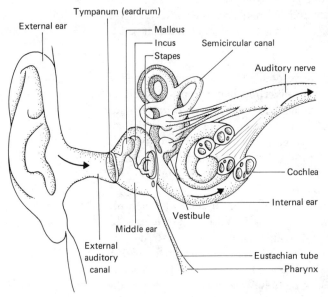

**Figure 26–10** A dissection of the human ear shows parts concerned with receiving sound and maintaining balance.

The **cochlea** is a coiled structure resembling a snail's shell. Within the **basilar partition** is the functional unit of the cochlea, the **organ of Corti**, with a normal response range between 20 and 20,000 cps. When the organ of Corti is displaced at various points, **hair cells** within the organ translate the physical displacement into nervous impulses. These are then transferred by the auditory nerve to the brain. When you speak, you hear yourself in two ways. What you primarily hear are sound waves in the air around your head. In addition to this, you hear a strong component from **bone conduction.** As your vocal cords vibrate, your entire mouth and nasal cavities reverberate the sound, and some of it is partially absorbed by your jaw and other facial bones. This accounts for the initial disbelief of many people when they hear a recording of their voice, since the recording does *not* contain the bone-conduction component.

## EQUILIBRIUM

The **membranous labyrinth** of vertebrates, from fish to human beings, are frequently considered part of the auditory system, since they appear as part of the inner ear. These organs are functional in an organism's special *orientation* and equilibrium.

The membranous labryinth of each human ear contains two sacs, the **utriculus** and the **sacculus.** These sacs contain **maculae,** which harbor cells similar to the lateral line neuromasts. **Otoliths** ("ear stones"), made of calcium carbonate, are found in both the utriculus and sacculus. These otoliths inform the animal of its vertical–horizontal orientation by responding to gravitational pull. The utricular otoliths are especially important in human equilibrium.

Three mutually perpendicular **semicircular canals** arise from the utriculus and comprise the rest of the membranous labyrinth. The semicircular canals, sacculus, and utriculus all contain **endolymph.** When the animal turns, the liquid endolymph exerts uneven pressure on the semicircular canals' cupulae. All this information is relayed to the brain, where integration and interpretation aid the organism in orientation.

# Hormones: Chemical Coordinators

We have seen how the nervous system acts as a coordinator of the responses necessary for animal behavior. The coordination of various sets of muscles and other organs require quick correlation; the nervous system is quite effective in regulating these activities.

In contrast, the **endocrine system** is a slower means of coordinating, usually acting over long periods of time. The endocrine system is composed of specialized tissues and, in some cases, ductless glands, which produce chemical substances called **hormones** (Fig. 26–11). Hormones pass directly into blood and other body fluids. They are chemical coordinators that may excite or inhibit cellular activity, thus affecting the behavior of the animal.

## HORMONES OF INVERTEBRATES

Hormones that occur in flatworms and higher phyla are highly varied in function. In crustaceans, squids, and octopuses, hormones control the expansion and contraction of chromatophores, enabling the organism to change body color.

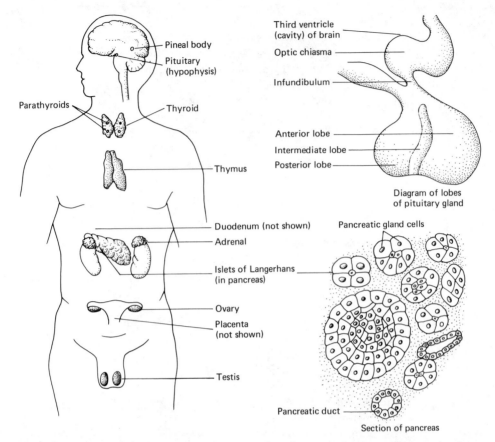

**Figure 26–11** Endocrine system. *Left:* approximate location of some endocrine glands in male/female composite human. The pineal body and thymus are not definitely known to be organs of internal secretion. *Right:* section of pituitary gland showing lobes and microscopic structure of the pancreas.

Another function of hormones is the control of reproductive activity in certain animal groups. In some worms, growth and regeneration are also under hormonal control.

Interest in the hormones of invertebrates has developed only in recent years; most of the research to date is on the arthropods. As more studies are undertaken, our understanding of the effects of hormones on all invertebrates is certain to increase.

## HORMONES OF VERTEBRATES

A great many investigations have been conducted to determine the effect of hormones on vertebrates. Because of the similarity of hormonal action in most vertebrates, the following discussion will focus on the human endocrine system.

The most important of the ductless glands in human beings are the thyroid, parathyroids, adrenals, pituitary, islets of Langerhans in the pancreas, and gonads. Hormones are also produced by membranes lining the intestines and stomach.

**THYROID.** The thyroid is a bilobed gland that lies anterior to the trachea below the larynx; it influences the general rate of metabolism in the body. The thyroid secretes a hormone called **thyroxin**, which contains about 65 percent iodine by weight. The production of this and other thyroid hormones may be decreased when there is insufficient iodine in the body. If too little thyroxin is secreted, the thyroid may enlarge to form a goiter.

When the thyroid fails to function properly in early life, children may become dwarfed and mentally deficient: this condition is known as *cretinism*. If the thyroid of an adult atrophies or is removed, myxedema occurs, which results in the swelling of the face and hands. Overactivity of the thyroid is known as *exophthalmic goiter* or *Graves' disease*. Thyroxin has been chemically isolated, crystallized, and synthesized.

**PARATHYROIDS.** The parathyroids are four small glands on the posterior surface of the thyroid. They produce a secretion containing **parathormone.** This hormone exercises a profound influence on the body's metabolism of calcium and phosphorus. A deficiency of parathormone results in low blood calcium and violent twitching of the muscles, a condition known as *tetany;* an excess, as in parathyroid tumors, may cause excessive withdrawal of calcium from the bones.

**ADRENALS.** The adrenals are small glands that sit atop the kidneys. Each consists of two parts, the (outer) cortex and (inner) medulla. The cortex secretes about 30 known compounds, chemically known as **steroids.** One of the best known steroids is **cortisone,** which, because it has antiinflammatory properties, is used to treat some types of arthritis and allergies. Removal of the adrenal cortex results in death. Injury to the adrenal cortex results in *Addison's disease*, characterized by anemia, low blood pressure, and intestinal disturbances: if not treated, this disease may be fatal.

The medulla secretes the hormone **epinephrine** (adrenaline), which comes into action at times of stress. Under emotional excitement, such as fear or anger, the sympathetic nervous system becomes activated, causing epinephrine to be secreted. This results in acceleration of the heart and blood vessel constriction, which in turn raises the blood pressure. In addition, the blood sugar is elevated by conversion of the liver glycogen. Epinephrine is widely used in treating asthma because of its relaxing effect upon the bronchioles of the lungs.

**PITUITARY, OR HYPOPHYSIS.** The pituitary gland is about the size of a pea and is located at the base of the brain in a depression of the sphenoid bone. It consists of anterior and posterior lobes.

The posterior lobe secretes two or more hormones, collectively called **pituitrin.** Injection of pituitrin into the circulatory system raises the blood pressure, decreases the amount of urine formed, and causes the contraction of smooth muscles (especially those of the uterus), and contractile elements in the mammary glands, causing the secretion of milk.

The anterior lobe secretes several hormones. **Prolactin** stimulates breast development and milk production; **growth hormone** affects the development of several organs and tissues; **TSH** (thyroid-stimulating hormone) stimulates thyroxin production; **ACTH** (adrenocorticotropic hormone) stimulates the adrenal cortex to secrete cortisol (a hormone that suppresses allergic and inflammatory reactions); **FSH** (follicle-stimulating hormone) stimulates the Graafian follicles in the ovaries to release eggs in females and stimulates spermatogenesis in the seminiferous tubules in males; and **LH** (luteinizing hormone) stimulates corpus luteum formation and ovulation in females and stimulates synthesis and secretion of testosterone in males.

An excessive amount of secretion of growth hormone in the young stimulates the development of long bones, resulting in **giantism;** when this occurs in adults, the bones become thickened, resulting in **acromegaly.** The pituitary also secretes a hormone that increases the sugar content of the blood, thus having effects opposite to those of insulin.

Since the pituitary gland influences other endocrine glands, it is referred to as the "master gland."

Several hormones, collectively called **estrogens,** are secreted by the ovary. Estrogens are formed in the Graafian (ovarian) follicles and contribute to the control of the menstrual cycle, sexual behavior, and development of accessory genital organs and secondary sex characteristics. **Progesterone** is secreted by the corpus luteum; it regulates the menstrual cycle, plays a part in the development of mammary glands, and prepares the female for pregnancy.

TESTIS. The interstitial cells of the testis secrete a hormone called **testosterone,** which influences sexual behavior and controls the development of the male accessory genital organs and secondary sex characteristics.

PANCREAS. The pancreas is a gland situated behind the stomach. The interspersed cells in the pancreas, known as the islets of Langerhans, secrete the hormones **insulin** and **glucagon.** Insulin is necessary for normal regulation of sugar metabolism; it stimulates uptake of glucose into the cells, accelerates synthesis of sugar to glycogen, and retards production of sugar in the liver from fat and protein. If the pancreas does not secrete a sufficient amount of insulin, an excessive amount of sugar occurs in the blood, resulting in the disease **diabetes mellitus.**

GASTRIC MUCOSA. There is considerable evidence that mucous membrane in the pyloric end of the stomach may, when food is present, produce a hormone called **gastrin.** This hormone is carried by blood to the gastric glands, where it stimulates the secretion of gastric juices.

INTESTINAL MUCOSA. Parts of the mucous membrane of the intestine produce a hormone called **secretin,** which was the first hormone discovered (1902). Secretin is carried by blood to the pancreas, where it causes immediate secretion of pancreatic juices and increases the flow of bile. A second intestinal hormone, **cholecystokinin,** causes the gallbladder to empty.

## Summary

Coelenterates represent the lowest level of animals possessing a true nervous system, with definite receptors, conductors, and effectors. Ascending the phyletic ladder, successively more refined nervous systems appear, with ganglia, specialized receptors, reflexes, and cephalization. The vertebrate nervous system consists of the central (CNS) and peripheral (PNS) nervous systems. The CNS is integrative and vastly more complex than the PNS, which serves mostly for conduction to and from the CNS. Another division of nervous systems is between somatic and autonomic, the former concerned mostly with voluntary functions, the latter with involuntary functions.

The autonomic nervous system is further divided into the sympathetic nervous system, used in emergency situations, and the parasympathetic nervous system, which returns the animal to a normal state. Neurons, the functional units of the nervous system, have axons, which conduct impulses away from the neuron, and dendrites, which carry impulses toward it. Neurons "connect" at chemical synapses.

The brain is divided into the cerebrum, cerebellum, and brainstem. The multilobed cerebrum includes the cortex and is responsible for thought, sensation localization, and interpretation. The cerebellum is responsible for most of the coordinating activity of the brain. The brainstem (medulla and pons) is concerned primarily with reflexes. All impulses entering or leaving the brain (with the exception of those occurring in the cranial nerves) pass through the brainstem. The reticular formation, within the medulla and pons, is responsible for activation of the organism as a whole.

A sensation may be defined as the discrimination of incoming stimuli or the interpretation of stimuli within the organism. Gustation (taste), olfaction (smell), vision, tactility (touch), and audition (hearing) occur in increasing complexity as we ascend the phyletic ladder.

Hormones are the chemical communicators within an organism. Minute concentrations of these chemicals affect target organs as either inhibitors or excitors. Many diseases are the result of hormonal imbalance, such as cretinism, tetany, Addison's disease, acromegaly, and diabetes.

# PART FIVE

# The Continuity of Life

# 27

# Reproduction and Development

In the early 1900s, the famous Scandinavian explorer V. Stefansson made some intriguing observations on Eskimo culture. Included in his study were reports on sexuality. According to Stefansson, sexual intercourse was enjoyed by all Eskimos, including adolescents, who freely participated in sexual episodes. In fact, it was considered an insult if a visiting man failed to have sex with his host's wife. Babies and children were looked upon favorably, but there was no connection made between sexual intercourse and pregnancy. Births were considered to be natural phenomena that happened to females *at random.*

We suspect that you find this amusing. Such a reaction is a result of training—in our present culture, we all know the "facts of life." A little later we will discuss the "birds and bees" with a more scientific approach than our parents (or peers) inculcated.

First we will study reproduction in the protist and animal kingdoms, and then turn our attention to the phenomena resulting from reproductive acts.

## Asexual Reproduction

**Asexual reproduction,** considered more primitive than sexual reproduction, is the "simplest" method of creating new individuals from preexisting ones. Five major methods of asexual reproduction are identified: binary fission, multiple fission, fragmentation, budding, and sporulation (Fig. 27–1).

**Binary fission** is a simple mitotic division where each new cell receives half the cytoplasm of the original cell. In nonradially symmetrical organisms, the **cleavage furrow** (line of cytoplasmic division) may be either longitudinal or transverse. This is the common mode of reproduction in protists and even occurs in a few members of the animal kingdom, including larval sponges, worms, and starfishes.

In **multiple fission,** nuclear division occurs *before* cleavage. After the nucleus divides, the cytoplasm cleaves into several units, each with one nucleus. Some amoebas, all the malarial parasites, many other sporozoans, and some other organisms reproduce by multiple fission.

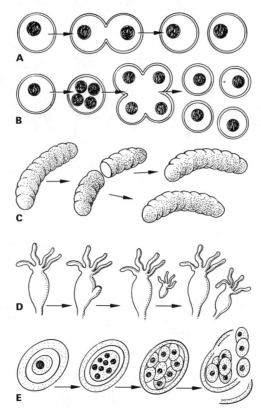

**Figure 27-1** Methods of asexual reproduction. (A) Binary fission. (B) Multiple fission. (C) Fragmentation. (D) Budding. (E) Sporulation. Note that sporulation is essentially the same as multiple fission except that it occurs within a cyst.

**Figure 27-2** Evolution of sex in Mastigophora. Light arrows represent asexual reproduction; dark arrows represent sexual reproduction.

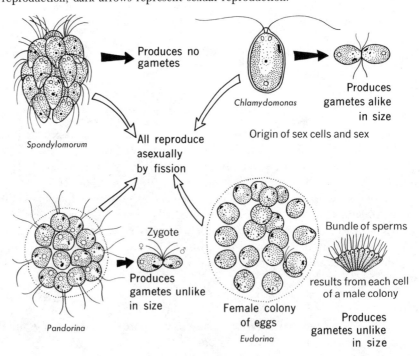

**Fragmentation** is the breaking off of parts of an organism. The parts then *differentiate* into complete, full-size organisms. Some flatworms, segmented worms, and spiny-skinned worms reproduce by fragmentation.

**Budding** consists of an unequal division of the organism's body—the smaller new individual is the "bud." Budding may be either internal or external. **Internal budding** (as in *Spongillidae*) occurs when several separate cells within the organism's body become encapsulated by dense coverings. These capsules are called **gemmules.** The parent organism's body disintegrates, and each gemmule forms a new individual. **External budding** is simpler—a part of the organism "grows out" of the parent's body, and may either remain attached (as in the colonial *Vorticella*) or separate and become morphologically independent (as in most *Hydra*).

**Sporulation** is (in zoology) a process whereby within a cyst many nuclear divisions and nuclear fission produce many cells. These cells may combine in pairs, forming zygotes, which secrete protective coverings. Since this phenomenon occurs mostly in parasitic organisms (as in the Sporozoa), the usual method of cyst dispersal involves ingestion of the cyst by a new host.

The evolutionary transition from an asexual to a sexual method of reproduction in species of Mastigophora is shown in Fig. 27-2.

## Sexual Reproduction and Fertilization

In **sexual reproduction,** a male cell, the sperm, fuses with a female cell, the ovum (egg) in a process called **fertilization.** Organisms with both sets of reproductive organs present in one individual are **monoecious** or **hermaphroditic.** Those with only one set of sex organs (**gonads**) are **dioecious.** The protists and many lower animals are monoecious; most higher organisms are dioecious. Although each monoecious individual is capable of producing both sperms and eggs, the majority of monoecious organisms **cross-fertilize,** thereby *exchanging* nuclear material. In *Paramecium*, although no gametes are actually formed, fertilization occurs almost exclusively through contact between two individuals. The earthworm, like most monoecious animals, is **self-sterile** and must cross-fertilize. **Self-sterilization** combined with cross-fertilization occurs in some *Hydra*, which release sperm into water. The gametes are carried by random water currents to other hydras and sometimes back to the emitters.

An interesting aspect of some monoecious and female dioecious animals is **sperm storage.** In the honeybee sexual cycle, the queen bee mates infrequently in her lifetime. She may use only a portion of the sperm immediately and hold the rest in reserve for future fertilizations. Some of her eggs become fertilized (diploid) and produce females, while many remain haploid and become males (drones). This production of viable individuals from unfertilized eggs also occurs in rotifers, crustaceans, and others, and is called **parthenogenesis.**

**Artificial parthenogenesis** may be induced in some normally nonparthenogenic organisms (such as the frog) by some types of egg membrane disturbances. Pinpricks, dilute acid solutions, warmth, and many other stimuli (including sperm penetration) have been known to induce embryo development.

The example of hydra's spewing sperm into its environment given above is illustrative of **external fertilization,** which also occurs in some vertebrates, particularly fish and amphibians. When frogs mate, the male mounts the female dorsally, stimulates her to release eggs, and proceeds to spread his sperm over the egg pile.

**Table 27–1** Gross Comparative Embryology—Some Basic Embryological Trends in Animalia*

| Structure | Mammals Viviparous | Reptiles and Birds Oviparous | Fishes Oviparous | Amphibians Oviparous |
|---|---|---|---|---|
| Yolk | Lacking | Telolecithal or oligolecithal; absorbed during development | Large amounts, nonsegmented | Oligolecithal |
| Cleavage | Complete | Discoidal | Meroblastic | Complete; uneven |
| Yolk sac | Hypoblast spreads under inner surface of trophoblast, encircling cytoplasm | Area opaca (periphery of blastodisk) spreads down; endoderm adheres to surface | Blastodisk spreads over yolk; endodermal portion does not go with it | Yolk is within gastrula |
| Body-yolk considerations | By means of body folds under and around embryo | Same as mammals | Same as mammals | Yolk is within gastrula; in early stages fills in entire gastrula except archenteron and blastocoel remnant |
| Amnion | Body folds fuse—later envelops embryo | Develops with chorion; fuses around embryo, produces amniotic cavity; function to prevent desiccation and shocks | | |
| Chorion | Outer surface of chorion is continuous with trophoblast | Develops with allantois, also called serosa | | |
| Allantois | Used exclusively for nutrition and $O_2$–$CO_2$ diffusion | Function in urinary bladder for $CO_2$–$O_2$ diffusion; between amnion, yolk sac and chorion; spreads evenly under chorion | | |
| Primitive streak/ Hensen's node/ neurulation | Present but shorter than in birds; ephemeral | Present; ephemeral | Neural plates formed but no tube formed | Neural tube eventually formed |

* The information in this table is highly simplified and generalized, and should only be used as a general guide. Note that entries are lacking under fishes and amphibians for amnion, chorion, and allantois. These organisms are **anamniotes,** without these structures. Fishes and amphibians develop in aqueous environments, and diffusional processes can readily occur without the aid of such structures.

**Internal fertilization,** generally characteristic of higher forms, provides a protected environment for the fertilization process, and hence helps to ensure reproductive success.

Embryo development takes place in one of three ways in Animalia (Table 27–1): **Oviparous** animals, such as birds, lay eggs that hatch outside the body of the mother. **Viviparous** animals, such as mammals, usually develop within the body of the mother and are nourished from her bloodstream through a placenta (see below). **Ovoviviparous** animals, such as sharks and reptiles, produce eggs that hatch within the mother's body but are not nourished by the mother's bloodstream.

## SPERMATOGENESIS AND THE MALE REPRODUCTIVE SYSTEM

This account refers mostly to mammals, especially human beings. The production of sperm occurs in the male gonads, the **testes.** Within the testes lie many **seminiferous tubules.** In young males, the tubules are simple and contain only immature sex cells (**spermatogonia**) and Sertoli cells. As the male matures, the spermatogonia develop into **spermatocytes,** which in turn become **spermatids.** The spermatids undergo a series of transformations, known as **spermatogenesis,** finally developing into sperm. Each **sperm** consists of a head, middle piece, and tail.

The head of the sperm cell becomes embedded for a time in the **Sertoli cells;** it is thought that the sperm here undergo a ripening process. The nourishing function of the Sertoli cells has led them to be called "sperm mother cells."

The **head** of the sperm is covered, in part, by the **acrosomal cap.** The main portion of the head is a glob of DNA (see Chapter 28). The **middle piece** is

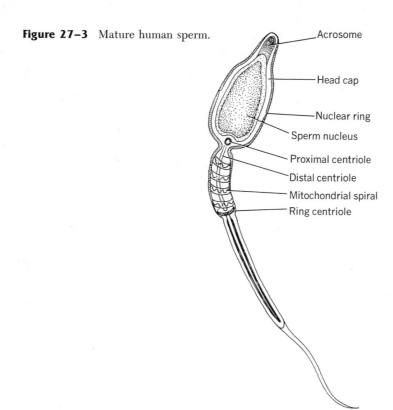

**Figure 27–3** Mature human sperm.

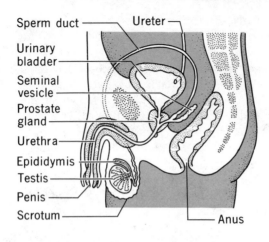

**Figure 27-4** Human male reproductive organs, shown in relation to urinary bladder and urethra.

separated from the head by two mutually perpendicular centrioles and is rich in mitochondria. The **tail,** an undulipodium, obtains energy from the mitochondria of the middle piece (Fig. 27-3).

In insects, the more mature sperm cells move toward the edges of the tubules while in human beings and other vertebrates, the more mature sperms are contained closer to the center of the tubules.

The seminiferous tubules unite to form the **epididymis,** a much-coiled tube about 6 meters in length. The epididymis leads into the sperm duct (**vas deferens**), which joins the duct of a **seminal vesicle** to form an **ejaculatory duct;** this opens into the **urethra.** Near the urethra is a gland about the size of a chestnut, the **prostate gland.** On either side of the prostate is a body about the size of a pea, the **bulbourethral (Cowper's) gland.**

Spermatozoa are formed in the testes and pass down the vasa deferentia. Secretions are supplied by the seminal vesicles, Cowper's glands, and the prostate gland; these are ultimately added to the sperm and constitute the **seminal fluid (semen).** These secretions provide a medium of transport, provide nutritive substance, allow activation of the sperm, and overcome acidity. The semen flows through the ejaculatory ducts into the urethra and then out of the body through the **penis** (Fig. 27-4).

## OOGENESIS AND THE FEMALE REPRODUCTIVE SYSTEM

In the female reproductive system are two **ovaries** that produce eggs and female sex hormones, two **oviducts** (or **Fallopian tubes**), about 10 cm long, a **uterus,** a **vagina,** and the **external genitalia** (Fig. 27-5). The ovaries are almond-shaped and contain at birth about 70,000 developing eggs (**ova**), most of which later disappear. Ova are discharged from the ovary into the coelomic cavity, enter the oviduct through the **ostia,** and are carried by peristalsis to the uterus. If an egg is fertilized, it remains in the uterus during embryonic development.

ESTROUS AND MENSTRUAL CYCLES. In mammals, females **ovulate** (release eggs or an egg from the ovaries) regularly. This cycle of hormonal and uterovaginal changes is called the **estrous cycle.** The characteristics of estrous cycles vary, depending on the organism under consideration. We will use the human female as an example of primate estrous and menstrual cycles (Fig. 27-6).

The first (**follicular**) phase of the estrous cycle begins with gradual matura-

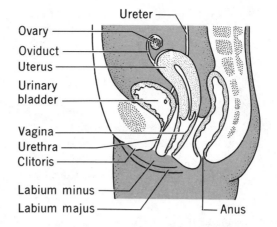

**Figure 27–5** Human female reproductive organs, shown in relation to urinary bladder and urethra.

tion of one or more **Graafian follicles** (each containing one oocyte) within the ovaries. **Follicle-stimulating hormone** (FSH) produced by the pituitary gland stimulates maturation and nutrition in the follicles. As the follicles mature, they secrete **estrogen**, a female sex hormone. **Luteinizing hormone** (LH) is also secreted by the pituitary gland; it almost immediately induces the follicle to

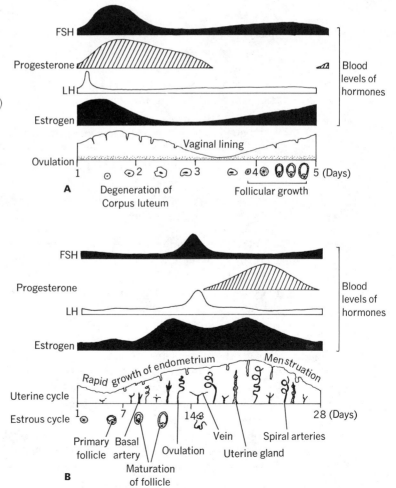

**Figure 27–6** Human female's estrous cycle (*top*) and menstrual cycle (*bottom*). (After R. A. Boolootian, *Human Reproduction*, slide sequence, New York: John Wiley & Sons, Inc., 1971.)

burst open, allowing the mature egg to escape (**ovulation**). The egg is drawn into the ostium of the Fallopian tubes by undulipodia, and begins its trek down to the uterus. Fertilization, when it occurs, usually takes place within the Fallopian tubes.

After the egg is emitted from the follicle, a dramatic change takes place. A yellow body known as the **corpus luteum** is formed. The corpus luteum begins prolific secretion of another hormone, **progesterone.** This marks the beginning of the **luteal phase** of the estrous cycle. If the egg is fertilized, the corpus luteum remains intact, continuously secreting its hormones, and the zygote becomes implanted in the wall of the uterus. If the egg is not fertilized, the corpus luteum *stops* its hormone production.

Owing especially to the cessation of progesterone secretion, the endometrial (interior) wall of the uterus begins to break down. In primates, the wall is sloughed off along with varying amounts of blood and is lost through the vaginal opening in a process called **menstruation.** For this reason, the estrous cycle in menstruating animals is often referred to as the **menstrual cycle.**

Nonprimate mammals go through estrous but not menstrual cycles. An estrous (or "heat") cycle is, among these organisms, a period of sexual fervor—the female may accept a male copulatory partner during this time, whereas at other times she may not only reject a male, but may even attack him.

## ZYGOTE FORMATION

After copulation, several million sperm approach the egg, but it is extremely rare for more than one to penetrate it. The acrosomal cap of the sperm cell contains lysosomal enzymes which penetrate the oocyte's outer membrane. As soon as the sperm gets through, the oocyte produces a protective membrane which keeps the other sperm out.

The nucleus of the oocyte is known as the **female pronucleus,** and the head of the sperm is the **male pronucleus.** After the sperm head has entered, the remainder of the sperm is cast off, and the male pronucleus merges with the female pronucleus. This forms the **zygote nucleus,** and the process of cleavage begins.

## CLEAVAGE

**Cleavage** is the mitotic division of the zygote into several connected cells. (If, in an early stage of cleavage, some of the cells separate, identical twins, triplets, etc., may result.)

After nuclear division comes the division of the zygote into two blastomeres (daughter cells). The daughter cells do not separate (as they do in the binary fission of paramecia), but remain attached to one another.

Several types of cleavage patterns are recognized (Fig. 27-7). In **radial cleavage** the furrows line up in successive tiers. In **spiral cleavage,** the furrows are at 90° angles to each other (Fig. 27-8). Many types of cleavage differ according to the yolk concentration (**lecithality**) in the egg. If the egg contains little yolk, the entire zygote divides into 2, 4, 8, etc., blastomeres. If these daughter cells are approximately equal in size, the process is known as **equal holoblastic cleavage,** characteristic of sea stars and *amphioxus*. If the daughter cells are unequal in size, **unequal holoblastic cleavage** proceeds, illustrated by the frog (except for the first two or three steps). If the egg contains a considerable amount of yolk, the entire egg is not divided into cells; only restricted

**413**
*Reproduction and Development*

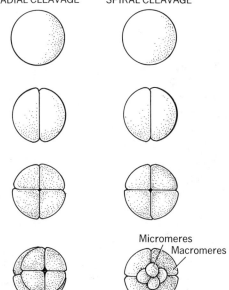

**Figure 27–7** Types of cleavage.

**Figure 27–8** Radial and spiral cleavage. Note that in spiral (and many other) cleavages, one tier (macromeres) has larger cells than the other tier (micromeres).

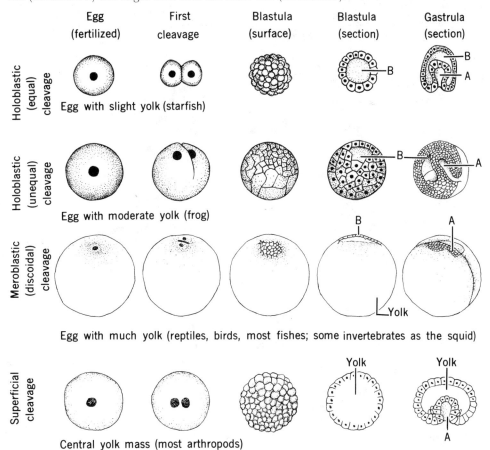

portions of the cytoplasm undergo cleavage. If cell division is restricted to a small cap or disk on one side of the egg (as in birds), it is **meroblastic** or **discoidal cleavage**. If it is restricted to an area of cytoplasm surrounding the egg (as in insects), it is **superficial cleavage**.

Cleavage may be characterized as either determinate or indeterminate. In **determinate cleavage**, each blastomere is destined to become a particular portion of the embryo, while in **indeterminate cleavage**, early blastomeres may develop into any part.

**BLASTULA.** As cleavage advances, a cavity becomes noticeable in the center of the egg, enlarging as development proceeds until the whole resembles a hollow rubber ball, the rubber being represented by a single layer of cells. At this stage the embryo is called a **blastula**, the cavity the **blastocoel** (**segmentation cavity**), and the cellular layer the **blastoderm**.

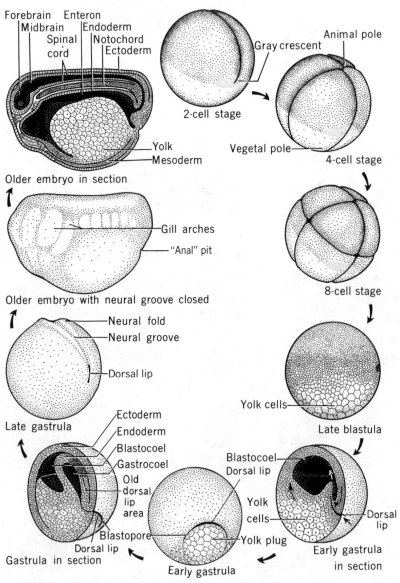

**Figure 27-9** Embryology of the frog. Note position and movements of dorsal lip of blastopore, which is destined to become the anus.

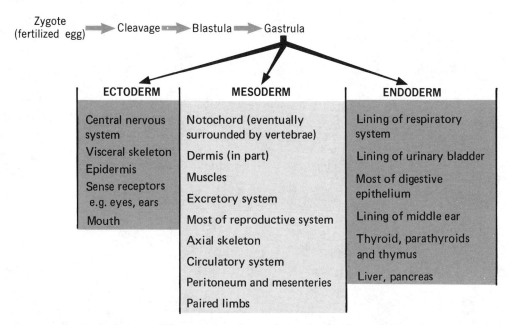

**Figure 27-10** Embryonic differentiation of germ layers in vertebrates.

## GASTRULATION

The cells on one side of the blastula may begin to invaginate into the interior. The blastocoel is gradually obliterated during invagination, while a new cavity, the **gastrocoel** (**archenteron**) is established, bounded by the invaginated cells. The developmental stage is now called the **gastrula;** its formation is called **gastrulation.** Figure 27-9 shows gastrulation and later stages in the embryonic development of the frog.

In the development of many species, invagination does not occur, but certain cells of the blastula divide and fill up the segmentation cavity, as in the *Hydra*. This results in a **stereogastrula**.

**GERM LAYERS.** All the tissues and organs that will develop are differentiated from **germ** (**embryonic**) **layers.** The gastrula of simple metazoans consists of two germ layers, an outer **ectoderm** and an inner **endoderm.** These phyla are said to be **diploblastic.** In the more complex animals, a third layer, the **mesoderm,** arises as a result of gastrulation and the organism is termed **triploblastic.** The structures arising from the germ layers are shown in Fig. 27-10.

## PLACENTATION AND EARLY DEVELOPMENT OF THE EMBRYO

We will now see how formation of the blastula and gastrula produce the structures necessary for embryo development; this section will be followed by a description of the development process of the human embryo.

In the primates, the **blastodisk** is formed of three major layers, the **trophoblast, epiblast,** and **hypoblast.** The **inner cell mass** is responsible for most of the formation of the embryo itself.

Below the blastocoel is a thin layer of cells called the **hypoblast,** and under that is the **subgerminal cavity.** The subgerminal cavity separates the hypoblast from the yolk to form the **yolk sac** (called such even if it does not contain yolk). (The blastodisk of fish lies directly over the blastocoel.)

**416**
*The Continuity of Life*

Under the blastocoel is the **periblast,** which does not function directly in embryo formation, only in metabolism of the yolk. Yolk sac formation in fishes is accomplished by the spreading of the blastodisk down and over the yolk. In amphibians, the yolk is usually passively enclosed in the process of gastrulation, and so there is no yolk sac per se.

In all groups mentioned except the amphibians, the body separates from the yolk by a process of **infolding** of germ layers, resulting in the embryo sitting atop the yolk sac and connecting to it by a **yolk stalk,** or, in higher vertebrates, by the **umbilicus.**

The infolding process is responsible for the **extraembryonic structures** in the

**Figure 27-11** Early development of human embryo, surrounded by protective membranes.

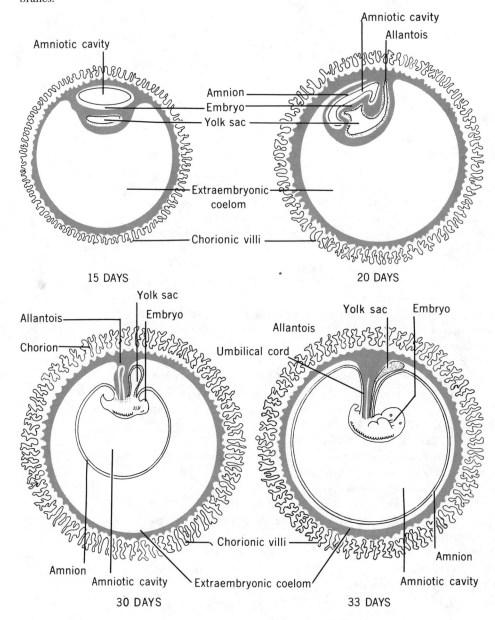

nonaquatic animals. These structures—the **amnion, chorion,** and **allantois**—serve for nutrition, waste excretion (through diffusion) and, in some cases, protection from shocks.

The amnion and chorion develop from the germ layers differently in mammals than in reptiles and birds. In mammals, the amniotic cavity is developed by cavitation (a type of splitting) either between the inner cell mass and the trophoblast or within the inner cell mass itself (Fig. 27-11).

The chorion is developed from the trophoblast, which, on contact with the maternal uterine wall, develops **villi** (increasing the surface area), which interject with maternal uterine tissues. The *combination* of maternal tissue and chorion is known as the **placenta** and is the means of nutrition and waste removal in mammals.

In reptiles and birds, the amnion and chorion are developed from the infolding process: the distal folds become the amnion, and the proximal ones the chorion.

In amphibians and fishes, their aquatic environment is usually sufficient for provision of gaseous exchange, waste removal, and so on, and therefore the extraembryonic membranes are usually absent.

## DEVELOPMENT OF THE HUMAN EMBRYO

Embryological development (**ontogeny**) begins similarly for most organisms and progressively becomes more characteristic of the particular organism. For that reason, space and scope do not allow for detailed descriptions of the ontogeny of each organism. We will confine our attention to the development of the human embryo (Fig. 27-12).

At approximately 30 hours after conception, the first cleavage has taken place in the developing human. The zygote is still within the Fallopian tubes of the mother. Some time between 40 and 60 hours, the second cleavage has taken place, and the third occurs at about 72 hours after conception.

Approximately 2 days after fertilization, the zygote leaves the oviduct and enters the uterus. At this point, it is composed of 32 cells. Soon thereafter the blastocyst is formed and within the blastocoel lies the inner cell mass. Opposite the mass, trophoblast has begun its proliferation, and at approximately day 6 the complete blastocyst is attached to the uterine wall. The mother–embryo connection is extended at about day 7, when the trophoblast invades the uterine wall and actually dissolves part of it. A number of spaces (**lacunae**) fill with blood and diffusionally communicate with the placenta. At about day 8 the embryo is fully embedded in the uterine wall. The inner cell mass splits, forming the **amniotic cavity.**

By day 18, much differentiation has occurred. A **primitive streak** (which gives rise to the notochord) has developed on the dorsal surface of the blastodisk. At the anterior end, **Hensen's node** represents the area of the **head process.** The mesoderm splits, and the area formed is the coelom.

By day 20, the nervous system has begun development. The mesoderm at the head process induces the overlying ectoderm to form the **neural tube.** Mesoderm is also divided into **somites,** which are to become muscle and bone tissue.

At the third week, the embryo is between 2 and 4 mm in length, with leg and arm **buds.** Between the amnion and chorion, **blood islands** appear, and the **vitelline network** of blood vessels is forming. By the fourth week, the circulatory system is functioning, as well as part of the nervous system.

At week 5, the head of the embryo is relatively large, and the ears are

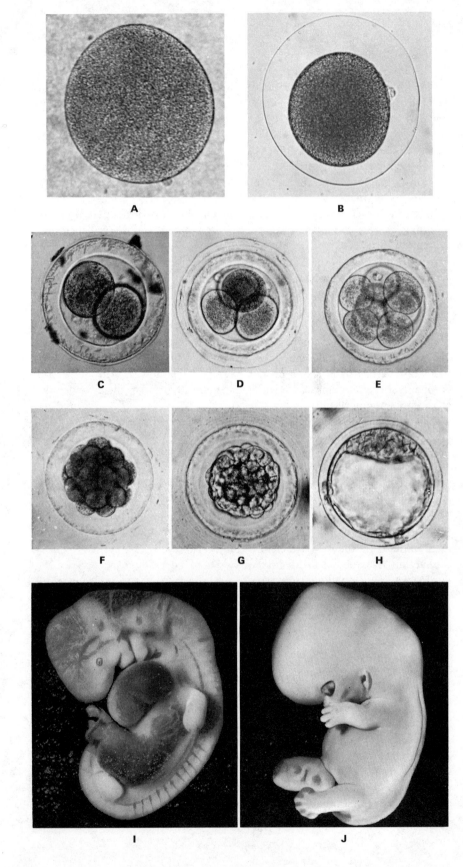

delineated. The embryo is about 8 mm long by week 6, and the limbs are evident, positioned across the abdomen. The hand and foot rudiments are visible, and some digitation is occurring. By week 7, the hands and feet are clearly differentiated. The tail is almost gone. At week 8, the embryo is about 23 mm long and appears humanoid. The organ rudiments continue development, and bone begins ossification.

At week 10, the intestines, which previously projected into the umbilicus, are drawn into the body. The spinal cord is definitive, and its internal structure is well on the way to completion. During the twelfth week, the sex of the embryo is visually obvious. The notochord begins degeneration, and the ossification of some bones is almost complete.

At the sixteenth week, the embryo or **fetus** may move about within the uterus, to a limited extent. Some hair appears on the head, which now grows more slowly as the body catches up to it in growth. The eyes, nose, and ears appear (externally) complete.

Between 20 and 40 weeks, the embryo increasingly approaches human form. In the male, the testes descend into the scrotum; in the female, the uterine glands appear. In earlier stages most of the blood was formed in the liver, but now the bone marrow is gradually taking over this function. The nose and ear ossify, although the ear is deaf until birth.

At birth, all organ systems are functioning fully. The sutures of the skull are not completely ossified, and the **fontanel** in the skull has yet to close. The brain is covered by a thick membrane at the locus of the fontanel. A hole in the interatrial septum of the heart (**foramen ovale**) closes shortly after birth.

Characteristics of the newborn organism are determined in part by its genes, which are inherited from both parents. A discussion of genetics and heredity is presented in the following chapter.

## Summary

Asexual reproduction may occur as binary fission, multiple fission, fragmentation, budding, or sporulation.

Sexual reproduction involves the fertilization of an ovum by a male sperm, producing a zygote.

The embryology of animals is at once strikingly similar and vastly dissimilar. Generally, groups of cells migrate in one way or another to loci within the prospective organism. Various sets of these cells affect one another, resulting in differentiation.

In the process of placentation, extraembryonic membranes (amnion, chorion, and allantois, and placenta) are formed. These membranes have different functions in different groups of animals.

Most higher animals' embryologies are similar in the earlier stages, and gradually diverge. In the case of human beings, it takes about 15 to 20 weeks before the embryo (or fetus) is easily recognizable as human.

**Figure 27–12** Development of mammalian egg. A–H, stages in development of rabbit's egg. (A) Unfertilized egg. (B) Fertilized egg. (C) Two-cell stage, 24 hours after fertilization. (D) Four-cell stage (29 hours). (E) Eight-cell stage (32 hours). (F) Morula stage (55 hours). (G) Trophoblast stage ($71\frac{1}{2}$ hours). (H) Blastocyst stage (90 hours). (I) Human embryo, 1 month old (6.7 mm) (J) Human embryo, 6 weeks old (19 mm). (C–H, courtesy P. W. Gregory; I–J, courtesy Carnegie Institute of Washington, Department of Embryology.)

# 28

# Genetics and Heredity

Thus far in our study we have concentrated on the chemistry and morphology of animal cells and the methods of cellular combinations (tissues, organs, and finally, organisms). Our attention has been focused on the physiological *systems*, which sum to form the organism. We are now ready to review the genetics underlying the similarities and differences of living creatures.

## THE CHROMOSOME AND DNA

In 1869, the German physiologist Friedrich Miescher made a discovery that revolutionized our understanding of every living organism; it explained how tissues, organs, and whole organisms came to have individual identity. Miescher found that cells' **chromosomes** had as a major component a substance called **deoxyribonucleic acid** (**DNA**). DNA's true significance was not known for some time, but continued research eventually solved the mystery of this intriguing molecule.

It was discovered that in a given organism, the amount of DNA in almost every cell is constant. It was also found that our characteristics, or **traits,** were determined by our chromosomes, and hence DNA. **Genes** are the hereditary units contained in the chromosome that are comprised of DNA.

DNA, as the unique substance in heredity, has a dual responsibility. It must contain the information for its own **replication** and for the **biosynthesis** of the other components of the cell.

The double-helix structure of the Watson–Crick model suggested that new molecules form directly, each one acting as a **template** for a new molecule or **strand.**

Each DNA molecule gives rise to **two daughter molecules,** each containing a parental strand and its newly synthesized complement. This form of duplication was called **semiconservative,** because even though individual parental strands were conserved during duplication, their association ended during duplication of the two daughter molecules. For each complementary strand to serve as a template for the next generation of DNA, the parental double helix must somehow unwind. Watson and Crick suggested that this occurred at the growing end of the chain at the time of formation of the two daughter helixes.

A chromosome is made of DNA and a protein coat. If we visually uncoil the DNA double helix, we may imagine it to look something like a ladder. In DNA,

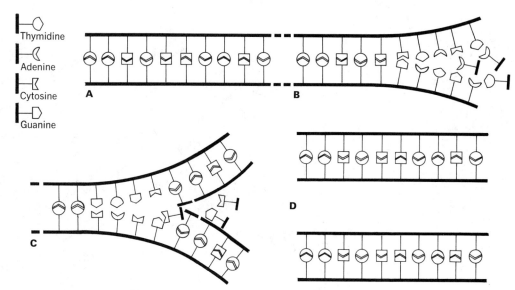

**Figure 28–1** Basic DNA replication. (A) Uncoiled section of DNA chain. (B) Deoxyribonucleotides (A, T, C, and G) link up with counterparts as the DNA chain "unzips." (C) As the "unzipping" process continues, two daughter chains (identical to mother chain) are formed. (D) Completed daughter chains.

there are four different **deoxyribonucleotides,** each with a phosphate–deoxyribose (sugar) backing: They are **adenine (adenylic acid** or **A), thymine (thymidylic acid** or **T), guanine (guanylic acid** or **G),** and **cytosine (cytidylic acid** or **C).** Hundreds of thousands of these molecules are lined up in pairs and form the "steps" of the ladder. It is important to note that cytosine can *only* pair with guanine and thymine can *only* pair with adenine in a DNA ladder; we will discover the importance of this **pair specificity** shortly.

The basics of **DNA replication** are shown in Fig. 28–1. A small portion of the intact DNA double helix is shown at the beginning of the S phase of the cell cycle (Chapter 2). As the chain starts to "unzip," A—T and C—G bonds break, and naked A, T, C, and G molecules, attached only by their phosphate sugar backbones, project into the karyoplasm. Within the nucleus are "free" deoxyribonucleotides with phosphorus–sugar backings. Aided by the enzyme **DNA polymerase,** these free nucleotides bond to free ends of their companion molecules on the old chain. The process continues until two *equivalent* DNA molecules have been formed and replication is complete.

## DNA, RNA, AND THE GENETIC CODE

Why is it so important that newly synthesized DNA maintain its integrity? What is the significance of adenine, thymine, cytosine, and guanine?

Although we, and all other animals, are made of different proteins, it would perhaps be more accurate to say that *we make* the proteins we are composed of. The synthesis of a protein is entirely dependent on the order of bases (nucleotides) in the DNA molecule. Since proteins are made of amino acids, the major function of chromosomes is to order successive amino acids so that the resulting sequence forms a precisely determined protein.

The four bases (A, T, C, and G) cannot individually "code" the over 20 amino acids used to build proteins. It has been determined that *sequences of*

**Table 28–1  Genetic Code**

| | | | |
|---|---|---|---|
| UUU ⎫ Phenylalanine<br>UUC ⎭<br>UUA ⎫ Leucine<br>UUG ⎭ | UCU ⎫<br>UCC ⎬ Serine<br>UCA ⎪<br>UCG ⎭ | UAU ⎫ Tyrosine<br>UAC ⎭<br>UAA ⎫ Term,°<br>UAG ⎭ | UGU ⎫ Cysteine<br>UGC ⎭<br>UGA ⎫ Tryptophan<br>UGG ⎭ |
| CUU ⎫<br>CUC ⎬ Leucine<br>CUA ⎪<br>CUG ⎭ | CCU ⎫<br>CCC ⎬ Proline<br>CCA ⎪<br>CCG ⎭ | CAU ⎫ Histidine<br>CAC ⎭<br>CAA ⎫ Glutamine<br>CAG ⎭ | CGU ⎫<br>CGC ⎬ Arginine<br>CGA ⎪<br>CGG ⎭ |
| AUU ⎫ Isoleucine<br>AUC ⎭<br>AUA ⎫ Methionine<br>AUG ⎭ | ACU ⎫<br>ACC ⎬ Threonine<br>ACA ⎪<br>ACG ⎭ | AAU ⎫ Asparagine<br>AAC ⎭<br>AAA ⎫ Lysine<br>AAG ⎭ | AGU ⎫ Cysteine<br>AGC ⎭<br>AGA ⎫ Arginine<br>AGG ⎭ |
| GUU ⎫<br>GUC ⎬ Valine<br>GUA ⎪<br>GUG ⎭ | GCU ⎫<br>GCC ⎬ Alanine<br>GCA ⎪<br>GCG ⎭ | GAU ⎫ Aspartic acid<br>GAC ⎭<br>GAA ⎫ Glutamic acid<br>GAG ⎭ | GGU ⎫<br>GGC ⎬ Glycine<br>GGA ⎪<br>GGG ⎭ |

° *Term.*, terminating codon.

three bases form the **genetic code** (Table 28–1). Each base triplet (**codon**) codes either for a specific amino acid or acts as a termination signal.

### RNA SYNTHESIS

Two very important points are relevant here: (1) proteins are not made directly from DNA and (2) protein synthesis occurs outside the karyoplasm. The intermediates necessary for protein synthesis are three forms of **ribonucleic acid (RNA)**. These are **messenger RNA (mRNA)**, **transfer RNA (tRNA)**, and **ribosomal RNA (rRNA)**. All three types are very similar to DNA with the following notable exceptions:

1. The sugar in DNA is deoxyribose; the sugar in RNA is **ribose**, which contains more oxygen.
2. In DNA the companion base to adenine is thymine; in RNA, adenine pairs with **uridine (U)**.
3. DNA replicates; RNA usually does not.
4. Different enzymes are used to catalyze DNA and RNA synthesis (**transcription**).
5. DNA remains within the nucleus, whereas all three RNA types travel to and perform protein synthesis in the cytoplasm.

### PROTEIN SYNTHESIS IN RNA TRANSLATION

Protein synthesis occurs in the cytoplasm in several steps. Once a tRNA molecule has been synthesized, it moves out of the nucleus and, with the aid of an enzyme and energy from ATP, bonds to an amino acid. There are about 25 different types of amino acid; the specificity of the bonds is imparted by the **active sites** of the tRNA molecule and the corresponding sites on the appropriate enzymes (Fig. 28–2). The template upon which the amino acids are united is provided by mRNA, synthesis of which is described below.

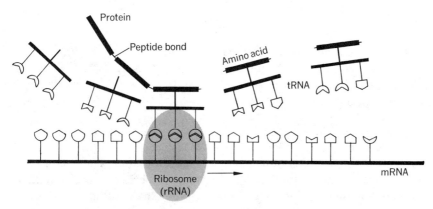

**Figure 28–2** Protein synthesis. Each tRNA is specified for a given amino acid. As the protein is formed, the tRNAs return to the cytoplasm to repeat their work.

## OPERON THEORY

A **gene** is the functional unit that comprises part of the chromosome and is responsible for the production of one protein. Since each chromosome contains thousands of genes, a control system must exist to determine when mRNA, hence synthesis should occur.

The controlling factors for mRNA synthesis are special genes which precede the codons for the amino acid constituents of the protein coded by the chromosome. Three types of **controlling sites** occur: the **regulator** (R), **operator** (O), and **promoter** (P) sites on the chromosome all control the **operon** (the set of codons that determine the actual amino acid sequence of the protein). In addition, an **inducer** may be present; its function will be discussed shortly.

The R continuously codes for **repressor** substance synthesis, which is diffused throughout the cell. When a cell is in a state that does *not* require the synthesis of a particular protein, the repressor binds the operator gene for that protein, and synthesis of mRNA does not occur. When an **inducer** is present (which occurs only when protein is lacking), it binds to the repressor and pulls it off the operator. This removes the inhibition on the operator, and synthesis occurs.

The process of synthesis is dependent on the factors above plus the characteristics of the **promoter** gene, which:

1. Binds to **RNA polymerase** (thus the enzyme "knows" which strand of DNA to use).
2. Determines the maximal rate of mRNA synthesis from the DNA.

The mRNA then travels from the nucleus into the cytoplasm. mRNA and tRNA are united at ribosomal sites to yield proteins. When the ribosome reaches the terminating codon on the mRNA (for which there is no tRNA), synthesis stops, but it may restart immediately if the operator is still unbound.

## ALTERATION OF THE CODE: GENE MUTATION

A change in the standard order of bases on the chromosome (and on the RNA made from it) is known as a **mutation,** and may be accomplished in several ways.

Three results of gene mutation are possible:

1. The mutation may be trivial and have no effect; this is extremely rare.
2. It may give rise to a different protein than originally coded for; if the new protein is used, it may have beneficial (rare) or harmful effects.
3. It may result in premature termination of synthesis.

Premature synthesis termination may have far-reaching effects. If a particular protein is not formed, a necessary bodily or cellular constituent may not be made, resulting in either impaired function or death to the organism. Also, a biochemical pathway, leading to the formation or breakdown of a particular substance, may be blocked (usually because a necessary enzyme is not formed). If a repressor is not coded for, a cell may go wild, synthesizing a protein that is not needed at that time. There are myriads of other results possible from gene mutation.

## Genetics and Heredity

The study of genetics deals with the origin of similarities and differences between parent and offspring. It is concerned with the nature of these similarities and differences, their sources, and how they develop. Briefly stated, genetics is a study of the transmission, from generation to generation, of developmental potentialities (genes) and how they come to expression.

We inherit from our parents the potentialities for various characteristics that can be observed and the determiners of many other characteristics that we never exhibit. When we discuss heredity, we do not refer to those traits that are common to all members of a group or species but to variations in body and mental characteristics that may be possessed by both parents and offspring. Thus we are not concerned with the fact that both a father and son possess hair, but with the color and texture of the hair.

In our study of genetics and heredity, certain terms will be used which are herewith defined.

Two genes at the same position (locus) in the homologous chromosomes are called **alleles**. For example, the gene responsible for color blindness is an allele of the alternative normal gene. An individual is **homozygous** for a trait when both the genes at corresponding loci in homologous chromosomes are identical. An individual is **heterozygous** for a trait if the two genes at any one locus in homologous chromosomes are different for a given trait. An organism may be homozygous for some pairs of genes and heterozygous for others.

In a heterozygote, the genes are usually of two types, **dominants** and **recessives**. A gene is dominant when the trait it represents appears in the individual; in such a case, its allele is said to be **recessive** (does not appear).

The term **genotype** is used to describe the sum total of genes inherited from both parents; these include both dominant and recessive genes. The term **phenotype** is used to describe the individual as shown by his expressed traits. For example, a person's genotype may include a gene for black hair (dominant) and one for blonde hair (recessive), and his phenotype will be black hair.

## REPRESENTATION OF THE GENOTYPE

Dominant genes are customarily represented by capital letters and recessive genes by lowercase letters. For example, if we are to study the trait of black hair in human beings, we might assign the trait the letter B. A homozygous dominant would have genotype *BB* (*two* letters, because a person receives a gene from each of his or her parents), a homozygous recessive would be symbolized as *bb*, and a heterozygote would be *Bb* (the dominant gene precedes the recessive gene in notation).

When two individuals are **crossed** (mated), it is represented as

$$P: BB \times Bb$$

In the notation above, the individuals crossed are a black-haired homozygote and a black-haired heterozygote. The P indicates that we are referring to the **parental generation.** Each successive generation is called a **filial generation** and is notated as $F_1$, $F_2$, and so on.

## MENDELIAN GENETICS

Gregor Mendel (1822–1884) was an Austrian monk who spent much of his time in what is probably the most famous garden in the world. Mendel raised pea plants, and through meticulous experimentation, laid the foundation for modern genetics.

One of his most famous experiments was concerned with height in the pea plants. Mendel **true-bred** several successive generations so that when any one late generation was **selfed** (mated or pollinated with others from the same generation), no phenotype changes were observed. He ended up with two different strains, one tall and one short. The selfing process eventually leads to a high probability of homozygosity, so we will assume that the tall and short plants were homozygous.

Mendel's cross, then, may be represented as:

$$P_1: TT \times tt$$
$$F_1: \text{all } Tt$$

The results were that all the offspring were tall. Mendel, however, did not stop there, for all he could do was observe phenotype; genotype is *inferred* from phenotype.

Mendel produced another generation by selfing the $F_1$ (thus producing $F_2$). The entire paradigm was:

$$P: TT \times tt$$
$$F_1: Tt \times Tt$$
$$F_2: TT, Tt, Tt, tt$$

Actually, Mendel inferred the above $F_2$ genotypes from the phenotypes tall and short. We see that the **phenotypic ratio** of the $F_2$ in this **monohybrid cross** (a cross in which only one trait is varied) is three tall to one short, commonly represented as 3:1.

In most crosses, the genotypic ratio is not the same as the phenotypic ratio. In the example above, the **genotypic ratio** was 1:2:1 (one *TT*, two *Tt*'s, and one

|  | ♀ Female | |
|---|---|---|
|  | p(0.5) T | p(0.5) t |
| ♂ Male p(0.5) T | TT | Tt |
| p(0.5) t | Tt | tt |

**Figure 28-3** Punnett square, crossing two *Tt* individuals. The probability of each potential zygote is computed by multiplying the marginal values. Thus the probability ($p$) of a *TT* is $0.5 \times 0.5 = 0.25$.

*tt*). These characteristic ratios (3:1 and 1:2:1) are found rather consistently in monohybrid crosses.

Through several series of such experiments, Mendel formulated two laws. The **law of segregation** states that there is a separation of the members of a pair of genes during maturation so that each gamete of an individual contains one gene from each pair. Thus, when two such gametes unite at fertilization, the two genes for each trait are brought together in the individual. Mendel's second law, the **law of independent assortment**, states that the distribution of each pair of genes to the gamete is entirely independent of the distribution of any other pair. [The second law, in light of more recent discoveries, was found to hold true only if the pairs of genes are on separate (nonhomologous) chromosomes. If Mendel had chanced to use traits that were linked on the same chromosome, many significant advances in genetics would have been substantially delayed.]

To compute the probability of obtaining any one type of genotype (or phenotype) in a genetic cross, one may use a **Punnett square**. Each possible gamete of one parent is listed across the top of the square, and the gametes of the other parent are listed along the left-hand side.

The Punnett square for the $F_1$ selfing example is shown in Fig. 28-3. Each square unit represents a possible zygote, and each, in this simple example, represents a unit of probability. Therefore, the probability of obtaining a *Tt* is 0.5, and for either *TT* or *tt*, it is 0.25.

## DIHYBRID CROSS

In the dihybrid cross, two traits are considered simultaneously. For our example, we will take the human traits of albinism (lack of integumentary pigmentation) and the ability to roll the tongue. Pigmentation is dominant over albinism (*A*, *a*), and rolling ability is dominant over nonrolling (*R*, *r*). Let us assume that the two parents are doubly heterozygous, that is, that both phenotypically show the dominant trait but both are **carriers**. The parents, then, must both be *AaRr*.

Mendel's laws tell us that each gamete produced by the parents has an equal probability of having either the dominant or recessive allele for each given trait. In the case of *AaRr*, that means that there is a 25 percent probability of *each* of the following combinations of genes in every gamete: *AR*, *Ar*, *aR*, *ar*.

The Punnett square can be used to compute the numerical probabilities of each genotype and phenotype by multiplying the probabilities of each parental contribution. A "4 × 4" square is used in a dihybrid cross since each parent can form four types of gametes (if they are heterozygous for both traits).

In this standard dihybrid cross, how many different *phenotypes* are there? In

|  | ♀ | | | |
|---|---|---|---|---|
|  | p(0.25) AR | p(0.25) Ar | p(0.25) aR | p(0.25) ar |
| p(0.25) AR | ① AARR | ② AARr | ③ AaRR | ④ AaRr |
| p(0.25) Ar ♂ | ⑤ AARr | ⑥ AArr | ⑦ AaRr | ⑧ Aarr |
| p(0.25) aR | ⑨ AaRR | ⑩ AaRr | ⑪ aaRR | ⑫ aaRr |
| p(0.25) ar | ⑬ AaRr | ⑭ Aarr | ⑮ aaRr | ⑯ aarr |

**Figure 28–4** Dihybird cross, a "4-by-4" Punnett square.

simple dominance, the heterozygote will exhibit only the dominant trait; hence all the following genotypes have the same phenotype, pigmented roller: AARR, AaRR, AaRr, AARr. Similarly, the phenotypes albino roller are aaRR and aaRr. Pigmented nonrollers are AArr and Aarr, and albino nonrollers are only aarr. Thus, there are four different phenotypes in a standard dihybrid cross (Fig. 28–4). The phenotypic ratio of this and other standard dihybrid crosses is 9:3:3:1.

## NON-MENDELIAN GENETICS

In all of Mendel's pea experiments, dominance was essentially complete, so there was no appreciable difference between the heterozygous and the homozygous individuals in the expression of a dominance trait: (*Tt*)tall peas could not be distinguished from (*TT*)tall peas. There are, however, many instances of lack of dominance; this is because the heterozygote is quite different from both the homozygous dominant and the homozygous recessive individuals.

One of the best illustrations of lack of dominance is provided by the blue Andalusian fowl. The "blue" appearance is due to very fine alternating white and black stripes. This type of fowl ($F_1$ in Fig. 28–5) is a hybrid (heterozygote from black and white-splashed parents). When blue Andalusians ($F_1$) are interbred, the offspring ($F_2$) are one white-splashed, two Andalusians, and one black. The number of visible types (phenotypes) here obtained is three instead of two (as would have occurred with complete dominance).

The expression of **incomplete dominance** or **partial dominance** is also used to describe cases of distinguishable heterozygotes, especially when the hybrids tend to resemble one parent much more than the other.

Among the phenomena that upset Mendel's dogma was the realization of **codominance**, such as exists in the MN blood-group systems of human beings, where two types of genes control the blood type (Table 28–2). In this instance, one gene does not preclude the expression of the other.

Another phenomenon that takes Mendel's dogma out for a walk is that of **incomplete penetrance**. Implicit in Mendel's assumptions is the statement that dominance is unfailingly expressed when it occurs. This is not always the case. In humans, the trait of polydactyly (extra fingers) is dominant over normalcy.

Incomplete penetrance is very important to our understanding of the interrelations of genes, phenotypes, and the environment, for it is the environ-

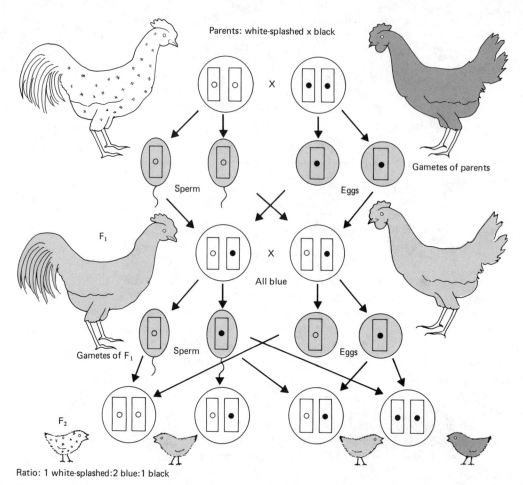

Ratio: 1 white-splashed:2 blue:1 black

**Figure 28–5** Lack of dominance. Cross between a white-splashed fowl and a black Andalusian fowl. The hybrid (heterozygote) is called a blue Andalusian.

ment which allows the expression of a gene—the actual inheritance of the gene is the inheritance of *potentiality*, not of a trait per se.

**Polygenic inheritance** is another non-Mendelian phenomenon. A prime example is weight in rabbits. If a Flemish giant rabbit of average weight 27 kg is mated with a Polish rabbit of 6 kg, the $F_1$ generation shows variability in weight in a range intermediate between the P generation's weights. The $F_2$ shows even more variation—the gradations approach continuity, and a Mendelian explanation is not possible.

The study of genetics is therefore still quite open to further investigations and discoveries. We will now study some of the areas where current knowledge of genetics have been put to use.

### MEIOSIS: SEX CELL GENETICS

**Meiosis** is cellular division that gives rise to the **sex cells.** You may recall that in mitosis, each cell divides to form two identical "daughter cells."

Meiosis differs from mitosis in three major respects:

1. Mitosis yields *two* cells for each original one. Meiosis yields *four* cells.

Table 28-2  Some Human Traits and Their Usual Mode of Inheritance*

| Trait | Dominant | Recessive | Several Genes |
|---|---|---|---|
| *Structural* | | | |
| Brown eyes | | | ● |
| Blue or gray eyes | | | ● |
| Premature grayness of hair | ● | | |
| Brachydactyly | ● | | |
| Skin color | | | ● |
| Albinism | | ● | |
| Polydactyly | ● R-P | | |
| Split hand "lobster claw" | ● | | |
| Extra teeth | ● | | |
| *Physiological* | | | |
| ABO blood groups | | | ● M-A |
| Ability to taste PTC (phenylthiocarbamide) | ● | | |
| *Psychological* | | | |
| Huntington's chorea | ● R-P | | |
| *Diseases†* | | | |
| Alkaptonuria | | ● | |
| Red-green color blindness | | ● S-L | |
| Deaf-mutism | | ● | ● |
| Nearsightedness (myopia) | ● | ● | |
| Absence of iris (aniridia) | ● | | |
| Bleeding (hemophilia) | | ● S-L | |
| Cancer of eye (retinoblastoma) | ● R-P | | |

* Key to the symbols: ●, inheritance of trait; R-P, reduced penetrance; S-L, sex-linked; M-A, several alternative genes; if more than one mode of inheritance is given for a trait, it means that in some families it is inherited in one way and in others in another.

† This is an artificial classification, because these diseases actually fall into structural or physiological categories.

2. Mitosis produces daughter cells with the same number of chromosomes as the mother cell (**diploid**). Meiosis produces daughter cells with *half* the chromosomes of the mother cell (**haploid**).
3. In mitosis only one division occurs. In meiosis there are *two* divisions, the first is a **reductional** division, the second being an **equational** division.

These differences explain why a zygote developed from a sperm cell and an egg cell will result in a new individual with the same number of chromosomes (rather than twice as many) as each parent. The reduction–division part of meiosis includes a random sorting of the four cells into pairs. This random assortment accounts for the variations that occur between siblings of the same parents.

## INHERITANCE OF SEX

In humans, two particular chromosomes are the determinants of sex: the X and Y chromosomes. A normal female has two X chromosomes and hence can produce only X eggs. A male human, however, has one X and one Y chromosome and can produce two types of sperm (with respect to sexual characteristics), X and Y. Generally, the probability of having a male or female child is thus 50 percent for each: 50 percent probability of the zygote receiving an X from the father, 50 percent of receiving Y from the father, and 100 percent of receiving X from the mother.

In many other organisms, sex determination is different from that in human beings. For example, in *Drosophila melanogaster*, a fruit fly used extensively in genetic studies, males are XY and females are XX *but* males may also be XO (the "O" means that no other sex chromosome is present). The latter results from fertilization of a **nullo-X egg**, one that contains no sex chromosomes. This was discovered in 1916 by C. B. Bridges, who correctly concluded that in *Drosophila*, the number of X chromosomes, not the presence of the Y chromosome, determined the sex of the zygote. The Y chromosome in *Drosophila*, however, does perform a sexually related function. It determines **fecundity** (fertility) in the male; an XO *Drosophila* will be a phenotypic sterile male.

## HUMAN SEX CHROMOSOMAL ABNORMALITIES

Sometimes, during meiosis, the sex chromosomes may not separate in either the first or second division. This **nondisjunction** may result in an egg containing

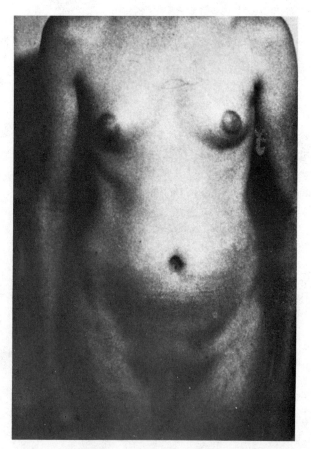

**Figure 28–6** Klinefelter's syndrome. Note the presence of both female and male characteristics in this XXY individual.

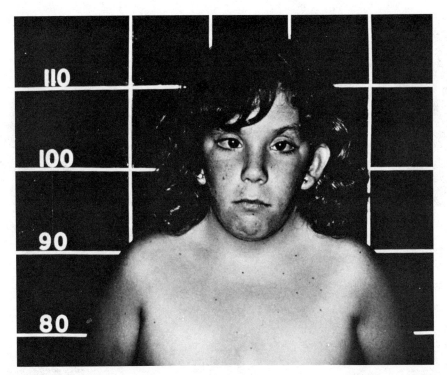

**Figure 28-7** Turner's syndrome.

two instead of one X chromosome, in a spermatid containing an X *and* Y (instead of either one or the other), or in an egg or sperm with no X or Y. The results of such disjunction may be manifested in Klinefelter's syndrome, Turner's syndrome, or others.

In **Klinefelter's syndrome** (Fig. 28-6), the individual gets two X chromosomes and one Y (XXY). The individual is phenotypically a male, since the Y chromosome in humans determines sexuality, but he is sterile and may show signs of secondary female characteristics (such as breast development). Mental retardation may be included among the symptoms resulting in this syndrome.

In **Turner's syndrome** (Fig. 28-7), an individual has only one X chromosome and no Y chromosome (XO). Phenotypically a female, a person with Turner's syndrome either lacks ovaries or has vestigial ones; additional symptoms may include an abnormally short stature, a webbed neck, and mental retardation.

Other sexual chromosome abnormalities also exist (Figs. 28-8 and 28-9), but will not be discussed here.

## SEX-LINKED INHERITANCE

The X chromosome, in addition to bearing genes influencing sex, also carries genes for many other traits. In cases of color blindness, for example, a color-blind father mated to a normal mother has no color-blind children. This is because the XY zygotes develop into normal males possessing one normal X chromosome and one Y chromosome, *which carries no gene for color blindness*. Similarly, the XX zygotes develop into normal females, since only one X chromosome bears the gene for color blindness, which is *recessive* to the normal condition. In the $F_2$ generation, however, half of the grandsons and half of the granddaughters are free of this defect; and the other half of the grand-

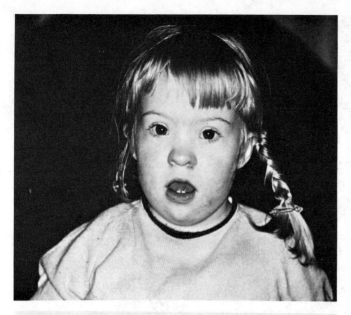

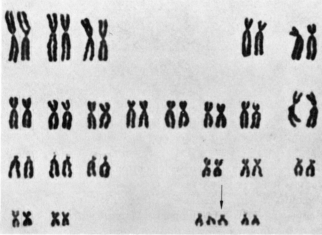

**Figure 28–8** Down syndrome. Note the three No. 21 chromosomes (arrow), making the total chromosome number 47. An error in meiosis, called nondisjunction, wherein an extra No. 21 chromosome is included in the genome produces this tragic syndrome.

daughters carry the gene for color blindness as recessive, whereas the other half of the males are color-blind, having no normal gene in the Y chromosome (Fig. 28–10).

There are many sex-linked traits; the genes of over 30 are known for human beings, and about 150 sex-linked genes have been found in the fruit fly.

## Population Genetics

This section deals with the *macro*scopic effect of individual variation. The mathematically based discipline of **population genetics** is attentive to the genetic variations within and among specific groups called *populations*. The common definition of a population is geographical—a set of boundaries is demarcated, and all organisms within the confines form a population.

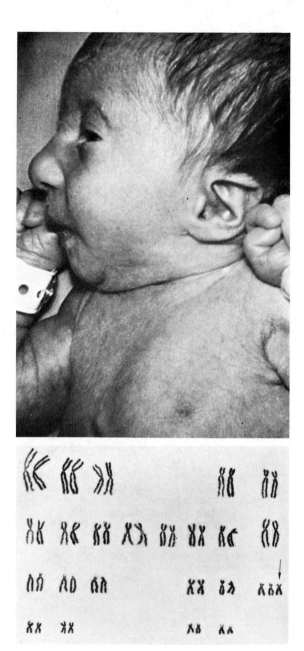

**Figure 28–9** Trisomy E syndrome. Note the large occipital lobe and the extra No. 18 chromosome (arrow).

The genetic definition of a population is much more specific. Several prerequisites must be met for a group to satisfy the stringent criteria of an ideal population:

1. The group must be capable of interbreeding.
2. The group must share a common **gene pool**—a group of all forms of every gene present in all gametes within the population. This implies that each member of the population must be able to combine its genes with any others present in the population and produce an offspring possessing that combination.

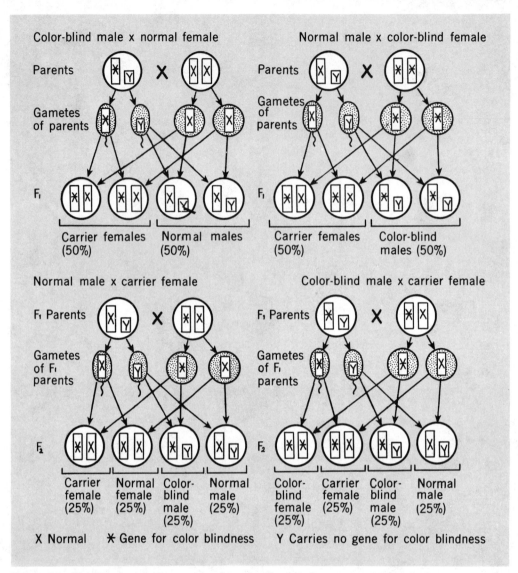

**Figure 28-10** Human inheritance of red-green color blindness over two generations.

3. A genetic population represents a level between the individual and the species.

## POPULATIONS AT GENETIC EQUILIBRIUM

The greatest interest in population genetics concerns change over time in the genetic character of the population. Before we can discuss that aspect, we must understand the population at **equilibrium**.

Certain prerequisites exist for an *ideal* population at equilibrium:

1. Either no mutation occurs, or the mutation *rate* is equal in both directions.
2. Members of each sex have equivalent mortality (death) rates.
3. Each adult member is capable of reproducing.

4. Mating is random (**panmixia**).
5. Zygote formation (fertilization) is at random.
6. Every zygote is equally viable.
7. Members of each sex are equally fertile.
8. Generations do not overlap (cannot simultaneously reproduce).
9. Segregation is Mendelian (nondisjunction, etc., do not occur).
10. The population approximates infinite size.

Obviously, these prerequisites cannot all be met. Although the study of ideal equilibrium populations is not of much *direct* use, it does provide some basic concepts of great interest to genetics. One of the more basic and important concepts is **gene** or **allele frequency**.

## ALLELE FREQUENCY

The allele frequency of the dominant allele (or gene) is traditionally given as the letter $p$, and the frequency of the recessive allele is given as $q$. By definition, then,

$$p + q = 1$$

Given a dominant–recessive dichotomy, the frequency of a particular allele can be determined by dividing the number of alleles of the $p$ or $q$ type in the population by the total number of alleles of both $p$ and $q$ types in the population.

## HARDY-WEINBERG LAW

The **Hardy-Weinberg law**, put forth by the English mathematician G. Hardy and the German physician W. Weinberg in 1908, is very useful in genetic studies. By using it, one can determine whether a trait is coded for by a recessive autosomal gene or by a sex chromosome.

If one has an estimate of the allele frequencies $p$ and $q$ and assumes that the population is at genetic equilibrium, percentages (probabilities) of genotypic ratios may be produced using the Hardy–Weinberg law. The basic formula used is

$$p^2 + 2pq + q^2 = 1$$

It is a simple matter to compare the predictions to the observations. If the deviations from computed expectations are statistically minimal, it is warranted to assume that the trait was coded for by a recessive autosomal gene.

# Methods and Advances in Genetics and Heredity Studies

## LINKAGE, CROSSOVER, AND CHROMOSOME MAPS

In *Drosophila*, the fruit fly, two of the many genes identified thus far are C, the gene for wing shape, and B, the gene for body color. Both represent cases of

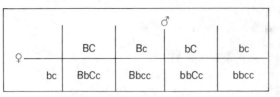

**Figure 28-11** Possible zygotes of *BbCc* × *bbcc* mating.

simple dominance: *C* codes for normal wings; *c*, its recessive counterpart, for curved wings; *B* codes for gray body, and *b* codes for black body.

Assume that a heterozygous individual, *BbCc*, mates with a homozygous recessive individual, *bbcc*. According to Mendelian theory, the heterozygote can make four kinds of gametes: *BC*, *Bc*, *bC*, and *bc*. The homozygous recessive can only make *bc*.

The Punnett square in Fig. 28-11 provides the possible zygotes. The four corresponding phenotypes, in order, are gray body, normal wing; gray body, curved wing; black body, normal wing; and black body, curved wing. Assuming equivalent allele frequencies, we would expect equal numbers of each of the four phenotypes to occur. In an *ideal* cross, however, drastically different results are found. *Only* the genotypes *BbCc* and *bbcc* (and their corresponding phenotypes, from which the genotypes are inferred) are found.

As mentioned earlier, genes are located on the chromosomes. This implies that homologous chromosomes contain the same genes (although in either dominant or recessive forms) and since chromosomes are essentially linear, a definite *order* must exist for the genes along the DNA strand. If we postulate

**Figure 28-12** Possible gametes of heterozygote and homozygous recessive, as described in text.

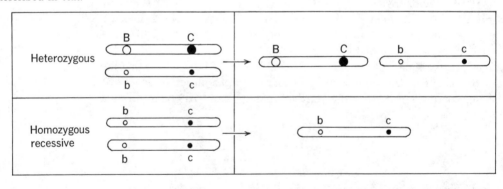

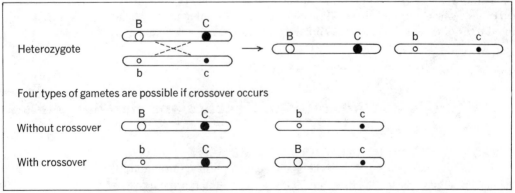

that genes *B* and *C* occur on the *same* chromosome, the seemingly deviant results above can be easily understood. In Fig. 28–12, we see that the double heterozygote can only produce two types of gametes upon anaphase chromosomal splitting, and the homozygote can only produce one type. Thus, there are only two combinations of gametes possible to form the zygote, *BbCc* and *bbcc*. The genes *B* and *C* are said to be **linked.**

A real (instead of ideal) cross between a heterozygote and a homozygous recessive, however, does not produce such neat results. Instead, all four genotypes are observed, but their proportions are not as Mendel would have predicted.

Generally, the vast majority of offspring are *BbCc* and *bbcc*, but some few turn out to be *Bbcc* and bbCc. This is explained by the phenomenon of **crossover.** The crossover leads to the production of two new gamete types. These may occur in addition to normal types, for not every pair will cross over. These four gamete types, in combination with the homozygous recessive's gamete type, will produce the observed kinds of offspring. Since the occurrence of crossover is much less frequent than its nonoccurrence, the observed greater portion of *BbCc* and *bbcc* is explained.

Another aspect of cross over is **crossover frequency.** Morgan and Sturtevant postulated that closer genes on a chromosome would experience crossovers between them less frequently than genes farther apart. Because of that, the percent crossover, monitored by the percentages of "unexpected" genotypes in offspring, would be directly related to the distance between genes on the chromosomes. Hence, gene positions could then be determined and drawn on a **chromosome map.**

The study of population genetics, therefore, can be a useful tool for calculating the accuracy of new genetic theories. Before closing, a few of the more recent advances in genetic studies will be presented.

## TWIN STUDIES

Twins occur about once in every 85 human births, and are of particular genetic interest. There are two types: fraternal and identical. *Fraternal twins* develop from two separate zygotes and hence have different genetic constitutions. *Identical twins,* on the other hand, arise from a single fertilized egg, which splits into two at some time during embryonic development. Since they arise from one egg, identical twins have the same genetic constitution; they are always of the same sex, and their characteristics, both physically and mentally, are remarkably similar.

Identical twins have therefore been used in various genetic studies; one is kept in a "normal" (*control*) state and the other is exposed to the variable in question. All differences that result can be assumed to be due to environment rather than heredity.

## THE TESTCROSS

A **testcross** is a genetic manipulation in which a *known* homozygous recessive (for the trait in question) is mated with an individual with unknown genotype. If the unknown is homozygous dominant, all offspring will be heterozygous for the trait. If the unknown is heterozygous, half the offspring will exhibit the dominant trait and half will not. If the unknown is homozygous recessive, all offspring will be identical to the parents with respect to the trait in question.

CELL FUSION

Recently, a new technique called **cell fusion** has proved useful in finding additional linkage groups in human chromosomes. In this method, mouse and human cells are made to fuse into a single hybrid cell by infecting them with a special virus that has been inactivated with ultraviolet light. In these hybrids, the DNAs from both the mouse and human nucleus remain active in synthesizing RNA and protein. However, as the hybrid cells undergo mitosis, some of the human chromosomes (and the genes they contain) are eliminated from the dividing cells. Investigators can determine by experiments which enzyme activities no longer occur in these cells and hence which group of genes that code for these enzymes has been lost.

Additional linkage groups are being explored in this manner.

# Eugenics

**Eugenics** is the science concerned with the application of the principles of genetics to the "improvement" of the human species. The principles of heredity are as applicable to man as to other organisms, but the practical difficulties of actually applying these principles, at least to the conservation of the human species, are very great.

The study of human genetics is difficult for several reasons: (1) the impracticability of experimental breeding, (2) the small number of offspring in a family, and (3) the relatively slow breeding of human beings.

Despite all the handicaps, significant achievements are being made in this field, as is evident from the fact that in the past 25 years, centers for counseling on heredity have been established as a practical way of giving information to people with genetic problems.

Physicians are realizing more and more that human genetics has an important bearing on clinical problems, and relatively rapid progress is being made in the accumulation of information, which, when applied, proves helpful in the prevention and diagnosis of disease.

# Summary

The DNA molecule is the carrier of genetic information. The RNA molecules, transcribed from DNA, translate the modified genetic code into amino acid sequences (protein).

The operon theory describes current thought on how protein synthesis is controlled. Regulator, operator, and promoter sites on the chromosomes interact to "decide" which genes are to be active at what times.

Genetics is the study of the transmission of genes, and of the genotypic and phenotypic results of such transmission. Pioneering work in genetics was done by Mendel. The identification of dominant–recessive relationships as well as the idealized mechanism of genetic transmission is attributable to him.

Mendel's laws have been shown to be incomplete by the subsequent discovery of partial dominance, codominance, sex inheritance, and other phenomena. They are still, however, the basics upon which the science of genetics has been built.

The mathematically based study of population genetics has great utility in most of the sciences. Populations at equilibrium, none of which really exist, provide the basis for the study of nonequilibrium populations.

Twin studies, the testcross, cell fusion, and the science of eugenics are currently extending our knowledge of genetic phenomena.

That extension underscores a general feature of human scientific intellect—with greater knowledge comes greater understanding, and understanding is the key to successful manipulation and acceptance of ourselves and our environment.

# 29

# Organic Evolution

Charles Darwin and Alfred Russell Wallace independently realized the observable facts of organic evolution—the development of existing plants and animals from their simpler ancestors. The scientific evidence supporting the organic evolution theory is overwhelming. Some previously dim areas of these mechanisms have been illuminated by recent advances in genetics, so there is little doubt that the main evolutionary mechanisms have been identified, although the extent of the role of each element in evolution is uncertain. Research from all areas of biology has helped to fit pieces into the evolutionary picture, and organic evolution is now universally accepted in the world of science.

## History of Early Evolutionary Thought

Where did the first animals and plants come from? How did they come to look and act as they do? These questions have been the subject of speculation throughout history. The most frequently proposed answers fall roughly into two categories: fixed creation and evolution.

The idea of **fixed creation** of life, or that life originated in its present forms usually under the influence of a supernatural force, has been shared by many myths and religions of the past, and into the present. The evidence in favor of this idea is obvious. Cows give rise to cows, dogs to dogs, and human beings to human beings. If one sees no marked change in a species in the lifetime of the human observers, or even after many generations of human beings, how could one reasonably expect members of one species to produce offspring of another species?

Fixed creation was not an unreasonable belief, for it was rooted in the everyday experiences of life. But, as science uncovered the secrets of nature, many commonly held beliefs were revised.

A more systematic approach to the question appears to have begun in the sixth century B.C. with the Greek philosopher Thales, who, noting that all living things contained water, concluded that animals arose in and from the sea. Aristotle attempted anatomically to classify organisms and noted the natural progression of organisms through the vivicum.

In the seventeenth century little was known of biology or the fossil record, and fixed creation was a scientifically reasonable hypothesis. When the hypothesis became part of Christian dogma, it was no longer considered a hypothesis subject to verification, and open investigation into the question thus became impossible; anyone bold enough to suggest the possibility of the alternative hypothesis, evolution, was branded a heretic. As the sanctions against such ideas were relaxed, some scientists advanced evolutionary theories. Georges-Louis Buffon (1707–1788), a French natural history writer, published evidence that species evolve from other species, as the result of inherited environmental modifications. He is also responsible for our present definition of a species: a group of organisms that can interbreed. Eramus Darwin (1731–1802), grandfather of Charles, expressed in poetry and prose a belief in organic evolution and the theory of inherited environmental modification.

Jean Baptiste Lamarck (1744–1829), a French naturalist, also proposed a theory of evolution. According to Lamarck, modifications due to use and disuse, or to direct action of environmental factors, are passed on to the offspring. Such inherited modifications would, for example, account for the long necks of giraffes, acquired after generations of foraging in the leaves of tall trees, and would explain the degeneration of certain organs, such as the eyes of cave-dwelling animals. Many investigators have attempted to obtain experimental evidence of inheritance of acquired characteristics without success.

## Darwin and Wallace

It might seem remarkable that two men, Charles Darwin (1809–1882) and Alfred Wallace (1823–1913), independently and from studies in different parts of the world, arrived at the same conclusions. It does, however, seem less remarkable if one keeps in mind that information about the physical and biological world had been accumulating, and it only remained for someone familiar with this knowledge to integrate and interpret it, although its actual presentation was the most dramatic new concept in modern history. We shall follow Darwin through his development of organic evolution theory, and return to Wallace later.

As a boy, Darwin liked to collect stones, insects, and plants. Later he turned to collecting rats and hunting game. His father, anxious over Charles's lack of interest in school, sent him to medical school. Charles detested the gore and was later sent to study for the clergy. Upon graduation he received an invitation on the H.M.S. *Beagle* as an unpaid naturalist. But his father offered great resistance: he wanted Charles to settle down in a career. Yet in 1831 Charles managed to sail away, reading Lyell's *Principles of Geology*. This book had a great influence on both Darwin and Wallace.

Lyell had rejected the prevailing biblical notions of the creation of earth, arguing that the earth's origins and present condition could be better understood in terms of *continuous, predictable, natural* causes, such as the action of rain, wind, volcanoes, and earthquakes. He further argued that these forces have been operating for millions of years and that the geological changes they produced led to the extinction of certain species.

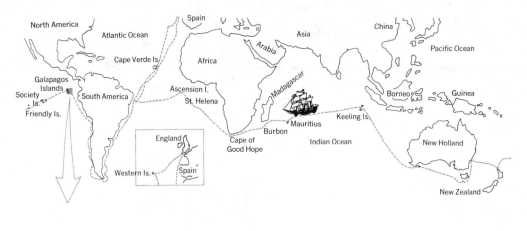
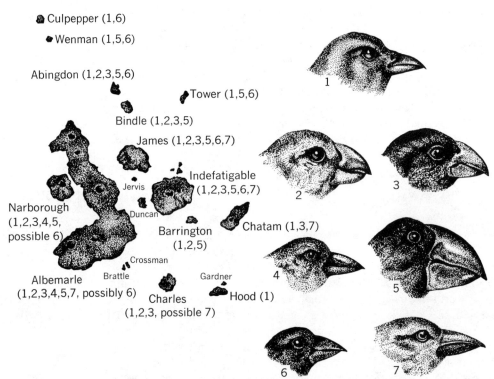

**Figure 29–1** Darwin's voyage to the Galapagos Islands on the H.M.S. *Beagle*. *Lower left:* numbers next to names of islands indicate which finches (*at right*) are inhabitants. The seven finches shown (out of an original 13) exhibit beak modification for diverse feeding habits.

During Darwin's five-year travels he kept detailed records on the nature and distribution of the local plants, animals, and fossils; he collected many specimens. His observations undermined his belief in fixed creation. If species were static, one would expect living species to be exactly like fossil ancestors, but in many cases they differed greatly. The distribution and different forms of organisms that he encountered in the Galapagos Islands also seemed unreasonable according to the fixed creation theory (Fig. 29–1).

Back in England, he carefully examined the finches he had collected and was struck by the fact that closely related but distinct species of finches, with a perfect gradation of structure, inhabited neighboring islands and resembled closely those of the nearest continent. If *one* mainland species had settled on an island and diversified into existing forms, the distribution of these finches could be understood. From such observations Darwin began to doubt the hypothesis that species were immutable. But if species were mutable, how did they change?

Darwin knew that animal breeders were able to produce new breeds of organisms by selecting and mating individuals that possessed the desired traits most fully. But in nature, what acts as the selecting force, what serves as the breeder? The answer came from Thomas Malthus's *An Essay on the Principle of Population* (Wallace was also influenced by this work). Malthus stated that the limited amount of food resources has continually kept the population in check by famine, war, and disease. Darwin realized that population pressure on resources must act on all organisms and that in the struggle for these resources, organisms with any advantage will more often succeed in reproducing. Variations that give an organism any advantage in its environment, and which are passed on to the offspring, will over a period of time give rise to a population quite different from its ancestors. In this way, the environment acts as the breeder, selecting organisms best suited for survival and producing a species better **adapted** to its environment.

Darwin realized this in 1838, but would not consider presenting his theory until he was certain it was correct and had substantiated every point with voluminous evidence. It was, in fact, this drive for an enormous compilation of data that separated his work—and his influence—from those who went before him. Others suggested, but Darwin marshalled the evidence and overwhelmed the massive opposition his ideas faced. In 1844, he wrote an essay on **evolution** in preparation for future ones which were to contain vast stores of evidence from the examples he gathered. But in 1858, his plan of presentation was hurried along when he received a letter from Wallace containing an evolutionary theory identical to his own. That same year, Darwin presented his paper jointly with Wallace's before the Linnean Society.

In 1859, Darwin's *On the Origin of Species* was published, a work he considered an "essay" or preliminary outline! Its bombshell effect echoed and reechoed throughout the Western world. As with previous revolutionary ideas, evolution proved to be deeply disturbing and frightening. It was rejected much as Copernicus's theory that the earth orbits the sun was rejected, because it implied that man's planet was not the center of the universe. Darwin's theory seemed to imply that it was not even man's planet—that man was simply here for no particular reason, and he was nothing but an animal. These philosophical implications have no bearing on the validity of the scientific theory, of course, but they were a focus of intense opposition on moral and religious grounds for many years, until the overwhelming evidence forced widespread acceptance of the evolution concept—although a recent resurgence of opposition on religious (or more properly, antiscience) grounds shows that we have not advanced as far from irrational behavior as we may have believed.[1]

---

[1] A recent California textbook ruling on "equal time" for pseudoscience (creationism) has resulted in new California high school biology books that scrupulously avoid even mentioning evolution.

# Evidence in Support of Organic Evolution

The compelling evidence for organic evolution has been derived from many scientific disciplines. The principles of evolution are based on a number of different types of evidence, some of which are presented briefly below.

## COMPARATIVE MORPHOLOGY

The study of comparative anatomy brings out similarities and dissimilarities in structure and offers much evidence in favor of organic evolution.

Vestigial organs, which are especially evident among vertebrates, furnish striking evidence of changes of ancestral conditions. Man's eye, for example, has a vestigial nictitating membrane, and the modern horse possesses splints in place of what in its ancestors were functional metacarpals and metatarsals.

## EMBRYOLOGY

The **biogenetic law** was formulated many years ago by the German biologist Haeckel, who observed that early stages of vertebrate embryos showed remarkable similarity. He concluded that the phylogenetic history of the species was represented in the various stages of the animal's embryonic development: **ontogeny repeats phylogeny.** This broad generalization was not entirely justified. A more accurate statement of this law is that the embryo of an animal may resemble *embryonic stages* of other animals on the phyletic scale (but not the adult form) (Fig. 29-2).

## COMPARATIVE BIOCHEMISTRY

Within recent years the discovery has been made that the biochemical characters of animals furnish convincing evidence in favor of the theory of organic evolution. For example, studies have been made of the crystals formed by the hemoglobin of blood, and comparisons show that the crystals of closely related species are more nearly alike in form than those of distantly related species.

Additional evidence of animal relationships is provided by the degree of similarity between the blood protein of various animals. Blood tests have shown that human beings are genetically closest to the great apes; next closest

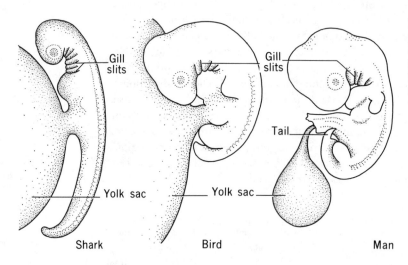

**Figure 29-2** Comparative embryology shows gill slits in three classes of vertebrates. They persist only in the adult fish.

in order are the Old World monkeys, the New World monkeys, and the tarsioids. Blood tests given to other animals give proof that dogs, cats, and bears are closely related, while sheep, goats, cows, antelopes, and deer form another closely related group. Sea lions and seals are more closely related to the carnivores than to other mammals, as anatomical observations also indicate.

## GENETIC INFLUENCES

Since Darwin's theory was universally accepted by scientists, certain questions have arisen which led the way toward a greater understanding of evolution. For example, how did variations arise in sufficient numbers to account for evolution? Assuming variations were diluted by blending effects in successive generations, Darwin postulated an impossibly high mutation rate.

Studies in genetics have shown that characteristics are not lost in blending as Darwin thought, but that they appear in later generations unchanged. Mathematicians have come to the aid of biologists and initiated a new era in evolutionary thought. The statistical treatment of genetics in populations has proved to be a powerful tool in gaining a more complete and detailed understanding of factors influencing evolution.

Evolution is now broadly seen as any change in allele frequency (gene frequency). Gene and chromosome mutations, and new combinations of these, produce the units of genetic variation but do not significantly alter the allele frequency. The frequencies of these new units are altered by other influences (**mutation pressure, meiotic drive, gene flow, genetic drift,** and **selection**).

# Modern Theory of Evolution

Much of Darwin's work provides a framework for the modern theory of evolution.

## SPECIATION

**Speciation** is the general term for the evolutionary change in species characteristics. Adaptations, adaptive radiation, and adaptive convergence are all types of speciation. Speciation, when complete, results in the formation of a new species.

ADAPTATION. That animals are adapted to their environment is obvious to anyone who has considered their structure, physiology, and habits. Structural adaptations are abundantly exhibited by all animals (Fig. 29–3). They must fit (adapt) a species for life in its environment, or that species cannot survive in the struggle for existence. The principle of organic evolution and the evidence in favor of that principle depend on the assumption that the degree of adaptability to the environment determines whether an animal shall live or die in the struggle for existence.

ADAPTIVE RADIATION AND ADAPTIVE CONVERGENCE. Darwin's finches on the Galapagos Islands present a beautiful example of **adaptive radiation** (Fig. 29–4). The birds are thought to have originally been one species when they came to one of the islands. Soon, some of them moved to nearby islands in the Galapagos group, and eventually several different forms emerged, each well suited to the particular island's peculiarities.

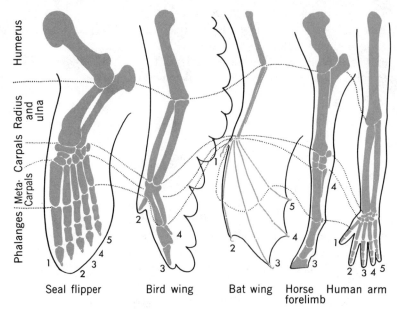

**Figure 29-3** Homology and adaptation of forelimb bones in vertebrates. The limbs share a fundamental structure based on common descent, but are adapted for special functions.

**Adaptive convergence** or **convergent evolution** may be illustrated by the streamlined body of a fish (shark), an extinct reptile (ichthyosaur), and a mammal (porpoise). Natural selection has favored in aquatic animals, whether they be fish or mammals, those characteristics that fit them for life in water. As far as body form is concerned, the ancestors of a fish and a porpoise were probably much less similar than these animals are today.

## MUTATION PRESSURE

Mutations must alter the gene frequency, thus exerting **mutation pressure.** For example, a mutation from A to B would certainly change the frequency of A and B. But since mutations occur rarely (usually less than 1 in 10,000), their effect on gene frequency, unless tied to selection, is probably negligible.

## MEIOTIC DRIVE

Through meiosis, a heterozygous male, $Aa$, will produce equal proportions of sperm containing $A$ and $a$ alleles. However, if the sperm from this male were to contain only the $a$ allele (with little or no $A$ sperm produced), *the frequency of $a$ would increase in the gene pool.*

Alleles exist (called *segregation distorters,* SD) that result in all sperm containing only the SD allele. One may think of SD genes as alleles that perpetuate themselves by eliminating competition. Although uncommon in natural populations, SD genes do occur. The extent of their role in producing evolutionary change is not known.

## GENE FLOW

Gene frequencies can be greatly changed by **immigration.** If members of one **deme** (breeding population) receive breeding guests from neighboring demes (of the same species), the change in gene frequency will be related to the proportion of breeding immigrants and the extent to which the immigrant's gene pool differs from that of the host deme's. The larger these differences are, the greater the changes in gene frequency will be. However, when genetic

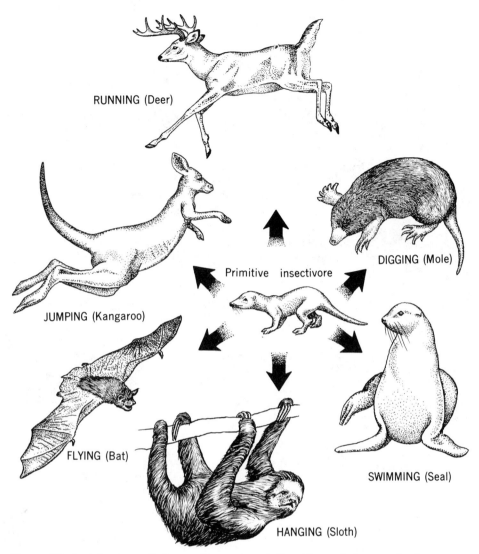

**Figure 29-4** Adaptive radiation or divergent evolution. Mammals have evolved from a common primitive insectivore ancestor (a five-toed land mammal). Note the adaptation of limbs to a wide variety of environments.

differences are too large (from mutations with increased frequencies), mating results in either sterile offspring, or in no offspring (thus speciation is complete).

## NATURAL SELECTION

Allele frequencies may also be changed when organisms with particular genotypes are prevented from being fully represented in the next generation. Organisms more successful in reproduction cause their genotypes to be more prevalent in the population, and hence they will alter the allele frequencies of the gene pool. The process that determines which organisms will reproduce and which will not is referred to as **natural selection.** Selection may result from many circumstances.

Let us consider a normal distribution curve for a trait, nose length, in a hypothetical population to illustrate some ways in which selection operates (Fig. 29–5).

Most organisms will have noses of an average length that we will arbitrarily call 5 cm. Fewer individuals will have extremes of 1 cm and 9 cm. If for some reason extremes are selected against, **stabilizing selection** will cause fewer individuals with extremes of nose length to become parents, in favor of more with the average nose. If the long nose is seen as attractive and is selected for, and the short nose is selected against, **directional selection** would shift the number of progeny individuals to the right of the graph to produce a population tending toward longer noses. If both short and long noses are selected for, **disruptive selection** would be in operation, resulting from a splitting of the curve between the extremes. Stabilizing selection may act to sustain a heterozygous condition, whereas disruptive selection can give rise to new species, especially when the diverging phenotypes reproduce selectively against themselves.

Any and all of these influences may interact to produce new genotypes and hence to produce evolutionary change. The interactions are complex and new

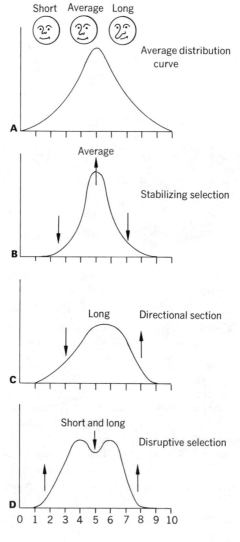

**Figure 29–5** Types of selection (arrows directed up are for selection, directed down, against). (A) Before selection, average distribution curve of nose length in hypothetical population. (B) After stabilizing selection (most individuals with average noses). (C) After directional selection (most with long noses). (D) After disruptive selection (most with either long or short, fewer with average, nose length).

models must be devised to get a total picture of evolution. As we turn to some of the examples of organic evolution, keep in mind the underlying genetic notions presented here and try to imagine what influences were operating to produce the observed phenomena.

## Fossil Animals: Historic Record of Evolution

From Darwin's time to the present, the remains of ancient animal life (fossils) have been regarded as strong evidence of evolution. Fossil remains of animals are usually *petrifactions*—parts have been replaced by mineral matter. The hard parts of the animals, such as bones, shells, and teeth, may be preserved intact; and parts of or entire animals may be preserved in ice (frozen mammoths in Siberia), amber (insects), asphalt (saber-toothed cats), and tar or oil-bearing soil (mammoths) (Fig. 29–6).

From evidence obtained from these fossils, paleontologists have constructed a table showing the geological periods, arranged in order of their succession, and the approximate time of origin of the different animal groups.

The most convincing evidence derived from paleontology results from investigation of the lines of descent of single groups, such as camels, elephants, and horses.

### EVOLUTION OF THE HORSE

One of the best paleontological lines of evidence of organic evolution is furnished by our knowledge of the evolution of the horse. The horses now living in America were descendants of domesticated animals brought to this country by the early settlers from Europe. In prehistoric times, the ancestors of our modern horse were native here, and some of the finest fossil remains of these ancestors have been found in America.

The evolution of the horse has been traced back through many distinct stages, extending through the Age of Mammals and the Age of Man. A brief

**Figure 29–6** Fossilized *Turritella* shells.

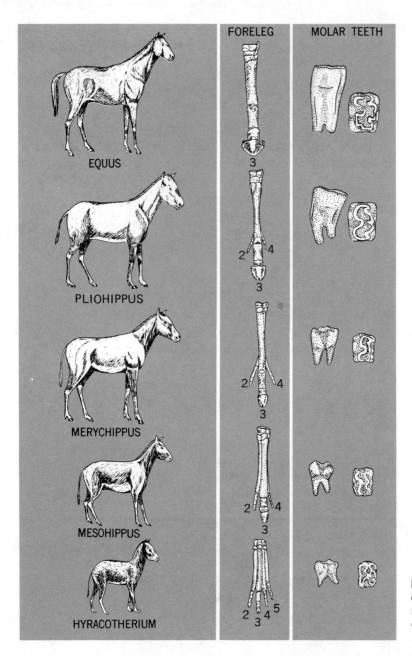

**Figure 29-7** Evolution of the horse. Digits or rudiments (splints) are designated by numbers 1-5.

description of five of these stages, together with Fig. 29-7, will serve to illustrate the principal changes that took place during this evolution.

**Eohippus** (dawn horse) was only 11 inches high and lived during the lower Eocene. Eohippus (also called Hyracotherium) lived in North America and Europe and was a browsing forest dweller. Its forefeet had four complete toes each but no trace of the first (thumb); the hind feet had three complete toes each and the rudiments (splints) of the first and the "little" (fifth) fingers.

**Mesohippus** (intermediate horse) belonged to the Oligocene period and was about the size of a sheep. Its forefeet possessed three complete toes each, and the fifth digit was represented by a splint; the hindfeet also possessed three complete toes each but no splint. All three toes touched the ground, but the middle toe was larger and bore most of the weight of the body.

**Merychippus** (ruminating horse) lived from the Miocene to the Pliocene periods. It marked the transition from the primitive horse to the modern horse. The milk teeth were short-crowned and fully cemented grinders, suited to the harsh vegetation of the plains. Both its forefeet and hindfeet possessed three toes each.

**Pliohippus** (Pliocene horse), from the Upper Miocene and Pliocene, was the first one-toed horse. Both forefeet and hindfeet were one-toed, and the second and fourth toes were represented by splints. The crowns of the upper molars were similar to those of the modern horse, but they did not possess as complex a pattern of ridges on the surface. Pliohippus had a height of some 40 inches, about the size of a modern pony.

**Equus** (horse) is the modern horse of the Pleistocene and recent epochs. The first and fifth digits are entirely absent, and the second and fourth are represented by splints. The third toe alone sustains the weight of the body. The crowns of the molar teeth are long, with complex enameled ridges well adapted for grinding dry, harsh vegetation.

The lengthened skull is accompanied by a larger and more complex brain. This horse is about 60 inches tall, considerably larger than any of its ancestors. The evolution of the horse has resulted in the development of an intelligent, long-legged, swift-running animal that is suited to live and feed on the open grasslands.

# Origin of Life

Several questions arise when studying evolution, the most common being that if all forms evolved from previous forms, where did the first life come from? And where is evolution in evidence today? This section will attempt to deal with these important questions.

## SPONTANEOUS GENERATION

Before the end of the seventeenth century, the theory of **spontaneous generation** or **abiogenesis** of animals was accepted by both scientists and philosophers. According to this idea, many species of animals were supposed to arise spontaneously from nonliving matter. Fly maggots, for example, were supposed to have come from decaying flesh.

The classical experiments of the Italian F. Redi in 1680 exploded this theory. Redi placed meat in wide-mouthed flasks that were either left open or covered with gauze or paper (Fig. 29–8). Flies entered the open vessels and laid eggs, which hatched into maggots. No larvae developed in the meat of the other vessels; but on the cloth of those covered with gauze, flies laid eggs that developed maggots. It was thus concluded from this experiment that maggots arose from the eggs of flies and not from decaying meat.

## BIOGENESIS

The theory of spontaneous generation or abiogenesis thus gave way to **biogenesis,** which maintains that all life arises from preexisting life. Since both

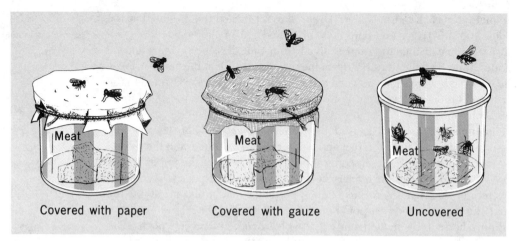

**Figure 29-8** Spontaneous generation was disproved by Redi's experiments with blow flies.

geologists and astronomers tell us that at one time life could not have existed on the earth, how did the world become populated?

A number of theories have been proposed to account for the origin of life on the earth during the period when the earth was cooling down from its original incandescence. There now seems to be general agreement among biologists that life originated through a progressive series of reactions in which atoms combined into molecules, molecules into simple compounds and compounds into more complex compounds. These complex compounds eventually became organized into living material.

The surface gases of early earth included hydrogen, carbon, nitrogen, and freed oxygen. Simple compounds can be made from these elements—ammonia ($NH_3$), methane ($CH_4$), water ($H_2O$), carbon dioxide ($CO_2$), hydrogen molecules ($H_2$), and hydrogen cyanide (HCN). Over a period of time, these compounds probably interacted with one another to yield new compounds. Such chemical interactions on early earth could have been activated and sustained by lightning and solar radiation.

In the 1950s, S. Miller demonstrated that simple life *can* be created in the manner described above. A mixture of methane gas, ammonia gas, and water was placed in a flask and electricity discharged through it for several days (to stimulate lightning). Subsequent analysis of flask contents revealed several amino acids, fatty acids, and other organic compounds. This classic experiment gave clear reason to think that simple gases and energy sources produced other compounds, which, through further interactions, gave way to more complex organic compounds, eventually yielding the basic material necessary for life on earth.

The remains of the earliest form of animal life are found in rocks that formed in oceans, and the body fluids of all animals contain certain salts found in seawater. It seems logical, therefore, to conclude that life began in the sea. The sea contains nutrients necessary and a place (the surface) where organic molecules could become concentrated.

We can only speculate on when life originated. The earth was probably formed between 5 and 7 billion years ago; 3 billion-year-old rocks contain fossil algae and bacteria. Photosynthetic plants made animal life possible by adding oxygen to the atmosphere. The first animal tests (shells) date from about 0.7 billion years ago.

## Current Evolution

A seemingly puzzling aspect of evolution is its inconspicuousness in our everyday life. Actually, evolution *can* be observed, but observation of current evolution is limited to its manifestation in rapidly reproducing organisms. This is because in organisms that reproduce more slowly, speciation may take about 1 million years. Evolution in the more rapidly reproduced animals produces data that can then be correlated to indicate the direction of evolutionary progression.

One example of current evolution is the **resistance** that often arises in insects and smaller organisms subjected to poisons. Experiments have been conducted to determine whether this increased resistance is due to exposure or to evolution (mutations quickly spread in the gene pool due to severe selection). The mutation hypothesis has not only been substantiated, but an actual observation of the evolution of one trait through indirect selection has been clearly made.

Other signs of evolution are seen everywhere. An insect-eating Galapagos finch has begun to explore a new food, blood from the rump of another bird (where it formerly picked ticks). Members of one species of American frog should now be considered two new species since the northern ones can no longer successfully mate with southern ones. Moths in industrial England have changed color as a result of allele frequency changes (see Chapter 28).

Short-legged mutant sheep appeared in a flock in Massachusetts in 1791 from which the Ancon breed developed. This mutation was of value to the farmer because these sheep could not jump over the low stone fences of New England. This breed became extinct about 90 years ago, but some 50 years later, a short-legged lamb appeared in the flock of a Norwegian farmer. From this, a new strain of Ancon sheep has been bred (Fig. 29–9).

**Figure 29–9** Ancon (short-legged) mutation in sheep (ewe in center, ram at right) compared with normal ewe (at left). (Courtesy *Life* Magazine, Time, Inc.)

Beginning with the earliest forms of life and taking into consideration the increasing variations and adaptations of successful life forms, the evolutionary perspective is a positive one. For the present, we must consider also that future potential depends upon the *existence* of a future, which in turn is dependent upon how all living organisms are interacting in the world today; this topic will be dealt with in Chapter 30.

## Summary

The basic theory of evolution was discovered independently by Darwin and Wallace. They observed that among species and individuals in nature, a wide range of variations existed and that species multiply by geometric progressions, yet populations remain constant over remarkably long periods of time. From that they concluded that not all embryos become adults; and not all adults reach reproductive maturity. Thus there is a struggle for existence. In this struggle, a process of natural selection is operative which results in the survival of the fittest. Thus, nature selects those individuals that are best adapted to their environment. These reproduce themselves, and their offspring presumably inherit the favorable variations responsible for their survival.

The modern theory of evolution has greatly modified Darwin's theory by incorporating knowledge from both classical and population genetics. Evolution is now seen as a change in allele frequency. Such changes can result from mutations, meiotic drive, gene flow, genetic drift, and natural selection.

The course of evolution appears to be either in the direction of adaptation (following on the heels of changing physical and biological demands), or in no particular direction (randomness that is a result of mutations, gene flow, and genetic drift), or it may be determined by interactions of the two. Based on the kinds of adaptations we have seen evidenced, the evolutionary perspective appears fairly optimistic in terms of the future potential for living organisms.

# Animal Ecology and Distribution

In Chapter 29, we saw the evolutionary development of animals. Concurrent with structural changes were changes in these animals' distribution throughout the globe. In this chapter, we will trace some major trends and theories culminating in the explanation of how present-day species arrived at their current locales, and their ecological interrelationship with their environment.

## Earth's Zoogeographical Regions

Approximately 29 percent of the earth's surface is covered with "dry" land. Certain sections of the land contain characteristic fauna, and it has become convenient to divide the earth into six zoogeographical regions: Palearctic, Nearctic, Neotropical, Ethiopian, Oriental, and Australian (Fig. 30-1). To gain some understanding of just why these regions have their current characteristics, it is necessary to look back in time.

The earth is approximately 7 billion years old. It is known that land masses of the earth changed in morphology over time. Several **paleogeographical theories** attempt to explain how this change occurred. The three most prominent theories will be presented here. It will be useful to refer to Table 30-1 in the course of this discussion.

## Paleogeographical Theories

**CONTINENTAL PERMANENCE**
Between 1830 and 1833, Sir Charles Lyell posited the theory of *continental permanence*, which contends that the continents have generally maintained their present positions, with only some minor changes. This theory assumes that narrow strips of earth sometimes connected the continents, across which animals were able to travel and disperse themselves.

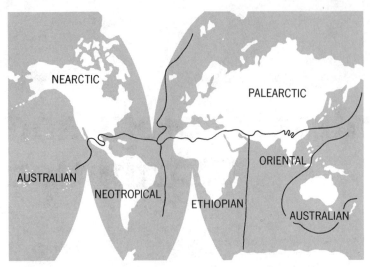

**Figure 30-1** Paleogeographical regions of the world. Palearctic is of temperate climate and contains a wide distribution of 28 mammalian families, most bird families, many amphibian, and few reptiles. Nearctic is also of temperate climate and contains 24 mammalian families, many reptiles, and three endemic (exclusive) families of rodents, among others. Neotropical climate is mostly tropical; there are 32 mammalian families, many reptiles, and almost all tailless amphibians; one-half of all bird families are endemic to this region. Ethiopian has a varied climate, from tropical to desert; it contains 38 mammalian families (4 are endemic). Australian region has tropical, arid, and temperate climates; 9 mammalian families, mostly marsupial; a varied bird population; and few amphibians; there are few vertebrate families and a conspicuous absence of placental carnivores.

**Table 30-1** Geographical Time Scale

| Era | Years × $10^6$ Since Beginning of Period | Period (Epoch) |
|---|---|---|
|  | 0 | Recent |
| Cenozoic | 1 | Pleistocene |
|  | 10 | Pliocene |
|  | 25 | Miocene |
|  | 40 | Oligocene |
|  | 60 | Eocene |
|  | 70 | Paleocene |
| Mesozoic | 135 | Cretaceous |
|  | 180 | Jurassic |
|  | 225 | Triassic |
| Paleozoic | 270 | Permian |
|  | 350 | Carboniferous |
|  | 400 | Devonian |
|  | 440 | Silurian |
|  | 500 | Ordovician |
|  | 600 | Cambrian |

One of the main points of the continental permanence theory was that the Mediterranean Sea was once (approximately 180 million years ago) part of the **Tethys Sea,** postulated to have existed between the northern extent of Africa and the southern region of Eurasia. The continental permanence theory has lost much of its appeal in recent years, owing largely to the new evidence in favor of the theory of continental drift (see below).

## LAND BRIDGES

The *land-bridge theory* suggests that large areas of dry land (*not* narrow strips) at one time connected the continents. The most famous bridge, according to this theory, was **Gondwanaland,** supposedly of nearly continental size. It is said to have existed during the early Cenozoic period, and toward the end of this period to have disappeared into the ocean.

The land bridge theory also postulated the existence of the Tethys Sea, in this case occurring between Gondwanaland and a northern mass that was composed of Eurasia, Greenland, and North America.

The land bridge theory flourished during the nineteenth century, but both it and the continental permanence theory have for the most part given way to the continental drift theory.

## CONTINENTAL DRIFT

In the early part of the twentieth century, A. L. Wegener put forth the theory of *continental drift.* According to this theory, all dry land was in one large mass (around the Carboniferous period or slightly before). Supposedly, the continents, lighter than the underlying earth crust material, gradually floated over it to their present positions.

According to this theory, by the Eocene period, South America lost its connection with Africa, and thus the Atlantic Ocean was formed. It is thought that Australia and South America were still joined by the Antarctic at that time.

Evidence for this theory is amassing constantly. Initially, the theory of continental drift is suggested by the outlines of our present continents. A quick glance at any map of the world will show that, for example, Africa and South America seem to "fit" each other if they are slid together. More scientific evidence has surpassed this simplistic (yet powerful) line of thought in support of this theory.

In the late 1950s, oceanogeographers identified a continuous ridge running midocean. Many earthquakes ("seaquakes") are associated with this area. In 1960, H. H. Hess of Princeton University set forth the possibility that the ocean floor might be in constant motion.

Perhaps one of the most important oceanogeographic discoveries was made by R. G. Mason, A. D. Raff, and V. Vacquier in 1961. The ocean floor just off North America was found to have had a striped pattern of magnetism. At about the same time, other scientists found that the earth's magnetic field had reversed itself several times during the history of the earth. Bringing a number of facts together, F. J. Vine and D. H. Matthews proposed that there was molten rock in past ages, which, when it cooled and solidified, took on the geomagnetic orientation of the world at those times. Hence, the geomagnetic striations in the ocean floor represent the various orientations of the earth's magnetic field at different times.

Further evidence for the continental drift theory came with the discovery that the midocean ridge changes direction abruptly at several loci, representing an axis for ocean floor **spreading**.

W. V. Morgan has postulated the existence of six **plates** (now called **tectonic plates**) which grow by the addition of *new* earth crust. Current theory of the tectonic plates is that, under the ocean floor and the continents, there is a conveyor-belt-like motion. Seismographic evidence lends credence to this theory. In addition, it has been found that the continents are *still* moving. The average rate is from about 1 to 6 cm per year away from the spreading axis. The Americas are moving west and Africa is moving east.

## Animal Distribution

We will now see how these land mass theories reflect the trends in animal distribution.

In the beginning of the **Cenozoic period** (the Paleocene), there was a great number of mammals in the northern hemisphere (Fig. 30–2). Many orders

**Figure 30–2** Distribution of major vertebrate groups throughout geological time. Changes in width of white areas indicate relative abundance; dashed lines suggest possible sources and time of origin. (After E. H. Colbert, *Evolution of the Vertebrates*, New York: John Wiley & Sons, Inc., 1955.)

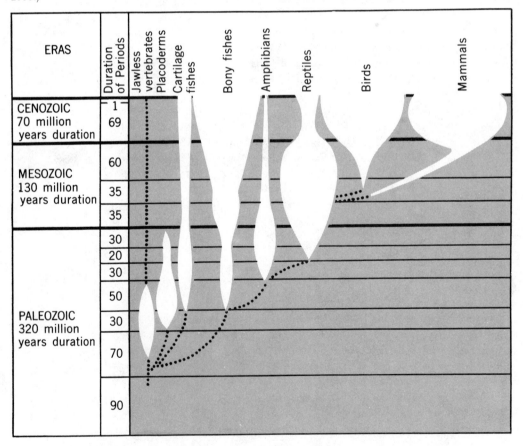

appeared that have since become extinct. Fossil evidence indicates that, for example, the Nearctic and Palearctic regions had extremely similar fauna in this and the previous Cretaceous periods. The trend seems to have been a general dispersal from the northern regions to the southern, with considerable east–west spread. Of course, some species moved very little or not at all.

In the **Eocene,** many new northern animals appeared, and there was an especially large increase in the number of northern mammalian families. In addition, it is thought that the northern regions lost any interconnections they may have had in the Eocene period.

The **Oligocene** is characterized by a rapid and pervasive spreading of Eocene animal populations. For that reason, it is thought that in this period there was probably a restoration of zoogeographical region connections that had existed in the early Eocene and previous to it. During this time, the Palearctic and Nearctic regions differed considerably in herbivore constitution; thus it may be assumed that they were not connected at that time. Fossil evidence indicates that in this period a migration of rodents, carnivores, and

**Figure 30–3** Environmental adaptations. Note the long ears and long bushy tail of the kit fox (*top*), a desert inhabitant. *Bottom:* the arctic fox; note shorter ears adapted to its cold habitat.

placentals occurred from the Palearctic to the Ethiopian region. Hence, Eurasia and Africa seem to have been connected at several points in the Oligocene.

It is thought that well into the **Miocene** a land bridge occurred between Siberia and Alaska, in the Bering Straits region. In this period and into the **Pliocene,** major earth movements threw up the Himalayas, according to land bridge theory. This caused a land barrier to exist between the Palearctic and the Oriental regions. By the end of the Pliocene, the Americas were joined, and they have remained so until the construction of the Panama Canal in Central America. The Pliocene is an important period, for the northern hemisphere began to cool during this time. This had profound effects.

Large numbers of carnivores (such as cats and bears) fled the dropping temperatures, moving south. The characters of the Nearctic and Neotropical regions changed radically. As the carnivores of the Nearctic headed toward the Neotropical region, the threat to herbivores in the Nearctic region decreased, and the herbivorous population of the Nearctic region increased. It is also possible that the herbivores adapted to the cold better than the carnivores. The Palearctic faunal complement also changed. Many of the animals there made a massive exodus south into the Ethiopian region.

By the arrival of the next period, the **Pleistocene,** most of the modern species had evolved. The African connections with Eurasia had become arid, and so there was only limited migration in that area. The most important phenomenon of this period was the Ice Age: the freezing of the greater portion of the northern hemisphere. At its peak, about 32 percent of the earth's surface was under massive glaciers. Many species died out, and the zoogeographical character of the world changed.

Several giant versions of certain species made their appearance throughout the non-iced parts of the world. Those animals maintaining life in the northern hemisphere usually were more protected from environmental extremes than their southerly counterparts (Fig. 30–3). Heavy fur coats were characteristic of much of the northern fauna. Mammoths and sabertooths died out because of the cold, and some organisms, such as tapir and "new" elephants, moved into southern areas.

# Ecology

Every living thing is continually influenced by, and continually influences, its environment. **Ecology** is the science of the interrelationships between organisms and their environments.

The study of ecology can begin by dividing environmental factors into those that are primarily physical (abiotic), such as light, temperature, chemicals, water, and substratum—and those that are primarily biological (biotic), such as the composition of plants and animals.

Another approach is to study our environment's organisms. The ecologist may start with a particular plant or animal population and direct his efforts toward the discovery of the various environmental conditions that can be tolerated by that particular group. Conditions that actively or passively promote the survival and welfare of the group, or adversely affect it are the concern of the ecologist.

Modern ecology is much less concerned with individual organisms than with groups. Ecologists have come to realize that it is the interaction of consistent environmental constants, changes, and cycles with relatively stable groups of organisms that is of importance to the orderly progress of evolution and survival.

Any integrated group of organisms of one or of several species that occupies a consistent environment may be called a **population.** The ecologist is well aware of the importance of the on-going relationship between the physical–chemical environment and populations. This is called the *ecosystem approach*. E. Odum defines an **ecosystem** as "any entity or natural unit that includes living and nonliving parts interacting to produce a stable system in which the exchange of materials between the living and nonliving parts follows circular paths."

# The Physical Environment

Before we can discuss the interactions between environments and populations, we must learn about what constitutes an environment. If the presence or absence of a particular factor in a certain minimal quantity is necessary for the success of a population, it is called a **limiting factor.** Abiotic factors in the environment that are often of great importance as limiting factors are light, temperature, wind, water currents, fire, soil texture, pH (acidity or alkalinity) of water and soil, presence or absence of certain inorganic salts, and the concentrations of oxygen, carbon dioxide, and other gases. Biotic factors may exist in relation to the food supply, predation, or parasitism. A few of the more important environmental factors follow.

## LIGHT

Without light, life on earth would be impossible.

Radiation from the sun (**solar radiation**) supplies both light and heat. The greatest importance of solar radiation is its energy contribution to photosynthesis. The photosynthetic organisms of the sea usually cannot exist any deeper than about 200 meters (and this is in almost-clear water) since below that depth the necessary spectral elements of the sun are lacking. Almost 90 percent of the ocean's photosynthetically derived energy is incorporated on the western seaboards of the continents, for it is there that upward swells bring fresh photosynthesizing material to the surface constantly.

By means of light, many animals are able to move about freely and carry on the necessary activities of life. Sense organelles, which detect light, are present in some protists, and complex light reception apparatus exist in most of the higher invertebrates and vertebrates.

The character of the lighting of an area has a profound effect upon the animals that live there. For example, the length of the period of daylight (photoperiod) appears to be one of the factors that stimulate the migration of birds. Certain fish that live in the ocean at depths where the light rays are few have large eyes; many deep-sea fishes manufacture their own light by means of luminescent organs. Many animals, such as owls, are nocturnal and also possess large eyes with pupils that will admit a great number of rays in dim light.

Ultraviolet light is both beneficial and harmful to animals. The earth's atmosphere acts as a partial shield; that is extremely important, for if all the UV light impinging upon the earth were to get to its crust, most organisms would die. UV light provides the impetus for synthesis of vitamin D in the skin, hair, or feathers of many animals. One's pigmentation has a profound effect on UV ray sensitivity. Light-pigmented people produce more vitamin D per unit area on the skin than do dark-pigmented people. Too much exposure to UV rays, however, may lead to skin cancer.

## TEMPERATURE

Variations in temperature generally take place over larger areas of the environment than variations in light. The world is, in fact, divided into zones (arctic, temperate, and tropical) largely on the basis of differences in temperature. Most terrestrial and aquatic animals react to differences in temperature, and each species has an optimum—the one at which they function best physiologically.

The effects of temperature on the presence or absence of animals in different habitats are varied. In some species, the adults die as winter approaches, but the race is maintained by means of "winter" eggs that can withstand the cold weather. Other animals escape the cold by hibernation or migration. Certain animals live normally in hot springs at a temperature that would soon kill others accustomed to more normal temperatures. Despite the variations that exist, all living things are adapted to a comparatively limited temperature range: in general, life exists at temperature of 10 to 45°C, the **biokinetic zone.**

Animals handle temperature variations in various ways. **Homeotherms** are animals that maintain a constant body temperature (such as human beings); they are usually **endothermic,** meaning that they produce their own heat, when necessary, and have some physiological mechanism (such as perspiration) to cool themselves down.

**Poikilotherms** are those organisms with variable body temperatures; they are normally **ectothermic,** which means that they obtain their heat directly from the environment. Most reptiles are poikilothermic, and hence they are usually not observed at night, when it is too cold for them to function with any degree of proficiency.

Some animals are **heterothermic.** They can switch over from homeothermy to poikilothermy. One manifestation of heterothermic animals is the ability to **hibernate.** Mostly in the northern regions of the world, some amphibians, reptiles, and some mammals such as ground squirrels and woodchucks can "switch off," lowering their basal metabolic rates, and become poikilothermic during a cold season. **Estivation** is another process involving dormancy but it is in response to heat instead of cold and is characteristic of several insects, spiders, and reptiles which live in hot, arid zones.

Animal size also seems to be related to adaptation to environmental temperatures. Many vertebrates of similar species differ according to the climate in which they live. External organs frequently reflect habitat, so that a species living in a colder climate will, for example, have shorter ears and/or shorter tails than a warmer-climate counterpart.

# CHEMICAL ASPECTS OF THE ENVIRONMENT

**WATER.** Water is in constant interchange among air, land, and sea, and between the living organism and its environment. It has an important bearing on the character of the animals that live in various types of habitat.

Because water is the most abundant component of animals, the chemical contents of water determine to a considerable extent its availability as a habitat for aquatic organisms. In deep water, less oxygen and more carbon dioxide are usually present than in surface water.

Animals that live in dry regions usually possess some method of preventing evaporation of water; the camel, by storing water in the reticulum of its stomach, can live for a week on dry food only, or for a month or more on green food. Certain species, such as the pronghorned antelopes, jackrabbits, and certain ground squirrels, can obtain all the water they require from green food.

Complete drying (**desiccation**) kills most animals, but members of some species can withstand it. This is particularly true of small organisms, such as some protistans, rotifers, and minute crustaceans, which escape death by encysting or by laying eggs with heavy shells that resist evaporation. The water bears (Tardigrada) are famous for their ability to withstand desiccation for several years.

**GASES.** The importance of oxygen ($O_2$) and carbon dioxide ($CO_2$) for photosynthesis and respiration in plants and for respiration in animals is primary. Air has an abundance of oxygen (210 $cm^3$/liter), especially when compared to fresh water (7.2 $cm^3$/liter) or salt water (5.8 $cm^3$/liter). Thus, aquatic animals must work harder for their oxygen then land animals.

There is a slowly increasing problem with carbon dioxide. About 50 percent of the $CO_2$ produced by industry, automobiles, and land animal exhilation is used by the oceans and green plants. The remainder accumulates in the atmosphere: the increase has been measured to be about 0.2 percent per year.

Industry and automobiles spew several types of toxic gases into the atmosphere. Some of them are sulfur dioxide, carbon monoxide (which is taken up more readily than oxygen by hemoglobin in blood and thus leads to suffocation if present in sufficient amounts), nitrogen oxides, and hydrocarbons. If these pollutants are allowed to continue indefinitely, their effects could drastically change our ecosystems.

## BIOGEOCHEMICAL CYCLES

The chemical elements which compose all organisms come from the environment and eventually return to the environment after the death of the organism. The constant interchange of these chemicals with plant and animal bodies results in natural cycles, three of which will be illustrated here.

**CARBON CYCLE.** All organic compounds in cytoplasm contain carbon (C). The carbon dioxide in air and/or water is synthesized into carbohydrate molecules through photosynthesis. These molecules, together with fats and proteins, make up plant tissues which are consumed by various animals. Following digestion and absorption, the carbon compounds are reorganized into animal tissues. These are passed through other animals (predators). Carbon dioxide is eliminated as a respiratory waste and is returned to the air or water.

**NITROGEN CYCLE.** Proteins constitute one of the most important animal foods. All proteins contain the element nitrogen (N). Free nitrogen in the atmosphere is directly combined into nitrates by nitrogen-fixing bacteria found in root nodules and soil. These nitrates are used by plants, which incorporate them into plant proteins. The nitrates can be returned to the soil by decay or are consumed by animals, which convert them into animal proteins. Through metabolism, these proteins result in nitrogenous wastes, including urea and uric acid. Bacterial action then converts the wastes to ammonia and nitrates. Further bacterial action yields free nitrates, which can be used by plants, and nitrogen, which is returned to the air.

PHOSPHORUS CYCLE. Phosphorus (P) is extremely important for all organisms; for the animals, its primary importance is in adenosine di- and triphosphate (ADP and ATP), the "energy molecules." Phosphorus from the earth washes into the sea during rains or tides. Marine algae take up the phosphorus and eventually either fishes or marine birds receive it. The fishes use it directly, while land animals rely on birds' feces; the phosphorus-rich fecal material is deposited again on land—frequently in areas where its presence is limited. Plants can then take up the phosphorus, and animals either get it directly from the plants or through carnivorous means.

### OTHER CHEMICAL CONSIDERATIONS

The ecology of a mineral can be detected by radioscopic **tagging**. Mercury (Hg) normally occurs in the environment in harmless "trace" amounts. Certain bacteria, however, are capable of transforming it into methylmercury, which is toxic, creating Minimata disease or **mercury poisoning** in animals. Fish may ingest the methylmercury-containing bacteria, and those fish may be eaten by human beings.

Green plants, necessary for our existence, depend on a number of **biogenic salts**, notably nitrogen and phosphorus. In addition, potassium (K), calcium (Ca), sulfur (S), and magnesium (Mg) are used by plants.

Seawater contains approximately 40 elements, the most abundant of which are chloride ($Cl^-$) and sodium ($Na^+$) ions. Other salt constituents (ions) present include calcium, magnesium, potassium, carbonate, sulfate, and bromide. The concentrations of these and other ions, inter- and intracellularly and extraorganismally, are important for osmoregulation in fishes.

# The Biomes

A **biome** is a unit resulting from the interaction of a regional climate, regional biota (animal and plant life), and substrate. The types of biomes are grasslands, deserts, coniferous forests, deciduous forests, tropical forests, tundra, freshwater, and marine environments (Fig. 30–4).

### GRASSLANDS

Every continent has one or more areas called **grasslands** (Fig. 30–5). The subtype of grassland biome depends mostly on tallness of grass. In South America, the two types are called **pampas** and **savanna**; in South Africa, it is the **veld**; in Russia, the **steppe**; and in western North America, tall grasslands are **prairies** and short grasslands are **plains**.

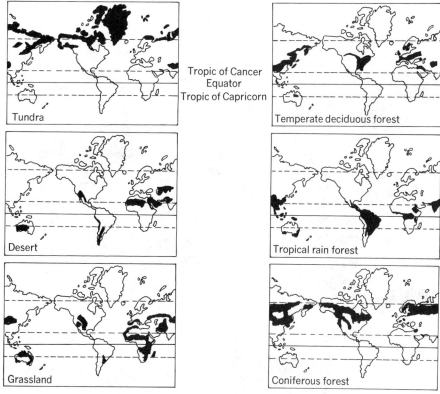

**Figure 30-4** Distribution of major biomes.

**Figure 30-5** Grassland biome in Wyoming.

**Figure 30–6**
Desert biome in Arizona.

**Figure 30–7** Coniferous forest in Boulder, Colorado.

## DESERTS

As with the grasslands, **deserts** are present in every major continent. Approximately one-fifth of the earth's dry land is desert. The extremes in temperature (hot days, cold nights) severely limit the possibilities for life. Plants called **xerophytes** are adapted to the arid environment; they can store water, and their seeds germinate rapidly in the presence of moisture. Rainfall in deserts is exceptionally low, usually less than 25 cm per year (Fig. 30–6).

## CONIFEROUS FORESTS

**Coniferous forests,** also referred to as **boreal forests** or **taiga,** are represented as a group of evergreen trees crossing Europe, Asia, and North America (Fig. 30–7). The fauna include the moose, rodents, lynx, and many birds. Winters are severe, and a relatively short growing season occurs, lasting from about 3 to 5 months.

## DECIDUOUS FORESTS

**Deciduous forests,** a biome characterized by a moderate (temperate) climate, is well suited to animal life. This zone is named for its deciduous trees, which lose all their leaves in the winter season (Fig. 30–8). The deciduous forest is replete with wildlife, including many large felines, rodents, predatory birds, and various other animals that are characteristic forest dwellers, such as the North American white-tailed deer.

**Figure 30–8** Deciduous forest biome.

**Figure 30-9** Tropical rain forest biome.

## TROPICAL FORESTS

**Tropical forests** usually occur around the equator and are rather varied in type. A major subtype is the rain forest, where over 200 cm of rain falls annually. The temperature, owing to the equatorial position, is usually high. The flora is restricted almost exclusively to trees; these are frequently dense enough to allow very little direct light to reach the ground.

Tropical forests are known for their high proportion of nocturnal and arboreal animals. Many species are represented, but there are relatively few individuals of each; hence, the tropical forest is one of the most diverse biomes (Fig. 30-9).

## TUNDRA

To get the feeling of **tundra**, listen to some high-register violin sections in Shostakovitch or some Prokofiev. The treeless ground is permanently frozen to a few centimeters of the surface. The tundra borders the Arctic Ocean (Fig. 30-10). The "growing" season is quite short, about 60 days.

The species' faunal complement is scanty, although in the summer there are some waterfowl, rodents, and lemmings, and insects are most abundant. Many inhabitants burrow and/or hibernate.

The various aqueous environments will be discussed next.

**Figure 30–10** Tundra biome in the arctic.

# Environmental Types

## FRESHWATER ENVIRONMENTS

The **freshwater** habitat is less extensive than any of the other aqueous environments, but it is probably better known to us. It may be divided into flowing water and standing water. The flowing-water habitat is divided into rapid streams and slow streams, and the standing-water habitat into lakes, ponds, and swamps. Each of these may be further divided—for example, the ponds are divided into those with bare bottoms and those whose bottoms contain vegetation. Each of these types of ponds contain a number of communities, such as those occupying the beach above the water, the shoreline, the shore water, the open water, and the bottom. All other habitats may be similarly divided.

Many of the animals we have studied live in freshwater communities; these include protists, rotifers, hydras, planarians, crayfishes, helminths, annelids, mollusks, and fishes.

## MARINE ENVIRONMENTS

**Marine** (or **saltwater**) environments differ from fresh waters in many aspects; their size, depth, and continuity make them more stable aquatic environments. Animals that live in salt water have a great deal of space to occupy, since 72 percent of the earth's surface is covered by the sea.

The marine environments may also be broken down into smaller subdivisions. The relatively shallow water of the continental shelf, called the **neritic** zone, is further divided into zones related to tidal activity. The deep waters beyond the continental shelf constitute the **oceanic region;** its subdivisions are vertical rather than horizontal. The **euphotic zone** is the upper 200 meters of the oceanic region in which effective photosynthesis takes place. Below it is the **bathyal zone** (200 to 2000 meters) and below this is the **abyssal zone,** which accounts for 87 percent of the ocean.

The number of animals is greatest near the surface, where the light penetrates; here they are subject to tides and currents. Along the shores live large numbers of barnacles, sea anemones, and limpets, which attach themselves to rocks; clams, snails, sea stars, and sea urchins cling to rocks or hide in crevices; and many species of worms and clams burrow into the sand or mud.

### TERRESTRIAL ENVIRONMENTS

The environments on land (**terrestrial**) are highly variable. They can be contrasted with the aquatic environments on the basis of a number of problems they offer to plant and animal populations.

On land, moisture is often a limiting factor. Air, as contrasted with water, permits rapid temperature variation but offers little variation in oxygen and carbon dioxide content. Air also lacks the supportive ability of water (buoyancy) and thus imposes upon many land animals a necessity for strong skeletal structure and special means of locomotion.

Terrestrial animals may live on the surface of the ground, under the surface, or out in the air. The types of animals that live in subterranean communities are largely determined by the character of the soil and the amount of water it contains.

The principal kinds of soil are clay and sand. Comparatively few species of animals live in pure clay soils. Clay soil rich in humus is usually well populated, since it contains an abundance of food and is easily penetrated by burrowing. Burrows in sandy soil collapse unless the walls are treated; hence, fewer species live in this type of soil. Soils that are too rocky for burrowing may provide animals with crevices that make good hiding places.

## Predisposing Factors in Animal Distribution

It was mentioned that the tendency for animals to spread out and utilize many of the locations we have just discussed was due to competition for living space and food. As you might suspect, things are somewhat more complex than that. Let us now take a look at the pressures that resulted in speciation.

### PHYSICAL BARRIERS

Some possible restrictions leading to isolation (partial or complete) of a population are purely geographical. The limiting effect of a physical barrier is, of course, the organism's ability to traverse it; a tapir may not be able to cross a river, but a bird can do so with ease. Some examples of common physical barriers are bodies of water, mountain ranges, deserts, forests, and, in some cases, concrete (cities).

## ANIMAL INTERACTIONS

There are four general ways in which animal interactions may affect speciation: courtship behavior, prey–predator relationships, competition, and parasitism.

If the **pattern of courtship** of two organisms is dissimilar, there will probably be no sexual interaction. Also, if a particular group of animals of the same species mate at different times (due to different estrous or "heat" cycles), no copulation will occur between these subgroups and their gene pools may become isolated.

The implications of **prey–predator relationships** are obvious, and tend to lend credence to Darwinian theories (see Chapter 29). Several predators may utilize the same prey; **competition** in the animal kingdom is usually for food, territory, or a mate.

The subject of **parasitism** has been treated throughout this text. If an animal cannot contend with its parasites, or rid itself of them, it will suffer some detriment, possibly enough to be fatal.

In summary, sexual and symbiotic relationships have diverse and important effects on speciation.

We will now study some of the more important principles involved in the study of animal ecology.

# The Ecosystem

An **ecosystem** is the conglomerate of all life in a given area (the **biotic** component), plus its interactions with the nonliving (**abiotic**) portions of the area. An **econiche** is the limited ecosystem of a particular group of organisms. Several econiches overlap and interact within an ecosystem (Fig. 30–11).

Abiotic components of an ecosystem include all the nonliving material, including air (or water), soil, minerals, and biogenic salts.

The biotic components can be divided into several categories. The **producers**—photosynthetic plants—rely on nothing organic for nourishment, utilizing the sun's energy directly, along with various minerals, salts, and water.

In an ecological community, it is a general rule of nature that for every step down, approximately 90 percent of the energy of the upper group is lost. This means that each successive group must eat approximately 10 times the mass the preceding group had to eat in order to gain the same amount of energy. For this reason, the producers, whose massive abundance dominates terrestrial ecosystems, are the largest group.

The next group, the **consumers,** are divided into primary, secondary, and so on, consumers. The primary consumers obtain nourishment directly from the producers, and are thus herbivores (or omnivores). Secondary consumers obtain nourishment from the primaries; and so on.

The **decomposers,** fungi and bacteria, obtain their energy mostly from dead tissues of the consumers. This group releases simple substances to the environment as by-products of their metabolism. The producers can then use these products, and hence the ecosystem itself is an essentially endless cycle.

The cycle is maintained by a **food web** (which incorporates the various and

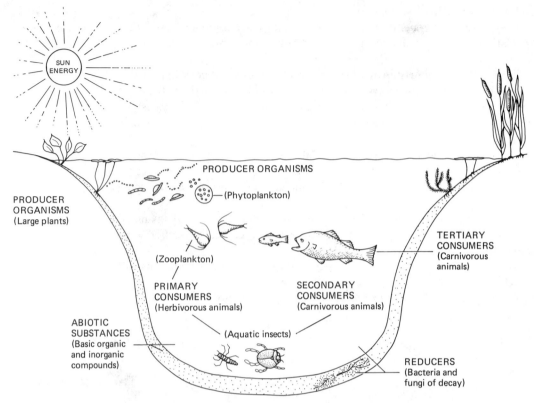

**Figure 30-11** Freshwater pond ecosystem. The ecosystem is the largest functional unit in ecology, including living (biotic) and nonliving (abiotic) environments which exchange materials in a circular path. (After E. P. Odum, *Fundamentals of Ecology*, Philadelphia: W. B. Saunders Co., 1953.)

more simple **food chains**). This is a complex system of interdependence; if one of the food "branches" were cut off, alternative forms of nourishment would continue to provide food for the ecosystem (Fig. 30-12).

A pyramid may be used to understand the interactions and results of a food web. Several types of pyramids have been constructed for a given ecosystem, but we will limit our discussion to two types.

In the **mass pyramid,** the base represents the *mass* (total) of the producers, and each successive tier stands for the mass of consumers (primary, secondary) supported. Since this pyramid indicates only mass, not energy, and since a very massive organism need not contribute a proportional amount of energy, this pyramid is of little use for our purposes.

Consequently, we can construct an **energy pyramid,** based on computing the energy in calories or kilocalories that each member of the ecosystem is worth, and drawing that proportionally in the form of a pyramid (Fig. 30-13).

# Principles Associated with Biotic Communities

A biotic community is a more-or-less complex group of plants and animals that occupies a particular area and influences the life of each member. They

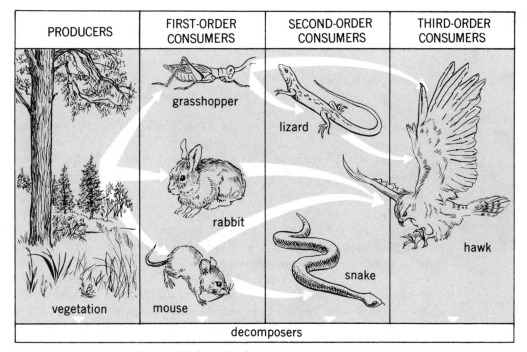

**Figure 30–12** Extremely simplified food web in forest community. Arrows point from the animal or plant eaten to the animal that feeds on it.

**Figure 30–13** (A) Mass pyramid for forest community. (B) Energy pyramid for forest community. (Values are approximate for both.)

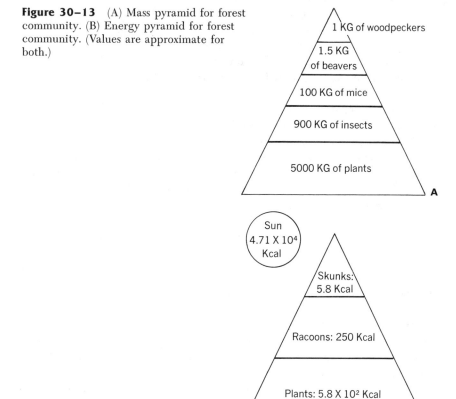

may be named after some conspicuous feature, such as the most conspicuous or numerous animal or plant species, or after some physical feature.

Environmental factors determine the types of plants and animals living in a biotic community. Areas with similar environments are likely to contain similar types of plants and animals. Freshwater ponds are never exactly alike, but are so similar in environment that we look for the same kinds of plants and animals in them.

Hundreds of different types of communities have been described and classified by ecologists. Study of these communities have led to a number of ecological principles; among these are **succession, dominance, stratification,** and **periodicity.**

## ECOLOGICAL SUCCESSION

Every biotic community has changed over time; these changes constitute **ecological succession.** For example, the overflow of a river might create a pond—a new habitat or **pioneer community.** At first, this pond may contain certain types of fish, including black bass and sunfish. In time, the sides of the pond may become overgrown with vegetation and the formerly clean bottom covered with deposits; the black bass and sunfish disappear because such an environment is not suited to them, but catfish may persist. Still later, just before the pond becomes a swamp, the mud minnow may replace the catfish. Finally, conditions become such that no fishes are able to live in the habitat. The final stage in this series is the **climax community;** at this time a dynamic equilibrium creates relative stability in the biotic community (Fig. 30–14).

We have just described an example of **primary succession.** Its phenomena are spread over relatively long periods of time, and it is normally not traumatic to the ecosystem. Rapid environmental changes, caused by volcanoes, floods, avalanches, and fires, produce **secondary succession** (Fig. 30–15). A brief discussion of fire succession will serve to illustrate the concept of secondary succession.

**Fire succession** is a cyclic series of changes. It has been determined that certain areas undergo fires periodically. After the fire, certain specific ecosystems, usually plant species, are the first to reappear, with other less fire-adapted species following. As time passes, the floral and faunal complements increase.

Thus, from an ecological viewpoint, fires are not as destructive as we have been led to believe. When a fire occurs, its presence stimulates microorganisms within the soil to increase their activity. Thus, fire succession is in part responsible for the increasing numbers of plants (as producers) and animals (as consumers), since plant growth depends upon nutritional substances produced by the decomposing action of the microorganisms.

## ECOLOGICAL DOMINANCE

In many communities where there are producers, consumers, and decomposers, certain organisms, because of their size and numbers, will have much to do with determinating the community's character. Such organisms are called ecological **dominants.** In land communities, these are usually plants: in a beech–maple community, these tall trees cast shade that limits the shorter plants to certain types, and these, in turn, limit the type of animal species that live there.

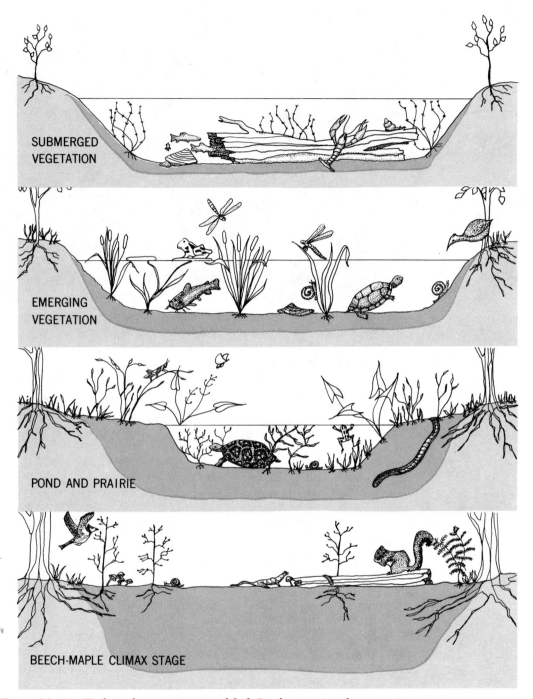

**Figure 30-14** Ecological successions, simplified. Pond succession, from practically bare bottom (pioneer) stage to relatively stable (climax) stage. Successive stages and climax will vary according to climate of environment.

**Figure 30–15** This fortuitous shot of an avalanche shows the snow just starting to fall. Avalanches are one of the most powerful natural phenomena.

## COMMUNITY STRATIFICATION

**Stratification** is the tendency for a community to arrange itself in vertical layers. This increases the number of habitats and thereby reduces competition between species.

In aquatic communities, this is usually due to a stratification of physical factors such as light, temperature, or oxygen content of the water. Thus, the organisms that live on the bottom of a deep lake will not be the same as those near the surface. Similar divisions occur in animal communities, usually based on the varying heights of vegetation, such as grass, shrubs, and trees.

## COMMUNITY PERIODICITY

**Periodicity** is the tendency for communities to exhibit rhythms or cycles. These involve recurring changes in the activities or movements of organisms. Thus, certain species may be active only during the day (**diurnal**) while others are active only at night (**nocturnal**).

In aquatic communities, animal plankton generally move toward the surface at night and return to deeper waters during the day; these are called **circadian rhythms.** Physical factors (e.g., temperature) may undergo **seasonal rhythms,** which bring about periodicity in populations. **Lunar rhythms** are well known in marine environments. Finally, there are rhythms which are not influenced by external factors but are inherent within a particular animal species.

## A FINAL WORD

We, as the species with the greatest capacity for rationalization and "intellect" (as proved thus far), must learn from the examples of other organisms how to interact in the various biotic communities that constitute our environment. This will be discussed more fully in the following, and final, chapter.

# Summary

There are six zoogeographical regions: Palearctic, Nearctic, Neotropical, Ethiopian, Oriental, and Australian. Zoological history follows the gross trend of a movement from the north to the south, with some east–west spreading; such motions were accelerated by the Pleistocene glaciers.

The major theories attempting to explain the present zoological state of the world are the theories of continental permanence, land bridges, and continental drift. The latter, which assumes that the continents are carried along by a moving substrate that operates like a conveyor belt, is currently the most widely accepted theory.

Ecology is the study of the interrelationships between organisms and their environments. Light, temperature, and chemical aspects all contribute to environmental diversity and the environment's differential effects on its inhabitants.

An ecosystem is composed of abiotic and biotic material. Biotic levels are those of the photosynthetic producers, the herbivorous or omnivorous primary consumers, the carnivorous or omnivorous secondary consumers, and the decomposers.

An understanding of the biotic communities is crucial if we as a species are to successfully interact with our environment.

# 31

# The Human Species

Our study of zoology would be incomplete if we were to exclude that unique mammal—the human species. Human beings have learned to evade predators, withstand extreme environments, travel distances no other animal could navigate, and obtain extraordinary amounts of food, enabling the support of enormous populations. Such adaptations are the result of a well-developed brain and an opposable thumb. These two anatomical characteristics ensure the human species a secure position in the world, much as the exoskeleton allows arthropods to adapt to a wide variety of habitats.

One attribute of the well-developed human brain is self-consciousness—our concept of ourselves. We do not know exactly when in our species' development this ability occurred, nor do we have any knowledge that it does not exist in other organisms. It may have been a gradual awakening of reasoning and self-awareness. Hopefully, the process is continuing.

As the human mind became functionally more alert, our ability to use the environment to our advantage increased at a tremendous rate. Driven by basic needs, our advancing technology eventually resulted in unexpected repercussions, such as pollution and overpopulation. Such surprises may stimulate a continuation of the awakening process. Thus may we become conscious of our place *within* nature and tune our technology to its pitch.

In this chapter, we will explore our species' history, technology, and impact on the environment, including a perspective of the species in terms of population growth and evolution. Biological principles that we have learned in the course of our study will be applied to discover limits and future possibilities for our continued existence.

## Origins of the Human Species

It is truly remarkable that evolution produced an animal aware of its own evolution. How is it that this one organism developed such great advantages that it may now control all others? Archeologists, evolutionists, geologists, and others have attempted to recreate our past in order to understand it. Much of the information about early man emerges from the study of fossil remains, and in particular the teeth and jaws that have fossilized better than other parts of the body. Workers have pieced together fossil remains and other fragments of evidence (such as tools, campsites, and other artifacts found with the fossils) and produced the following picture of our past.

The first fossil ape, *Propliopithecus*, lived about 30 million years ago and apparently was a predecessor of apes and man. Through many years of evolution the ape stock developed larger canines (for defense or sexual selection) and tended to be herbivorous, while in man's ancestors the canines reduced as he became dependent on tools for defense and became more omnivorous.

### *Ramapithecus*

The next most significant ancestor was dated at about 14 million years B.P. (before present). At about this time a continuous belt of forest linking Africa and Asia began to die off and recede. As the forests became sparce, this small African ancestor, **Ramapithecus wickeri,** spent increasingly more time on the ground. Some of his **arboreal** (tree-dwelling) characteristics were used in novel ways. His ability to swing freely through the trees was dependent on several factors: agile arms with full circle rotation at the shoulder; depth perception through three-dimensional stereoscopic vision (which required a reduced nose); branch holding through dextrous hands with opposable thumbs; corresponding increases in the areas of the brain concerned with these functions; and coordination of the eyes and hands, through further development of the brain. When *R. wickeri* descended from the trees, his reduced nose was next to useless in perceiving the ground-hovering scents, but standing up he was able to see over the tall grass with stereoscopic vision. Predators wary of this strange posture were repelled by rocks and sticks thrown by dextrous hands—hands that used tools!

*R. wickeri's* teeth were characteristic of herbivores; however, there is some evidence that they were omnivorous. Smashed bones found near their remains along with crude stone implements indicated that these tree-dwelling animals probably scavenged carcasses left by predators, using the marrow and brains as one food source.

Their bipedal posture (upright, two-legged) also had residual effects. Evolutionary changes in the pelvic girdle narrowed the birth canal, necessitating smaller (hence less mature) babies, resulting in longer parental care. Increased parental care undoubtedly altered behavioral patterns, further separating male and female roles.

### *Australopithecus*

Ascending through the fossil records, we encounter **R. punjabicus** in India around 9 million years B.P. Around 2 to 3 million years B.P., four species of **Australopithecus** appeared. Two of these species **A. boisei** and **A. robustus** (Fig. 31–1), were specialized for a herbivorous existence, with massive jaws, large teeth, and powerful jaw muscles required to eat rough vegetation. These muscles were attached to a ridge on top of the head, which limited the brain case. The other two species, **A. africanus** and **A. habilis,** were less specialized omnivorous hunters. Since there was no need for large head muscles, the brain case was not limited in size. Indeed, since existence depended on interactions of hand and eye, and possibly imagination (they made tools), there might have

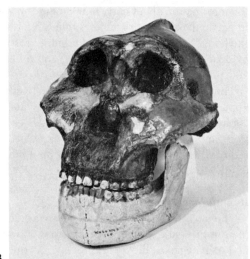
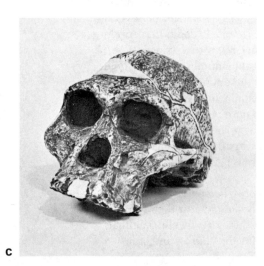
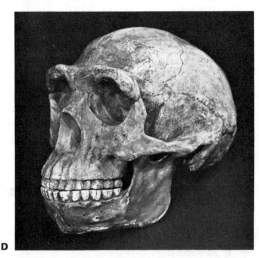

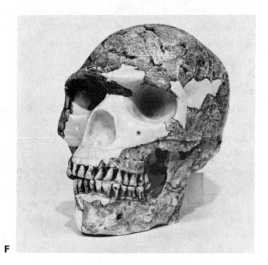

**Figure 31-1** Skulls of humans and our ancestors. There is an evident progression from the apelike skulls, with large sagittal crests and jaws, to the modern human being, with pronounced forehead, chin, and reduced jaw. (A) *Australopithecus robustus*. (B) *A. boisei*. (C) *A. africanus*. (D) *Homo erectus* (Peking man). (E) *Homo sapiens neanderthalensis* (Rhodesia man). (F) *H. sapiens* (intermediate). (G) *H. sapiens*, modern human (Cro-Magnon).

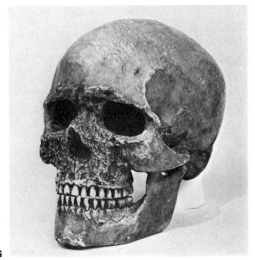

G

been selective pressure for a larger brain and its inescapable corollary, a larger brain case.

The remains of omnivorous australopithecines indicate that although *A. africanus* did make tools, those of *A. habilis* were superior. *A. habilis* fossils are also found in close proximity to stone rings (possibly shelters) and the bones of large mammals (such as elephants and rhinoceroses) that were apparently chased into the mud and killed—evidence of socialization. As in the social insects and wolves, *A. habilis* found that in groups they could overcome large prey more easily than as individuals. Here we may see the beginnings of a culture. Individuals brought together for some activity exchanged information. In this manner methods of tool making could pass from one generation to the next, with gradual improvements along the way.

Estimates have placed the australopithecine population at about 0.25 to 0.50 million.

*Homo erectus*

The next major fossil group was that of ***Homo erectus***, a group that flourished about 0.9 million years ago. Their bones have been found all over the Old World, some in layers directly above those of *A. habilis*, indicating a possible evolutionary relationship between the two. It is thought that a population of about 1 million *H. erectus* lived in ancient times. Like the australopithecines, members of *H. erectus* were probably very good hunters, since they were one of the few diurnal predators and had effective weapons. *H. erectus* had an additional weapon—fire. For this reason they were probably a superpredator, whose exploits may have led to the extinction of several animal species.

Fire could be used as a weapon not only against animals but against the cold as well. Another defense against the cold used by *H. erectus* was animal skins—clothes. Now he could travel into previously hostile environments—his horizons were broadening.

Fire was used in several important ways. Peking man used it to frighten away cave bears that competed with him for the shelter of caves. In time, he excluded the bear from the cave entirely—man was developing better ways to maintain fire. Fire could also cook food previously too tough to eat. Cooked food generally required less use of the teeth and jaw, leading to reduction of both. This is an excellent example of interrelationship of cultural and morphological evolution.

### Homo sapiens

*Homo sapiens* appeared about 0.3 million years ago. The earliest fossils are found in Europe and Africa (it is not yet clear that the latter are *H. sapiens*). The *H. sapiens* skull showed further reduction of the jaw and a more prominent forehead (a larger brain). The European population was probably driven south by icy sheets, and at the next interglacial period this first European group was replaced by the puzzling Neanderthal man.

Neanderthal man had more primitive characteristics, such as an enlarged jaw, a shallow long brain case, and no chin. At about the same time that Neanderthal man lived (0.1 million years B.P. or less), similar populations lived in other parts of the world; Solo man in Southeast Asia, Rhodesia man in Africa, and Palestinus in Asia.

After the next glacial period, modern human beings appeared all over the Old World, radiating from the Middle East and North Africa. About 50 thousand years B.P., *H. sapiens* migrated across the Bering Straits and entered North America. Nomadic groups of *Homo* probably interbred, resulting in several relatively similar populations in which there may have been cultural exchange.

## Cultural Development

Cro-Magnon man decorated his caves with renderings of animals, an indication that culture and communication had progressed—man used symbols. He also decorated his body with clothes and jewelry, a sign of self-consciousness.

The morphological evolution of our species was giving away to our cultural evolution; our very survival depended on our culture. But this same culture also threatened to destroy us. Hunting techniques were so successful that some prey species became extinct, which probably stimulated development in agriculture and animal domestication. Agriculture and animal domestication (beginning with the goat) seems to have first developed in Southwest Asia about 10,000 years ago.

The development of agriculture had profound effects on *H. sapiens*. Our species lost the freedom to roam as our nomadic gatherer–hunter ancestors had, for the crops had to be tended and harvested. Larger stationary groups were formed—the beginnings of urbanization. Agricultural communities could support more life than previous hunting societies, allowing population increases, which in turn created new problems. Dense populations provided a new niche for disease organisms; increases in population demanded larger food supplies; the removal of forests and planting of homogenous (hence unstable) crops allowed insects into the crops; and deforestation, combined with the grazing of domestic animals, led to favorable selection of weed plants—and weed animals as well, such as the rat and mouse, and the bedbug.

We have long been unaware of the complexities of nature and, in our ignorance, have compounded problems with short-term solutions. This is easily seen by following subsequent cultural advances. We fought diseases resulting from urbanization, using medicines and pesticides, which led to overpopulation, poisoning the environment (through the food chain), and encouraging the evolution of hardier disease vectors. We prevented floods that occurred after deforestation, using dams to hold back nutrient-rich silt, resulting in sterile soil below. To remedy the situation, we used fertilizers that polluted the water supply, while the still water behind the dams provided excellent breeding grounds for insects. We see countless examples of short-sighted solutions to lingering problems in the present. Some of these are considered below. We are now approaching an important new stage in our cultural development: ecological foresight. With the increasing ability to assess our activities' impact on our environment, we can hopefully make more rational decisions about our actions.

Environmental changes resulting from our activities will partly determine our future evolution. An example is our creation of malaria-free environments through elimination of mosquitoes, which altered the selective pressure (and hence the allele frequency) of the sickle cell trait which, in a heterozygous condition accorded protection against the disease. When penetrated by *Plasmodium falciparum*, the delicate red blood cells would lyse, giving the parasite no roosting place.

## Impact on the Environment

We have a greater impact on our environment than any other animal. Evolution of both a highly developed brain and dextrous hands enabled us to use tools culminating in modern technology. With powerful tool technology, many of the problems of survival were overcome. This power, in a world of delicate interactions, has resulted in imbalances, some of which are irreversible. Ecological sense is not the mourning of a single squashed ant, it is concern for the environment upon which we all depend. The survival of the human species is not assured unless we use our technology wisely (Fig. 31–2).

Our survival depends on meeting basic biological requirements for life: food, water, air, and living space. When populations were small, the environment could easily provide these essentials and absorb the wastes we produced.

**Figure 31–2** Future human being? With technological developments, many new adaptations become necessary; these may be morphological and/or cultural.

Increasing populations, with their growing demands on resources and careless attitudes toward nature, have resulted in pollution of our food, water, and air. We will now explore the ways in which we have mishandled our water resources and some possible solutions to the resulting problems. The student interested in other environmental problems should refer to the selected references.

## WATER

Earth's water is confined mostly to the oceans (97 percent). The fresh water available is from lakes left by receding glaciers and rain. Water evaporated from the ocean surface by solar radiation creates warm tropical air. When this warm air encounters cold continental air, water condenses into rain. Rainwater usually finds its way to rivers and flows back to the oceans. We will never exhaust our supply of water, but we may pollute it to such an extent that it becomes useless to us and many other organisms.

Water is absolutely necessary for human life (we need approximately 1 liter per day per person). In the United States, domestic consumption accounts for about 72 billion liters of water per year. It is used to bathe, drink, prepare food, flush toilets, water gardens, and wash cars. All domestic uses represent only a small fraction of the water used for agriculture, power production, and industrial uses (cooling and waste disposal). Some of the effects of industrial uses are deoxygenation, heat loading, and toxicity.

**DEOXYGENATION.** Bacteria in water break down organic wastes, using oxygen in the process. Their numbers are usually limited by the amount of organic substances available. When the amount of organic wastes increases (and we will see how our species produces this effect), the bacteria proliferate, using up the oxygen that fish need to respire. When the oxygen levels fall off considerably, the fish die (Fig. 31–3). In addition, a lack of oxygen will cause certain organisms to emit foul-smelling and combustible wastes.

Among the many creators of organic wastes are human sewage, paper mills, petroleum refineries, and food processors (canneries and frozen food packagers). Nutrients such as phosphorus enter the water supply through agricul-

**Figure 31–3** Water pollution takes its toll of alewife fish in Lake Michigan.

**Figure 31–4** Algae, fed by some pollutants, form large mats on the surface of the water (above). In addition to disruption of food webs and habitats, such pollution renders lakes and rivers useless for drinking, bathing, and recreation (below).

tural fertilizers and detergents and cause population booms of algae that form mats on the water surface. These organisms use oxygen at night and may deoxygenate the water enough to kill fish. The mats often wash up on the shore and decompose, releasing hydrogen sulfide and making the lake unfit for any use. If the algae die on the water surface, they sink and feed the oxygen-depleting bacteria (Fig. 31–4).

Lake Erie is an example of what can happen when wastes are cast indiscriminately into the water supply. Large cities and industry have dumped wastes into the lake to such an extent that in many areas instead of a clear lake, one sees heaps of decaying algae on the shores, algae mats, and scum floating on sewage and water. Few people would care to go boating or swimming in such a lake, much less drink from it—and one must be careful not to light a match near it, lest combustible wastes explode! Even in the unlikely event that pollution of the lake were to stop immediately, conditions might not improve, since deoxygenation has removed a layer of iron, releasing phosphorus and other nutrients that were beneath the layer in the mud.

Since new freshwater supplies through the desalination of seawater are too energetically expensive to be carried out on a large scale, our supply of usable water is limited; it is only good sense not to ruin this life-sustaining material. It appears that the best way to prevent water contamination is to recycle the water and to treat water that is returned to the environment. Much water could be saved by retaining bath water for use in washing cars and flushing toilets.

Sewage treatment is another way to save water, and in some places is in use. Suspended particles and solids are removed in primary treatment. In secondary treatment facilities, another step is taken—organic wastes are removed using bacteria, chlorine, and filtration. Tertiary treatment removes nutrients such as phosphorus; this could be accomplished using holding ponds to grow algae, which could then be used as food for some animals.

**HEAT LOADING.** Warm water cannot hold as much dissolved oxygen as cold water can. **Heat loading** thus has a deoxygenating effect on water, causing the problems cited above. Many industries use water for cooling purposes; the worst offender is the electric power industry.

Electric power is generated by turning blades of turbines that turn generators. Water flowing rapidly over dams was once used to provide the power to turn the turbines. Larger demands for electricity resulted in using steampower instead of waterpower to run the turbines. Steam was produced by heating water with fossil fuels (oil, coal, and gas). As technology advanced, nuclear fuels replaced oil, coal, and gas. After it is used for power, the steam is recondensed by being passed over pipes that contain cool water (coolant); this process heats up the coolant before it is returned to the environment.

The fossil-fuel plants used approximately 50 percent less coolant in generating one unit of electricity than the nuclear plants do. The more recent nuclear breeder plants recover some heat energy but are still bigger heat polluters than the fossil-fuel plants.

Although nuclear plants do not contribute to air pollution (as did the burning of fossil fuels), nuclear power is by no means clean. In addition to producing more thermal pollution than petroleum plants, it adds another problem—disposal of radioactive wastes. The boom in construction of nuclear plants reflects their low fuel costs and the fact that fossil-fuel supplies are beginning to dry up. They will not help the environment as long as they heat up our water and produce radioactive wastes.

Some efforts have been made to curb heat pollution, the most effective (and expensive) of which is the **dry tower,** which passes cool air over pipes containing hot water, transferring the heat to the air.

Radioactive wastes, however, present a tougher problem. Nearly 1 billion liters of highly radioactive wastes are presently stored in underground tanks. These wastes will take hundreds of years to decay. The large amount of wastes produced in nuclear plants (breeder reactors included), even if condensed into solids, will require vast areas of land for disposal. The reactors themselves leak small amounts of radiation into the environment, but are insignificant compared to the amounts of radioactive dust resulting from the mining and processing of uranium.

The dust finds its way into rivers and lakes and eventually into crops, livestock, and man. The breeder reactors use large amounts of plutonium, a small fraction of which, if brought together through any number of accidents, could form a critical mass (the amount of plutonium necessary to blow up). Although there are many safeguards to prevent this from occurring, it can happen.

Clearly, alternative energy sources should be explored. Experiments using deflectors to derive energy from the sun have resulted in a small number of private households which are currently run using only solar energy. Other areas being explored are tidal energy (using the difference between high and low tides), geothermal energy, and improvement of our storage batteries.

**Toxins.** Certain metallic wastes, such as mercury and lead, interfere with vital processes and are hence poisonous to organisms that live in water with an abundance of these substances.

Plating industries, among others, dump metals that create these wastes directly into waters. Other sources of mercury are discarded batteries, coal smoke, dental wastes, and more; these all eventually enter water supplies.

Although there has always been some mercury in our environment, the levels are rising. The average weekly human intake of mercury is about 10 times greater than it was 40 years ago in developed countries. In the 1950s, over 52 deaths and 3 times as many illnesses due to eating mercury-contaminated fish were reported. In the United States, a large percentage of swordfish and some tuna were seized by the government because it exceeded legal limits of mercury concentration (0.5 parts per million).

We have explored some aspects of one way in which the human species has influenced the environment. We will now turn to a discussion of how our actions directly affect us as a species, both now and in the future.

# Population Size

Evolution results in species that ensure their own continuation by producing excess offspring. In an ecosystem, the population size is limited by available food, predators, and living space. The population size stabilizes around this limit, the **carrying capacity** ($K$) of the ecosystem. When a species invades a new econiche, the population increases rapidly until $K$ is reached. If changing conditions alter $K$, the population size will correspondingly change.

It has been demonstrated, both in laboratory and field studies, that when $K$ is raised, the population increases at a characteristic rate (Fig. 31-5). The rate depends upon several factors: the original number of individuals, the birth rate, and the death rate. The rate of increase is simply the number of individuals multiplied by the births minus the deaths.

Using this method, it has been calculated that by the year 2005 (if the carrying capacity does not intervene and the birth rate is not controlled), our population size will have doubled.

Since the size of the earth is finite, this kind of growth cannot continue for

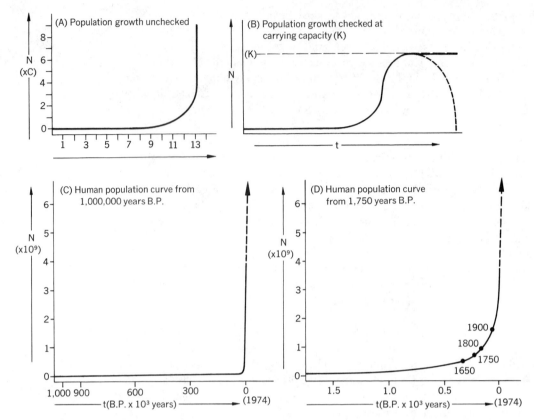

**Figure 31–5** Population growth curves. Vertical axes at left indicate the number of individuals (N) when multiplied by number indicated. Horizontal axes indicate time.

very long. Without obvious limitations, the human race would outweigh the visible universe in about 5000 years.

It is in everyone's interest to stop population increase, and even decrease it to some extent; according to some estimates, the earth can comfortably carry a population of 1 billion humans. Population increases are greatest in South America and Asia, although it is also continually increasing in the United States, Europe, and other parts of the world.

Advising people to practice birth control is not the only solution, since each American consumes many times the resources that an individual in an agricultural country does. Part of our species' problem, in fact, lies in the unequal distribution of the total amount of resources at our disposal.

The technological means for birth control, while not perfected, do exist; the pill, intrauterine devices, sterilization, abortion, and other means have proven effective against conception (Fig. 31–6). Some governments, religions, and traditions, however, are obstructing their effective use. Abortion is now the most widespread form of birth control used in the world. Where it is legal, it is safer than childbirth; in other, less enlightened countries, it is often performed by amateurs, forcing women to take extreme risks.

Will the grim predictions of Malthus and the like, who contend that our population will eventually increase to the point where our food supply could not possibly sustain us, come about? Is there hope for us as a species despite our traditional abuse of our environment?

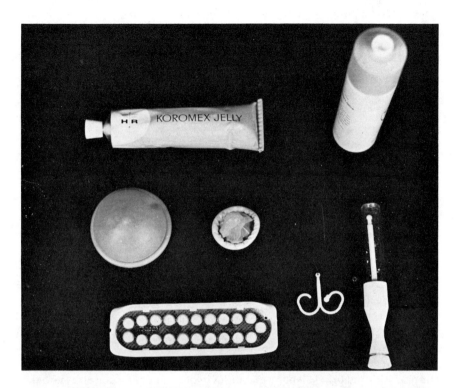

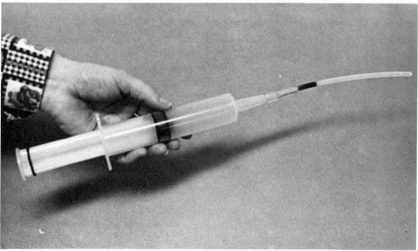

**Figure 31-6** Birth control devices. *Top:* spermicidal jelly and diaphragm (that fits over cervix), prophylactic or "rubber" (that fits over penis), spermicidal foam and applicator, I.U.D. and birth control pills (the last two are the most effective.) Surgical means of birth control (not shown) include vasectomy, where the male vas deferens is cut and tied to prevent sperm flow during ejaculation; and tubal ligation, where the female fallopian tubes are severed and tied. *Bottom:* Karmin cannula, used to gently extract fetus during early pregnancy; the syringe's plastic tube is small and passes through the cervix, providing suction. After the fifth week of pregnancy, a suction machine is usually used for abortions.

If our species were to recognize themselves as such, and our separate governments worked together for a better understanding of how we as the human species can successfully relate to nature, then our shared knowledge of ecology could open the door to a stable global, and possibly universal, existence.

## Summary

The human species are unique animals. No other species has had such a tremendous impact on the environment. The lives of every organism in the vivicum are affected by our actions. The evolutionary history of the species has resulted in a highly developed nervous system, which has enabled the development of a powerful technology.

The technological power of human beings has far outpaced our evolutionary restraints, so that we now sit precariously perched atop a barricade against nature. We prolonged life through the control of disease, but are now feeling the squeeze of overpopulation. Our method of food production has permanently upset ecosystems. We have built industrial empires, squandering irreplaceable resources to produce products, and fouling our nest with pollution in the process.

As a species, we cannot ignore nature forever. Unless we drastically change our priorities and bring all our capabilities to bear on tuning ourselves into the world, our barriers will fall to the continuous onslaught of nature, or crumble under their own weight in a catastrophe of epic proportions.

Scientists have proposed technical solutions to some of the problems resulting from technological power guided by animal impulses. But what is obstructing us from solving our problems may be our concept of ourselves as apart from and superior to nature. If we can overcome this notion of separateness, limit our population size, and act with ecological awareness, we may live a life of biological peace.

# Index

All numbers in this index refer to pages; those in bold face type contain figures or illustrations of the entries. Mention may also appear in the text on pages in bold face. Scientific names of genera and species are in italics.

## A

Ardvark, 306
A band, 355, **356**
Abalone, 201
Abdomen
 arthropod, **153**–155, 157, 162, **163**–**164**, 173–174, 188
 mammal, 307
Abdominal
 appendage, **156**
 artery, **156**
 cavity, 45, 251, 309
 vein, **252**, **257**, **271**
Abducens nerve, 231, **258**, **259**, **311**, **384**, **392**
Abductor muscle, 250, **251**
Abiogenesis, 451
Abiotic portion of ecosystem, 461–464, 471, **472**
ABO group, **377**
Abomasum, **323**
Aboral
 end, 43
 pore, 95–**96**
Absorption, 361
Abyssal zone, 470
*Acanthaster planci*, 210
Acarina, **163**
Accessory
 genital organs,
 glands, 177
Accommodation, 296, 395

Acellular organisms, 53–82
Acetabulum, 115
 (see Ventral sucker)
Acetabulum (pelvic), **248**
Acetylcholine, **387**, 388
Acid, 16, 29
Acipenseridae, **234**
Acoelomata, 45, 112–120
*Acontia*, **106**–108
Acordate, **213**
Acorn barnacles, 162
Acorn worm 214–**215**
Acromegaly, 400
Acrosomal cap, **409**, 412
Actin, 355
*Actinophilus*, **56**
Action potential, 387
Activation energy, 10–11
Active sites
 muscle, 355
 tRNA, 422
Active transport, 20
"Adam's apple," 379
Adaptation(s), 443, 445, 454, **459**, 461–464
 behavioral, 329–330
 future human, **483**
 limb, **446**, **447**
 mammal, 316, **317**
Adaptibility, 333
Adaptive
 coloration, 179, 222, 245, 288, 339 (*see* Coloration)

convergence, 446
radiation, 445, **447**
Addison's disease, 399
Adductor brevis, **249**
Adductor longus, **249**, **250**, **251**
Adductor magnus, **249**, **250**, **251**
Adductor muscle
 clam, **192**
 human, **354**
 vertebrate, **249**, **250**, **251**
Adenine (adenylic acid, A), **421**, **422**
Adenosine diphosphate (ADP), 464
Adenosine triphosphate (ATP) 20, 26, 464
 and muscle contraction, 355
 and protein synthesis, 422
Adhesive pad, **104**
Adipose tissue, **35**–**36**
 brown, 36
 white, 36
Adrenal gland, 37, 233, 260, 296, **398**, 399
Adrenaline, 385, 399 (*see* Epinephrine)
Adrenocorticotropic hormone (ACTH), 399
Adult form, 47, 177–**179**, 207, 263
*Aeolosoma*, 143
Afferent arteriole, 368
African elephant, **322**

491

African honey guide, 341
African sleeping sickness, 60, 186
Agama, **268**
Agamidae, **268**
Agglutination, 377–378
Agglutinins, 377
Agglutinogens, 377
Aggression, 338, 340
*Agkistrodon*, **278**
Agnatha, **214**, 220–222, **223**–227, 241
Agriculture, 482
Air
  pollution, 463
  sac, 175–**176**, 205, **295** (see Alveoli)
  tube, **379**
Air bladder, 233, 241
Albatross, 286, **287**, 298, 300
Albinism, 426, **429**
Albumin, **297**, 363
Alewife fish, **484**
Algae, 464, 485
Alkaptonuria, **429**
All-or-none law, 387
Allantois, **408**, **416**, 417
Allele, **424**
  frequency, 435, 445, 447
Allergic reactions, 376, 399
Alligator, 280, 281, 285
  American, 281
  brain, **392**
  Chinese, 281
Alligator gar, **234**
Alligatoridae, **268**
Altitudinal migration, 343
Altricial birds, **297**, 298
Alula, 292
Alveolar duct, 379
Alveoli, 253, 311, **379**, 380 (see Air sac)
Ambulacral groove, 203, **206**
Amblyopsidae, **234**, 239
*Ambystoma tigrinum*, 264
Ambystomidae, 243
Amino acids, 15, 363, 421, **423**
  codons for, **422**
  essential, **14**, 15
Amino end, 15
Amitosis, 78
Ammonia, 13, 464
Ammophila, 335
Amnion, **408**, **416**, 417
Amniotic cavity, **416**, 417
Amoeba, 47
  asymmetry, 43–44
  shell, 346
Amoeba proteus, 43, 63–**67**
  feeding, 59, 64–66

locomotion, 65
morphology, **64**–65
Amoebic dysentery, 68
Amoebocytes, 49, 85–**86**, 88–89, 203, 205, 366
Amoeboid
  cells 85–**86**, 88–89
  white corpuscles, 236, 254
Amoeboid movement, **65**
Amoebotaenia, 149
Amphibia, 49, 213, **214**, 216, 220, 221, 242–266
  age of, 264
  arboreal, **264**
  brain, 258
  evolution, 240
  fossil, 264
  heart, **255**
  legless, 264
  physiology, **252–261**
  regeneration, 266
  relations to man, 265–266
Amphiblastula, 90
Amphioxus, **217**–219
  external anatomy, **217**
  internal anatomy, **218–219**
Amphipoda, **161**
Amphiumidae, **243**
Ampullae
  echinoderm, **205, 206, 209, 210**
  *paramecium*, 77
Ampullae of Lorenzini, 296
*Amyda*, **283**
Amylase, 138, 363
  salivary, 361, 363
Anabolism, 5, 12
Anal
  canal, 361
  opening, 120, 136, 305
  pit, **414**
  sphincter, **360**
Analogous organs, 45, 360
Anaphase, **27–28**, 67
Anatomy, **6**
Anchovies, 240
Ancon sheep, **453**
*Ancyclostoma duodenale*, 126, 127
Anemia, **364**
  pernicious, **364**
  sickle-cell, 381, 483
Anguidae, **268**
Anguillidae, **234**
Animal interactions, 471
Animal pole, **414**
Animalia, 47–49
Animals
  basic characteristics of, 5
  classification of, 3, 40–49

relationship to man, 3–4
study of, 3–4
Anion, 9
Aniridia, **429**
Anisogametes, 62
Anisomyaria, **190**
Ankle joint, 293
Annelida, 48, **134**–150, 183, 188, 200
  origin of, 148
  relations to man, 148–149
Annular ligament, **354**
Anodonta, 191–195
*Anolis*, **273**
*Anopheles*, 70
Anoplura, **171**, 186
Anostraca, **161**
Anseriformes, **287**
Antagonistic muscle pairs, 354
Antennae, **153**–155, 159–160, 170, 172–**173**, 177, **184**, 383
Antennules, **153**, 156, 157, 160
Anthozoa, 43, 93, 105–108
Anthropomorphism, 329
Antibodies, 375, 377–378
Anticoagulant, 149
Antigens, 375, 377–378
Antimeres, 43
Antivenom, 280
Antlion, **185**
Ants, **180**, 182
Anus, 46
  amphioxus, **217**, 218
  annelid, 144, **146**, 147
  arrow worm, **132**
  arthropod, **156**, 174, **175**
  bonyfish, 236, **237**
  bryozoan, 130–**131**
  cell, **74**
  crocodilian, 280
  echinoderm, **206, 209, 210**
  embryonic development of, 46
  human, **360**
  lamprey, 226
  mollusk, **193, 196, 197**
Aorta
  arthropod, 164, 175
  bird, 293, **294**
  bonyfish, **237**
  cat, **310**
  dogfish shark, **230**
  dorsal, 219, 220, 229, **230**, **237**, 255, **257**, **271**, 293, **294**, **310**, **374**, **375**
  frog, **254–257**
  human, 371, **374**, 375
  mollusk, 193–194
  turtle, **271**

ventral, **218**, 219, 229, **230**, **237**, 375
Aortic arches, 138–**139**, 255–257, **271**, 293
Apes, anthropoid, **324**
Aphid, 182, **185**
Aphislion, **185**
Apical organ, **146**
Aplacophora, **190**
Apoda, 242
Apoenzyme, 13
Apopyle, **87–88**
Appendages, 220, 349
   adaptation, **446**, **447** (*see* Limbs, type of appendage)
   arachnid, 162
   biramous, **154–155**
   cartilaginous fish, 228
   crayfish, **154–155**
   foliaceous, 155
   lateral, 220
   pelvic, 220
   snake, 274
   and taxonomy, 46
   thoracic, 220
   uniramous, 155
   vertebrate, 220
Appendicular skeleton, 228, 236, **246-248**, 349, **351**, 352
Appetitive behavior, 335
Apposition image, **158–159**
Apteria, 291
Apterygiformes, **287**
Aquatic
   animals, 446
   birds, 288, 298, **300–301**, 304
   carnivores, 319, **320**, **321**
   mammals, 319, **320**, **321**, 322
   reptiles, 268, 283
Aquatic (*see* Fish)
Aqueous environments, 5, 469–470
Aqueous humor, **395**
Arachnida, **151**, 162–169, 184
Arachnoid, **389**, 390
Araneae, 163
Arbor vitae, 390
Arboreal characteristics, 479
Arcella, 68, 346
*Archaeopteryx*, 298, **299**
*Archaeornis*, 298, **299**
Archaeornithes, 298, **299**
Archenteron, **408**, 415
Archeocytes, 89, 91
Arches
   gill, 235
   hyoid, 235
   mandibular, 235

   paired, 235
Archiannelida, **134**, 146–147, 150
Arctic fox, **459**
Arctic tern, 343
Areola, 155
Argon, **9**
Aristotle, 3, 40, 441
Aristotle's lantern, 208, **209**
Armadillo, **306**, **317**
Arms
   human, **446**
   oral, **105**
   *Ramapithecus*, 479
Arrow worms, 130, **132**, 133
Arteries, 37
   abdominal, 165
   bird, 293-**294**
   branchial, 219, 229, **230**, 236, **237**
   cat, **310**
   caudal, 165
   frog, 254, 255
   histology, **372**
   human, 372, **374**, 375
   muscle layer, **372**
   villi, **362**
   (*see* Blood vessels; specific arteries)
Arterioles, 373
Arteriosclerosis, 373
Arthritis, 24, 399
Arthropoda, 48, 151–188
   origin, 183–184
   relations to man, 184–188
Artificial selection, 330
Artiodactyla, **306**
Ascaphidae, **243**
*Ascaris lumbricoides*, 123–126
Ascending colon, **360**, 361
Aschelminthes, 48, 121–129
Ascon, 86–**87**
Asexual reproduction, **55**, 90, 405–406
Asiatic elephant, **322**
Asiatic land salamander, **243**
Assimilation, 66
Association neurons, 35, 140, **141**, 332
Assortment, independent, **426**, 429
Aster, 24, **27–28**
*Asterias*, 203–208
Asteroidea, 203, **204**, 208
Asthma, 381, 399
*Astrangia*, 93, **107**
Asymmetry, 43–**44**, 47
Atlas, **246**, **292**, **307**, **350**
Atmosphere
   early earth, 452–453

   pollution, 463
Atoms, 8–9, 29
Atomic bonds, **10**
Atomic weight, 8
Atrial cavity, **216**
Atriopore, **217**, 219
Atrioventricular
   node, 371
   partial block, **373**
   valve, **256**, **371**, 372
Atrium, amphioxus, **217–219**
Atrium (heart), **237**, **254–256**, **271**, 293, **294**, **310**, **371**-**374**
Atrophy, 333
Audition, 266, 296–297
Auditory
   bones, 312, 349, 350
   canal, 226, 232, 296
   capsule, 228, 247, 259
   nerve, **231**, **258**, **259**, **311**, **384**, **392**, **396**
   organ, 177
   sac, 174
*Aurelia*, 93, **104–105**
Auricle (heart), **193–194**, 371
Auricles (ear), 113–114
Australian region, **455**, **456**
*Australopithecus*, 479, **480**, 481
*A. africanus*, 479, **480**, 481
*A. boisei*, 479, **480**
*A. habilis*, 479, 481
*A. robustus*, 479, **480**
Autogamy, 80
*Autolytus*, 145
Autonomic
   ganglion, **259**
   nerve trunk, **259**
   nervous system, 37, 176, 226, 259, 384–385, 390
Autophagy, 24
Autotomy, 166, 203, 208, 273
Aves, **214**, 220, 221, 286–304
   (*see* Birds)
Avoiding reaction, **77–78**
Axes
   morphological, 42–44
Axial cells, 83–84
Axial skeleton, 228, **246-248**, 307, 349, **351**
Axis, **246**, **292**, **307**, **350**
Axolotl, 264–265
Axon, **385**, **386**, **387**, **388**
Azygos vein, **310**

# B

Baboon, **306**, **323**
Bacteria, 471, 472

**493**

Bacteria (cont.)
  nitrogen-fixing, 463
Balancing organs, 105, 109, 157, 195, 226, 234, 238, 260, **396**, 397
*Balantidium*, 82
*Balanus*, 162
Bald eagle, **301**
Barbs, 290–**291**
Barbules, 290–**291**
Barn owl, 302
Barn swallow, **287**
Barnacle, **161**–162
Basal
  body, 75
  disk, 95–**96**, 97, 99, **106**, 107
  granules, 82
  membrane, 31, **141**
  pigment cells, 157–158
Base, 16, 29
Base (tunicate), **216**, 217
Basement membrane, 31, **113**, 152, 172
Basilar partition, 397
Basket star, 208
Basophil, **35**, 354, **376**
Bass, **234**, 240
Bat, little brown, 316
Bat, true vampire, 316
Batesian mimicry, 339
Bath sponge, 90–**91**
Bathyalzone, 470
Bats, 316, **317**, 341
Beaded lizard, **284**
Beard worm, 132
Beaver, 316, **317**
Beadbug, **188**
Bee, **151**, 180–182
Beetle, **171**
Behavior, 329–344
  aggressive, 338
  *Amoeba*, 66
  appetitive, 335
  arthropod, 159–160, 180–183
  communicative, 338–339
  courtship, 339–340
  earthworm, 143
  frog, 260
  homeostatic, 330
  homing, 343
  innate (instinctive), 334–335
  mating, 340
  *Paramecium*, 77–**78**
  precopulatory, 340
  sexual, 400
  social, 340–341
  territorial, 338
Bell toad, **243**
Beri-beri, **364**
Bering Strait, 460

Bicarbonates, 381
Biceps, **249**, **251**, **308**, **354**
  femorus, **308**
Bicuspid valve, **371**
Bicuspids, 360
Bidder's organ, 242
"Big game animals," 322
Bile, 229, 253, 270, 309, 361, 363, 367, 400
  duct, 229, **253**, 270, 293, **360**
Bill, **288**, **290**, 291, 295, 301, 302
Binary fission, 58–59, 62, 66–67, 78–**79**, 114–115, 405, **406**
Binomial nomenclature, 41
Biochemistry, comparative, 444–445
Biogenesis, 451–453
Biogenetic law, 46, 444
Biogenic salts, 464
Biokinetic zone, 462
Biological drives, 332
Biome, 464
  coniferous forest, **465**, **466**, 467
  desert, **465**, **466**, 467
  grassland, 464, **465**
  temperate deciduous forest, **465**, 467
  tropical rain forest, **465**, **468**
  tundra, **465**, 468, **469**
Biosynthesis, 420
Biotelemetry, 330
Biotic communities, 472, 474–476
Biotic portion of ecosystem, 464–470, 471–**473**
Biotin, **364**
Bipedal locomotion
  bird, 293
  early man, 479
  primate, 324, 325
Bipinnaria larva, 206–**207**
Biramous appendage, **154**
Bird, 49, 213, **214**, 216, 286–304
  aerial, 286
  ancient, 298, **299**
  aquatic, 288, 298, 300–301, 304
  bills, **288**, 290, 291, 295
  brain, **392**, 393
  coloration, 288
  development, **297**–298
  domesticated, 303
  ducklike, 301
  eggs, **297**–298
  extinct, 298
  falconlike, 301–302
  feathers, 290–**291**
  feet, **288**, 290, 293, 298

  flight, 286
  flightless, 298, **300**
  form and function, 286, 288
  fossil, 298, **299**
  heart, **255**
  march, 301
  metabolism, 293, 295
  migration, 289, 341, 343
  nest, **296**–298
  origin, 286, 298
  perching, 288, 298, 303
  pest control, 304
  of prey, 288, 295, **301**–302, 304
  relations to man, 303–304
  song and call notes, 288–289
  tail, 286, 290
  terrestrial, 286
  wading, 288, **300**, **301**
  wings, 286, 290–293
Birth control, 488–**489**
Biting louse, **171**
Bittern, 300
Black
  andalusian fowl, 427, **428**
  ant, little, **188**
  snake, 275
  widow spider, 162, **167**
Bladder, 38, **366**
  bony fish, **237**, 238
  cat, **309**, 311
  frog, **252**, 253
  human, 367–369, **410**, **411**
  rotifer, 121
  turtle, **272**
    accessory, **271**, 272
Bladder worm, 119
Blastocoel, 146, **408**, **414**
Blastocyst, 417, **418**, 419
Blastoderm, 238, **239**, 414
Blastodisc, **297**, **308**, 415
Blastomeres, 412
Blastopore, 46, **414**
Blastopore, 131
Blastostyle, 102–**103**
Blastula, 46, **101**, **103**, **105**, 142, 207, 413, 414, 415
Blending, genetic, 445
Blind cavefish, **234**, 239
Blood, 31, 35, 37, 375
  arthropod, 165
  bony fish, 236
  clothing, 254, **364**, **376**, 377
  constituents of, 375–377
  earthworm, 138
  flow, 370
    frog, **256**, 257
    crayfish gill, **157**
    human heart, **371**
    human circulation, **374**

frog, **254**
functions of, 375, 380
groups, 377–378
  Rh, 323, 377
  ABO, **377, 429**
  human, 377–378
  other, 377
  MN, 427
human, 375–378, 382
mollusk, 197
respiratory function of, 380
sugar level, 399, 400
transfusions, 377
vessels, 37, **136,** 137, 138–**139**
Blood flukes, 199
Blood islands, 417
Bloodletting, 149
Blood-vascular lymphatic system, 36–37 (see Circulatory system)
Blue Andalusian fowl, 427, **428**
Blue racer, 275
Blue whale, 322
Boa constrictor, 274–275, **276**
Bobwhite, 286
Body
  cavity, 45, 137
  column, 95
  form, 446
  plan, 151, 189, **191,** 202, **213,** 220
  wall, 124, **137,** 189, 208, **217,** 251 (see Coelom)
Body lice, 187
Bodidae, **268**
Bolus, 361
Bond, chemical, 10–11, 29
Bone(s), **31,** 34
  auditory, 312, 349, 350, **396**
  cartilage, 307
  conductor, 396–397
  development, 347–348, 356
  growth, **348**–349, 356
  hollow, **295,** 304
  marrow, 38, 254, **348,** 356, 376
  membrane, 307
  sasamoid, 307
  skull, 352 (see Skull)
  vertebrae, 219, 220
Bony
  basal plate, 228
  plates, 223, 227, 233, 267, 269, **270,** 280, 345
  shield, **223**
Bony fishes, 49, 222, 233–241
  brain, 391, **392**
  circulatory system, 236, **237**
  development, 238
  digestive system, 236

excretory system, 238
external anatomy, 233–234
internal anatomy, 234–238
muscular system, 236
nervous system, 238
parasites of, **195**
reproduction, 238
respiratory system, 236
sense organs, 238
skeletal system, 234–236
Book gills, **168**
Book lungs, 48, **164,** 165
Booklouse, **171,** 188
Boophilus, **168**
Boreal forests, 467
Boring sponges, 91
Bowman's capsule, **367,** 368
Box turtle, 283
Brachial
  nerves, **259**
  artery, 255, **257,** 310, 375
Brachiolaria, 207
Brachiopoda, 131, **132**
Brachydactyly, **429**
Braconid fly, **185**
Braconid fly, **185**
Brain, 37, 401,
  amphioxus, **217,** 219
  annelid, **136,** 140, **141,** 147
  arrow worm, **132**
  arthropod, **156,** 165, **175, 176**
  aschelminth, **122**
  bird, 295
  bony fish, **237,** 238
  cat, **311**–312
  dogfish shark, **231**–232
  frog, **258**–260
  human, 384, 390–**392,** 478
    ventral surface, 391, **392**
  human ancestors, 481
  lamprey, **225,** 226, 227
  platyhelminth, **113**–114
  turtle, 272
  vertebrate, **220,** 391, **392,** 393
Brain case, **247,** 352, 479, 481
Brainstem, 390, 401
Branchial
  arches, 228, 231
  arteries, 219, 229, **230,** 236, **237**
  basket, 225
Branchiocardiac grooves, 153
Branchiostegites, 155
*Branchiostoma lanceolatus,* **217**
Breathing, 380, 381
Breeder reactors, 486–487
Breeding population, 446
Breeds, 330
Bristles, **132,** 135

Brittle star, 203, **204,** 208
Bronchial tubes, 37
Bronchioles, **379**
Bronchus, 253, 271, **295,** 311, **379**
*Brontosaurus,* 281
Brood pocket, **300**
Brood pouch, **195**
Brown recluse, 162
Brownian motion, 17
Bryozoans, 130–**131,** 133
Bubonic plague, **186**
Buccal
  cavity, 137–138
  funnel, 224–**225**
Budding, 90, 100, 102, 107, **406,** 407
  external, 407
  internal, 407
*Bufo americanus,* **244,** 263
*Bufo marinus,* 265, 266
*Bufo terrestris,* 263
Bufonidae, **243,** 263
Bufotoxins, 265
Bulbourethral glands, 312, 410
Bulbus arteriosus, **237**
Bullfrog, **243,** 244 (see Frog)
Bumblebees, 180
Burrowing owl, 302
Bursa fabricii, 293
Butterfly, 178–**179**

# C

Caddisly, **171**
Caecilians, 242, **243**
Caeciliidae, **243,** 264
Caiman, 280
Calcaneus, **239,** 351
Calcarea, 85
Calcareous plates, 203, 208, 210
Calciferous glands, 138
Calcium, 347, 362, 399
California sea lion, 319, **321,** 338
Call notes, 288–289
*Cambarus,* 153–161
Cambrian period, 130, **456**
*Camerina,* 69
*Cambarus bartoni*
  classification of, 42
Camouflage, 222, 245, 288, 339
*Campodia Staphylinus,* 177
Canal
  central, **218,** 219
  digestive, 220
  radial, 88, **104**–105, 203, **205, 206,** 209
  systems, 86–88

**495**

Canal of Schlemm, 395
Canaliculi, 348
Cancellous bone tissue, **348**
Canines, **313**, **314**
Capillaries, 35, 37, 138, 372, 373
    frog, 256
    gill, 236
    head, **237**, 374
    human, 372, 373, 374
    intestinal, 219, **237**, 374
    kidney, **237**, 374
    lower body, **374**
    lung, **374**, 379
    lymph, 375
    tail, 237
Carapace
    arthropod, 153, **157**, **168**
    turtle, 269, **270**, 347
Carbaminohemoglobin, 381
Carbohydrates, 14, 363, 369
Carbon, 9
    cycle, 463
    isotopes, **9**
Carbon dioxide, 11, 37, 355, 365, 378, 379, **380**, 381, 463
Carboniferous period, 264, **456**
Carboxyl function, 14–15
*Carcharodon carcharias*, 229, 233
Cardiac
    glands, 361
    orifice, 361
    stomach, 155–**156**, 205, 253
Cardiac muscle, 38, 248, 250, 354–355, 371
    tissue, **31**, **33**
Cardinal
    sinuses, 229, **230**
    veins, **230**, 237
Caribou, 341
"Carnival," 340–341
Carnivora, **306**, 316, **319**
Carnivores, 358
    annelids, 148
    aquatic, 319, **320**, **321**
    arthropods, 155, 165, 169
    bony fish, 236
    coelenterates, 93
    ctenophores, 110
    dentition of, **359**
    echinoderms, 205
    lamprey, 224, 225
    mammals, 305, 316, **319**
    mollusks, 190, 197
    placoderm, **227**
    shark, 233, 246
    snakes, 274–**277**
Carotid, 255

arch, **256**
    common, 255, **257**, 310, 375
    external (lingual), 255, **257**, 310, 375
    gland, **256**, **257**
    internal, 310, 375
Carp, **234**
Carpal ligament, **354**
Carpals, **246**, **249**, 292, **307**, 308, **351**, **352**
Carpet beetle, **188**
Carpometacarpus, **292**
Carrier
    coenzyme function, 13
    membrane, 20
Carriers, 426
Carrying capacity (K), **487**, **488**
Cartilage, 347
    articular, 352
    bird, 295
    costal, 352
    dogfish shark, 228
    elastic, 34
    fibro, **34**
    frog, 247
    hyaline, **31**, **34**
    lamprey, 224, 225
    squid, **196**
    vertebrae, 219, **220**
Cartilaginous fishes, 227–233, 241
Cassowary, **287**, 298
Castes, **181**–**183**
Castings, 135, 148–149
Casuariiformes, **287**
Cat, domestic, 305–312, 316, 325–326
    brain, **311**, 312, **392**
    circulatory system, 309–311
    digestive system, **309**
    excretory system, 311
    external anatomy, 305, 307
    foot posture, **314**
    internal anatomy, 307–312
    muscular system, **308**–309
    nervous system, **311**–312
    reproductive system, 312
    respiratory system, 311
    sense organs, 312
    skeletal system, **307**–**308**
Catabolism, 5, 12, 361
Catalyst, 12, 362
Cataract, 395
Caterpillar, 178–**179**
Cation, 9
Cattle grub, **187**
Caudal
    artery, **230**, **294**, 310
    vein, **230**, 204, 310
Caudal fins, 132

Caudal muscle, **308**
Caudal vertebrae, 291, 292, **307**
Caudata, 242
Caves, 481
Cavitation, 417
Cecum, 130–**131**, **164**, 165, **196**, 197, 293, 309, **360**, 361
    hepatic, 205
    intestinal, 205, **206**
    pyloric, 205, **206**, 236, **237**
Cell, 5, 8
    amoeboid, 88
    axial, 83–**84**
    chemistry, 11–16
    collar, 85–**89**
    damaged, 26
    daughter, 27, 29, **116**, 412
    division, 27–29, 30
    fusion, 438
    hybrid, 438
    hydra, 95–**97**
    membrane, 17–**21**, 29, 30
    morphology, 16–26
    number, 46
    sex, 29, 428
    size, 17
    somatic, 83–**84**
Cellular respiration, 26, 378, 381
Cellular specialization, 85, 93
Cellulase, 138
Cellulose, 361
Cement, **313**, **314**
Cement glands, 121–**122**
Cenozoic era, 314, **456**, **458**–459
Centipede, **151**, **169**, 184
Central cavity, 85–89
Central nervous system (CNS), 35, 37, 140, 156, 384, 400 (*see* Nervous systems)
Central suicus, **391**
Centrarchidae, **234**
Centrioles, **18**, **24**, **27**–29
    sperm, **409**, 410
Centromere, 28–29
Centrosphere, **18**, 24
Centrosome, 24, 28
*Centruroides sculpturatus*, 168
*Cephalaspis*, **223**
Cephalization, 45, 112, 145, 149–150, 383
Cephalochordata, 215, 217–219
Cephalopods, 189–**191**, **196**–**198**
    evolution of, 200
    parasites of, 83–84
Cephalothorax, **153**, 162, **163**–**164**, 188
*Ceratium*, 60

Cercariae, 116–117
Cere, **290**
Cerebellum, 391–393, 401
  alligator, **392**
  bird, 295, **392**
  bony fish, **237**, 238, **392**
  cat, 311, 312, **392**
  dogfish shark, **231**, 232, **392**
  frog, **258**
  human, 390, **391**, 392
  lamprey, 226, **392**
  turtle, 272
Cerebral
  cortex, 390
  ganglion, **136**, 140, **141**; 144, 147
  hemispheres, 226, **231**, 258, 272, 311, 390, **391**
  vesicle, 219
Cerebropleural ganglion, **194**
Cerebrospinal fluid, 390
Cerebrovisceral connective, **194**
Cerebrum
  alligator, **392**
  bird, 295, **392**
  bony fish, **237**, 238, **392**
  cat, 311, **392**
  dogfish shark, **231**, 232
  frog, **258**, **392**
  human, 390, **391**, 392
  lamprey, 226, **392**
Cervical groove, **153**
Cervical vertebrae, 291, **292**, **307**, **350**, **351**
Cestoda, 112, 117–119
Cetacea, **306**, **320**, 322
Chaetoderma, **190**
Chaetogaster, **134**
Chaetognatha, **132**
Chaga's disease, 60, **186**
Chalaza, **297**
Chamaeleonidae, **268**
Chamber, sponge, **87**
Chambered nautilus, **190**, **198**
Chameleon, **268**, 273, 339
  American, **273**
Charadriiformes, **287**
Cheese skipper, **188**
Cheetah, 316
Chelicerae, 162, 163, **164**, 166, **168**
Chelicerata, 162–169
Chelidae, **268**
Cheliped, **153**, 155, **156**, 159
*Chelonia*, **269**, 284
Chelydridae, **268**
Chemical bond, 10–11, 29
  covalent, **10**
  ionic, **10**
Chemical reactions, 10, 13, 29

types of, 15–16
Chemical(s)
  environmental, 452–464
  reactions to, 66, **77–78**
  sense, 160
Chemoreceptors, 393
Chemosensory apparatus, 383, 393
Chicken louse, **187**
Chicken mite, **168**, 184
Chickens, 303
Chigger mite, **168**
*Chilodonella*, **81**
*Chilomastix mesnili*, 61
*Chilomonas*, 59
Chilopoda, **151**, **169**
Chimeras, 228
Chimney swift, **297**
Chimpanzee, **324**, 335, 340–341
Chinchilla, 316, **319**
Chipmunk, 316
Chiroptera, **306**
Chitin, 153, 346, 347
Chitinous
  bristles, 135
  cuticle, 123
  jaws, 144
  ossicles, 155
  rods, 123, 194, **195**
  shell, 91, **101**, 130
  teeth, 148, 155, 190
  tube, 347
Chitin-like jaws, 121
Chitons, 189–**191**, 200
*Chlamydomonas*, 60, **406**
Chloragogue cells, **137**, 138, 140
*Chlorohydra viridissima*, 94
Chlorophyll, 58, 358
Chloroplast, **57**–58
Choanocytes, 85–89
Choanoflagellata, 92
Cholecystokinin, 400
Cholesterol, 363
Chondrichthyes, **214**, 220, 222, 224–233
Chondrin, 34
Chordae tendineae, 371, 372
Chordata, 49, 213–221
  body plan, **213**
  characteristics, 213, 217, 220
  evolution, 209, 213–216, 219, 221
Chorion, **408**, **416**, 417
Choroid coat, 394, **395**
Chromatids, 28–29
Chromatin, **18**, 21
Chromatophores, 58, 59, 159, 198, 222, **245**, 260, 397
Chromosomes, 21, 420

diploid number, 29, 429
haploid number, 29, 429
homologous, 424, 436
in mitosis, **27**–29
replication, 28
sex, 430
(3°) abnormalities, **430–431**
Chromosome maps, 437
Chrysalis, **179**
Chyme, 361
Ciconiiformes, **287**
Cilia, 54, 56, **74–76**, 122, 191, 205
  structure, **25**
Ciliary body, 395
Ciliary muscle, **395**
Ciliary ring, **146**
Ciliated pit, 48
Ciliates, 47, 54, 73–82
  freshwater, 73–80
  parasitic, 82
Ciliophora, 47, 56, 73–82
  parasitic, 56, 82
Circadian rhythms, 476
Circular DNA, 26
Circular muscle
  earthworm, **137**
  intestinal, **362**
  hydra, **96–97**
  planarian, **113**–114
Circulatory system, 36–37, 38, 370–378
  amphioxus, 219
  arthropod, 165, 174, 175
  bird, 293–295
  bony fish, 236, **237**
  cat, 309–**310**, 311
  chordate, 221
  closed, 197
  dogfish shark, 229
  earthworm, 138–**139**
  frog, 254–**257**, 266
  human, **371**–378, 381
  invertebrate, 370
  mollusk, 193–194, 197
  open, 165, 175, 370
  turtle, **271**
  vertebrate, 371
Circumesophageal connective, **175**
Circumoral nerve ring, 206
Circumpharyngeal connective, 104, **136**, **141**, 157
Cirri, 144, 218
Cirripedia, **161**
Cladocera, **161**
Clams, 49, 189–**191**, 201
  freshwater, 191–195
    anatomy, **192**, **193**
    circulation, 193–194

Clams (cont.)
  (3°) digestion, 192–193
  (3°) dispersal, 195
  (3°) excretion, 194
  (3°) larva, **195**
  (3°) life cycle, **195**
  (3°) nervous system, 194–195
  (3°) physiology, 192–195
  (3°) reproduction, 195
  (3°) respiration, 194, **195**
  (3°) sense organs, 195
  (3°) shell, **192**, **193**
Claspers, 228, **232**–233
Class, 41
Classical conditioning, 333
Classification of organisms, 40–47, 49
  natural and artificial, 41
Clavicle, **307**, 308, **351**, 352
Clavobrachialis, **308**
Clavotrapezius, **308**
Claws
  arthropod, 163, 169, **173**–174
  bird, **288**, 290, 293, 298, **299**, 301, 302
  mammal, 307, 313
  onychophoran, **184**
  poison, 169
  turtle, 269
Cleavage, 29, 46, 312, **408**, 412, **413**, 414
  annelid, 142, 145
  arthropod, 159
  bony fish, 238, **239**
  complete, **408**
  determinate, 414
  discoidal, **408**, **413**, 414
  holoblastic, 46
    (3°) equal, 412, **413**
    (3°) unequal, 412, 413
  hydra, **101**
  indeterminate, 414
  meroblastic, 46, **408**, **413**, 414
  radial, 412, **413**
  spiral, 145, 149, 412, **413**
  superficial, 159, **413**, 414
Clevate furrow, 405, 412
Climatic migration, 343
Climax community, 474, **475**
Clitellum, 135, **147**
Clitoris, 312, **411**
Cloaca, 121–**122**, 123, 124, 229, **230**, **232**, **252**–253, 270, **271**, **272**, 293, 296
Cloacal opening, 121–122, 123, 229, **230**, 233, 245, **252**, 253, **260**, **261**, **271**, **272**, 280, 293
Clothes, 481

Clothes moth, **188**
Clypeus, 172–**173**
Cnidaria, 47, 93–109, 111 (see Coelenterata)
Cnidoblast, **98**
Cnidocil, **98**
Cnidospora, 55, 59, 73, 82
  parasitic, 73
Coachwhip snake, **268**
Coat-of-mail shells, 200
Cobra, **277**
  venom, 276
Cobra de capello, 276
Coccidians, 73
Coccygeoiliacus, **249**
Coccygesacralis, **249**
Coccyx, **350**, **351**
Cochlea, 396–397
  bird, 295
  cat, 312
  human, **396**–**397**
Cockroach, **188**, 340
Cocoon, 142, 148, 166, 178, **185**
Cod-liver oil, 241
Codfish, 241
Codling moth, **185**
Codominance, 427
Codon, **422**
Coelacanth, **234**
Coelacanthidae, **234**
Coelenterata, 47, 93–109, 111
  origins of, 108–109
  (see specific classes and animals)
Coeliac
  artery, **230**, 257, 271, 294
  axis, **310**
Coelom, 45, 48
  amphioxus, **218**
  annelid, 135, **137**, 144, 145, 149–150
  arthropod, 155
  echinoderm, **206**, 209, 210
  hemichordate, 214
  minor coelomate phyla, 130–133
  molluscan, 189
  and taxonomy, 45
  vertebrate, **220**
Coelomates, minor phyla, 130–133
Coenosarch, **103**
Coenzyme, 13, 363
Cold-blood animals (see Poikilotherms)
Coleoptera, **171**, **185**, 188
Collagen, 33–34
Collar cells, 85–89
Collared lizard, **268**
Collecting tubule, **367**, 369

Collembola, **171**
Collenchyma, 33
Colloblasts, 110
Colloid, 17
Colon, 37, 174, **175**, 253, 309, **360**, 361
  ascending, **360**, 361
  descending, **360**, 361
  sigmoid, **260**, 361
  transverse, **360**, 361
Colonial organisms
  coral, **107**–**108**
  hydroid, 102–**103**
  insects, 180–183
  *Volvox*, 54–**55**
Colorblindness, 424, 431–432
  red-green, **429**, **434**
Coloration
  amphibian, 245, 263, 266
  bird, 286, 288
  coral, 109, 207
  ctenophora, 109
  fish, 222
  insect, 179
  mimicry, 339
  mollusk, 397
  reptile, 273, 275, 278, 283
  skin, **429**
  sponge, 90
  warning, 263, 265
Colubridae, **268**
*Columba livia*, 303
Columella, 259
Columnar epithelia, 31–32
Comb jellies, 48, **109**–**110**, 111
Comb plates, 48, **109**–**110**
Commensalism, 341, **342**
Communication, 338–339
  aquatic mammal, **320**, **321**
  auditory, **320**, **321**, 338
  bee, 182, 338–**339**
  human, 482
Community
  periodicity, 476
  stratification, 476
Compact bone, 348
Compatibility, blood group, 377–378
Competition, 471
Compound eyes, 157–159, 176, 338, 394
Compounds, 10
  in the cell, 11–15
Concentration gradient, 20
Conditioning, 333–334, 344
  classical, 333
  operant (or instrumental), 333–334
Conducting nerve cells, 97 (see Motor nerve cells)

Conductivity, 383
Conductors, 383
Cones, 394
Coniferous forest, **465, 466,** 467
Conjugation, 78–80
Connective neurons, 386 (*see* Interneurons)
Connective tissue, **31,** 33–34, 36, 39
Connectives, 195
  circumesophageal, **175**
  circumpharyngeal, **136,** 140, **141,** 157
  cerebrovisceral, **194**
Constrictor muscle, **279**
Consumers, 471, **472, 473**
Continental drift, 457–458, 477
Continental permanence, 455, 457
Contraction, 31, 33
Contraction, muscle, 355–356
Controlling sites, 423
Conus arteriosus, 254–**256**
Convergent evolution, 446
Convoluted tubules, **367**
  distal, 368, 369
  proximal, 368
Convolutions, **311**
Cooper's hawk, 302
Coot, **287**
Copepoda, **161**
Copper, 363
Copperhead snake, **278**
Copulation, 140, **142,** 148, 233, 272, 296, 340, 412
Corals, 48, 93, 107–109
  atolls, **108,** 109
  islands, 108
  polyps, **107**–108, 346
  reefs, 108
  uses of, 109
Copulatory
  organ, 267
  sac, **113**
Coracoid process, 291, **292,** 308
Coral king snake, 275
Coral snake, **268,** 275, 287
  Arizona, 278
Cormorant, 303
Cornea, 157–**158,** 176, 238, 394, **395**
Corona, **122**
Coronary arteries, 310, 375
Corpora quadrigemina, 393
Corpus luteum, 399, 400, **411,** 412
Corpuscles, 138, 375
  human blood, **376**
  red (*see* Red blood corpuscles)
  white, 175, 236, **254,** 310
Corrodentia, **171, 188**
Cortisol, 399
Cortisone, 24, 299
Costal cartilages, 352
Cottonmouth, **278**
Cotylosaurs, 264
Cougar, 316, **319**
Courtship behavior, 339–340
  cockroach, 340
  flicker, 340
  mantid, 340
  and speciation, 471
  spider, 166, 340
  turtle, 272
Covalent bond, 10
Cow, 322, **323**
Cowbird, **296**
Cowper's glands, 312, 410
Coxa, **173**–174
Coxal glands, 48, 165
Crab louse, **171**
Crabs, 151, 184, 240
  fiddler, 162
  hermit, 162
  king, 162
  shore, **161**
  spider, 162, 166
Cranial nerves, **384**
  bony fish, 238
  cat, **311,** 312
  distribution, **384**
  dogfish shark, **231,** 232
  frog, **258,** 259
  function, **384**
  human, 391, **392**
  lamprey, 226
  origins, **384**
  turtle, 272
Cranium, 216, 347, 349
  bird, 291
  dogfish shark, 228
  frog, **247**–248
  human, 249, 351
  lamprey, 225
Crayfish, 48, 152, **153**–161
  behavior, 159–160
  development, 159
  digestive system, 155
  endocrine glands, 159
  excretory system, 156
  external anatomy, **153**–155
  fertilization, 159
  as food, 184
  internal anatomy and physiology, 155–159
  locomotion, 160
  muscular system, 159
  nervous system, 156
  reaction to stimuli, 160
  regeneration, 160–161
  reproduction system, 159
  sensation, 160
Cretaceous period, 282, **456,** 459
Cretinism, 399
Cribellum, 166
Cricket frog, 264
Crimson-spotted newt, 263
Crinoidea, 203, **204,** 208
Cristae mitochondriales, **25**–26
Critical mass, 487
Cro-Magnon man, **481,** 482
Croaking, 254
Crocodile, **268,** 280
  African, 280–281
  American, **280**
  heart, 271
Crocodilians, 267, **280**–281, 285, 371
Crocodylidae, **268**
Crop, **136,** 138, 147–**148, 290,** 293
Cross, 330, 425
  dihybrid, 426–**427**
  monohybrid, 425
Cross-fertilization, 407
Crossover, 436, 437
  frequency, 437
Crotalidae, **268**
Crustacea, **151,** 152–162, 184, 188
Cryptobranchidae, **243,** 264
*Cryptobranchus alleganiensis,* 264
*Cryptomonas,* 60
Crystalline cone, 157–**158**
Ctenoid scales, 235–**236**
Ctenophora, 48, 93, 109–111
Cuboidal epithelia, **31**–32
Cubshark, 240
Cuckoo, 296
"Cues," 335
Culture
  evolution, 482–483
  origins, 481
Cupula, 396
Cuspids, 360
Cutaneous
  artery, 255, **257**
Cutaneous abdominis, **249**
Cuticle, 347
  annelid, 136, **137,** 141, 144–145
  arthropod, **152,** 153, 159, 172, 174, 175
  aschelminth, 121–**122,** 123–**124**
  platyhelminth, 115–116
Cuticulin, 347
Cutting plates, 126

**499**

Cuttlefish, 201
Cyanocobalamin, **364**
Cycloid scales, 235–**236**
*Cyclops viridis,* **161**
Cyclosis, 76–77
Cyclostomata, 221, 222, 223–227, 241, 391, **392**
Cyprinidae, **234**
Cyst, **406**, 407
   division in, **58**–59
   *Opalina,* 62–**63**
   platyhelminth, 115–117
   sporozoan, 69–70, 72
   wall, 59, 69–**70**
Cysticercus, **118**–119
Cytokinesis, 29
Cytology, **6**
Cytopharynx, **57**–58
Cytoplasm, 5, **18**, 27–28
Cytosine (cytidylic acid, C), **421**, **422**
Cytostome, 57, 76

# D

Daddy longlegs, 162
Dams, 483
Damselfly, **171**
Dance
   bee, **338**–339
   chimpanzee, 340–341
Daphnia, **161**
Darwin, C., 40, 440, 441–443
Daughter
   cell, 27, 29, **116**, 412, 428
   molecules, 420, **421**
Deaf-mutism, **429**
Decapoda, **161**, 162
Deciduous dentition, 314
Deciduous forest, **465**, 467
Decomposers (reducers), 471, **472**, **473**
Decussation, 390
Deer fly, **186**
Defecation, 361
Deforestation, 482, 483
Dehydration, 16
Deltoid, **259**, **308**, **354**
Deme, 446
Demospongiae, 85
Dendrites, 333, 385
Dentritic
   spines, 333
Dentine, 228, **313**
Dentition, **358**–359
   (*see also* Teeth)
Deoxygenation, 484–486
Deoxyribonucleic acid (DNA), 21, 420, 438

double-helix structure, **420**
   mitochondrial, 26
   replication, 420, **421**
Deoxyribonucleotides, **421**
Deoxyribose, 422
Deplasmolysis, 17, **19**
Depolarization, 386–387
Depressor mandibularis, **249**
Depressor muscle, **250**
Dermacentor, **168**, 184
Dermal
   barbels, 238
   branchiae, 205, **206**
   epithelium, 85
   papillae, 290, 313, **346**
   plica, 245
Dermanyssus, **168**
Dermaptera, **171**
Dermatitis, **364**
Dermis, **346**
Dermochelidae, **268**
*Dermochelys,* 283
Dermoptera, **306**
Desalination, 486
Descending colon, **360**, 361
Desert, **465**, **466**, 467
Dessication, 267, 463
Deuterostomes, 46, 131
Development, 412, 419
Development
   *Amoeba,* 66
   arthropod, 159, 177–**179**
   bird, 296–298
   bony fish, 238–**239**
   cat, 312
   earthworm, 142
   frog, 261–263
   hydra, **101**
   lamprey, 227
   mollusk, **191**
   skeletal, 347–349
Devonian period, 228, 264, **456**
Destral, 198
Diabetes mellitus, 400
Diamondback terrapin, **284**
Diaphragm, 45, 308, **309**, 310, 311, 380
Diaphysis, **348**, 349
Diarrhea, **364**
Diastole, 373
Dibranchia, **190**
*Dicyema,* 83–84
*Diemictylus viridescens,* 263
Diencephalon, 226, **231**, **258**, 311
*Difflugia,* 68, 346
Diffusion, 17, **19**
Digestion
   extracellular, 99, 111, 359
   intracellular, 65, 76, 89, 99,

111, 114, 359
   vacuoles, 24
Digestive filament, **106**
Digestive glands, 155–**156**, **157**, **193**, 253
Digestive system, 37, 38, 369
   *Amoeba,* 65–66
   amphioxus, 218
   annelid, 137–138, 144, 148, 149
   arrow worm, **132**
   arthropod, 155–156, 165, 174
   aschelminth, 121–**122**, 123–**124**
   bird, 293
   bony fish, 236
   brachiopod, 131
   bryozoam, 130, **131**
   cat, **309**
   dogfish shark, 228–229
   echinoderm, 205
   frog, 252–**253**, 266
   human, **260**–264
   hydra, 99–**100**
   lamprey, **225**–226
   mollusk, 192–193, 197
   nemertine, 120
   *Paramecium,* **74**, 76–**77**
   platyhelminth, **113**–114, 115
   sponges, 89
   turtle, 269–270, 285
   vertebrate, 220
Digestive tract, 38
   presence and taxonomy, 46
   (*see* Digestive system)
Digitation, 419
Digitigrade, 307, **314**
Dogits, 269, 280, 292, 313, 314
Dihybrid cross, 426–**427**
Dimorphism
   chromatic, 228
   seasonal, 179, 288
   sexual, 179, 288
*Dinobryon,* 60
Dinoflagellate, 59
Dinophilus, **134**
Dinosaurs, 281–**282**, 285
Dioecious organisms, 47, 407
   annelid, 144, 145, 150
   arthropod, 159, 165
   aschelminth, 121, 123
   cephalochordate, 219
   coelenterate, 100, 107
   echinoderm, 206, 210
   mollusk, 190, 195
   nemertine, 120
   vertebrate, 226, 233, 238, 260
Diploblastic phyla, 46, 415
Diploid
   chromosome number, 29, 429

Diplopoda, **151**, 169–170
Dipolar moment, 11
Dipole, 11
Diptera, **171**, **185**, **186**, **187**, **188**
Directional selection, **448**
Disaccharides, 363
Discoglossidae, **243**
Disruptive selection, **448**
Distal convoluted tubule, 368–369
Distribution, animal, **458**–460
  factors in, 470–471
Distribution, discontinuous, 184
Diurnal animals, 476, 481
Diurnal birds, 301
Divergent evolution, **447**
Division of labor, 180
DNA polymerase, 421
Dodo, 298
Dog
  ascarid, 128
  teeth, **313**
  hookworm, 126
Dogfish shark, 222, 228–233
  circulatory system, 229
  digestive system, 228–229, **230**
  external anatomy, 228
  nervous system, **231**
  notochord, **220**
  respiratory system, **231**
  sense organs, 231–232
  skeletal system, 228
  urogenital system, **232**–233
Dolphin, **320**, 322
Domesticated birds, 303
Domestication, 358, 482
Dominance, 424, 427
  incomplete, 427
  lack of, 427, **428**
  partial, 427
Dominant-subordinate hierarchies, 340
Dominants, ecological, 474
Dorsal lip, **414**
Dorsal blood vessel, 37, **136**, **137**, 138–**139**
Dorsal root, 258, **389**, 390
  ganglion, **389**, 390
Dorsalis scapulae, **249**
Dorsolateral fold, 245
Down's Syndrome, **432**
Dracunculus medinensis, **127**
Dragon lizard, 273
Dragonfly, **178**
Drones, 180–182
Drosophila melanogaster, 430, 435–436
Dry tower, 486
Duck, 288, 301, 303

Duck-billed platypus, 315
Ducklike birds, 301
*Dugesia tigrina*, 112
Dumdum fever, 60
Duodenum
  cat, 309
  frog, **253**
  human, 161, **398**
Dura mater, **389**, 390

# E

Ear, 226, 232, 238, 259, 305, **396**,
  drum (*see* Tympanic membrane)
  human, 396–397
  inner, 259–260, 312, **396**–397
  mammalian, 396
  middle, 259, 312, **396**
  opening, **290**
  ossicles, 312, 349, 350
  outer, 312, **396**
  stones, 238, 397
  vestibule, **396**
Earth
  age, 452
  creation of, 441
  early atmosphere, 452
  morphology, 455–458
  zoogeographical regions, 455, **456**
Earthquakes, 457
Earthworm, 48, 134, 135–143
  behavior, 143
  circulatory system, 138–**139**
  development, 142
  digestive system, 137–138
  excretory system, 139–140
  external anatomy, 135–137
  grafting, 142
  internal anatomy, **137**
  learning, 143
  nervous system, 140, **141**
  parasites of, 69–**70**, 149
  reaction to stimuli, 143
  regeneration, 142
  reproductive system, 140, 142
  respiration, 139
  sense organs, 140, **141**
Earwig, **171**
Eccles, J., 333
Echidna, **315**
Echinodermata, 49, 203–210, 213
  origins, 209
  relations to man, 210
Echinoidea, 203, **204**, 208
Echiuroids, 132–**133**

Ecological dominance, 474
Ecological succession, 474, **475**
Ecology, **6**–7, 460–470, 477
Econiche, 471
Ecosystem, 461, 471–472
  abiotic portion, 461–464
  biotic portion, 464–470, 471–473
  freshwater pond, **472**
Ectoderm, 46, **94**, 101, **414**, **415**
Ectoplasm, **64**–65, **69**, **74**
Ectoprocts, 130
Ectotherms, 462
Edema, 375
Edentata, **306**, 316, **317**
Eels, 222, **234**, 239
Effectors, 35, 140, 332, 383, 384
Efferent arteriole, 368
Egestion, 65–66, 99, 114, 205
Egg, 38, 407
  annelid, 142
  arthropod, 159, 166, 177, 180
  *Ascaris*, 124
  bird, 296, **297**, 298
  bony fish, 238, **239**
  cat, 312
  cleavage, 29, 46, 412, **413**, 414 (*see* Cleavage)
  echinoderm, 206–**207**, 210
  fertilization, 407 (*see* Fertilization)
  frog, 242, **261**, **262**
  human, 410, 412
  hydra, 100–**101**
  lamprey, 226–227
  mammal, 315
  mollusk, 190–191, 195
  nullo-X, 430
  *Obelia*, 102–**103**
  platyhelminth, **116**–**118**
  reptile, 267, 272, 283
  sponge, 90
  summer, 123
  winter, 123, 462
Egg funnel, **136**
Egg-laying mammals
Egg sac, **136**, 140
Ejaculatory duct, 124, 177, 410
Elapidae, **268**
Elasmobranchs, 227, 233
Elastic cartilage, 34
Elastic fibers, 33–34, 372
Elastin, **34**
Electric eel, 233
Electric power plants, 486–487
Electrical potential, 355
Electrocardiogram (EKG), **373**
Electroencephalogram (EEG), 336, **337**
Electrolytes, 10

**501**

Electron, 8–9
  shells, 9
Electronegativity, 11
Electroreception, 226, 232, 396
Elements, 9
  in cells, 11–12
Elephant, 306, 322, 460
  Indian, 322
  African, 322
Elephant seal, 321, 338
Elephant's ear sponge, 91
*Emberiza citrinella*, 340
Embiid, 171
Embioptera, 171
Embryo, 46
  annelid, 142
  Aschelminth, 124, 127
  bony fish, 238, 239
  comparative, 408, 409
    (3°) and evolution, 444
  differentiation of germ layers, 415
  early development, 414, 415–417
  frog, 414
  human, 418
    (3°) development, 416, 417–419
  hydra, 101
  and phylogeny, 46
  six-hooked, 119
Embryology, 6–7, 203, 415, 412–419
  comparative, 408, 409
  (3°) and evolution, 444
Embryonic development, 408, 409, 415–419
  and phylogeny, 46
  differentiation of germ layers, 415
Emo, 298, 300
Emulsification, 363
Enamel, 313
Endangered species, 331
  whale, 320, 322
Endergonic reaction, 11
Endocrine glands, 37, 397–400
  (*see* specific gland)
Endocrine system, 37, 38, 397–400, 401
  arthropod, 159
  bird, 296
  frog, 260
  human, 398–400
  invertebrate, 397–398
  lamprey, 226
  vertebrate, 398
Endocuticle, 347
Endocytosis, 20
Endoderm, 46, 94, 101, 414, 415

Endolymph, 397
Endometrium, 411, 412
Endopeptidase, 114
Endoplasm, 64–65, 69, 74, 76
Endoplasmic reticulum, 21–22
Endopodite, 152, 154, 155
Endoskeleton, 37, 47, 49, 346, 356
  bony fish, 235
  chordate, 213
  development, 347–349
  echinoderm, 203, 209, 210
  frog, 246–247
  human, 349–352
Endosome, 57–58
Endostyle, 218, 226
Endothelium, 372
Endotherms, 462
Energetics, 11
Energy, 5
  activation, 10
  chemical bond, 10
Energy pyramid, 472, 473
Energy resources
  fossil-fuel, 486
  geothermal, 487
  nuclear, 486
  solar, 487
  tidal, 487
*Entamoeba*, 68
*Enterobius vermicularis*, 127, 128
Enteron, 95–96, 99, 104–105, 106–107, 414
Enteropneust, 214–215
Enterozoans, 46
Entomology, 172, 187
Entoprocts, 129
Environment
  aqueous
    freshwater, 469
    marine, 469–470
  biological, 464–470, 471–473
  digestive, 37, 138, 155–156, 165, 205, 218, 361, 362, 363
  and evolution, 443
  human impact on, 483–487
  lysosomal, 24, 412
  physical, 461–464, 474
  pollution of, 483–487
  and protein synthesis, 422
  studies, 331
  terrestrial, 470
  trait development and, 427–428
Enzymes, 12–14, 29
  digestive, 37, 138, 155–156, 165, 205, 218, 361, 362, 363

  lysosomal, 24, 412
  and proteinsynthesis, 422
Eocene period, 456, 457, 459
Eohippus, 450
Eosinophil, 254, 376
Ephemeroptera, 171
Ephyra, 105
Epiblast, 415
Epicuticle, 347
Epidermal gland, 224
Epidermis, 93, 95, 101, 104–106, 107, 110, 115, 136, 137, 141, 152, 159, 172, 174, 245, 346
Epididymis, 312, 410
Epiglottis, 309, 379
Epigynum, 164
Epinephrine, 260, 385, 399
Epiphragm, 198
Epiphyseal plate, 348, 349
Epiphysis, 348, 349
Epipodite, 154
Epithelia, 30–32, 36
  moist membranes, 36, 379
  sensory, 394
  villi, 362
  (*see* Epithelial types)
Epitheliomuscular cells, 95, 97, 98
Equilibrium
  organs of, 105, 109, 157, 195, 226, 234, 238, 260, 396, 397
Equilibrium constant, 15
Equus, 450, 451
Erepsin, 363
Eryops, 243
Eryopsidae, 243
Erythrocytes, 31, 35, 220, 236, 254, 309, 376
  and malaria, 70–72
  sickle-cell, 381
  and tonicity, 19
Esophageal gland, 144, 145
Esophagus, 37
  annelid, 136, 138, 144–146
  arthropod, 155, 164, 165, 174–175
  aschelminth, 121, 123, 125
  bird, 293
  bony fish, 236, 237
  bryozoan, 130
  cat, 309
  cow, 323
  dogfish shark, 228, 230
  echinoderm, 205, 209, 210
  frog, 252
  human, 360, 361, 379
  lamprey, 225
  mollusk, 193, 196, 197

turtle, 270, **271**
Estivation, 269, 462
Estrogens, 400, 411
Estrous, cycle, 410, **411**, 412
Ethiopian region, 455, **456**, 460
Ethmoid bone, **349**
Ethology, 329, 330, 344
Eucoelomata, 45
*Eudorina*, **406**
Eugenics, 438
*Euglena vividis*, 54, 57–59
Euglenoid movement, **58**–**59**
Eulamellibranchia, **190**
*Eunice*, 146
Euphotic zone, 470
*Euplectella*, **91**–**92**
*Euplotes*, **80**
*Eupomatus*, **146**
Eustachian tube, 259, 269, 295, 396
Eutrombicula, **168**
Evolution, 6–7, 30
 and taxonomy, 40–41
Evolution, 440–454
 cultural, 482–483
 current, 453–454
 defined, 445
 evidence for, 444–445
 fish, 223, 227
 fossil record, 449–451
 history of thought, 440–445
 horse, 449, **450**, 451
 human, 478–482
 modern theory of, 445–449
Excitatory synapse, 388
Excretion, 365–369
Excretory
 canals, **118**–119, 366
 pore, 115, **125**, **154**, **156**, 366
 tube, 121–**122**, **125**
Excretory system, 38
 *amoeba*, 65–66
 amphioxus, 219
 annelid, 137, 139–140, 148, 149
 arthropod, 156, 165, 174–175
 aschelminth, 121–**122**, 123–**125**
 bird, 293
 cat, 311
 dogfish shark, **232**–233
 Echinoderm, 205
 frog, 253
 human, **367**–369
 hydra, 99
 lamprey, 226
 mollusk, 194
 *paramecium*, 76
 platyhelminth, **113**–114, 115, 119
 sponge, 89

turtle, **272**
Excurrent canal, 87–88
Exercise, 356
Exergonic reaction, 11
Exoccipital bones, **246**–247
Exocotidae, **234**
Exocuticle, 347
Exopodite, **152**, **154**, 155
Exopthalmic goiter, 399
Exoskeleton, 37, 47, 48, 346–347, 356
 arthopod, **152**, 153–155, 159, 172, 188
 bony fish, 234–235
 mollusk, 189
 vertebrate, 347
Expiration, 380
Extensor
 carpi radialis, **308**
 carpi ulnaris, **308**
 communis digitorum, **308**
 cruris, **250**, **251**
 lateralis digitorum, **308**
 longus digitorum, **308**
Extensor muscles, **250**, **251**
 of hand, **354**
External oblique, **259**, **308**, **354**
Extracostal muscles, 311
Extraembryonic structures, **416**–417
Exumbrella, **104**
Eyes, 394
 annelid, 144, **145**, **146**, 147
 arthropod, 157–159, 164, 170, 172, 383
 bird, 290, 295–296, 302
 bony fish, 238
 cancer of, **429**
 cat, 305, 312
 color, **429**
 compound, 157–**158**, 159, 172, 176, 338, 394
 dogfish shark, 232
 frog, 244, 259–260
 human, **394**–395
 mollusk, **196**, 197, 198
 lamprey, 226
 pineal, 226
 platyhelminth, **113**, 115, 394
 simple, 164, 165, 170, 172, 176
 tunicate, 219
 turtle, 269, 272
 vertebrate, 394
Eye worm, **127**
Eyeball, 260, 395
Eyelashes, 305
Eyelid, 260, 269, 290, 305, **395**, (*see* Nictitating membrane)

Eyespots, 48, **57**–59, **113**, 115, **206**, **217**, 219, 394

# F

Facets, 157–**158**, 176
Facial bones, 291, **349**
Facial nerve, **231**, **258**, **259**, **311**, **384**, **392**
Facilitated transport, 20
Fairy shrimp, **161**
Falconiformes, **287**
Falconlike birds, 301–302
Fallopian tubes, 410, **411**, 412
Family, 41
Fangs, **279**
Fascia lata, **249**
Fat droplet, **18**
Fatigue, muscle, 355, 356
Fat(s), 13, 363, 369
 tissue, **35**–36
 (*see* Lipids)
Fatty acids, 13–14
Feather, 38, 290–**291**, 303, 304, 345
 archaeornithes, 298, **299**
 contour, 290–291
 disk, **302**
 down, 290–**291**
 filoplumes, 290–**291**
 flight, 292
 pigments in, 288
 preening, 288
 primaries, 292
 secondaries, 292
 tertiaries, 292
 tracts, 291
Feces, 361, 367
Fecundity, 430
Feeding
 *amoeba*, **64**–66
 amphioxus, 218
 annelid, 138, 144, 147–148
 anthozoa, 107
 arthropod, 155, 165, 172
 birds, 293
 bony fish, 236
 bryozoa, 130–131
 ctenophora, 110
 echinoderm, 205, **207**, 210
 frog, 252
 hydra, 99–**100**
 hydroid, 102–**103**
 lamprey, **224**, 225
 mollusk, 190, 192
 *paramecium*, **74**, **76**–**77**
 planaria, 114
 platyhelminth, 121, 123, 126, 128

Feeding (*cont.*)
  scyphozoa, 105
  shark, 240
  snakes, 274–**277**
  sponges, 89
  turtle, 269
Feet
  bird, **288**, 290, 293, 298
  webbed, **288**, 298, 301, 316, 319
*Felis catus*, 305
  (*see* Cat, domestic)
Female reproductive system, 410, **411**, 412
Femoral
  artery, **271**, **294**, **310**, 375
  vein, **257**, **271**, **294**, **310**
Femur, 173–174, **246**, 248, 249, 292, 293, **307**, 308, **351**, 352
Fertility, 430
Fertilization, 24, 29, 38, 80, 407
  amphioxus, 219
  arthropod, 159, 166
  *ascaria*, 124
  bird, 296
  bony fish, 238
  cat, 312
  cross, 115, 142, 407
  echinoderm, 206–**207**
  external, 145, 266, 408
  frog, 261, 266
  human, 412, 417
  hydra, 100–101
  internal, 409
  lamprey, 226–227
  mollusk, 195
  self, 80, 407
Fetal membranes, 312
Fetus, 419
Fibrin, 254, 377
Fibrinogen, 377
Fibrocartilage, **34**, 307, 352
Fibula, **292**, 293, **307**, 308, **351**, 352
Fiddler crab, 162
Filter feeders, 85–86, 89, 93, 190, 200, 209, 223
Filtration, 368, 369
Fin, 227, **235**, 241
  anal, **228**, 233, **235**
  caudal, 132, **217**, **228**, 233, **235**
  dorsal, **217**, **223**, **228**, 233, **235**
  lateral, 132
  origin, **228**
  pectoral, **228**, 233, **235**
  pelvic, **228**, 233, **235**
  ray, **217**, **218**, **234**, **235**

squid, 197
ventral, **217**
Final Consummatory Act, 335
Finches, Galapagos, **442**–443, 445, 453
Finger sponge, **91**
Fingerprints, 313
Fire salamander, 263
Fire succession, 474
Fire use, 481–482
Fishes, 213, **214**
  bony, 222, 233–241, (*see* Bony fishes)
  brain, 391, **393**
  cartilaginous, 227–233, (*see* Cartilaginous fishes)
  coloration, 222
  deep sea, 239
  food, 240–241
  general characteristics, 222
  heart, **255**
  jawless primitive, 214, 220–222, **223**–227
  relation to man, 240–241
Fishmeal, 241
Fission
  binary, 58–59, 62, 66–67, 78–**79**, **114**–115, 405, **406**
  longitudinal, **58**–59, 107
  multiple, 71–**72**, 405, **406**
  transverse, 143
Fixed creation, 440–442
Flagella, **25**, 54
Flagellates, 47, **54**–55, 57–61
Flame bulb, 366
Flamingo, 300
Flatworms, 48, 112–120
Flea, **171**, 187
Flicker, 340
Flight, bird, 286, 304
  adaptations for, 290–293
Flightless birds, 298, **300**
Flipper
  bird, 298, **300**
  seal, **446**
Flame cell, 48, **366**
  amphioxus, 219
  aschelminth, 121–**122**
  nemertine, **119**
  platyhelminth, **113**–114, 115
Flexor carpi, **308**
Flexor muscles, 250, **251**
  of hand, **354**
Floating ribs, 352
Flounder, 222, 241
Flukes, 48, 112–117, 199
Flying fish, **234**, 239
Flying lemur, **306**
Flying mammals, 316, **317**
Folic acid, **364**

Follicle-stimulating hormone (FSH), 399, **411**
Follicular phase, 410–**411**
Fontanel, 419
Food, 5, 358, 369
  capture, 358–359
  chains, 54, 56
  choice, 369, 393
  cooked, 481–482
  cup, **65**
  digestion of, 37, 361–364 (*see* Digestive system)
  hydra, 95, 97
  ingestion, 359
  to maintain individual, 6
  molluscan, 48, 189, **191**, 192, 193, 197, 198, **199**, 200, 202
  nutrients in, 362–364
  preferences, 393
  resources, 443
  rotifer, 121–**122**
  vacuole, 65, **74**, 76–**77**, **97**, 359
  web, 471, **473**
Foramen ovale, 419
Foramen magnum, **247**
Foraminifera, **68**–69
Foregut, 174
Forelimbs
  adaptation, **446**
  amphibian, 245, 248
  bird, 292, 298
  cat, 307, 308
  horse, **450**, 451
  reptile, 280
Forgetting, 333
Forming face, 23
Fossil
  amphibia, 264
  ancestral man, 479, **480**, **481**, 482
  ape, 479
  birds, 298, 299
  crinoids, 208
  foraminifera, 68–69
  horse, 449–451
  mammals, 314, **315**
  placoderm, **227**
  record, 449, 459
  reptiles, 281–282
  *Turritella*, **449**
  (*see* Living fossils)
Fossil fuels, 486
Fossilization, 449
Fowl, common, 291, **292**, 303
Fragmentation, 107, 120, **406**, 407
Freshwater
  amphibians, 242

annelids, 135, 147, 149–150
bryozoa, 130
ciliates, 73–80
clam, 191–195, 200–**201**
environments, 5, 469
fish, 233–241
flatworms, 112–115
hydrozoan, 94
pond ecosystem, **472**
snail, 198–199
sponges, 85, **91**
Frog, **214**, 242–263, 264–266
behavior, 261
blood, **254**
brain, **258**–260, 391, **392**
circulation, **254**–**257**
digestion, 252–**253**
endocrine glands, 260
excretion, 252–253
external anatomy, 244–245
feeding, 252
heart, 251, 254, **255**, **256**
internal anatomy, 251–261
life cycle, 261–263
lymphatic system, 257–258
metamorphosis, 261–263
muscular system, 248–**251**
nervous system, **258**–260
relations to man, 265–266
reproductive system, **260**–**261**
respiration, 253–254
sense organs, 259–260
skeleton, **246**–**248**, **249**
skin, 242, **245**, 256, 259, 260
sound production, 254
urogenital system, **260**–**261**
Frons, 172–**173**
Frontal
bone, **308**, 349
Frontal lobe, 390, **391**, **392**
Fruit fly, 430, 432, 435–436
Functional groups, 15
Fundic glands, 361
Fungi, 471, **471**
Funiculi, 388
Funnel, **196**, 197
Funnel-eared bat, **306**
Fur, 305, 316
under, 313
Fur-bearing mammals, 316, **319**
Fur, seal, 319
Fusiform shape, 286
Fusion, cell, 438

# G

Gaboon viper, **268**
Galapagos Islands, **442**, 445
Gallbladder

bony fish, 236, **237**
cat, **309**
dogfish shark, 229, **230**
frog, **252**–**253**
human, **360**, 361, 400
turtle, **271**
Galliformes, **287**
*Gallus gallus*, 303
*Gambusia*, 241
Gamete, 29, 62, 69, **70**, **436**, 437
Gametocytes, 71–**72**
Ganglia, 37, 116, 259
cephalopod, 197–198
cerebropleural, 194
cerebral, **136**, 140, **141**, 144, **147**
nerve cord, **147**, 149
pedal, 194
subesophageal, 144, 147, **175**–176
subpharyngeal, 140, **141**
supraesophageal, 156–157, **175**–**176**
visceral, 194
Ganoid scales, 235–**236**
Garden snail, **199**
Garden spider, **163**, 166
Garter snake, 275, **276**
Gas exchange, lung, 379
Gastric
artery, **257**, 271
vein, **271**
Gastric filaments, **105**
Gastric mill, 155
Gastric mucosa, 400
Gastric pouches, **105**
Gastrin, 400
Gastrocnemius, **249**, **250**, 251, 308, 354
Gastrocoel, **414**, 415
Gastrodermis, 93, 95, 101, **104**–**105**, 107, 110
Gastropoda, 189–**191**, 198–199
evolution, 200
pulmonate, **190**, 198–199
Gastrotricha, 128
Gastrocascular
canal, **110**
cavity, 47, 93, 95–96, 99, **104**–**106**, 113, 359
Gastrula, 101, **105**, 142, **207**, 408, **413**, **414**, **415**
Gastrulation, 415
Gavial, 280
Geckos, 273
Gekkonidae, **268**
Gel, 17
Gemmules, **90**, 91, 407
Genae, 172–**173**

Gene, 420
flow, 446–447
frequency, 435, 445
pool, 433
Generation, filial and parental, 425
Genetic
code, 421, **422**
equilibrium, 434–435
studies, 435–438
Genetics, 6–7, 420, 424–439
Mendelian, 425–427
non-Mendelian, 427–428
population, 432–435, 439
sex-cell, 428–429
Genital
opening, 164, **175**, 177, 305
organs, accessory, 400
pore, **113**, 115, 118, 123, **209**, 226
Genital chamber, **113**
Genotype, 424, 425
Genotypic ratio, 425
Genus, 41
Geographical isolation, 470
Geological time scale, **456**
Geomagnetic striations, 457
Germ cells, 97, 100, **116**
Germ layers, 46, **415**
differentiation, 415
infolding, 416
Germinal disk, 238, **239**
Gestation, 312
Giant
anteater, 316, **317**
fibers, 140, 141
octopus, 198, 202
salamander, 264, 265
squid, 198
tortoise, 283
Giantism, 400
*Giardia lanblia*, **61**
Gibbon, **324**
Gila monster, **268**, 284
Gill(s), 37, 254, 378–379
annelid, 144, 149
arch, 235, 236, **414**
arthropod, 48, 155, **157**, 159
bailers, 159
bar, **217**–219, 227
bony fish, 236–**237**
capillaries, 236, **237**
channels, 379
chordate, 49, 213, 214
dogfish shark, 231
filaments, **194**, **195**, **225**, **231**, 236, 379
lamellae, 194, **195**
mollusk, 190, **193**, **194**, **196**–198

**505**

Gill(s) (*cont.*)
  mud puppy, **265**
  plates, 192, 194
  pouches, 214, 221, **225**, 226
  raker, **231**, 235
  ray, **231**
  slits, 213–**215**, **217**–221, **225**, 226, 228, **230**, **231**, 236, **444**
  tadpole, 261, **262**, 263
Girdles, limb, 349
  bird, 291, 292
  bony fish, 236
  cat, 308
  dogfish shark, 228
  frog, **247**–**248**
  human, 352, 479
  pectoral, 228, 236, **247**–**248**, 291, 292, 308
  pelvic, 228, 236, **247**–**248**, 292, 308
Glaciers, 460
Gland cells, **97**, 99, 107, **113**, 141
Glands
  endocrine, 37, 397–400
  skin, 313, 345–**346**
  (*see* specific gland)
Glass rope sponge, **91**
Glass snake, **268**, 273
Glaucoma, 395
*Globigerina*, 69
Globin, 376
Glochindium larva, **195**
Glomerular filtrate, 368–369
Glomerulus, **367**, 368
Glossobalanus, **215**
Glossopharyngeal nerve, **231**, **258**, **259**, **311**, **384**, **392**
Glottis, **252**, 253, 269, 271, 295, 309
Glucagon, 400
Glucose, 14, 363, 368
  blood, 400
Gluten, 363
Gluteus, **249**
Gluteus, maximus, **308**, **354**
Glutinants, 98–99
Glycerol, 363
Glycogen, 14, 22, 355
*Glyptodon*, 314
Gnathostomata, 221, 222, 227, 241
Gnawing mammals, 316, **317**
Goblet cell, **362**
Goiter, 362, 399
Golden eagle, 302
Goldfish, 238–239
  (*see* Bony fishes)
Golgi apparatus (or bodies), **18**, 22–23
Gollum, 418
Gonad-stimulating hormone, 260
Gonadotropic hormone, 260
Gonals, 38, 407
  amphioxus, **217**, 219
  arthropod, 155, 159
  bird, 296
  echinoderm, **206**, **209**, **210**
  hydra, **96**, 97, 100–101
  hydroid, 102–**103**
  lamprey, 226
  mollusk, 195
  vertebrate, **220**
Gonangia, 102–**103**
Gondwanaland, 457
*Gonionemus*, 102
Gonochoristic organisms, 47
Gonotheca, **103**
*Gonyaulax*, 59
*Gonyaulax catenella*, 59
Goose, 301, 303
Goose barnacle, **161**
Gopher, 316
Gorilla, **324**
Gorilla rib, 352
Graafian follicles, 312, 399, 411
Gracilis, **308**, **354**
  major, **249**, **250**, 251
  minor, **249**, **250**, 251
Grafting, 102, 115, 142
Grasshopper, **171**, **172**–**177**
  circulatory system, 175
  digestive system, 174
  excretory system, 174–175
  external anatomy, 172–174
  head, **173**
  internal anatomy, 174–177
  life cycle, 177
  mouthparts, **173**
  muscular system, 174
  nervous system, 175–**176**
  physiology, 174–177
  reproductive system, 177
  respiratory system, 175–**176**
  sense organs, 176–177
  sound production, 172, 177
Grasslands, 464, **465**
Graves' disease, 399
Gray crescent, **414**
Gray matter, 258, 388, **389**, 390
Gray whale, **320**
Great auk, **287**, 298
Great blue heron, **301**
Great horned owl, 289, **302**
Great white shark, 233
Green glands, 155–**156**, 165, 366–367
Green turtle, **269**, 283–284

*Gregarina blattarum*, **71**
Gregarines, **73**
Gizzard, **136**, 138, 174, 293
Grizzly bear, 316, **319**
Ground beetle, **185**
Growth, 5
  bone, 348–349, 356
  curve, population, **488**
  amoeba, 66
Growth, hormone, 260, 348, 399–400
Gruiformes, **287**
Guanine (guanylic acid, G), **421**, **422**
Guano, 293, 303
Guest, 341
Guinea worm, **127**
Gullet, **105**, **106**–107
Gustation, 393
*Gymnodinium*, 59
*Gymnodinium brevis*, 59
Gyri, 390

# H

H band, 355, **356**
Habitat, 5
  (*see* Environment)
Habituation, 334
Hagfishes, 223, 241
Hair, 38, 305, 312–313, 345, **346**
  bulb, **346**
  guard, 312
  muscle, **346**
  papilla, **346**
  root, **346**
  wooly, 313
Hair cells, 397
Halibut, 241
Hallux, **292**, 293, 303
Hammerhead, 233
Hand, 245
  eye coordination, 479
  *ramapithecus*, 479
  split, **429**
Haploid,
  chromosome number, 29, **429**
Hardy-Weinberg law, 435
Harlow, H., 331
Harvestman, 162, **163**
Haversian canal system, **348**, 356
Hawk, **287**, **288**
Hawksbill turtle, 284
Head
  bony fish, 233
  cephalopod, 197
  end, 44–45

insect, 172, 188
pigeon, 290
turtle, 269
vertebrate, 233
Head (sperm), **409**
Head process, 417
Hearing, 165, 177, 266, 272, 295
    range, 331
Heart, 37
    arthropod, **164**, 165
    beat, 372–373
    bony fish, 236, **237**
    brachiopod, 131
    cat, **310**
    contraction, 371–373
    dogfish shark, 229, **230**
    dorsal, 193
    earthworm, 138–**139**
    frog, 251, 254, **255**, 256
    four-chambered, 271, 371
    human, **371**, **374**
    mammalian, 371
    mollusk, 193–194, **196**
    reptile, 267, 271
    sounds, 372
    three-chambered, 254–**256**, 266, 285, 371
    univentricular, 254–**256**, 266, 285, 371
    valves, **371**, 372
    ventral, 49, 220
    vertebrate, **220**, **255**
Heat
    in chemical reactions, 10
    loading, 486–487
Heat-sensitive organ, 278–279
Heath hen, 298
Hedonistic motives, 332
*Helix*, 199
Hellbender, **243**, 264
Helodermatidae, **268**
Hematin, 376
Heminchordates, 214–**215**
Hemiptera, **171**, **186**, 188
Hemocoel, 45, 48, 174, 175, 183, 370
Hemoglobin, 138, **254**, 367, 376
    in evolutionary studies, 444, 445
    respiratory function of, 380
Hemolymph, 174, 175
Hemophilia, 377, **429**
Hen, common, 303
Hensen's node, **408**, 417
Hepatic
    artery, **257**, **310**, **374**, 375
    ducts, 155
    sinus, 229
    vein, 219, **237**, **257**, 271, **294**, **310**, 374

Hepatic portal system, 49, 229, **230**, 256, **257**, 271, 310
Hepatic portal vein, 219, **294**, **310**, **374**, 375
Herbivores, 190, 269, 358, 479
    dentition of, **359**, 479
Heredity, 420, 424
    and behavior, 330
    counseling, 438
Hermaphroditic organisms, 47, 407
    annelid, 135, 140, 148, 149, 150
    arrow worm, 132
    bryozoa, 130
    ctenophore, 110
    flatworm, 115, 116
    mesozoa, 84
    mollusk, 190, 195
    sponge, 90
    vertebrate, 226
Hermit crab, 162
Heron, **287**, **288**, 300, **301**
Herring, 240
*Heterakin gallinae*, 128
Heterocercal, 228
Heteromorphosis, 161
Heteronomous segmentation, 45
Heterotherms, 462
Heterozygous, 424, 448
Hexacanth, 119
Hexactinellida, 85
Hibernation, 269, 462
Himalayas, 460
Hindgut, 174
Hindlimbs
    amphibian, 245, 248
    bird, 293
    cat, **307**, 308
    reptile, 280
Hinge, **192**, **194**
Hippopotamus, **306**
Hirudin, 149
Hirudinea, **134**, **147**–148, 149, 150
*Hirudo medicinalis*, 148
Histamine, 376
Histology, **6**, 30, 39
Hive, **181**
H. M. S. *Beagle*, 441, **442**
Holoblastic cleavage, 46, 142
    equal, 412, **413**
    unequal, 412, **413**
Holothuroidea, 203, **204**, 208
*Homarus americanus*, 153
Homeostatic behavior, 330, 332
Homeotherms, 37, 298, 305, 462
Homing, 343
Homing pigeon, 343
Hominidae, 324

*Homo erectus*, **480**, 481–482
*Homo sapiens*, 324–325
*H. sapiens* (intermediate), **480**
*H. sapiens* (modern human), **481**
*H. sapiens neanderthalensis*, **480**, 482
Homologous
    appendages, **154**–155
    organs, 45
Homology, 45
    forelimb, **446**
    serial, 155
Homonomous segmentation, 45
Homoptera, **171**, **185**
Homozygous, 424
Honey, 184
Honeybees, 180–182
Hoof, 313, 314, 345
Hoofed mammals, 322
Hooklets, 290–**291**
Hooks, 117–**118**, 128
Hookworms, 126, **127**
Hormones, 37, 260
    human, 398–400, 401
    invertebrate, 397–398
    vertebrate, 398–400, 401
Horned lark, **296**
Horns, 38
Horns (spinal), 388
Hornworm, **185**
Horse, 322
    evolution, 449, **450**, 451
    foot posture, **314**
    teeth, **313**
Horse botfly, **186**
Horsehair worms, 128
Horseshoe crab, 162, **163**, **168**, 169
Host, 6, 341
Host cells, 98
Hot springs, 462
Hondini, H., 381
Housefly, **171**, **186**–187
House spider, 166
Human beings, 324–325, 478–490
    birth control, 488–**489**
    blood, 375–377
    brain, 390, **391**, 392
    circulatory system, 371–378, 381
    culture, 482–483
    dentition, 358, **359**
    development, 412–419
    digestive system, **360**–364
    disease vectors, **186**–187
    endocrine system, 398–400
    excretory system, **367**–369
    evolution, 444–445, 478–482, 490

Human beings (cont.)
  heart, 371
  impact on environment, 483–487
  integumentary system, 345–346
  lymphatic system, 375, 381
  muscular system, 354
  nervous system, 384–391
  parasites of, 60, 61, 68, 70–72, 82, 115, 117–118, 123, 126–128, 129
  population size, 487–488, 490
  reproduction, 409–412
  respiratory system, 379–381
  skeletal system, 349–352, 357
Human itch mite, 168
Humerus, 246, 249, 292, 307, 308, 351, 352
Hummingbirds, 288
Hunger, 332
Hunting societies, 482
Huntington's chorea, 429
Huxley, A. F., 355
Huxley, H. E., 355
Hyaline cartilage, 31, 34, 352, 353
Hybrid, 427
  cells, 438
Hydra, 93
  behavior, 99–100
  body plan, 94
  body wall, 97
  buddings, 95–96
  cell types, 95–97
  development, 101
  differentiation, 96, 101
  digestion, 99
  embryology, 101
  excretion, 99
  feeding, 99–100
  grafting, 102
  histology, 95–97
  locomotion, 98, 99–100
  mesoglea, 94–97
  morphology, 94–95
  muscular system, 96–97
  nematocysts, 96–98, 99
  nervous system, 96–97, 100
  regeneration, 96, 102
  reproduction, 100–101
  respiration, 99
  response to stimuli, 99–100
Hydranths, 102–103
Hydrocauli, 102
Hydrochloric acid, 363
Hydrogen
  ions, 16, 381
  molecule, 10
Hydroids, 48

  colonial, 102–103
  polyp, 94
  (see Hydra)
Hydrolases
  lysosomal, 24
Hydrolysis, 16, 362
Hydronium ions, 16
Hydrophiidae, 268
Hydrorhiza, 102
Hydrozoa, 93–104
  (see specific animal)
Hyena, 306
*Hyla andersonii*, 264
*Hyla versicolor*, 264
Hylidae, 243
Hymenoptera, 171, 185, 188
Hynobiidae, 243
Hyoid
  arch, 228, 235, 236, 247–248
  bone, 307, 350
Hyperbranchial groove, 218
Hyperpolarization, 388
Hypertonicity, 17, 19
Hyperventilation, 381
Hypnotoxin, 98
Hypoblast, 408, 415
Hypobranchial groove, 218
Hypoglossal nerve, 311, 384, 392
Hypopharynx, 172, 173, 175
Hypophysis, 37, 391, 392, 398, 399–400
Hypostome, 95–96, 97
Hypothalamus, 390
Hypotonicity, 17, 19
Hyracoidea, 306
Hyrax, 306

# I

I band, 355, 356
Ice Age, 282, 460
*Ichthyosaurus*, 281–282
Iguanidae, 268
Ileocecal valve, 361
Ileum, 174, 175, 253, 270, 361
Iliac
  arteries, 230, 257, 294, 310, 374, 375
  veins, 294, 310, 374
Ilium, 246, 248, 292, 307, 308, 352
Immigration, 446
Imprinting, 335, 336
Incisor, 313, 314, 360
Incomplete penetrance, 427–428
Incubation, 296, 298, 300
Incurrent pores, 86–87

Incus, 350, 396
Indian elephant, 322
Inducer, 423
Inert gases, 9
Inflammation, 24
Infraorbitals, 309
Infundibulum, 226, 258, 392, 398
Ingestion, 359
  amoeba, 65–66
  anthozoa, 107
  hydra, 99–100
  *paramecium*, 76
  planarian, 114
  sponge, 89
  (see Feeding)
Inheritance
  human, 429, 434
  polygenic, 330, 428
  sex, 430
  sex-linked, 431–432
Ink sac, 196, 197
Innate behavior, 334
Innate releasing mechanism, 335
Inner cell mass, 415, 417
Innominate
  arteries, 271, 293, 294, 310
  bone, 308, 351, 352
  vein, 257
Inorganic compounds, 11
Insect-eating mammals, 316, 317
Insectivora, 306, 316, 317
Insectivores, 316, 317, 358
  primitive ancestral, 447
Insects, 48, 151, 170, 171–183, 188
  carnivorous, 187
  control, 187
  disease vectors, 186
  helpful, 184
  household, 188
    pests, 188
    predacious, 185
  respiration, 165, 175–176, 378
  scavengers, 184
  social, 180–183
Insertion, muscle, 250, 251
Insight, 334
Inspiration, 380
Instinct, 334–335, 344
Insulin, 260, 400
Integument
  sensory, 238, 259, 272
  (see Skin)
Integumentary system, 38, 345–346, 356
  bird, 290
  frog, 245, 256, 260, 264

human, 345–346
mammal, 312–313
reptile, 267, 273–274, 283
vertebrate, 345
Intensity, stimulus, 331
Intercartilaginous development, 347–349
Intercilliary fibril, **75**
Intercostal muscle, **308**, 380
external, 380
internal, 380
Interference cells, 245
Interfilament bridges, 355
Interlamellar partitions, 194, **195**
Intermedin, 260
Intermembranous development, 347
Internal oblique, **308, 354**
Interneurons, 35, 385, 388, **389**, 390
Interphase, **27**–29, **67**
Interspersed globule model, 20
Interstitial cells
hydra, 96–**97**, 101, 109
testicular, 400
Interventricular septum, 271, 371
Intervertebral
disks, 307, **389**
ligaments, 307
Intestinal gland, **262**
Intestinal mucosa, 400
Intestine, **36**
amphioxus, **217**, 218
annelid, 138, 144, **145**, 147
arthropod, 156, **157**, 165
aschelminth, 121–**122**, 123–**125**
bird, 293
bony fish, 236, **237**
bryozoan, 130–**131**
cat, **309**
dogfish shark, 229, 230
echinoderm, 205, **209, 210**
frog, 251–**253**
human, **360**, 361
lamprey, **225**, 226
mollusk, **193, 194**
platyhelminth, **113**–114
turtle, 270, **271**
Intrafusal fiber, **389**
Intraocular pressure, 395
Invertase, 363
Invertebrates
body plan, 213
circulatory system, 370
development, 412–413, 415
digestion, 359
endocrine system, 397–398

excretory system, **366**–367
nervous system, 383
reproduction, 407
respiratory system, 378–379
sense organs, 394 (*see* Sense organs)
skeletal systems, 346–347
(*see* specific animal)
Involuntary muscle
(*see* Smooth muscle, Cardiac muscle)
Iodine, 362, 399
Iodopsin, 394
Ions, 9
hydrogen, 16
hydronium, 16
Ionic bond, 10, 16
Iris, 272, 394, **395**
absence of, **429**
Ischium, **246, 248, 292, 307, 308,** 352
Islets of langerhans, 296, **398, 400**
Isopoda, **161**
Isoptera, **171**
Isotonicity, 17, **19**
Isotopes, 9
Irritability, 5, 34, 383
Ivory-billed woodpecker, 298

# J

Jaw
annelid, 144, **147**–148
arthropod (*see* Mandibles)
cartilages, 349
chitin-like, 121
crocodilian, 281
fishes
bony, 236
cartilaginous, 227–**229**
primitive, **214**
frog, **247**–248
mollusk, 190
snake, 274, **279**
Jawless vertebrates, **214**, 220–227
Jellyfish, 48, 93–**94**, 104–**105**
Jejunum, 361
Joints, 352–353, 357
cartilaginous, 352
diarthroid, 353
fibrous, 352
synovial, 353
Jugular
veins, **257, 271, 294,** 295, **310**
Jumping spiders, 166
June beetle, **185**
Jurassic period, **456**

# K

Kala-azar, 60–61, **187**
Kangaroo, 316, **317**
Karyoplasm, 17, 21, 27
Keel, 291, **292**
Kidneys, 38, **367**–369
bird, 293
bony fish, **237**, 238
cat, **309**, 311
dogfish shark, **230, 232**–233
frog, 251, **252,** 253
lamprey, **225**, 226
mollusk, 194
turtle, **272**
vertebrate, **220**
Killer shark, **240**
Killer whale, **320**
Kineses, 335, 344
Kinesthetic stimuli, 331–332
King crab, 162, **163, 168,** 169
King insect, **182**
King snake, 275, **277**
Kingdom, 41
Kinorhycha, 128
Kissing bugs, 60, **186**
Kit fox, **459**
Kiwi, **287**, 298
Klinefelter's syndrome, **430**, 431
Knee jerk reflex, 332
Koala, 316, **318**
Komodo dragon, 273

# L

La Brea Tar Pits, 281
Labium, 172–**173**
Labium (vulva)
minus, **411**
majus, **411**
Labrador duck, 298
Labrum, 172–**173**
Labyrinth, membranous, 238
Lacertidae, **268**
Lacewing, **171**
Lacrimal bone, **292, 308,** 349
Lacrimal glands, 313
Lactase, 363
Lacteal vessel, **362, 374**
Lactic acid, 355, 356
Lacunae (bone), 347, **348**
Ladybird beetle, Australian, **185**
Lagomorpha, **306**
Lake Erie, 485–486
Lamarck, J., 441
Lamellae (gill), 194, **195**
Lamellae (skin), 273
Lamellus, **348**
Lampshells, 130, 131–**132**, 133

**509**

Lamprey, 222, 223–227, 240, 241
  brain, **392**
  circulatory system, 226
  digestive, system, **225**–226
  endocrine glands, 226
  external anatomy, 223–**224**
  muscular system, 225
  nervous system, 226
  relation to man, 240
  reproduction, 226–227
  respiratory system, 226
  sense organs, 226
  skeletal system, 224–225
  suctorial mouth, 223–**224**, **225**
  urogenital system, 226
Lancelet, 215
Land bridges, 456, 459, 460
Langurs, 323
Lappets, **105**
Large intestine, 37
  cat, **309**
  human, **360**, 361
Larva
  aschelminth, 124, 126–**128**, 129
  axoloti, 264–265
  bipinnarias, 206–**207**
  coelenterate, 93, 102–**103**, **105**, 107
  ctenophoran, 110
  cydippid, 110
  echinoderm, 206–**207**
  glochidium, **195**
  insect, 178–**179**, **181**, **185**, 186
  lamprey, 227
  mesozoan, 84
  mollusk, **191**, **195**, 200
  planula, 102–**103**, **105**
  platyhelminth, 116–117, **118**–119
  polychaete, 144–**146**, 148,
  sponge, **90**
  and taxonomy, 47
  trochophore, 144–**146**, **191**, 200
  tunicate, **216**–217
  veliger, **191**
Larynx, 37, 247, 253, 254, 271, **309**, 311, **360**, 379
Lateral fins, 132
Lateral line, 224, 232, 238, 396
Latissimus dorsi, **249**, **308**
Latitudinal migration, 343
Law of independent assortment, 426
Law of priority, 42
Law of segregation, 426

Leadership, 340
Leafhopper, **171**
Learning, 332–334, 344
  abridged, 334
  chemical theories of, 332–333
  earthworm, 143
  structural theories of, 333
Leatherback turtle, **268**, 283
Lecithality, 412
Lecithin, 363
Leeches, 48, **134**, 147–148, 149, 150
Legs
  arthropod, 153–155, 163, 168, 169–170, **173**–174
  walking
    (3°) arachnid, 163, 168
    (3°) crayfish, **153**–155
Legumin, 363
Leishmanias, 60–**61**
Lemurs, 322
Lens
  arthropod, 157–**158**, 176
  bony fish, 238
  frog, 260
  human, crystalline, 394–**395**
Leopard, 316
Leopard frog, 242, **244**, 265 (see Frog)
Lepidoptera, 171, 178–**179**, **185**, **188**
Lepisosteidae, 234
Lepodactylidae, **243**
Leucon, **87**
*Leucosalenia*, 85–87
Leukocytes, **31**, **35**, 254, 258, 375, **376**
Levator muscle, 250
*Libinia emarginata*, 162
Life
  origin of, 451–453
Life cycle
  anthozoan, 107
  cestode, **118**–119
  ctenophora, 110
  echinoderm, **207**
  hydroid, **103**
  insect, 117–**179**, 180–183, 186
  mesozoan, 84
  mollusk, 191, **195**
  *opalina*, 63
  schyphozoan, **105**
  sporozoan, **70**–72
  trematode, 116–117
  *trichinella*, **128**
Life spans, **325**
Ligaments, **34**, 352, 353
Light, 461–462
  reactions to, 59, 143

  receptors, 143, 461
  sensitive spot, 105
Limb buds, 417
Limbs
  adaptation, **446**, **447**
  bird, 292, 293
  cat, **307**, 308
  crocodilian, 280
  horse, **450**, 451
  human, **351**, 352
  turtle, 269, 283
  (see Appendages; Forelimb; Hindlimb; Leg)
Limiting factor, 461
Limulus, **163**, **168**, 169
Lines of growth, **192**
*Lineus socialis*, 120
*Lingula*, 131
Linkage, 435, 437
Linnaeus, C., 40–41
  *systema natural*, 41
Lion, 305, 316
Lipase, 138, 363
Lipids, 13, 22, 363, 369
Lipoids, 363
Lips
  cat, 305
  roundworm, 123, **125**
Liver, 367
  amphioxus, **218**
  bird, 293
  bony fish, 236, **237**
  cat, **309**
  dogfish shark, 229, **230**
  frog, 251, **252**, **253**, 256
  human, **360**, 361
  lamprey, **225**, **226**
  mollusk, 193, **196**, 197
  spider, **164**, 165
  turtle, 270, **271**
Liver flukes, 115–**117**, 199
"Living fossils," 131, 162, **168**–169, 282
Lizards, 267, 273, 285
  poisionous, **284**
*Loa loa*, **127**
Lobefinned fish, **234**
Lobster, 48, **151**, 153, 184, 240
Localization, receptor site, 395–396
Locomotion, 5, 38
  *amoeba*, **65**
  arthropod, 160
  bipedal, 293, 324
  bird, 286, 288, 291, 293, 298
  bony fish, 233–234
  ctenophora, 109
  echinoderm, 203, 205, 208–209

**510**

*euglena vividis*, 57–58
hydra, 98, 99–100
lamprey, 223,
mulluscan, **191**
*opalina*, 63
*paramecium*, 74, 76
primate, 324
snake, 274
Locus, 424
*Loligo*, **196**–197
Longitudinal fission, **58**–59, 107
Longitudinal fissure, **392**
Longitudinal muscle
   *ascaris*, **124**
   earthworm, **137**
   echinoderm, **210**
   hydra, **96**–**97**
   intestinal, 362
   planarian, **113**–114
Loons, 298
Loop of Henle, **367**, 368
Lophiidae, **234**
Lophophore, 130, **131**
Lorenz, K., 329, 334, 335
Lugworm, **134**, 149
Lumbar enlargement, **259**
Lumbar vertebrae, **307**, **350**, **351**
*Lumbricus terrestris*, 135–143
   (*see* Earthworm)
Lumen, **137**
Luminescent organs, 239, 462
Lunar rhythms, 476
Lungs, 37, 367, 379
   bird, **290**, **295**
   cat, **309**, 311
   frog, 251, **252**, 253–254
   human, **379**
   turtle, **271**
   vertebrate, 220
Lungfish, 233, **234**, 240
Lungworm, 149
Luteal phase, 412
Luteinizing hormone (LH), 399, **411**
Lyell, C., 441, 443, 455
Lymph, 35, 37, 258, 375
   capillaries, 375
   duct, **374**
   hearts, 257
   nodes, 293, **310**, 311, **374**, 375
   spaces, 257
   vessels, 257, 311, **362**, **374**, 375
Lymphatic systems, 37
   cat, 311
   frog, 257–258
   human, **374**, 375, 381

Lymphocytes, **254**, 310, 375, **376**

# M

Macaques, 323
Mackerel, 240
Macromeres, **413**
Macronucleus, 74, 78–**79**, 81
Maculae, 397
Madreporite, 203, **205**, **206**, **209**, **210**
*Malaclemys*, **284**
Malaria, 70–72, 241
Male reproductive system, 409–**410**
Mallard, **287**
Malleus, 350, **396**
*Mallomonas*, 60
Mallophaga, **171**, **187**
Malpighian tubules, 48, **164**, 165, 174–**175**, **366**, 367
Maltase, 361, 363
Malthus, T., 443
Mammals, 49, 213, **214**, 220, 221, 305–326
   age of, 314, 449
   aquatic, 319–322
   brain, **392**, 393
   carnivorous, 316, **319**
   claws, hoofed, 313
   egg-laying, **315**
   flying, 316, **317**
   foot posture, 314
   fossil, 314–**315**, 449
   gnawing, 316, **317**
   hair, 312–313
   heart, **255**
   hoofed, 322
   insect-eating, 316, **317**
   pouched, 316, **317**, **318**
   primates, 322–325
   skin glands, 313
   teeth, 313–314
   toothless, 316, **317**
Mammary glands, 305, 307, 313, 315, **318**, 325, 345, 399, 400
Mammillary body, **392**
Mammoths, 449, 460
Man
   age of, 449
   (*see* Human beings)
Manatee, **306**
Mandible, 155–**156**, 169, 170, 172–**173**, **175**, 184, 247–**248**, **292**, **308**, **349**
Mandibular arch, 235

Manganese, 363
Mantid, 340
Mantle, 48, 189–190, 192, **193**, 200, 216
   cavity, 189, 190, 192, 200
   collar, 197
   line, **192**
Manubrium, **104**
Marginal lappets, **105**
Marine
   annelids, 144, 146
   chordates, 214, 216, 217
   coelenterates, 93
   crustaceans, 161, 162
   ctenophores, 93, 109
   environment, 5, 469–470
   fish, 224, 228, 239, 240
   minor coelomate phyla, 130–133
   mollusks, 197–198, 199–200
   sponges, 85, 87, 90–91
   turtles, 283
Marsh birds, 301
Marsupialia, **306**, 316, **317**
Marsupials, 316, **317**, **318**
Marrow, bone, 38, 254, **348**, 376
Mass pyramid, **472**, **473**
Masseter, **249**, **308**
Mastodon, **315**
Mastax, 121–**122**
"Master gland," 400
Mastication, 37, 361
*Mastigamoeba*, 59
Mastigophores, **54**–**55**, 57–61
   parasitic, 60–61
Mating
   behavior, 340
   in birds, 296
Matriarchal leadership, 340
Maturation face, 23
Maxillae, 154–155, 163, 170, 172–**173**, **246**–**248**, **292**, **308**, **349**
Maxillipeds, **153**–155
Mayfly, **171**, 177
Meatus, **396**
Mechanism, 329
Mechanoreceptors, 395
Mecoptera, **171**
Mediterranean Sea, 457
Medulla oblongata
   bony fish, **237**, 238, 391
   cat, **311**
   dogfish shark, **231**, 232
   frog, **258**
   human, 390, **392**
   lamprey, 226
   turtle, 272

**511**

Medusae, **94, 102–103, 104**–106, 111
Meiosis, 29, 428–429, 446
Meiotic drive, 446
Meissner's corpuscles, 395
Melanophores, **245**
Membrane
  basement (or basal), 31
  cell, 17–**18, 20**–**21**, 29, 366
  mitochondrial, **25**
  moist, 36
    (3°) mucous, 36
    (3°) serous, 36
  nuclear, **18**, 21, **27**–**28**
  organelle, 17–**18**, 21, 22, 25
  semipermeable, 17, **19**, 29
  structure, 20–21
  transport, 20
Membrane potential, 386
Membranous labyrinth, 397
Memory, 332–334
  associative, 258
  experiential, 334
  latent, 143
  psychological, 334
  psychophysiological, 334
  recognition, 334
Mendel, G., 425, 426
Mendelian genetics, 425–427
Mendel's law, 426
Meninges, 390
Menstruation, **411**, 412
Menstrual cycle, 400, 410, **411**, 412
Mercury poisoning, 464, 487
Meroblastic cleavage, 46, **408**, **413**, 414
Merozoites, **71**–**72**
Merychippus, **450**, 451
Mesaxon, **386**
Mesenchyme, 88, 111, **113**–**114**, **146**, 347, 348, 356
Mesenteric
  artery, **230, 257, 271, 294**, 310
  vein, **294, 310**
Mesenteric filaments, **106**–107
Mesentery, 106, **209, 210**, 251
Mesoblastic bands, 142
Mesoderm, 45–46, 112, 114, 142, 149, **414, 415**
Mesoglea, 85–86, 93–94, 95–97, 101, 104–**105**, 110, 111
Mesohippus, **450**–451
Mesonephric ducts, 238, 253
Mesothelium, 36, 45
Mesothorax, **173**–174
Mesozoa, 83–84
Mesozic era, 227, 281, **456**

Messenger RNA (MRNA), 422
  synthesis, 423
Metabolic wastes, 365
Metabolism, 5, 12, 16, 29, 358
  bird, 293, 295
  hormone regulation of, 260, 398, 400
Metacarpals, **246, 249**, 292, **307, 308, 351**, 352
Metacercaria, **116**–117
Metachronal waves, 76
Metagenesis, 102
Metamere, 44, 134
Metamerism, 44–45, 48–49, 134–135, 149, 183, 220
  external (or superficial), 44, 148
  internal, 44
Metamorphosis
  echinoderm, **206**–**207**
  frog, 261–263, 264–265
  insect, 48
    (3°) complete, 178–**179**
    (3°) gradual, 177–**178**
    (3°) incomplete, 177–**178**
  lamprey, 227
  mollusk, 191
  retrogressive, 217
Metaphase, **27**–**28**
  plate, 28
Metapleural fold, 217, **218**
Metastrongylus, 149
Metatarsals, **246**, 249, 293, **307, 308, 351**, 352
Metathorax, **173**–174
Metazoans, 83, 85, 92
Methyl mercury, 464
*Metridium*, 93, **106**–107
Mice, 316
Micelles, 20
*Microciona*, 90
Microhylidae, 243
Micromeres, **413**
Micronucleus, 74, 78–**80**
*Microsporidia*, **78**
Microtubules, **24**–**25**
Midbrain, 336, 390, **414**
Midgut, 155, 174, 218
Midocean ridge, 457, 458
Meischer, F., 420
Migration, 341, 343, 462
  altitudinal, 343
  bird, 289, 341–**342**, 461
  bony fish, 238, 341
  Cenozoic, 459–460
  climatic, 343
  early human, 482
  insect, 341
  lamprey, 226–227

latitudinal, 343
mammals, 341
reproductive, 343
reptile, 341
Milk dentition, 314
Milk glands, 307
(*see* Mammary glands)
Miller, S., 452
Millipede, **151**, 169–**170**, 184
Milt, 238
Mimicry, 339
Mineral salts, 362, 369
Minimata disease, 464
Mink, 316
Miocene period, **456**, 460
Miracidia larve, **116**–117
Missing link, 183, 200, 214
Mites, 48, 151, 162, **168**, 184
Mitochondria, **18**, **25**–26
Mitochondrial spiral, **409**
Mitosis, 27–29, **67**, 428–429
*Mnemiopsis,* 43, **109**
Moist membranes, 36, 39
  mucons, 36, 39, 361, 394
  respiratory, 379
  serous, 36, 39
Molars, 313, 314
Molecular polarity, **11**
Molecules, 10, 29
  in the cell, 11–15
  polar, **11**
Moles, 316, **317**
*Molgula,* **216**
Mollusca, 48, 189–202
  origin of, 200
  relations to man, 200–202
Molting, 48, 152, 159, 177, 291, 347
Monarch butterfly, 339, 341
Monitor, **268**
Monkeys, 323
*Monocystis lumbrici,* 69
Monocytes, **376**
Monoecious organisms, 47, 100, 407
Monohybrid cross, 425, **426**
Monoplacophora, 189–**190**, 199–200, 202
Monosaccharides, 14, 363
Monotremata, **306**
Morphological terms, 42–43
Morphology, 5
  cell, 16–26
  and taxonomy, 40–47
Morula, **70**
Mosiac image, **158**–159, 176
Mosquitos, 187
  and malaria, 70–72, 241, 483
"Moss animals," 130–**131**

Moth, 171, 178–179, 453
Mother-of-pearl, 192, **193**
Motility, 53
Motives, 332
Motor
  cells (neurons), 35, **97**, 140, 141, 385–386, **389**, 390
  fibers, 140, **141**
Mountain lion, 316, **319**
Mouth, 37
  amphioxus, **217**, 218
  annelid, **136**, 137–138, 144, 147
  anthozoan, **106**
  arthropod, 153, **164**, 165, 172–**173**, 174, **175**
  aschelminth, 121–123, 125
  bird, 293
  bony fish, 236, **237**
  bryozoan, 130–**131**
  cat, **309**
  cavity, 244, **252**, 293, **309**, 360
  ctenophoran, 109–**110**
  dogfish shark, 228, **230**
  echinoderm, 205, **207**, 210
  embryonic development of, 46
  frog, 244, **252**
  human, **360**, 361
  hydra, 95–**96**, 99, 101
  hydroid, **103**
  hydrozoan medusa, **104**
  lamprey, 223, **224**, 225
  mollusk, 190, **191**, 192, **193**, **196**
  platyhelminth, **113**–114
  suctorial, 223, **224**, 225
  turtle, 269
Mouthparts, 172–173
Movement, 5
  amoeboid, **65**
  euglenoid, **58**–59
  spontaneous, 99, 258
Mucosa, **362**
  gastric, 400
  intestinal, 400
Mucous
  glands, 233, 245, 345
  membranes, 36, 39, 361, 394
  secretion, 114, 141, 198, 233, 247, 309
  "slime road," 114
Mucus-secreting gland cells, 96–97
Mud puppy, 242, **243**, 265
Mud turtle, **268**
Müllerian mimicry, 339
Multicellular organisms, 30, 83, 85, 92

Multiple fission, 71–72, 405, **406**
Muscular system, 38, 353–356, 357
  arthropod, 159, 172, 174
  *ascaris*, **124**
  bird, 293
  bony fish, 236
  cat, **308**–309
  clam, **192**
  earthworm, **137**
  frog, 248–**251**, 266
  human, **354**
  hydra, **96–97**
  lamprey, 225
  molluscan, **192**
  platyhelminth, 113–114, 116
Muscle
  abductor, 250, **251**
  adductor (*see* Adductor)
  antagonistic, 354
  bands, 355, **356**
  cardiac, 38, 248, 250, 354–355, 371
  circular (*see* Circular muscle)
  constrictor, 279
  contraction, 355–356
  control mechanism, **389**
  depressor, **250**
  dorsoventral, **113**–114
  extensor, **250, 251**
  fibers
    smooth, 31, 353
    striated, 31, 353
  flexor, 250, **251**
  hair, **346**
  insertion, 250, **251**, 354
  levator, 250
  longitudinal (*see* Longitudinal muscle)
  origin, 250, **251**, 354
  protractor, 225
  radial, **210**
  retractor, 192, **210**, 225
  rotator, 250
  skeletal, 38, 353
  smooth (involuntary), 38, 248, 250, 353
  striated (voluntary), 38, 174, 248, 250, 353
  tissue, **31, 33**, 39, 353
  ultrastructure, 355
Muscularis, **362**
  mucosa, **362**
Musculocutaneous vein, **257**
Musk organs, 280
Musk turtle, 283
Mussels, 189, 201

  freshwater, 200–201
Mustelids, 316
Mutation, 423–424, 445
  pressure, 446
  sheep, **453**
Mutualism, 341, **342**
Myelin, 385, **386**
Myelination, 385, **386**
Mylohyoideus, **249, 308**
Myocytes, 90
Myofibrils, 33, 353, 355
Myofilaments, 355
Myomeres, 236
Myopia, **429**
Myosin, 355
Myotis, 316, **317**
Myotome, **217**–219
Myxedema, 399
*Myxidium*, 73

# N

Nacre, **193**
Nacreous layer, 192, **193**
Naiad, 177–**178**
Nails, 345
*Naja naja*, 276
Nares
  external, 244, **252**, 253, 269, **309**
  internal, **309**
Narrow-mouthed toad, 243
Nasal
  capsule, 228
  cavity, 253, 259, 312, 379, 394
  passages, 271
Nasal bone, **247, 292, 349**
Natural selection, 447–449
"Nature versus nurture," 330–331, 344
*Nauphoeta cinerea*, 340
Nautiluses, 189–**190, 198**
Navigation, 343
Neanderthal man, **480**, 482
*Neanthes vivens*, 144–**145**
Nearctic region, 455, **456**, 459, 460
Nearsightedness, **429**
*Necator americanus*, 126
Neck, 49, 117, 220, 269, 290, 305
*Necturus*, 265
Nematocysts, 48, 94, **96–98**, 99, 104, 107, 110
Nematodes, 48
Nematomorpha, 128
Nemertinea, 119–120

**513**

Neoceratodontidae, **234**
*Neophron percnopterus*, 335
*Neopilina*, **190**, 199–200
Neornithes, 298
Neoteny, 264
Neotropical region, 455, **456**, 460
Nephridia, 48, **366**
  amphioxus, 219
  annelid, 137, 139–140, **145**–**147**, 149
  arthropod, 183
  mollusk, **193**, **194**, 197
  platyhelminth, **116**
Nephridiopore, **116**, **137**, 140, **147**, **366**
Nephron, **367**, 368–369
Nephrostome, **137**, 140, **366**
Neritic zone, 469–470
Nerve
  fibers, 140, **141**, 386, 388
  net, 94, **96**–**97**, 383
  plexus, 94, **96**–**97**, 383
  ring, 97, **104**, 206
  tracts, 123
  transverse, **113**, 115
Nerve cell, 34–35, 37
  (*see* Neuron)
Nerve cord
  amphioxus, **217**–219
  annelid, **136**, **137**, 140, **141**, 144, **145**, 147
  bony fish, **237**, 238
  dogfish shark, **231**–232
  dorsal, **213**, 214, **217**–221
  echinoderm, 205–**206**
  lamprey, **225**, 226
  platyhelminth, **113**–**115**, **118**–119
  radial, 205–**206**
  ventral, **136**, **137**, 140, 144, **145**, 147, 157, **175**–**176**
Nerves
  cranial (*see* Cranial nerves)
  dorsal, 219
  motor, 35, **97**, 140, **141**, 219
  sensory, 35, 140, **141**, 219
  spinal (*see* Spinal nerves)
  ventral, 219
Nervous
  conduction, 383, 386–387
  impulse, 386–387
  stimuli, 355
Nervous system, 38–393, 400–401
  amphioxus, 219
  annelid, 140, **141**, 144, 149
  arrow worm, **132**
  arthropod, 156–157, 165, **175**–**176**

*ascaris*, 123–**124**
autonomic, 37, 176, 226, 259, 384–385, 400
bird, 295
bony fish, 238
cat, **311**–312
central (CNS), 35, 37, 384
chordate, 221
dogfish shark, **231**–232
dorsal, 384
echinoderm, 205–206, 209, 210
frog, **258**–260, 266
human, 384–391
hydra, **96**–97, 100
invertebrate, 384
mollusk, **194**–**195**, 197
parasympathetic, 385
peripheral (PNS), 37, 384
platyhelminth, **113**–115, **116**, **118**–119
subectodermal, 209, 210, 214
sympathetic, 176, 384–385
turtle, 272, 285
vertebrate, 384
visceral, 157
Nervous tissue, 34–36, 39
Nest
  bee, 180
  bird, **296**–298
  wasp, 180
Neural
  arch, 225, 227, 350
  canal, **350**
  fold, **414**
  groove, **414**
  pathways, 333
  plates, **408**
  spine, **350**
  tube, 217, **408**, 417
Neuromast, 396
Neuron, 34–35, 37
  alpha motor, **389**
  association, 35, 140, **141**, 332
  bipolar, **385**, 386
  connective, 386
  gamma motor, **389**
  interneuron, 35, 385, 388, **389**, 390
  motor, 385–386
  multipolar, **385**
  sensory, 35
  (3°) bipolar, 386, 394
  spindle receptor, **389**
  sympathetic nervous system control, **389**
  tendon receptor, **389**
  voluntary cortical control, **389**
Neuroptera, 171

Neurulation, **408**
Neutron, 8–9
Neutrophil, **254**, **376**
Newt, **243**
  poisionous, 265
Niacin, **364**
Nictitating membrane, 244, 260, 269, 290, 305
Night blindness, **364**
Nine-banded armadillo, 316, **317**
Nitrates, **464**
Nitrogen, **380**
  cycle, 463–464
  fixation, 463–464
Nitrogen compounds, 363
Nitrogen-fixing bacteria, 463
Noble gases, 9
Nocturnal animals, 476
  eyes, 394, 462
Nocturnal birds, 302
Nodes of Ranvier, 385
Nomenclature (taxonomic), 41–42
  binomial, 41
  trinomial, 41
Non-Mendelian genetics, 427–428
Nondisjunction, 432
Noradrenaline, 388
Norepinephrine, 388
Nostrils, 37, 244, 253, 269, **290**, 295, 305, **379**
Notochord, 49, 349
  amphioxus, **217**–219
  chordate, 213, 221
  development, **414**
  dogfish shark, 228, **230**
  hemichordate, **214**–**215**
  lamprey, 224, **225**
  presence and taxonomy, 46, 213, 214
  vertebrate, 219–**220**
Nuclear membrane, **18**, 21, 27–28
Nuclear plants, 486
Nuclear ring, **409**
Nuclear sap, 21
Nucleolus, **18**, 21, 28–29
Nucleoplasm, 17–**18**, 21, 27
Nucleotides, 421
Nucleus, atomic, 8
Nucleus, cell, **18**, 21
  *amoeba*, 54, **64**–65
  ciliophora, 56, **74**, **81**
  cnidospora, **73**
  hydra, interstitial, 97
  during mitosis, 27–29
  *opalina*, 54, **62**
  *paramecium*, **74**, 78–79

sperm, **409**
sporozoa, **69**
Nudibranch, **190**
Nullo-X egg, 430
Numerical taxonomy, 40
Nurse cells, 177
Nut clam, **190**
Nutrition, 358
  carnivorous, 358
  cell, 20
  *euglena viridis,* 58, 358
  herbivorous, 358
  holophytic, 58, 358
  holozoic, 58, 358
  insectivorous, 358
  modes of, 358
  omnivorous, 358
  saprophytic, 58, 358
Nutritive cells, 96–**97**
Nymph, 177–**178**

# O

*Obelia,* 39, 102–**103**, 347
Obstetrical toad, **243**
Occipital
  bone, **292**, **308**, 349
  condyle, **247**, 291
Occipital lobe, 390, **392**
Ocean floor, 457–458
  geomagnetic striations, 457
  spreading, 458
Oceanic region, 470
Ocelli, 172–**173**, **175**, 176
Ocelot, 316, **319**
Octopus, 49, 189, **197**–198, 201
  giant, 198, 202
Oculomotor nerve, 231, **258**, **259**, **311**, **384**, **392**
Odonata, **171**
Oil glands, 288, 345
Old world monkeys, **323**
Olfaction, 266, 393–394
  (*see* Smell)
Olfactory
  bulb, **231**, **311**, 312, 391, **392**, **394**
  capsule, 247
  hairs, 170
  lobes, 226, 231–232, **258**, 259, 311, 391, **392**
  nerves, **231**, **258**, **259**, **311**, **384**
  organs, 198, 259, 393, 394
  pit, 177, 219
  pouch, 394
  sacs, **231**, 232, 238, 294

tract, **231**, **311**, **392**
Oligocene period, **456**, 459–460
Oligochaeta, **134**, 135–143, 148, 149–150
Oligodendrocytes, 385
Oligolecthal eggs, **408**
Omasum, **323**
Ommatidium, 157–**158**, 176, 394
Omnivores, 269, 358, 479
  dentition of, **359**, 479
Ontogeny, 417
  recapitulates phylogeny, 46, 444
Onychophora, **183**
Oocyst, **72**
Oocytes, 177
Oogonia, 29, 177
Ookinete, 71–**72**
*Opalina,* 62–**63**, 81
Opalinids, **54**, 62–**63**, 82
  parasitic, 62
Operant conditioning, 333–334
Operator, 423
Operculum, 236, 261–**262**
Operon, 423
Operon theory, 432, 438
Ophiuroidea, 203, **204**, 208
Opisthobranchia, **190**
Opossum, 316, **318**
Opposable
  great toe, 322, 325
  thumb, 322
Optic
  chiasma, **258**, **311**, 391, **392**, 398
  lobes, 226, 231–232, **237**, 238, **258**, 295, 391, **392**
  nerve, 157, **158**, **231**, **258**, **259**, **311**, **384**, **392**
  tract, **392**
  ventricle, **258**
  vesicles, **239**
Oral
  arms, **105**
  end, 43
  funnel, **224**, **225**, 226
  groove, **74**, 76
  hood, **217**, 218
  lobe, **104**
  papilla, **184**
  spear, 123
  tentacles, **217**–218
Orangutan, **324**
Orb-weaving spider, **166**
Orbits, 260, 291, **292**
Order, 41
Odrovician period, **456**
Organ(s)
  analogy, 45

homology, 45
system level, 36–39, 49, 112, 135
vestigial, 282, 444
Organ of corti, 397
Organic compounds, 11, 463
  abiogenic synthesis of, 452
Organic wastes, 484–486
Organism, 38
Organology, 30, 39
Organs of Jacobson, 274
Oriental rat flea, **186**
Oriental region, 455, **456**, 460
Oriental zone, 60, **187**
Orientation, 331, 335, 338, 343
Origin, muscle, 250, **251**
Ornithodoros, **168**
Orthoptera, **171**, **185**, **188**
Os coxae, **351**, 352
Osculum, 85–90
Osmoregulation, 233, 293, 365–367, 369
Osmosis, 17, **19**
Osmotic pressure, **19**
Osphradium, 195
Osseous tissue, **31**, 34, 347–349
Ossicle, **206**
Ossification, 347–348, 419
Osteichthyes, **214**, 221, 222, 227, 233–240, 241
Osteoblastic processes, 347
Osteoblasts, 347–348
Osteocytes, 347, **348**
Ostia, 47, **106**, 165
Ostium (oviduct), **261**, 312, 410
Ostrich, **287**, 298
Ostium, **232**–233
Ostracod, **161**
Ostracoda, **161**
Ostracodermi, 221–**223**, 241
Otocyst, 146
Otoliths, 238, 397
Otter, 316, **321**
Oval window, 396
Ovarioles, 177
Ovary
  annelid, **136**, 140
  arthropod, 159, 164, 166, **175**, 177
  *ascaris,* **124-125**
  bird, 296
  bony fish, 238
  cat, 312
  dogfish shark, **232**–233
  frog, 251, **261**
  human, 398, 400, 410, **411**
  hydra, **96**, **101**
  platyhelminth, **113**, **116**
  turtle, 272
  vertebrate, 400

Overpopulation, 483, 487–488, 490
Oviduct
  annelid, **136**, 140
  arthropod, 159, **175**, 177
  *ascaris*, **124–125**
  bird, 296
  cat, 312
  dogfish shark, **232–233**
  frog, 251, **261**
  human, 410, **411**, 412
  platyhelminth, **113**
  turtle, 272
Oviparous, 266, 272, 409
Ovipositor, **173–174**, **175**, 177
Ovoivparous, 409
Ovulation, 399, **411**, 412
Ovum, 312, 407, 410
  (*see* Egg)
Owls, **302**, 462
Oxidation reactions, 16
Oxidized
  definition, 16
Oxygen, 37, 358, 378, 379, **380**, 381, 463
Oxyhemoglobin, 376, 380–381
Oysters, 49, 189, 190–191, 201, 210

**P**

Pacemaker, 371
Pacinian corpuscles, 395
Paddles, 298, **300**
Pads, digital, 307, 313
*Pagurus*, 162
Pain, 395
  phantom limb, 395–396
Pair specificity, 421
Painted turtle, **283**
Paired arch, 235
Palate, 309, **360**
Paleartic region, 455, **456**, 459, 460
Paleocene period, **456**, 458
Paleogeographical theories, 455, 457–458
Paleontology, **6**, 400
Paleozoic era, 200, 264, **456**
Palestinus, 482
Palmaris, longus, **308**
Palmitic acid, 14
Palolo worm, 146
Palp, 144–**145**
  labial, 192–**193**
Pampas, 464
Pancreas
  bird, 293, 296
  bony fish, 236, **237**

cat, 309
dogfish shark, **230**, 299
frog, **252–253**, 260
human, 360, 361, 363, **398**, 400
mollusk, 197
turtle, 270, **271**
vertebrate, **220**
Pancreatic
  duct, 270, 360, **398**
  gland cells, **398**, 400
  juices, 361
Pandorina, **406**
Panmixia, 435
Panther, 316, **319**
Pantothenic acid, **364**
Papulae, 205
Paradoxical sleep, 336
*Paramecium*, 56
Paramecium caudatum, 73–**80**
  behavior, 77–**78**
  cyclosis, 76
  digestion, **74**, 76
  locomotion, **74**, 76
  morphology, 74
  offense and defense, 76
  regulation of water content, 76–77
  reproduction, 78–**80**
  respiration, 76
  spiral path, **74**, 76
Paramylum, **57**, 58
Parapodia, 135, 144–**145**, 149–150
Parasites, 6, 47, 48, 341
  annelids, 144, 147, 148, 150
  arthropod, **168**, 184, **185**
  cestoda, 112, 115, 117–119
  cnidospora, 73
  cyclostome, 227, 241
  roundworm, 121, 123–129
  sarcomastigophora, 54, 60
    (3°) mastigophores, 60–61
    (3°) opalinids, 62
    (3°) sarcodines, 68
  sporozoa, 69–73
  trematoda, 112, 115–**117**, 199, 202
Parasitism, 341, 471
Parasympathetic nervous system, 385
Parathormone, 399
Parathyroids, 37, **398**, 399
Parenchyma (connective tissue), 33, 114, 115
Parenchyma (organ tissue), 36
Parental cave, 322, 325, 330–331, 479
Parietal
  bone, **308**, 349
  eye, 282

layer, 36
lobe, 390
vessel, 138–**139**
Parotids, 309, **310**
Parthenogenesis, 123, 407
  artificial, 407
Passenger pigeon, 298
  passeriformes, **287**, 303
Patella, **307**, **351**, 352
Patellar reflex, 332
Pathology, **6**
Patterns
  of behavior, 329
  of courtship, 471
Pavlov, I., 333
Peacock, 286, 303
Peanut worms, 132
Pearl, **193**, **201**
Peck order, 340
Pectin, 21
Pectoral
  girdle, 228, 236, 247–**248**, 291, 292, 308, 352
  vein, **294**
Pectoralis major, 293, **354**
Pedal ganglion, 194
Pedicellaria, 203, **209**
Pedipalpi, 163
Pedipalps, 163, **164**, 166, **168**
Pectoralis minor, 293, **308**, **354**
Peduncle, 131, 164
Peking man, **480**, 481–482
Pelecypoda, 189–**191**, **192–195**
  evolution, 200
Pelican, 300
Pellagra, **364**
Pellicle, **57**, 74–76, 347
Pelobatidae, **243**
*Pelomyxa*, 67
Pelvic
  girdle, 228, 236, 247–**248**, 292, 308, 352, 479
  vein, **257**, **271**
Pen, **196**, 197
Penetrants, **98–99**
Pengiun, 298, **300**
Penial spicules, 123
Penis
  bird, 296
  cat, **309**, 311, 312
  human, **410**
  platyhelminth, **113**, **116**, **118**
  turtle, **272**
Pentamerous divisions, 49
Pepsin, 138, 363
Peptide bonds, 15, **423**
Peptones, 363
*Perca flavescens*, 233–238
  (*see* Fishes, bony)
Perch, 222, 233–238
  (*see* Fishes, bony)

Perching, 286
  birds, 288, 298, 303
Percidae, 234
Periblast, 416
Pericardial cavity, 45, 189, **193**, **194**, **220**, 236, **237**
Pericardium, 36, 165, 251, 371
Perichondrium, 34
*Peridinium*, 60
Perihemal canal, **206**
Periodicity, 476
Periosteum, 348, 356
Periostracum, 192, **193**
*Peripatus*, 183
Peripheral nervous system, 37, 140
  (*see* Nervous system)
Perisarc, **103**, 347
Perissodatyla, **306**
Peristalsis, 138, 353, 361
Peristomium, 144–**145**
Peritoneum, 36, 45, 130, 133, **137**, 144, **206**, 251, 261, **362**
Periwinkles, 198
Permeability, 17, **19**–20
Permeable membranes, 17, **19**–20, 29
Permain period, **456**
Peroneal artery, **257**
Peroneus, **249**, **251**, **354**
Perspiration, 38
Pesticides, 483
Pests
  agricultural, 123, 187, 266, 304, 482
  control, 266, 304
  household, 188, 482
Petrifactions, **449**
Petromyzon marinus, 223–227
  (*see* Lamprey)
pH, 16
  blood, 381
Phagocytes, 254, 376
Phagocytosis, 20
Phagosome, 24, 65
Phalanges, **246**, **249**, **292**, **307**, 308, **351**, 352
Phalangida, **163**
Phalanx, **292**
Phantom limb pain, 395–396
Pharyngeal gills
  (*see* Gills)
Pharyngeal sac, **216**, 217
Pharyngeal slit, **216**
Pharynx
  amphioxus, **217**–219
  annelid, **136**, 138, 144, **147**–148
  arthropod, 165

aschelminth, 121–123
bony fish, 236, **237**
cat, **309**
ctenophore, **110**
dogfish shark, 228, **230**
human, **360**, 361, **379**
lamprey, **225**
mollusk, 190, 197
platyhelminth, **113**–116
turtle, 269
Pheasant, **287**, **288**
Phenotype, 424
Phenotypic ratio, 425
Phenylthiocarbamide PTC, ability to taste, **429**
Pholidota, **306**
Phoronids, **132**
*Phoronis*, **132**
Phosphate-deoxyribose backing, 421
Phospholipids, 14
Phosphorus, 362, 399
  cycle, 464
  pollution, 484–485
Photoperiod, 461
Photoreception
  amphioxus, 219
  annelid, 140, 143
  ctenophora, 109
  molluscan, 195
  platyhelminth, **113**, 115
  scyphozoa, 105
Photosynthesis, 12, 358, 453, 461, 463, 471
  in *euglena viridis*, 58
Phototropism, 335
Phyla, 41
  synopsis of major, 47–49
Phylogeny, 46, 444
  ontogeny recapitulates, 46, 444
  and taxonomy, 40–41
*Physalia*, 102
Physical barriers, 470
Physiology, **6**
Pia mater, **389**, 390
Pigeon, **289**–296, 304
  brain, **392**
  circulatory system, 293–295
  digestive system, 293
  domestic, 303
  endocrine glands, 296
  excretory system, 293
  external anatomy, 289–291
  internal anatomy, **290**, 291–296
  muscular system, 293
  nervous system, 295–296
  reproductive system, 296
  respiratory system, **295**
  sense organs, 295–296

  skeleton, 291–293
Pigment
  cells, 157–**158**, 198, 222, 245, 260
  granules, 222, **245**
  human skin, 462
  photosensitive, 394
  respiratory, 138, 254, 376
*Pinctada*, 201
Pineal body, 226, **231**, **258**, **392**, **298**
Pinnae, 396
Pinnules, 208
Pinocytosis, 20
Pinworm, **127**, 128
Pioneer community, 474, **475**
Pipidae, **243**
Piping plovers, **297**
Pisces, 221, 222
Pit organs, 232, 278–**279**
Pit viper, 278
  venom, 275
Pituitary
  gland, 37, 226, **258**, **260**, 296, **311**, 391, **392**, **398**, 399–400
  anterior, 226, 260, **398**
  posterior, 226, 260, **398** 399
  sac, 225–226
Pituitrin, 399
Placenta, 305, 312, 316, **398**, 417
Placentation, 415–417, 419
Placodermi, **214**, 220, **227**, 241
Placoid scales, 228
Plaice, 241
Plains, 464
Planarians, 48, 112–115
  anatomy, 112–114
  digestive system, **113**–114
  excretory system, **113**–114
  feeding, 114
  learning experiments, 332–333
  muscular system, **113**–114
  nervous system, **113**–115
  reproductive system, **113**–115
Planes
  morphological, 42–44
Plankton, 476
Plantigrade posture, **314**
Plants
  in food web, 471
  gases necessary for, 463
  insects injurious to, 187
  nitrogen incorporation, 464
  pollenation of, 184
  roundworm parasites of, 123
  weed, 482
Planula larva, 102–**103**, **105**

**517**

Plasma
   blood, 35, 175, 254, 258, 375
Plasma membrane, 17–**18**, 20–21, 29, 366, 378
   (*see* Membrane, cell)
Plasmagel, 64
Plasmasol, 64
*Plasmodium,* 70–**72**, 483
Plasmolysis, 17, **19**
Plastron, 269, **270**, 347
Platelets, **35**, **376**, 377
Platyhelminthes, 48, 112–120
Plecoptera, **171**
Pleiotropic studies, 330
Pleistocene period, **456**, 460
   glaciers, 460, 477
Plethodontidae, **243**
Pleura, 36, 380
Pleurisy, 380
Pleuron, **152**, 153, 174
Pliocene period, **456**, 460
Pliohippus, **450**, 451
Plumage
   breeding, 288, 291
   juvenile, 288
   natal, 290
   owl, **302**
   winter, 288
*Plumatella,* 130–**131**
Plutonium, 487
*Podophyra,* 81
Poikilotherms, 267, 285, 462
Poison
   claw, 169
   environmental (*see* Pollution)
   fangs, **279**
   glands, **164**–166, **245**, 247, 263, 265, 266, **279**, 284
Polar cap, **84**
Polar capsule, **73**
Polar filaments, **73**
Polar molecule, **11**, 20
Polarity (molecular), 11
Polarity (morphological), 45
Polarization, light, 338
Pollen, 180–183
Pollination, 184, **342**
Pollution, 490
   air, 468
   mercury, 464, 487
   thermal, 486
   water, 59, 60, 200, 201, 483, **484**–487
Polychaeta, **134**, 144–146, 148, 150
Polydactyly, 427, 429
*Polydora,* 149
Polygenic inheritance, 330, 428
Polymorphism, 94, **103**, 105–106, 111, 179
Polypeptides, 363

Polyplacophora, 189–**190**, 200
Polyps, **94**, 111
   coral, **107**–108
   hydroid, **94**, 102–**103**
Polypteridae, **234**
Polysaccharides, 363
Polyzoans, 130–131, 133
Pongidae, **324**
Pons, **311**, 336, 390, **392**
   varolii, 390
Popliteal artery, 375
Population, 432–434, 461
   breeding, 446
   control, 443, 488–**489**
   genetics, 432–435
   human, 482–483, 487–488
   growth rate, 487–**488**
   isolation, 470
   size, 487
Porcupine, **306**, 316
Pore(s)
   aboral, 95–**96**
   cells, 89
   cuticle, 136, **141**
   dorsal, 137
   excretory, 115, **125**
   genital, **113**, 115, **118**, 123, 209, 226
   membrane, 17–20
   sponge, **86**–89
   water, 194, **195**
Porifera, 47, 85–92
   amoeboid cells, 85–**86**, 88–89
   canals, 86–88
   classification, 85
   freshwater, 85, **91**
   morphology, 85–89, 90
   movement, 89
   origin, 92
   physiology, 89–90
   regeneration, 90
   relations to man, 90–91
   representative, 91
   reproduction, 90
   skeletons, 85, 89, 91–92, 346
   (*see* Sponge)
Pork
   tapeworm, 117–**118**
   trichina worms, 126–**128**
Porocyte, **86**, 89
Porpoise, 322
Portuguese man-of-war, 102
Postsynaptic inhibition, **388**
Posture, foot, **314**
Potential
   action, 387
   membrane, 386
   resting, 386
Pouched mammals, 316, **317**, **318**
Power generation, 486–487

Prairies, 464
Prawns, 162
Praying mantis, **185**
Precapillary sphincter, 38
Precocial birds, **297**, 298
Precopulatory behavior, 340
Predacious arthropods, 165, 166–168, **185**
Pregnancy, 400
Premaxillae, 236, **246-248**
Premolars, **313**, 314
Presynaptic inhibition, **388**
Prey, birds of, 288, 295, **301**–**302**, 304
Prey-predator relationships, 471
Primaries, 292
Primary succession, 474
Primates, **306**, 322–325, 338
Primitive
   amphibians, bony fish, **214**
   jawed fish, **214**, **227**
   jawless fish, **214**, **223**
   insectivore ancestor, **447**
   reptiles, **214**, 264
Primitive streak, **408**, 417
Prismatic layer, 192, **193**
Proboscidea, **306**
Proboscis, 120, 128, **215**
Procellariiformes, **287**
Procornea, 274
Producers, 471, **472**, **473**
Progenital opening, 226
Progesterone, 400
Prolottids, 117–119
Prolactin, 399
Promoter, 423
Pronator teres, **308**
Pronotum, 174
Pronucleus,
   female, 412
   male, 412
Prophase, **27**–28, **67**
*Propliopithecus,* 479
Prosobranchia, **190**
Prosopyle, **87**–88
Prostate gland, 312, **410**
Prostomium, 135–**136**, 141, 144–**145**, 150
Proteidae, **243**
Protein, 421, 463–464
   cell, 14
   coat, 420
   digestion, 363, 369
   synthesis, 22, 422, **423**
Protein matrix, 347
Proteoses, 363
Prothorax, **173**–174
Prothrombin, **364**
Protista, 47, 53–82
*Protoceratops,* 281
Protochordates, 215–217

Proton, 8–9
Protonephridia, **218**, 219
Protoplasm, 16
Protopodite, **152, 154,** 155
Protostomes, 46, 131
Protozoa, 47, 53–82
Protractor muscles, **225**
Protura, **171**
Proturan, **171**
Proventriculus, 174, 293
Proximal convoluted tubule, 368
Pseudocoel, 45, 48, 121, 123–**124,** 129
Pseudocoelomates, 121–129
Pseudopodia, 47, 54, **64–66,** 68, **97**
Pseudoscorpion, **163**
Pseudoscorpionida, **163**
Pseudostratified epithelia, **31–32**
Psychogenic stimuli, 331
Psychophysics, 331
Ptarmigan, 288
*Pteranodon*, 281
Pterobranchia, 214
Pterosaurs, 281
Pterygoideus, **249**
Pterylae, 291
Ptyalin, 361, 363
Pubis, **248, 292, 307,** 308, 352
Pulmocutaneous
  arch, **256**
  artery, 255, **257**
Pulmonary
  artery, 255, **257, 271,** 293–294, **371,** 374, 375
  veins, **256, 257, 271,** 293, **310, 371,** 374, 375
Pulmonata, **190**
Pulp cavity, **313,** 314
Pulse, 373
Pulvillus, **173–174**
Puma, 316, **319**
Punishment, 334
Punnett square, **426, 427, 436**
Pupa, 178–**179, 185,** 186
Pupil, 238, 260, 272, 394, **395**
Purkinje fibers, 371
Pycnogonida, **163**
Pygostyle, 291, **292**
Pyloric
  cecum, 205, **206,** 236, **237**
  glands, 361
  sphincter, 253
  stomach, 155–**156,** 205, 253
  valve, 229, 270
Pylorus, 360
Pyorrhea, 68
Pyramid
  energy, 472, **473**
  mass, 472, **473**
Pyrenoid, 58
Pyridoxine, **364**
Pyrgota fly, **185**
Pyriformis, **249**
Python, **268,** 274–275
  Indian, **275**

## Q

Quadratojugals, **247**
Quadriceps femoris, **354**
Quadruped, 305
Queen, 180–183
  conch, 198
  insect, 180–183

## R

Rabbit, **306**
Radial arteries, 375
Radial canals, 88, **104–105,** 203, **205, 206, 208, 210**
Radial chambers, 106
Radiating canals
  *amoeba*, 77
Radioactive wastes, 486–487
Radiolaria, **68**
Radioulna, **246, 249**
Radius, **292, 307,** 308, **351,** 353
R. punjabicus, 479
Radula, 189–190, 197
*Ramapithecus wickeri*, 479
*Rana pipiens*, 242
*Rana catesbeiana*, 244
Random assortment, 429
Range, hearing, 331
Ranidae, **243**
Rapid eye movement, 336
Rat, 316
Rattlesnake, **268, 274,** 278, **279,** 284
Ray (echinoderm), 203, 205, **206,** 208
Rays, 228
Reabsorption, 368, 369
Reaction, chemical, 10, 13, 29
  types of, 15–16
Reaction to stimuli
  *amoeba*, 66
  arthropod, 160
  coelenterate, 93, 99–100, 105
  earthworm, 143
  *euglena viridis*, 59
  frog, 260
  negative, 66, 77
  *paramecium*, 77–78
  positive, 66, 77
  sponge, 90
Receptor, 35, 140, **141,** 195, 332, 383, 386, 393
  site localization, 395–396
  threshold response, 331
Recessive, 424
Rectal gland, 229, **230**
Rectum, 37, 123, **147, 164,** 165, 174–**175,** 193, **194, 196,** 197, **209,** 270, **272, 309, 360,** 361
Rectus abdominus, **354**
Rectus anticus femoris, **249**
Red-backed salamander, **243**
Red-bellied newt, **263**
Red blood corpuscles, 31, **35,** 220, 236, **254,** 309, 376
  and malaria, 70–72
  sickle-cell, 381
  and tonicity, **19**
Red fox, 316
Red racer, 275
"Red tide," 59
Redi, F., 451, **452**
Reduced
  definition, 16
Reduced penetrance, 429
Reduction division, 429
Reduction reactions, 16
Reefs, 108, 149, 210
Reflex, 140, 258, 266, 332
  arcs, 383, **389,** 390
  muscle stretch, **389**
  patellar, 332
  spinal, 332, **389,** 390
Refractory period, 387
Regeneration, 90, 96, 102, 398
  amphibian, 266
  arthropod, 160–161
  earthworm, 142
  echinoderm, 203, 208
  hydra, 96, 102
  planaria, 142
  sponges, 90
Regulator, 423
Reinforcement
  negative, 333
  positive, 333
Relapsing fever vector, **168,** 186
Releasing stimulus, 335
REM sleep, 336
Renal
  artery, **271, 294, 310, 367,** 368, **374,** 375
  corpuscles, **367**
  cortex, **367**
  medulla, **367**
  pelvis, **367,** 369
  vein, **237, 257, 271, 294, 310, 367,** 374
Renal portal system, 229, **230, 256, 257, 271, 294,** 295

**519**

Renal portal vein, 237, **257,** **271, 294**
Renettes, 366
Replication, 420, **421**
Repressor, 423
Reproduction
 asexual, **55,** 90, 405, **406**
 budding, 90, 100, 102, 107
 conjugation, 78–**80**
 fission, 405–**406** (*see* Fission)
 fragmentation, 107, 120
 metagenesis, 102–**103**
 sexual, **55,** 407
 sporulation, **406**
Reproductive migration, 343
Reproductive organs, 38
 accessory, 38
 (*see* Gonads)
Reproductive system, 37–38, 47, 405–412
 *amoeba,* 66–**67**
 amphioxus, 219
 annelid, 140, 142, 145–146, 148, 149
 anthozoan, 107
 arthropod, 159, 165–166, 177
 *ascaris,* 123–**125**
 bird, 296
 bony fish, 238
 cat, 312
 dogfish shark, **232**–233
 echinoderm, 206–**207**
 entroproct, 129
 *euglena viridis,* **58**–59
 frog, **260**–261
 human, 409–412
 hydra, 100–101
 hydroid, 102–**103**
 lamprey, 226–227
 mesozoan, 84
 mollusk, 190–191, 195
 *opalina,* **62**–63
 *paramecium,* 78–**80**
 platyhelminth, **113**–115, **116**–117
 scyphozoan, **105**
 sponge, 90
 sporozoan, 69–**72**
 *volvox,* **55**
Reptiles, 49, 213, **214,** 221, 267–285
 age of, 281
 brain, **392,** 393
 "feathered," 286
 fossil, 281–**282**
 general characteristics, 267
 heart, **255**
 primitive, **255**
 relations to man, 284–285
Reservoir, 57

Resistance, 453
Respiration, 37, 378
 cell, 26
 control, 381
 cutaneous, 190, 198, 256
 external, 37, 378
 internal, 37, 378
 movements, 380
 rate, 381
Respiratory system, 37, 38, 378–381, 382
 *amoeba,* 66
 amphioxus, 219
 annelid, 139, 148
 bird, **295**
 bony fish, 236
 cat, 311
 dogfish shark, **231**
 echinoderm, 205
 frog, 253–254, 256, 266
 human, **379**–381
 invertebrate, 378–379, 382
 lamprey, 226
 mesozan, 84
 mollusk, 190, 194–**195,** 198
 *paramecium,* 76
 turtle, 271–272
 vertebrate, 379
Respiratory tree, **210**
Response to stimuli, 332–334
 conditioned, 333
 kineses and taxes, 335
 local, 99
 nonlocalized, 99
 unconditioned, 333
 (*see* Reaction to stimuli)
Resting potential, 386
Reticular formation, 336, 390–391
Reticulum, **323**
Retina, 176, 260, 394, **395**
 pure-rod, 394
Retinoblastoma, **429**
Retinular cells, 157–**158**
Retractor, 192, **210, 225**
Retroperitoneal, 251
RH blood factor, 323, 377–378
 incompatibility, 378
Rhabdite, **113**
Rhabdome, 157–**158**
Rhagon, **87**
Rheas, 289
Rhesus monkeys, 322, 331
*Rhineodon typus,* 233
Rhinoceroses, 322
*Rhinodrilus,* 143
Rhodesia man, **480,** 482
Rhodopsin, 394
Rhynchocephalia, 282
Ribbon worms, **119**–120

Ribonucleic acid (RNA), 422
 and learning, 332–333
 synthesis, 422
Ribose, 422
Ribosomal RNS (rRNA), 422
Ribosomes, 21–22, **25**–26, **423**
Rib(s), 269, **307,** 308, 349
 bird, 291
 cage, 380
 cervical, 291
 flase, 352
 floating, 352
 gorilla, 352
 human, **351,** 352
 thoracic, 291
 true, 352
Rickets, **364**
Rickettsial organisms, 184
Ring canal, **105,** 203, **205, 209, 210**
Ringhals, **277**
RNA polymerase, 423
Robber frog, **243**
Robin, 289
Rocky Mountain spotted fever, 184
Rodentia, **306,** 316, **317**
Rods, 394
Root nodules, 464
Roots
 dorsal (sensory), **389,** 390
 ventral (motor), 388, **389**
Rostrum, **153,** 156
Rotator muscle, 250
Rotatoria, 121–123
Rotifers, 48, 121–123
Rough endoplasmic reticulum, **22**
"Round dance," **339**
Round window, 396
Roundworms, 121, 123–129
"Royal jelly," **181**
Rumen, **323**
Ruminants, 322, **323**

## S

Sabertoothed cats, 449, 460
Sacculus, 238, 397
*Sacoglossus kowalevskii,* **215**
Sacral vertebrae, **246, 307, 350**
Sacrum, **350, 351**
*Sagitta hexaptera,* **132**
Sagittal crests, 479, **481**
Salamander, 242, 263
 aquatic, 265
 giant, 264, 265
 *necturus,* **265**
 poisonous, 265
 tiger, 264–265

*Salamandra salamandra*, 265
Salamandridae, **243**
Salientia, 242
Salinity, 5
Saliva, 37, 361, 363, 393
Salivary glands, 37, 174, **175,** 197, 309, 360, 361
Salmon, 240, 341, 343
Salmonidae, **234**
Salts, 10, 29
  biogenic, 464
  defined, 10, 16
  mineral, 362
  seawater, 464
Saltwater environment, 469–470
Sand dollars, 49
Sandflies, 60
Sandworm, 48, 144–145
Sarcodines, **54,** 63–69, 82
  and geology, 68–69
  parasitic, 68
Sarcomastigophora, 47, 54–55, 57–69, 82
  parasitic, 54, 60–62, 68
Sarcomeres, 33, 355, 356
Sarcoplasm, 33
Sarcoptes, **168**
Sargassum fish, **234**
Sartorius, **249, 250, 251,** 308, **354**
Savanna, 464
Scab mite, 184
Scales, 38, 144, 179, 222, 233–**236,** 273, 290, 345, 347
  ctenoid, 235–**236**
  cycloid, 235–**236**
  ganoid, 235–**236**
  placoid, 228
Scallops, 189–**190,** 201
Scaphopoda, 189–**191,** 200
Scapulae, **292, 307,** 308, **351,** 352
Scavengers
  bird, 302
  insect, 184, 187
  molluscan, 200
Scent glands, 170, 313
Schizogony, 71–**72**
Schizont, **71–72**
Schwann, cell, 385, **386**
Sciatic
  artery, **294**
  vein, **257, 294**
Sciatic nerve, **259**
Scientific method, 4
Scincidae, **268**
Sclerites, 172, 174
Scleroblasts, **88**–89
Sclerotic coat, 394, **395**
Scolex, 117–119

Scorpion, 151, 162, **163,** 167–168
Scorpionfly, **171**
Scorpionida, **163**
Screech owl, 302
Scrotal sacs, 312
Scrotum, 307, 312, 410
Scud, **161**
Scurvy, **364**
Scutes, 269, **270**
*Scutigera*, 169
*Scypha*, 87–90
Scyphistoma, **105**
Scyphozoa, 93, 104–105
Sea (*see* Marine environment)
Sea anemone, 48, 93, **106**–107, 335, **337**
Sea cucumbers, 49, 203, **204,** 208, **210**
Sea gooseberries, 48, 109
Sea gull, 286, **341**
Sea lamprey, 222, 223–227, 240 (*see* Lamprey)
Sea lilies, 203, **204,** 208
Sea otter, 316, **321**
Sea snake, **268**
Sea spider, **163**
Sea squirts, 215
Sea star, 49, 203–208, 210
  commensualism, **342**
  digestion, 205
  excretion, 205
  external anatomy, 203
  internal anatomy, 203
  life cycle, **207**
  nervous system, 205–206
  regeneration, 208
  reproduction, 206
  respiration, 205
  sense organs, 205–206
  water-vascular system, 203, **205,** 210
Sea turtle, 283, 341
Sea urchin, 203, **204,** 208, **209,** 210
Sea walnuts, 48, 109
Seahorse, **234,** 239
Seaquakes, 457
Seasonal rhythms, 476
Sebaceous glands, 312, 313, 345, **346**
Sebum, 345
Secondaries, 292
Secondary succession, 474
Secretin, 400
Secretion
  hormone, 37, 260
  mucus, 96–**97,** 114, 141, 198, 233, 247, 309
  (*see* Endocrine system)

Segment, 44, 134–135, 172
Segmentation, 44, 134
  annelid, 134–135
  chordate, 213, 219
  heteronomous, 45
  homonomous, 45
  molluscan, 200
Segmentation cavity, 414, 415
Segregation, 426, 435
Segregation distorters, SD, 446
Selection, **448**
Selective permeability, 20
Self-consciousness, 478, 482
Self-fertilization, 407
Self-sterile, 407
Selfing, 425
Semen, 410
Semicircular canals, 226, **231,** 232, 238, 397
Semiconservative duplication, 420
Semilunar valve, **256,** 371
Semimembranosus, **249, 250, 251**
Seminal
  fluid, 312, 410
  grooves, 142
  receptacle, **136,** 137, **140,** 142, 159, **164,** 166, **175,** 177
  vesicle, 124, 136, 140, 142, **260, 261,** 296, **410**
Seminiferous tubules, 399, 409
Semipermeable membranes, 17, **19,** 29
Semitendinosis, **249,** 308
Sense hairs, 140
Sense organs, 37
  annelid, 140, **141,** 143
  amphioxus, 219
  arthropod, 157–159, 160, 176–177
  bird, 290, 295–296
  bony fish, 238
  cat, 312
  ctenophoran, 109
  dogfish shark, 232
  echinoderm, 206
  frog, 259–260
  human, 393–397
  lamprey, 226
  mollusk, 195, 198
  plathelminth, **113,** 115, 116
  scyphozoan, 105
  turtle, 272
Senses, 37, 393–397
  audition, 396–397
  chemical, 383, 393–394
  electroreception, 226, 232, 396

Senses (cont.)
   equilibrium, 397
   gustation, 393
   olfaction, 393, 394
   tactility, 395–396
   vision, 394–395
Sensory
   canals, 232
   cells (neurons), 35, **97**, 113, 140, **141**, 195, 219, 383, **389**, 390, 396
   fibers, 140, **141**
   hair cells, 232
   hairs, 165, 396
   nerves, 219
   receptors (see Receptor)
   stimuli, 331
Septa, 45
   annelid, **136**, 137
   sea anemone, **106**
Serial homology, 155
Serosa, **362, 408**
Serous membranes, 36, 39
Serpentes, 273
Serratus anterior, **354**
Sertoli cells, 409
Sessile, 43, 47, 85, 92, 105, 209
Setae, 48, 135, **137**, 144–145, **152**, 160
Sewage treatment, 486
Sex
   chromosomal abnormalities, **430–431**
   inheritance, 430
   roles, 479
Sex-linked inheritance, 431–432
Sexual characteristics
   secondary, 400
   and taxonomy, 47
Sexual reproduction, **55**, 407
   evolution of, **406**
Sexuals, **182**
Shaft, 290–**291**
Shark, 214, 228, 233, 240
   dogfish (see Dogfish shark)
   great white, 233
   hammerhead, 233
   jaw and teeth, **229**
   killer, **240**
   liver oil, 240
   whale, 233
Sheep scab mite, 184
Sheep, short-legged, **453**
Sheep tick, **187**
Shell, 48, 347
   aschelminth egg, 123, 124
   bird egg, 297
   bivalve, 131, 189, 200
   brachiopod, 131–**132**
   bryozoan, 130

echinoderm, 208–**209**, 210
gland, 233
hydra, **101**
molluscan, 189, **191, 192, 193, 196**, 197, 198, **199**, 200, 202
reptile egg, 267, 272
sarcodine, 68–69
sponge, 91
tortoise, 284
turtle, **269, 270**, 283, 347
Shells, electron, 9
Shelter, 481
Shipworms, **201**–202
Shore crab, **161**
Shorebirds, 298, 301
Short-term solution, 482–483
Shrew, **306**, 316
Shrimp, 184
   fairy, **161**
Sickle cell anemia, 381, 483
Sigmoid colon, **360**, 361
Silk, 184
Silk glands, 166
Silurian period, **456**
Silverfish, **171**, 188
Sinistral, 198
Sinoatrial node, 371
Sinus gland, 159
Sinuses, 370
Sinus venosus, 229, **230, 237**, 254–**256, 257**, 271
Siphon
   dorsal, 191
   excurrent, 191–194, **216**
   funnel, **196**, 197
   incurrent, 191–194, **216**
   ventral, 191
Siphonaptera, **171**, 186
Siphonoglyph, **106**
Siren, 242, **243**
Sirenia, **306**
Sirenidae, **243**
Skates, 228
Skeletal axis, 213, 307
Skeletal muscle, 353
   tissue, **31**, 33, 353
Skeletal systems, 34, 37, 38, 346–353, 356, 357
   amphioxus, 218
   arthropod, **152, 153**–155, 159, 172
   bird, 291–293, 295
   bony fish, 234–**235**
   cat, **307–308**
   development, 347–349
   dogfish shark, 228
   echinoderm, 203, 208–**209**, 210
   frog, **246–248**

human, 349–352, 357
invertebrate, 347–349
lamprey, 224–225
molluscan, 189
snake, **274**
sponge, 85–89, 91–92, 346
and taxonomy, 46
turtle, 269, **270**
vertebrate, 220, 347–349
Skeleton
   appendicular, 228
   axial, 228, 307
   cartilage, 228
   ossified, 267
   somatic, 349
   visceral, 235, 247, 349
Skin, 38
   bird, 290
   bony fish, 233
   cancer, 462
   color inheritance, **429**
   dogfish shark, 228
   as excretory organ, 345
   frog, 242, **245**, 256, 260, 265
   glands, 313, 345–**346**
   human, 345–**346**
   lamprey, 224
   reptile, 267, 273–274, 283, 285
   as respiratory organ, 190, 198, 256, 378
   salamander, 263, 265
   as sensory organ, 238, 259, 272
   (see Tactility; Touch)
   in thermoregulation, 345–346
   toad, 242, 263, 265
   vertebrate, 345
   vitamin D synthesism, 462
Skink, **268**
Skinner, B.F., 333
Skull, 347
   bird, 291
   bony fish, 235
   cat, **307, 308**
   dogfish shark, 228
   frog, **247-248**
   human, **349, 351, 480, 481**
   human ancestors, **480**
Skunk, 316
Sleep, 336, **337**
Sliding filament model of muscle contraction, 355, **356**
Slime
   eels, 223
   glands, 198
   road, 114
Sloth, 316
Slow-wave sleep, 336

Slugs, 189
　land, 198, 202
Small intestine
　cat, **309**
　cow, **323**
　frog, **252**, 253
　human, **360**, 361
　turtle, 270, **271**
Smell, 160, 165, 226, 259, 272, 312
Smooth endoplasmic reticulum, **22**
Smooth muscle, 38, 353
　invertebrate, 353
　tissue, **31**, **33**, 353
　vertebrate, 248, 250, 353
Snails, 49, 189–**191**, 198–**199**, 201
　freshwater, 198–199, 202
　land, **190**, 198–**199**
　marine, 198
Snake-necked turtle, **268**
Snakebite, 280, 285
　antivenom, 280
　first-aid measures, 280
Snakes, **214**, 267, 273–280, 285
　venomous, 274, 276–280, 285
Snapping turtle, **283**
Sneeze, 332
Social behavior, 340–**341**
Social insects, **180**–183
Social interactions, 330
Sociology, 6
Sodium chloride, **10**
　formation of, 16
Soft-shelled turtles, **283**
Soil, 470
Sol, 17
Solar azimuth, **343**
Solar radiation, 461–462, 471, **472**
Soldiers, **182**–183
Sole, 241
Solenocytes, 219
Soleus, **308**, **354**
Solo man, 482
Somatic cells, 83–**84**
Somatic nervous system, 384
Somatic skeleton, 349
Somite, 44, 134–135, 239, 417
Song sparrow, 286
Songs, 288–289
Sound production
　cat, 311
　fishes, 222, 241
　frog, 254, 264
　grasshopper, 172, 177
　human, 380
South American monkeys, 323
Sow bug, **161**

Spadefoot toad, **243**
Sparrow, **288**
Sparrow hawk, 302
Spawning, 223, 226–227, 238, 343
Speciation, 447, 470–471
Species, 6, 41
Sperm, 29, 38, 71, 407, **409**
　bundles, **55**, **406**
　duct, **113**, **116**, **118**, **410**
　formation, 409–410
　funnel, **136**, 142
　human, **409**–410
　nucleus, **409**
　penetration, 24
　receptacle, 180
　storage, 407
　structure, **409**–410
Sperm oil, 322
Sperm whale, **306**, 322
Spermatheca, 177
Spermatid, 409
Spermatocytes, 409
Spermatogenesis, 399, 409
Spermatogonia, 29, 409
*Sphenodon punctatus*, 267, 282, 285
Sphenidintidae, **268**
Sphenoid bone, **349**
Spherical symmetry, 43
Spicules, 85–89, 91–92, 346
Spider crab, 162, 166
Spider, 48, **151**, 162–**167**
　circulatory system, 165
　courtship behavior, 166, 340
　digestive system, 165
　excretory system, 165
　external anatomy, 163–**164**
　internal anatomy, **164**–166
　nervous system, 165
　physiology, 165–166
　reproductive system, 165–166
　respiratory system, 165
　web, 162, **166**, 334
Spinal accessory nerve, **258**, 311, 392
Spinal cord, 37
　bony fish, **237**, 238
　dogfish shark, **231**–232
　cat, **311**, 312
　frog, **258**, **259**
　human, 234, 388, **389**, 390
　vertebrate, 234
Spinal nerves, 37, **231**, 232, 238, 258, **259**, 311, 312, **389**, 390
Spinal reflex, 332
　arc, **389**, 390
Spindle, **389**
Spindle cells, 254

Spindle fibers, 24, **27**–29
Spines, **206**, 208, **209**, 235
Spines, dendritic, 333
Spinnerets, **164**, 166
Spinning organs, 166
Spiny anteater, **306**, 315
Spiny-head worms, 128–129
Spiracles, **164**, 173–**176**, 228, 378
Spiral rings, 175–**176**
Spiral valve, 229, **230**
Spleen
　bird, 293
　bony fish, **237**
　cat, **309**, 310
　dogfish shark, 229, **230**
　frog, **252**, 254
　human, 374
　turtle, **271**
　vertebrate, **220**
Splenic
　artery, 375
　vein, **257**
Splints, 444, **450**
*Spondylomorum*, **406**
Sponges, 47, 85–92
　ascon, 86–**87**
　bath, 90–91
　boring, 91
　elephant's ear, **91**
　finger, **91**
　freshwater, **91**
　glass rope, **91**
　marine, 85, 87, 90–91 (*see* Porifera)
　origin, 92
　rhagon, **87**
　sycon, **87**, 89
　venus's flower basket, **91**
*Spongilla lacustris*, 91
Spongillidae, 85, **91**
Spongin, 85, 87–88, 346
Spongoblasts, 87
Spongocoel, 85–89
Spongy bone tissue, 348
Spontaneous generation, 451, **452**
Spores, 47, 55, **69–73**
　wall, **69**, **73**
Sporocyst, **70**, **116**–117
Sporozoa, 47, 55, 69–73, 82
　parasitic, 69–73
Sporozoites, **69**–**72**
Sporulation, **406**, 407
Spotted fever tick, **168**
Spring peeper, 264
Springtail, **171**
Sprue, **364**
*Squalus acanthias*, 228–233
Squamous epithelia, **31**–**32**

Squids, 49, 189–191, **196**–198, 201
  giant, 198
Squirrel, **214**, 316
Stabilizing selection, **448**
Stalk, 95–**96**, 208, 217
Stapes, 350, **396**
Starfish, 203–208, 210
Starling, 343
Statoblasts, 130, **131**
Statocyst, **104**, 109, 157, 195, 198
Statolith, 109, 157, 195
Stearic acid, 14
Stegocephalis, 264
*Stegosaurus*, 281
Stellar navigation, 343
*Stentor*, 81
Steppe, 464
Sterocoral pocket, **164**, 165
Sterogastrula, 415
Stereoscopic vision, 479
Sternocleidomastoideus, **354**
Sternomastoid, **308**
Sternum, **152**, 153, 164, 174, 247–**248**, 291, **292**, 307, 308, **351**, **352**
Steroids, 13, 399
Sterols, 363
Stethoscope, 372
Stickleback, 335
Stigma, 58
Stimuli
  reaction to (*see* Reaction to stimuli)
Stimulus, 331–332
  conditioned, 333
  external, 331
  intensity, 331
  internal, 331
  kinesthetic, 331–332
  psychogenic, 331
  releasing, 335
  sensory, 331
  threshold, 331
  unconditioned, 33
Stimulus-response, 331–332, 344
Sting, 167–168
Sting capsules
  (*see* Nematocysts)
Stomach, 37
  annelid, 144, 146, **147**
  arthropod, 155, 165, 174–**175**
  aschelminth, 121–122
  bord, 293
  bony fish, 236, **237**
  bryozoan, 130–131
  cat, **309**
  cow, 323

ctenophore, **110**
dogfish shark, 229
echinoderm, 205, **206**, **207**, **209**, 210
four-chambered, **323**
frog, 251–**253**
hormone secretion, 400
human, **360**, 361, 400
mollusk, **193**, **196**, 197
ruminant, 323
turtle, 270, **271**
Stomodeum, 106, 110
Stone, canal, 203, **205**, **209**
Stone rings, 481
Stonefly, **171**
Strainer, 155
Stratification, 476
Stratified epithelia, **31**–32
  in mucous membranes, 36
Strepsiptera, **171**
Stress, 399
Striated muscle, 38, 353–354
  arthropod, 174, 353
  tissue, **31**, 33, 353
  vertebrate, 248, 250, 353
Stridulation, 177
Strigiformes, **287**
Strobila, **105**
Stroma, 36
*Strongylocentrotus purpuratus*, 208
*Strongylus vulgaris*, 128
Struthioniformes, **287**
Sturgeon, **234**
*Stylonchia*, 81
Subcellular organelles, 17–**18**, 21–26, 29
Subclavian
  arteries, **230**, 257, **271**, **310**, 374, 375
  veins, 257, **374**
Subdermal cavity, 87
Subesophageal ganglion, 144, 147, **175**–176
Subgerminal cavity, 415
Subintestinal vein, 219
Sublinguals, 309
Submaxillaries, 309, **310**
Submucosa, **362**
Subneural vessel, 138–**139**
Subpharyngeal ganglion, 140, **141**
Subscapular vein, **257**, **310**
Subspecies, 41, 330
Substrate, 12
Subumbrella, **104**
Succession, ecological, 474, **475**
Sucker, 117
  cephaloped, **196**, 197

leech, 147–148
oral, 115–**116**, 225
ventral, 115–116
Suctorians, 47
Sugars, 14, 363
Sulcus, **311**
  central, **391**
Sunfish, **214**
Supportive tissue
  (*see* Connective tissue)
Suprabranchial chamber, **194**
Supraesophageal ganglion, 156–157, **175**–176
Suprascapula, **246**, **248**
Surface-to-volume ratio, 30
Surinam toad, **243**
Surrogate parent, 331
Survival of the fittest, 454
Suture, **349**, **352**, 419
Swallow, 286
Swallowing, 359
  cat, 309
  human, 361
  snake, 361
Swan, 301
Sweat glands, 38, 313, 345–**346**, 367
Swimmeret, **152**, 153, **154**, 155, **156**, 159
Sycon, 87, 89
Symbiosis, 54, 341
Symmetry, 43–44
  asymmetry, 43–44, 47
  bilateral, 43–**44**, 48, 49
  biradial, 43–**44**, 47, 48
  radial, 43–**44**, 47, 49
  spherical, 43–44
Sympathetic
  ganglia, **389**, 390
  trunk, **389**, 390
Sympathetic nervous system, 176, 384–385
Symphysis pubis, 353
Synapse, 35, 386, **387**–**388**
  excitatory, 388
  inhibitory, 388
Synaptic
  cleft, **387**–**388**
  knobs, **387**–**388**
  transmitters, 388
  vesicles, **387**–**388**
Syncytium, **124**, 355
Syngamy, **62**
Symgnathidae, **243**
Synonyms, 42
Synocial
  fluid, **352**, 353
  joints, 353
  membrane, **352**

Synthesis reaction, 16
*Synura*, 60
Syrinx, **295**
Systole, 373

# T

Tactile
  disks, 395
  hairs, 177
  organs, 295, 395
  sensation, 266
Tactility, 395–396
Tadpole, 261, **262**
Tagging, radioisotopic, 464
Tagmata, 152
Tagmosis, 152, 188
Taiga, 467
Tail
  bird, 286, 290, 291
  bony fish, 233
  postanal, 49
  prehensile, 323
  vertebrate, 220
"Tail-wagging dance," **339**
Tapetum lucidum, 296, 312, 394
Tapeworms, 48, 112, 117–119
Tapirs, 322, 460
Tarantula, 166–**167**
Tarsals, **246**, **249**, **307**, 308, **351**, 352
Tarsometatarsus, **292**, 293
Tarsus, **173**–**174**
Taste, 160, 165, 266, 393
  buds, 393
  organs, 177, 226, 259
  pore, 393
  receptors, 393
Taxes, 335, 344
Taxodonta, **190**
Taxon (taxa), 41, 49
Taxonomy, 3, 6, 40–47, 49, 273
Teats, 307, 316
  (*see* Mammary glands)
Technology, 478, 483
Tectonis plates, 458
Teeth, 37, 345
  ape, 479
  archaeornithes, 298
  bony fish, 236
  cat, **308**
  chitinous, 148, 155, 190
  early man, 479, 482
  extra, **429**
  frog, 242, **247**, 252
  horse, **450**, 451
  human, 358, **359**, 360, 361

mammals, 313–314
  shark, **229**
  snake, 274, **279**
Telolecithal eggs, **408**
Telophase, **27**, **29**, **67**
Telson, **153**, 155, **156**, 160, **168**
Temperate decidous forest, **465**, 467
Temperature, 462
  zones, 462
Template, 420
Temporal
  bone, **308**, **349**
  lobe, 390, **391**, **392**
  muscle, **249**, **308**
Tendon, 31, 34
  achilles, **249**, **354**
Tensor fascia lata, **308**, **354**
Tentacular
  bulb, 104
  pouches, 110
Tentacles, **94**-**96**, 99–100, 101, 102–**106**, 110, 130, 200, 206, **210**
Tentaculocysts, 105
Tergum, **152**, 153, 174
Terminal nerve, **231**, **384**
Termite, **171**, **182**–**183**
Terrestrial environments, 6, 470
Territoriality, **338**, 340, **341**
Tertiaries, 292
Test, 208–**209**, 216
Testcross, 437
Testes, 38, 400
  annelid, **136**, 142
  arthropod, **156**, 159, 177
  aschelminth, 124
  bird, 296
  bony fish, 238
  cat, **309**, 312
  dogfish shark, **232**–233
  frog, 251, **252**, 260
  hydra, **96**, 100–101
  human, **398**, 400, 409, **410**
  platyhelminth, **113**, **116**, **118**
  turtle, **272**
Testosterone, 399, 400
Testudindae, **268**
*Testudo*, 283
Tetany, 399
Tethys sea, 457
Tetrabranchia, **190**
*Tetrahymena*, 65, **66**, 81
Tetrapoda, 221, 222, 242
Texas cattle tick, **168**
Thalamus, 390
Thales, 440
Thermoreceptive organ, 278–279

Thermoregulation, 37, 345, 346, 367, 375, 462
  bird, 295
  mammal, 313
Thiamine, **364**, **365**
Thoracic
  cavity, 45, 309, 311, 380
  pressure changes, 380, 382
  vertebrae, **307**, **350**, **351**
Thorax
  human, 352
  insect, 172–174, 188
  mammal, 305, 308
Thought, 390
Threshold, stimulus, 331
Thrips, **171**
Thrombin, 254, 377
Thrombocytes, **35**, **254**
Thrombokinase, 377
Thromboplastin, 377
Thumb, opposable, 322, 478
Thymine (thymidylic acid, T), **421**
Thymus, 37, 293, **398**
Thyroid gland, 37, 226, 260, 296, **398**–399
Thyroid-stimulating hormone (TSH), 399
Thyroxin, 260, 362, 398–399
Thysanoptera, **171**
Thysanura, **171**, 187
Tibia, 173–174, **307**, 308, **351**, 352
Tibial artery, **257**, 375
Tibialsi
  anterior, **308**
  anticus, **249**, **250**, **251**, **354**
  posticus, **250**, **251**
Tibiofibula, **246**, **249**
Tibiotarsus, **292**, 293
Tick paralysis, 184
Ticks, 48, 151, 162, **163**, **168**, 184, 186
Tiedman's bodies, 203, **205**
Tiger, 305, 316
Tiger beetle, **185**
Tiger salamander, **243**, 264–265
Tinbergen, N., 329, 334, 335
Tissue, 30–36, 39
  embryonic, 45–46, 112
  level, 92, 93, 119, 111
  and organs, 30, 36–39
  types of, 30–36, 39
    connective, **31**, 33
    epithelial, 30–**32**
    osseous, **31**, 33
    muscle, **31**, 33
    nervous, **31**, 34
    vascular, **31**, **35**

Tissue (cont.)
    adipose, 35
Toad, 242, 263
    American, **243**, 244
    poisonous, 263, 265
    relations to man, 265–266
Toadfish, 368
Toes, 307, 308
Tomato sphinx moth, **185**
Tongue
    bird, 295
    bony fish, 236, **237**
    cat, 309
    extensile, 252
    human, **360**, 361, 393
    lamprey, **224**, **225**
    rasplike, **224**, **225**
    as sense organ, 259, 274, 295, 393
    snake, 274, **279**
    turtle, 269
Tongue worm, 214–**215**
Tonicity, 17, **19**
Tonsils, 309
Tonus, 354
Tool
    making, 481
    use, 335, 479
Tooth shells, 189–**191**, 200
Toothed whales, **320**, 322
Toothless mammals, 316, **317**
Topminnows, 241
Tortoise, 267, 284, 285
    giant, 283
    shell, 284
Tortoise armadillo, 314
Touch, 160, 238, 266, 272, 395
*Toxocara canis*, 128
Trabeculae, 347–348, 356
Trachea, 37
    bird, **295**
    cat, **309**, 311
    human, **379**
    turtle, **271**
Tracheal (arthropod), 48, **164**, 165, 175, 183, 378
Tracheal rings, 379
Tracheoles, 175–**176**, 378
Traits, 420, 424
    human, **429**
Transcription, 422
Transfer reaction, 16
Transfer RNA (tRNA), 422, **423**
Transfusions, blood, 377
Transitional stratified epithelia, **31**
Translation, 422, **423**
Transport, membrane, 20
Transverse
    canal, 205

colon, **360**, 361
    muscle, **249**, **308**
Transverse fission, 143
Trapezius, **308**, 354
Tree frog, **243**, **264**
Trematoda, 112, 115–117, 199
Trench fever, **186**
Triassic period, **456**
Triceps
    brachii, **249**, **250**, **308**
    femoris, **249**, **251**
Trichima worms, 126–128
*Trichinella spiralis*, 126–**128**
Trichinosis, 126–**128**
Trichocysts, **75**–76
*Trichomonas*, 60–**61**
Trichoptera, **171**
*Trichuris trichiura*, **127**
Tricuspid valve, **371**
Tricuspids, 360
Trigeminal nerve, **231**, **258**, **259**, **311**, **384**, **392**
Trimorphism, 179
Trinomial nomenclature, 41
Triploblastic organisms, 46, 48–49, 112, 415
Trisomy E syndrome, **433**
Trochanter, **173**–174
Trochlear nerve, **231**, **258**, **259**, **384**, **392**
Trochophore larva, 144–146, **191**, 200
Trogon, **287**
Trogoniformes, **287**
Trophoblast, **408**, 415, 417, **418**, 419
Trophozoite, 55–**56**, 63, 69–**72**
Tropical rain forest, **465**, **468**
Tropisms, 335
Trout, **234**, 240
True-breeding, 425
Truncus arteriosus, 255–**256**
Trunk, 220, 233
Trypanosomes, **60**
Trypsin, 38, 363
Tsetse fly, 60, **186**–187
Tuatara, **268**
Tubal excretion, 368
Tube feet, 49, 203, **205**–210
Tubercles, 164
Tubes, 130
    water, **194**, **195**
*Tubifex tubifex*, 143
Tubuli dentata, **306**
Tularemia, **186**
Tuna, 240
Tundra, **465**, **468**, **469**
Tunic, **216**, 217
Tunica propria, **362**
Tunicates, 215–217

internal structure, **216**
Turbellaria, 112–115
Turbot, 241
Turkey, 303
Turner's syndrome, **431**
Turtle, 267, 269–272, 283, 285
    circulatory system, 271
    digestive system, 269–270
    external anatomy, 269
    internal anatomy, 269–272
    nervous system, 272
    relations to man, 272
    respiratory system, 271–272
    sense organs, 272
    shell, 269, **270**, 283, 285
    skeletal system, 269, **270**
    urogenital system, **272**
Tusk shell, **190**, 200
Twins, 412, 437
    fraternal, 437
    identical, 412, 437
Twisted wing, **171**
Two-toed amphiuma, **243**
Tympanic
    cavity, 290
    membrane, 174, 177, 244, 259, 269, 312, 396
Type specimen, 41
Typhlosole, **136**, 138, **194**
Typhus, **186**
*Tyrannosaurus*, 281

## U

*Uintatherium*, 314, **315**
Ulna, **292**, 307, 308, **351**, 352
Ulnar arteries, 375
Ultraviolet light (UV), 462
Umbilicus, **416**
Umbo, **192**
Unicinate process, 291–**292**
Undulipodia, 47, 54, 56, 58, 62–66, 74, 81, 88, 109, 114, 120, 121–122, 183, **191**, **195**, 205, 219, 261
    structure, 24–25
Unguligrade, **314**
Unicellular organisms, 30, 53
Unit membrane model, 20–**21**
Urbanization, 482, 483
Urchins, 49, 203, **204**, 208, **209**, 210
Urea, 13, 38, 368, 369, 464
*Urechis caupo*, **133**
Ureters, 238, **272**, 293, **309**, 311, 367, 369, **410**, **411**
Urethra, 38, **309**, 311, 312, 365, 367, **410**, **411**
Uric acid, 293, 464

Uridine (U), **422**
Urinary duct, 226, **230, 232,** 233, **237,** 238, **252,** 253, **260, 261**
Urination, 367, 369
Urine, 13, 38, 311, 367–369
  composition, **368**
Urochordata, 215
Urogenital
  opening, **237,** 238, 307, 312
  sinus, 226, 233
Urogenital system
  dogfish shark, **232–233**
  frog, **260–261**
  lamprey, 226
  turtle, **272**
*Uroglenopsis*, 60
Uropod, **153, 154,** 155, **156,** 160
Urostyle, **246–247**
Uterus
  *ascaris*, **124–125**
  vat, 312
  dogfish shark, **232–233**
  frog, **261**
  human, 410, **411,** 412
"Uterus externus," 166
Utriculus, 397

# V

Vacuole, **18**
  contractile, 57–58
  digestion, 24
  food, **65, 74,** 76–**77,** 97, 359
Vagina, 124–**125,** 177, 312, 410, **411**
Vagus nerve, **231, 258, 259, 311,** 384, **392**
Valves
  ear and nose, 316
  shell, **192**
Valves, heart, 138–**139,** 256, **371,** 372
  atrioventricular, **256, 371,** 372
  bicuspid, **371**
  semilunar, **256, 371**
  spiral, **256**
  tricuspid, **371**
  venous, **372,** 375
Vane, 290–**291**
Varanidae, **268**
*Varanus*, 273
Vas deferens, 124, **136,** 142, **156,** 159, 177, **232,** 233, 261, **272,** 296, 309, 312, **410**
Vas efferens, **136,** 142, **232**–233
Vascular tissue, **31, 35,** 36, 39

Vastus
  externus, **249**
  internus, **249**
  medialis, **354**
Vectors, disease, 162, **186**–187
Vegetal pole, **414**
Veins, 37, 255
  bird, 293–295
  cat, **310**
  frog, 254, 255–**257**
  histology, **372**
  human, 372, **374,** 375
  turtle, **271**
  villi, **362**
  (*see* Circulatory system; specific veins)
Velar tentacle, **217**–219
Veld, 464
Veliger larva, **191**
Velum, **104,** 191, 218, **225**
Venae cavae, 255–**257**
  anterior, 255–**257, 271, 294**
  bird, 293, **294**
  cat, **310**
  frog, 255–**257**
  human, **371, 374**
  inferior, **371, 374,** 380
  reptile, **271**
  superior, **371, 374**
Venom, 274, 279
  cobra, 276
  duct, **279**
  gila monster, 284
  gland, **279**
  pit-viper, 275
Vent, 293
Ventilation, 381
Ventral root, 258, **388, 389**
Ventral vessels, 138–**139**
Ventricle, **193**–**194, 237,** 254–**256, 271,** 293–**294, 310, 371,** 373, **374,** 375
Ventricles (brain), **258,** 390
Ventriculus, 174, 293
Venules, 373
Venus's flower basket, **91**–92
Vermiform appendix, 324, **360,** 361
Vertebrae, 216, 219–221, 349
  bony fish, 235
  cat, **307**
  caudal, 291, **292,** 307
  cervical, 291, **292,** 307
  frog, 247
  human, **350, 389,** 390
  lumbar, **307, 350, 351**
  sacral, **246, 307,** 350
  thoracic, **307, 350, 351**
  turtle, 269
Vertebral column, 49, 213, 219

  cat, **307**
  dogfish shark, 228
  frog, **246**–248
  human, 349, **350, 389**
  turtle, 269
Vertebrates, 219–221
  body structure, 220
  brain, 391, **392,** 393
  circulation, 371
  development, **408,** 412–419
  digestive system, 360
  distribution through time, **458**
  endocrine system, 398
  excretory system, **367**–369
  evolution, **214,** 215–216
  nervous system, 384
  oldest known, 223
  organization, 220–221
  reproduction, 409
  respiration, 379
  sense organs, 393–397
  skeletal systems, 349
Vertebrochondral (false) ribs, 352
Vertebrosternal (true) ribs, 352
Vessels
  blood, 37, **136, 137,** 138–**139**
  lymph, 257, 311, **362, 374,** 375
Vestigial organs, 282, 444
Vibrassae, 305, 338
Viceroy butterfly, 339
Villi
  chorionic, **416,** 417
  intestinal, 361, **362**
Viperidae, 268
Visceral
  arches, 247
  ganglion, 194
  layer, 36
  muscle (*see* Smooth muscle)
  skeleton, 235, 247, 349
Visceral organs, 384
  muscles in, 353
Vision, 160, 394
  arthropod, 157–159, 176
  bird, 295–296
  close-up, 395
  color, 394
  dim light, 394
  distant, 395
  frog, 266
  human, 394–395
  lamprey, 226
  stereoscopic, 479
Vitamins, 13, 363–365
  A, 240, 241, **364, 365**
  $B_1$ (thiamine), **364, 365**
  $B_6$ (pyridoxine), **364**

**527**

Vitamins (cont.)
  $B_{12}$ (cyanocobalamin), **364**
  C (ascorbic acid), **364**
  D, 214, **364**, **365**, 462
  deficiency symptoms, **364**, **365**
  E, **364**
  H (biotin), **364**
  K, **364**
  M (folic acid), **364**
  niacin, **364**
  pantothenic acid, **364**
  sources, **364**
Vitellarium, **122**
Vitelline network, 417
Vitreous humor, **395**
Vitrodentine, 228
Vivicum, 42
Viviparous, 305, 409
Vocabulary, 338
Vocal
  cords, 254, 311, 379, 380
  organ, 295
  sacs, 254, 264
  signals, 338
Volatile substances, 394
Voluntary muscle (see Skeleton muscle)
Volvants, 98–99
*Volvox,* 44, **54–55**
Vomeronasal organs, 274, 394
Von Frisch, Karl, 182
*Vorticella,* **81**
Vulture, **302**, 335
Vulva, 123–**125**

# W

Wading birds, 288, 300, **301**
Walking legs, 153–155, 163
Wall gekko, **268**
Wall lizard, **268**
Wallace, A., 440, 441, 443
Walaroo, 316, **317**
Walrus, 319
Warbler, **288**, 343
"Warm-blooded" animals (see Homeotherms)
Wasp, **171**, 180, 335
Wastes
  organic, 484–486
  metallic, 487
  radioactive, 486–487
Wastes, excretion of, 365–369, 370
  (see Excretory system)

Water, **11**, 362, 463, 484
  cellular, 13
  chemistry, 11
  consumption, 194, **195**
  contamination, 59–60, 200
  deoxygenation, 484–486
  desalination, 486
  heat loading, 486–487
  pollution, 484–487
  pores, 194, **195**
  recycling, 486
  tubes, **194**, **195**
Water-expulsion vesicle, **57**, 64–65, **74**, 76–77
Water moccasin, **278**
Water strider, **171**
Water-vascular system, 49, 203, **205**, 210
Watson-Crick model, 420
Wax, 184
Waxes, 13
Weasel, 316
Webbed feet, **288**, 298, 301, 316, 319
Webs, 162, **166**
  orb, **166**
  sperm, 166
Wegener, A., 457
Whale, 228, 322, 341
  blue, 322
  gray, **320**
  killer, **320**, 322
  sperm, 322
  toothed, **320**, 322
Whale shark, 233
Whaling, **320**, 322
"Wheel animals," 121–123
Whelk, **190**, 198, 335, **337**
Whip scorpion, **163**
Whipworm, **127**
Whiskers, 305
White corpuscles, 175, 236, **254**, 310, **376**
White matter, 388, **389**, 390
White-splashed Andalusian fowl, 427, **428**
Wild rock dove, 303
Wild cats, 305
Wing
  bat, **446**
  bird, 286, 290–293, 295, 298, 300, 301, 304, **446**
  grasshopper, **173**–174
  termite, **182**–183
  veins, 174, 177
Wingless birds, 298, **300**
Workers, 180–183

Worms, 48, 123–129, **132**, 134–150
  segmented, **134**–150
  unsegmented, 123–129, **132**
  (see type of worm)
*Wuchereria bancrofti,* **127**

# X

X-gland, 159
Xerophthalmia, 364
Xerophytes, 467

# Y

Y-organ, 159
Yellow-bellied sapsucker, 304
Yellow fever, 241
Yellowhammer, 340
Yellow mealworm beetle, **188**
Yellow perch, 222, 233–238, 240 (see Fishes, bony)
Yolk, 38, 238, **239**, 297, 408, 413, 414
  cells, **116**, 414
  concentration, 412
  duct, **116**
  glands, **113**, **116**
  plug, 414
  sac, 238, **408**, 415, **416**
  stalk, 416

# Z

Z band, 355, **356**
*Zalophus californianus,* 338
*Zea Mays*
  cell, **18**
Zebra, **306**
Zoogeographical region, 455, **456**, 477
Zoogeography, **6**
Zoological history, 458–460, 477
Zoology, 3–4
  relevance to human problems, 7
  subdivisions of, 6–7
Zoos, 331
Zonules of Zinn, 394–395
Zygomatic arch, **308**, 349
Zygote, 29, 62, 69, **70**–71, 102–103, 107, 415, 429
  formation, 412
  nucleus, 412

GORDON MEMORIAL LIBRARY

3GORL000225480

Gordon Memorial Library
917 N. Circle Drive
Sealy, TX. 77474

DEMCO